D0415320

Statistics in Society

Y

Arnold Applications of Statistics Series

Series Editor: **BRIAN EVERITT**
Department of Biostatistics and Computing, Institute of Psychiatry, London, UK

This series offers titles which cover the statistical methodology most relevant to particular subject matters. Readers will be assumed to have a basic grasp of the topics covered in most general introductory statistics courses and texts, thus enabling the authors of the books in the series to concentrate on those techniques of most importance in the discipline under discussion. Although not introductory, most publications in the series are applied rather than highly technical, and all contain many detailed examples.

Other titles in the series:

Statistics in Education Ian Plewis

Statistics in Civil Engineering Andrew Metcalfe

Statistics in Human Genetics Pak Sham

Statistics in Finance David Hand and Saul Jacka

Statistics in Sport Jay Bennett

Statistics in Society

The arithmetic of politics

Edited by
Daniel Dorling
and
Stephen Simpson

A member of the Hodder Headline Group
LONDON • NEW YORK • SYDNEY • AUCKLAND

First published in Great Britain in 1999 by
Arnold, a member of the Hodder Headline Group,
338 Euston Road, London NW1 3BH

http://www.arnoldpublishers.com

Co-published in the United States of America by
Oxford University Press Inc.,
198 Madison Avenue, New York, NY10016

Whilst the advice and information in this book are believed to be true and
accurate at the date of going to press, neither the authors nor the publisher
can accept any legal responsibility or liability for any errors or omissions
that may be made.

British Library Cataloguing in Publication Data
A catalogue record for this book is available from the British Library

Library of Congress Cataloging-in-Publication Data
A catalog record for this book is available from the Library of Congress

ISBN 0 340 71994 X

1 2 3 4 5 6 7 8 9 10

Commissioning Editor: Nicki Dennis
Production Editor: Julie Delf
Production Controller: Priya Gohil
Cover Design: M2

Typeset in 10/11pt Times by J&L Composition Ltd, Filey, North Yorkshire
Printed and bound in Great Britain by J W Arrowsmith Ltd, Bristol

What do you think about this book? Or any other Arnold title?
Please send your comments to feedback.arnold@hodder.co.uk

Contents

List of contributors   x

Preface   xxv

1   Introduction to statistics in society   Daniel Dorling and
        Stephen Simpson   1
        Statistics matter

Part I: Collecting Statistics   7

2   The census   Ian Diamond   9
        The biggest and best data source in the UK?

3   Government household surveys   Charlie Owen   19
        Using government household surveys in social research

4   Confidentiality of official statistics: an excuse for secrecy
        Angela Dale   29
        *Truly open government would reduce concerns about
        confidentiality of data*

5   Who pays for research? The UK statistical picture   Ian Miles   38
        *Research and development produces knowledge, but is hidden in
        official statistics*

6   Working with government to disseminate official statistics
        Michael Blakemore   46
        The contentious relationships in the dissemination of official statistics

Part II: Models and Theory   53

7   **Eugenics and the rise of mathematical statistics in Britain**
     Donald MacKenzie   55
     Even the mathematical theory of statistics can be shaped by social
     circumstances and by political aims

8   **Science, statistics and three 'psychologies'**   Daniel B Wright   62
     Scientific claims are too easily made and too easily believed in
     'psychology'

9   **Using statistics in everyday life: From barefoot statisticians to**
     critical citizenship   Jeff Evans and Ivan Rappaport   71
     Not every user of statistics has to be a professional statistician

10   **Resources for lay statisticians and critical citizens**
     Ivan Rappaport and Jeff Evans   78
     Help is at hand to understand social statistics

11   **Qualitative data and the subjectivity of 'objective' facts**
     Ian Parker   83
     Research fails unless it engages with subjectivity

12   **Untouched by statistics: representing and misrepresenting**
     other cultures   David Sibley   89
     Statistical methods on their own do not provide a way to
     understand the world

13   **Models are stories are not real life**   Jane Elliott   95
     Statistical models are the means by which researchers construct
     meanings for variables

Part III: Classifying People   103

14   **Missing subjects? Searching for gender in official statistics**
     Diane Perrons   105
     How gender issues are emerging within official statistics

15   **Playing politics with pensions: legitimating privatisation**
     Jay Ginn   115
     Misleading statistics have helped to reinforce gender inequality
     in later life

16   **Ethnic statistics: better than nothing or worse than nothing?**
     Waqar Ahmad   124
     Standard categories of ethnicity are often not appropriate

17   **The religious question: representing reality or compounding**
     confusion?   Joanna Southworth   132
     There should not be a question on religion in the census

18 **Measuring eating habits: some problems with the National Food Survey**   Mary Shaw   140
Is there such a thing as an average diet?

19 **Measuring international migration: the tools aren't up to the job**   Ann Singleton   148
A case study of European cross-national comparisons

Part IV: Counting Poverty   159

20 **Poverty and disabled children**   David Gordon and Pauline Heslop   161
Households with a disabled child are found to be among the 'poorest of the poor'

21 **Where are the deprived? Measuring deprivation in cities and regions**   Peter Lee   172
Deprivation is not easy to measure: it depends on what you want to find

22 **The limitations of official homelessness statistics**   Rebekah Widdowfield   181
Homelessness counts are getting further from reality

23 **The use and abuse of statistics on homeless people**   Walid Abdul-Hamid   189
We have a misconception of the mental health of the homeless because of flawed statistical studies in the past

24 **Are crime and fear of crime more likely to be experienced by the 'poor'?**   Christina Pantazis and David Gordon   198
Crime has a greater impact on poorer people, although they lose less property

Part V: Valuing Health   213

25 **The racialisation of ethnic inequalities in health**   James Y Nazroo   215
Ethnicity does not reflect presumed genetic or cultural attributes

26 **What do official health statistics measure?**   Alison Macfarlane and Jenny Head   223
Monitoring health administration is not the same as measuring health or health services

27 **Making sense of health inequality statistics**   Mel Bartley, David Blane and George Davey Smith   234
Careful measurement shows that health inequalities are rising nationally

28   **Poverty and health**   Anthony Staines   244
      Local statistics can be used to show the widening gap between
      the health of rich and poor people in Britain

29   **Statistics and the privatisation of the National Health Service**
      and social services   Alison Macfarlane and Allyson Pollock   252
      Using statistics to investigate the hidden privatisation of health
      and social care services

30   **Industrial injury statistics**   Theo Nichols   263
      Official statistics on industrial injury are not a valid measure of
      safety performance

Part VI: Assessing Education   271

31   **What's worth comparing in education?**   Ian Plewis   273
      Debates about standards are often fruitless because they focus
      on impossible tasks

32   **Performance indicators in education**   Harvey Goldstein   281
      Knowing about uncertainty in performance indicators prevents
      their misuse

33   **Can trends in reading standards be measured?**   Pauline Davis 287
      The use of the National Curriculum's Standard Assessment Tasks
      (SATs) for assessing trends in reading standards

34   **Inspecting the inspection system**   Nicola Brimblecombe   294
      OFSTED inspection statistics give a misleading picture of schools
      in this country

35   **Special statistics for special education**   Cecilio Mar Molinero   302
      Special education statistics measure willingness to provide rather
      than need

Part VII: Measuring Employment   311

36   **Problems of measuring participation in the labour market**
      Anne E Green   313
      To understand how labour markets operate, a range of measures
      of participation are needed

37   **The politics and reform of unemployment and employment**
      statistics   Ray Thomas   324
      Problems and solutions concerning two of the UK's best-known
      data series

38   **Unemployment and permanent sickness in Mid Glamorgan**
      Roy Davies   335
      Local statistics are difficult to compile but can produce important
      results

39 **Voodoo economics: 'art' and 'science' in employment forecasting**   Jamie Peck   340
Forecasting employment is pointless

40 **Working with historical statistics on poverty and economic distress**   Humphrey Southall   350
Statistics change with society but, with care, comparisons can be made

Part VIII: Economics and Politics   359

41 **Measuring the UK economy**   Alan Freeman   361
When reworked, UK national accounts show a lack of investment by profit-makers

42 **Household projections: a sheep in wolf's clothing** Richard Bate   369
Statistics should inform debate on planning policy, not substitute for it

43 **The statistics of militarism**   Paul Dunne   376
Military statistics serve the interests of the military industry

44 **Counting computers – or why we are not well informed about the information society**   James Cornford   384
The information industry largely determines the information that we have about it

45 **The British electoral system and the British electorate** Ron Johnston, Charles Pattie and David Rossiter   391
Votes that count and votes that don't

46 **Illuminating social statistics**   Graham Upton   400
There are better graphics than pie charts and bar charts!

47 **Conclusion: statistics and 'the truth'** Stephen Simpson and Daniel Dorling   414
Social statistics are a social product

Appendix: a list of Radical Statistics publications   421

References   424

Author index   465

Subject index   473

List of Contributors

Walid Abdul-Hamid is a psychiatric Senior Registrar at the Maudsley Hospital, London, SE5. He was previously a research fellow at the Department of Public Health, UCL Medical School, where he completed a PhD on Homeless People and Community Care.

Waqar Ahmad was the Director of the Ethnicity and Social Policy Research Unit at Bradford University. He is the editor of *Race and Health in Contemporary Britain* and (with Karl Atkin) *Race and Community Care* (both published by Open University Press). He is currently the Professor of Primary Care Research, and Director of the Centre for Research in Primary Care, at the University of Leeds.

Mel Bartley is a contract researcher in medical sociology, presently employed at the Department of Epidemiology and Public Health, University College. Her research interests include health inequalities, women's health, and the social construction of medical knowledge.

Richard Bate is a partner in the planning and environment firm Green Balance. He is special advisor to the Housing Inquiry of the House of Commons Select Committee on the Environment, Transport and Regional Affairs, Session 1997–98.

Michael Blakemore is Professor in the Department of Geography, University of Durham. Since 1986 he has been the Executive Director of Nomis, the Office for National Statistics' online labour market statistical database, based at the University of Durham. Since January 1995, he has been Executive Director of the Economic and Social Research Council (ESRC) Research Centre, r•cade (Resource Centre for Access to Data on Europe).

David Blane is Reader in Medical Sociology at Charing Cross and Westminster Medical School, where he teaches sociology as a basic medical science. His research interests include health inequality, measurement issues in health studies, general practice research and the study of the life course.

Nicola Brimblecombe has a background in social policy. Previously a Researcher in Education, she is currently a Research Associate at the University of Bristol, researching geographical inequalities in health, part of the ESRC Health Inequalities Programme.

James Cornford is a Senior Researcher at the University of Newcastle's Centre for Urban and Regional Development Studies (CURDS). His research concerns the implications of information and communication technologies for the development of cities and regions.

Angela Dale is Director of the Cathie Marsh Centre for Census and Survey Research (CCSR) and Professor of Quantitative Social Research at the University of Manchester. CCSR provides access to the Samples of Anonymised Records from the 1991 Census and also conducts research using census and survey data.

George Davey Smith was a member of the noise terrorism outfit Scum Auxillary in the early 1980s. As commercial and critical success completely eluded the group he has had to work as a social epidemiologist since then. (George is currently Professor of Clinical Epidemiology at the University of Bristol – eds.)

Roy Davies has a background in teaching industrial relations and social studies in Further Education and adult education. He was Dean of Management and Business Studies at S. Glamorgan Institute from 1980–88. He has written several articles on health issues and employment.

Pauline Davis is a Researcher in Education at the University of Manchester. She is also Visiting Lecturer in Research Methods at Oxford Brookes University where she used to be employed on a full-time basis. Her main research interests are reading and teacher communication styles in the Further Education sector. Previously she spent six years teaching mathematics at Cardiff's Tertiary College.

Ian Diamond is Professor of Social Statistics at the University of Southampton. He has worked extensively on census analyses in many countries and is currently working with the Office for National Statistics on the development of a One Number Census for 2001.

Daniel Dorling is Lecturer in Human Geography in the School of Geographical Sciences, Bristol University, following several years spent in full-time research as a Joseph Rowntree Foundation Fellow and subsequently British Academy Fellow at Newcastle University. During 1996/7 he was joint editor of the Radical Statistics Newsletter with Stephen Simpson.

Paul Dunne is Professor of Economics at Middlesex University Business School. He has published widely on the economics of military spending and conversion and recently co-edited *The Peace Dividend*.

Jane Elliott is a Research Fellow at the Cathie Marsh Centre for Census and Survey Research, University of Manchester. She is a sociologist with a particular interest in the analysis of life histories using both qualitative and quantitative methods.

Jeff Evans teaches social statistics and research methods as Middlesex University. His research interests include: public understanding of mathematics and bringing together quantitative and qualitative methodologies. He is a founding member of Radical Statistics, co-editor of *De-mystifying Social Statistics* (Pluto, 1979), and author of *Adults and Numeracy: Mathematical Thinking and Emotions in Context* (Falmer, in press).

Alan Freeman lectures in Economics at the University of Greenwich and is co-organiser of the International Working Group on Value Theory. He may be contacted at a.freeman@greenwich.ac.u.k.

Jay Ginn is currently employed in the Sociology Department at Surrey University, researching the changing pensions mix in Britain. She has published widely on gender and pensions. Books (co-authored/co-edited with Sara Arber) include *Gender and Later Life* (1991) and *Connecting Gender and Ageing* (1995).

Harvey Goldstein has been Professor of Statistical Methods at the Institute of Education, University of London, since 1977. He is a chartered statistician, and has been editor of the Royal Statistical Society's Journal, Series A. He was elected a member of the International Statistical Institute in 1987, a fellow of the Royal Society of Arts in 1991 and a fellow of the British Academy in 1996. His research interests include the use of statistical modelling techniques in the construction and analysis of educational tests, and the methodology of multilevel modelling. The major text on multilevel modelling is his book *Multilevel Statistical Models* (Arnold, 1995).

David Gordon combined his background in biology and geology with anti-poverty policy, while helping to find safe public water supplies in the South Pacific. He currently works for the School for Policy Studies at the University of Bristol, UK. He has published extensively in the fields of scientific measurement of poverty, crime and poverty, childhood disability, area-based anti-poverty measures, the causal effects of poverty on ill health, housing policy and rural poverty. He recently co-edited *Breadline Britain in the 1990s*, with Christina Pantazis.

Anne Green is a geographer with particular research interests in spatial dimensions of economic, social and demographic change. Recent research projects include analyses of the geography of unemployment and non-employment, and investigations of individual and household location and mobility decisions. She is a Principal Research Fellow at the Institute for Employment Research, University of Warwick.

Jenny Head is Lecturer in Statistics in the Department of Epidemiology and Public Health at University College London. Her interests include health surveys and official health statistics.

Pauline Heslop comes from a background of paediatric nursing in Britain and overseas. Her work has included setting up a maternal and child health clinic in India, evaluating health care services for Tibetan refugees in Nepal and teaching computing within the NHS. She is currently a research student at the University of Bristol completing a PhD on childhood factors associated with adulthood disablement.

Ron Johnston is a Professor in the School of Geographical Sciences at the University of Bristol, having previously been Professor of Geography at the University of Sheffield and Vice-Chancellor at the University of Essex. He has published widely on various aspects of the UK's electoral geography, and has recently completed a book with David Rossiter and Charles Pattie on *The Boundary Commissions: Redrawing the UK's Map of Parliamentary Constituencies* (Manchester University Press).

Peter Lee is a lecturer in Urban and Regional Studies at the University of Birmingham. He has recently been involved in debates concerning targeting deprived areas and identifying areas in 'need' for regeneration purposes.

Alison Macfarlane joined Radical Statistics when it was founded in 1975 and has been active in its Health Group ever since, contributing to many of its publications. She works as a statistician at the Perinatal Epidemiology Unit in Oxford.

Donald MacKenzie studied applied mathematics, and then went on to become a historian and sociologist of science and technology, especially of their quantitative aspects. His books include *Inventing Accuracy* (MIT Press, 1990) and *Knowing Machines* (MIT Press, 1996). (He can be found at the Department of Sociology, University of Edinburgh, but edited this chapter while visiting Harvard – eds.)

Cecilio Mar Molinero is a Reader in Operational Research at the Department of Management of the University of Southampton. He graduated in Engineering at the University of Barcelona, Spain. He is interested in education management, particularly the effects of policy decisions on deprived schools. He has published extensively in international academic journals. He is Chairman of the Educational Systems Study Group of the Operational Research Society and a member of the Community OR Network.

John Martyn is a lecturer in research methods and Senior Research Fellow in the Sociology and Social Policy School at Roehampton Institute London, Southlands College, 80 Roehampton Lane, London SW15 5SL. An experienced market researcher and a fellow of the Royal Statistical Society, he represents the Market Research Society on the Statistics Users' Council.

Ian Miles is Professor of Technology and Social Change at the University of Manchester. His publications include over 100 published journal articles and book chapters, 15 books (three as contributing editor) and 16 published reports – in the fields of innovation studies, social aspects of information technology, technology foresight, research evaluation and social indicators.

James Nazroo qualified from St George's Hospital Medical School in 1986. In 1989 he obtained an MSc in Sociology of Health and Illness from Royal Holloway and Bedford New College. He is now Senior Lecturer at University College London. For the past nine years he has been involved in research into inequalities in health, with a particular focus on ethnic and gender inequalities.

Theo Nichols is Professor of Sociology at the University of Bristol. He is author of *The Sociology of Industrial Injury*, and of a wide range of other contributions to economic sociology.

Charlie Owen is a Research Officer in the Thomas Coram Research Unit, Institute of Education, University of London. He specialises in the secondary analysis of large datasets, such as the household surveys.

Christina Pantazis is a researcher for the Centre for the Study of Social Exclusion and Social Injustice in the School for Policy Studies at the University of Bristol, with interests in poverty, crime and criminal justice. She has previously researched the criminalisation of female poverty, and inequalities in the criminal justice system.

Ian Parker is Professor of Psychology and Director of the Critical Psychology MSc Programme at Bolton Institute, author of books on psychological discourse including *Psychoanalytic Culture* (Sage, 1997), and a member of Psychology Politics Resistance.

Charles Pattie is Senior Lecturer in Geography at the University of Sheffield. He has published widely on electoral geography and political campaigning.

Jamie Peck is Professor of Geography and Director, International Centre for Labour Studies, University of Manchester. He is author of *Work-Place: the social regulation of labor markets* (Guilford, NY, 1996). He is currently researching labour markets and welfare reform.

Diane Perrons is a lecturer in Human Geography at the London School of Economics and Political Science. Her research focuses on economic and social cohesion in Europe, specialising in regional and gender dimensions of inequality. She has recently directed a cross national qualitative research project on flexible working and the reconciliation of paid work and family life for the European Commission.

Ian Plewis is Senior Lecturer in Statistics at the Institute of Education and Senior Research Officer, at the Thomas Coram Research Unit, Institute of Education. He is the author of *Statistics in Education* (Arnold, 1997), an intermediate textbook showing how modern statistical ideas can be applied to educational data.

Allyson Pollock is a Senior Lecturer in public health medicine at St George's Hospital Medical School, University College London and a consultant in public health medicine working for Merton Sutton and Wandsworth Health Authority. She has published widely on a number of areas including health policy, rationing, cancer epidemiology and long-term care. She and other colleagues are currently analysing the implications of acute and community service strategies for health care needs in the context of the private finance initiative and other government policies.

Ivan Rappaport moved from working in semi-conductor physics to statistics, seeing this as a bridge between science and society. Statistical interests range from social surveys to medical applications (including survival analysis). He teaches statistics to a wide variety of students at Middlesex University.

David Rossiter is a Research Fellow in the School of Geographical Sciences, at the University of Bristol. His main research interest is the geography of voting and, together with Ron Johnston and Charles Pattie, he has recently completed a major study of electoral redistricting in Britain.

Mary Shaw has a PhD in Sociology from the University of Queensland, and is currently a researcher on the ESRC Health Variations Programme, at the University of Bristol.

David Sibley teaches human geography at Hull University. He is currently working on problems associated with the representation of space and place as they affect and are affected by transgressions. His interests include animals in cities, nomads and transgressive ideas in the social sciences.

Stephen Ludi Simpson is currently employed at Bradford City Council and at Manchester University. During 1989–1992 he administered the Radical Statistics Group. His interests lie in demography and the use of statistics by communities and governments.

Ann Singleton works in the Eurostat Migration Team, on secondment to Luxembourg from Bristol University's Centre for the study of Social Exclusion and Social Injustice, School for Policy Studies. She has worked with international migration statistics since 1992. Her previous position was Senior Research Fellow in the Migration Research Unit at University College London. Her research interests include the use of migration statistics in policy development, migration as crime and the political economy of the management of response to disasters.

Humphrey Southall is Reader in Geography at Queen Mary and Westfield College, University of London. He has been researching the linked histories of unemployment, labour markets and welfare systems for many years. He is currently leading a major project to construct a historical Geographical Information System for Britain, and to create from this a new social atlas to mark the bi-centenary of the first census in 1801.

Jo Southworth is currently a postgraduate student of geography at the University of Bristol, where she is researching into the issues surrounding the ethnic and proposed religious questions on the UK census. She has recently completed an MSc and a Research Assistant post, both at the University of Bristol.

Anthony Staines's main academic passions are spatial stuff and poverty. He believes that public health can change things, if we get our act together. Helping this process is his real job.

Ray Thomas is a Research Fellow in the Social Sciences Faculty at the Open University, Affiliate Lecturer at Cambridge University, and a regular contributor to *Radical Statistics*. His research focuses on the nature and influence of the categorisations used in official statistics.

Graham Upton was educated at Leicester and Birmingham Universities. After five years at the University of Newcastle upon Tyne, he moved to the University of Essex in 1973. He is currently Head of Department and Reader in Applied Statistics.

Rebekah Widdowfield completed a PhD at the University of Newcastle-upon-Tyne, examining the allocation of council housing to lone parent families. She then moved to the University of Bristol where she is a Research Associate on an ESRC-funded project looking at rural homelessness.

Daniel Wright is a lecturer in psychology at the University of Bristol and has interests in memory (especially eyewitness testimony), methods (surveys) and statistics. He is author of *Understanding Statistics* (Sage, 1997), and currently Chair of the British Psychology Society's Mathematics, Statistics and Computing Section.

Preface

The idea for this book, and its making, arose from the members and activities of the Radical Statistics group.

What is Radical Statistics?

Radical Statistics is a group of statisticians and others who share a common concern about the political assumptions implicit in the process of compiling and using statistics, and an awareness of the actual and potential misuse of statistics and its techniques. In particular, we are concerned about:

- the mystifying use of technical language in order to disguise social problems as technical problems;
- the lack of control by the community over the aims of statistical investigations, the way these are conducted and the use of the information produced;
- the power structures within which statistical workers are employed and which control their work and the uses to which it is put; and
- the fragmentation of statistical questions into separate specialist fields in ways that can obscure common problems.

Our history

Radical Statistics was formed in January 1975 and is proud to have been part of the radical science movement. This movement dates back to before the Second World War. Its most influential expression was in JD Bernal's book *The Social Function of Science*. He argued that science was the motor of human progress and history. The bombing of Hiroshima and Nagasaki and other events led to the disillusionment and eventual collapse of this pre-war movement. Some years later, in 1969, involvement in the anti-Vietnam War campaigns led a new generation of young radical scientists to found the British Society for Social Responsibility in Science (BSSRS).

The idea that statistics can be used as a tool for social change has a much longer history and lay behind statistical developments in the mid-nineteenth century.

Some of these ideas surfaced anew in the 1970s in the form of a heightened interest in social statistics in general and social indicators in particular. Radical Statistics rejected the idea that statistics were solely for measuring the 'economic' well-being of the state. We felt that statistics should and could be used for 'radical' and 'progressive' purposes. Statistics should be used to identify 'social' needs and to underpin rational planning to eliminate these needs. Many of these ideas were expressed in a book *Demystifying Social Statistics*, published in 1979.

These movements grew together through the 1970s. At its height, BSSRS had over 1200 members and a number of affiliated organizations including the Politics of Health Group, Radical Statistics Group, Radical Statistics Health Group and the Radical Science Journal Collective. Sadly, only Radical Statistics managed to retain an active membership and survive the numerous political defeats of the 18 years of Conservative Party rule from 1979 to 1997. BSSRS finally collapsed in the early 1990s, leaving only its charitable arm, the Science and Society Trust. Since the demise of BSSRS, Radical Statistics has not been affiliated to any other organization.

Our activities

Apart from our Annual Conference and occasional conferences on single issues, most of our activities are focused on producing publications, which are often used by campaigning groups, journalists, politicians and others. These have included books, pamphlets, broadsheets and articles in political and topical journals and, of course, our newsletter, *Radical Statistics*, which appears three times a year. Often our work has reached a wider audience anonymously. For example, an influential Channel 4 documentary, *Cooking the Books*, which set in train much-needed changes in government statistics, showed 10 examples of misleading use of statistics by the then Conservative government. Most of these examples came from the Radical Statistics Health Group's book *Facing the Figures*, published in 1987, and from other material produced by Radical Statistics.

Many of our activities are still centred around subject-based working groups. Some, notably those on health and education, have had a long existence. Others, such as groups on Nicaragua, nuclear issues, race, surveys for pressure groups and the poll tax, have been formed to respond to issues which were topical at the time. Regional and local subgroups have also campaigned on issues as diverse as food safety, community planning, economic statistics and women's rights. At the time of writing, the Radical Statistics Health Group is about to complete a new edition of its *Unofficial Guide to Official Health Statistics*, which was originally published in 1980. This book too, will be published by Arnold, with a new title: *Official Health Statistics: An Unofficial Guide*.

What is radical about statistics?

The members of Radical Statistics believe that statistics can be used as part of radical campaigns for progressive social change. We have always seen our role as belonging to a spectrum of campaigning organizations rather than as an academic or professional organization. Working with Radical Statistics is unlikely to help anyone build their career.

Although we have no 'party line', most of us share the view that the needs of the community can never be met fully by competition. The pursuit of profit alone will not eliminate the problems of poverty, inequality and discrimination. Only rational, democratic and progressive planning can tackle the manifest injustices of our present society and help the least 'powerful' groups to realize their full potential. Meaningful statistics are needed for this process.

To paraphrase the old Marxist adage, the purpose of statistics in general and Radical Statistics in particular is not only to describe the world but also to change it.

If you wish to join us, please write to Radical Statistics, c/o 10 Ruskin Avenue, Heaton, Bradford BD9 6EB.

David Gordon and Alison Macfarlane
October 1998

New plans for government statistics
A note provided by John Martyn, Roehampton Institute, London

It is timely for this book that the government issued in 1998 a consultative Green Paper *Statistics: A Matter of Trust*, in fulfilment of its pledge to develop the accuracy and independence of official statistics and ensure that they are reliably and honestly presented by people who understand their limitations.

The Green Paper accepts that reliable official statistics are a cornerstone of democracy and are essential to good government and proper accountability. It presents four main options, ranging from 'strengthening existing arrangements' (the minimum change option), the setting up of an independent Statistical Commission, or for a Committee of the House of Commons to assume direct responsibility for the National Statistics programme. Thus there may soon be developments towards an independent National Statistical Service able to withstand political manipulation.

However, Ministers and politicians in general are not widely trusted, and it has been suggested that the Head of the new National Statistical Service should report direct to the Prime Minister, rather than the Treasury as at present, and that the new, overseeing Statistical Commission should be well-respected and independent, rather than a creature of government. The size of the budget and how this is determined, as well as spent, will also act as constraints on what can be done.

Dr John Martyn, Roehampton Institute London

1

Introduction to Statistics in Society

Daniel Dorling and Stephen Simpson

Statistics matter

Have you ever heard it said that:

A third of children in Britain live in poverty.

Most people find a lover within a mile of their home.

More than half of all present marriages will end in divorce.

You are 27 times more likely to kill yourself than be killed.

Most people go blind before they die, almost all are disabled.

Ninety per cent of the population believe all statistics presented on television.

These statements may all be true. Many people would accept them at first sight. Statistics are pervasive and powerful; every day of your life you are directly or indirectly affected by them. In this book we argue that without evidence you should not necessarily believe everything you read. And even when evidence is presented you should always question it.

Most people in your road, college or town will know of you only indirectly. Most of them you will know only through statistics about society. How you treat and view others will depend to a large extent on the statistics you have read, and which you believe, because statistics have become the language of politics and persuasion. Social statistics are at the heart of many generalisations, stereotypes and prejudices. On the other hand, if statistics are understood, created and acted upon, they can be used to change society.

Lives led by statistics

Let's start from the beginning. How were you first described? 'It's a girl/boy, 6 lb 10 oz ...' Why were you described that way? Why does birthweight matter?

Who decides and where did this information go? Perhaps you didn't have much of a personality to record at birth, but later, surely, people treated you differently? For services that did not cost very much they may have, but once money is at stake statistics matter.

As soon as a service, a possession or a position involves money, statistics appear. For most of us, the clothes we can buy depend on a mixture of the results of fashion surveys, the cost of fabrics and their manufacture into clothes, and the economic forecasts of clothing markets. To clothing manufacturers and retailers you are a size 12 or 16, a 32- or 38-inch waist, a set of numbers. These measures of your size and weight, the statistics you are told about the dimensions of 'normal' people, statistics about health and food have direct effects on your behaviour and self-perception. The public services that you use are shaped even more strongly by statistics. For instance, you went to school. Why was it there? Why did you go there? How many teachers were there? How did they treat you when you went? League tables, reading tests, educational assessments, population forecasts, local authority spending assessments and many other statistics played a large part in the counting and controlling of your education.

What you have got in life and where you have ended up are strongly affected by statistics that are kept about you, such as exam results, National Insurance records, crime and debt records; and by how those statistics about you are presented and interpreted by others. You may be at college, for instance, only because somebody argued a case for increasing student numbers with the result that your place was funded. Decisions about whether a bank gives mortgages, or whether the government builds a million new homes, are always made with reference to statistics. If you rent from a council or a housing association, decisions about how many houses are provided in your area, and who can live in them, are determined by the interpretation of statistics. You need 'points' to climb the housing waiting list. The housing points you get depend on statistics about you and others. Your credit rating for loans depends on statistical models. Often they use statistics about yourself which other people have provided. The decisions that were taken to introduce statistics such as these were themselves based on statistical assessments.

Statistics and politics

Social statistics in their broadest sense were first collected when states needed to raise armies and taxes from people's work and wealth. The Domesday Book was the first national social survey in England. It was compiled in the eleventh century by the Normans to assess the land that they had conquered, and its owners. War and international commerce became the guiding force behind statistics. Trade in alcohol was measured, if not more accurately than the population, then at least with more effort, in order to monitor the collection of duty. British censuses of Ireland in the nineteenth century provided better counts of pigs, cows and horses than of people during the famine years. The national censuses of Britain itself began only in 1801 when manufacturing and labour concentrated in towns. Only in the past century have national statistics begun to focus on individuals and families, and statisticians have started to produce what would be recognisable now as social statistics. Indicators of social trends are a recent invention; they arose in many developed countries in the 1970s, partly as a reaction to the inadequacy of economic statistics to acknowledge social needs. Even more recently, concern at environmental hazards has demanded the integration of economic and social statistics. Indicators of sustainability are now on the agenda. But the agenda is changing rapidly. This book is part of that process of revising our priorities, methods, assumptions and measures.

Like the statistics themselves, methods of statistics have also always been political and are also changing quickly. In this book Donald MacKenzie (Chapter 7) describes how the academic discipline itself, barely a hundred years old, was founded on studies which assumed that people's intelligence could be related to measurements of parts of their bodies. This kind of politics can be difficult to take part in as it requires numeracy to participate, as Jeff Evans and Ivan Rappaport show in Chapter 9. But you should not be intimidated, as only a basic level of numeracy is really required, not the mathematical acrobatics so often presented to mystify the uninitiated.

To understand percentages, tables and the arguments requires a feel for numbers. Many people do not think they have this feel and so do not take part in this politics. They simply accept every statistic they hear, or accept what they have heard many times. Other people are inclined to dismiss all statistics and are excluded from this politics. The technical jargon used is a foreign language to most people, often to confuse more than to clarify, just as Latin was used in religious services in the past and the colonial power's language was used to conduct politics in former colonies. Technical language often mystifies and curtails debate, restricting it to those who claim to be fluent in the jargon.

This book provides a wide range of examples of how statistics are used in society. It aims to describe the nature of the major government surveys and censuses, and give examples of the way in which social statistics are used to make decisions in our lives, to demystify the use of statistics, illustrating how misleading statistics can be identified and challenged. By doing this the book investigates applications of statistics in society and aspects of the production of statistics and statistical methods. We hope this will allow you to judge the real meaning and reliability of everyday statistics in society as well as statistics used in social science research.

We do not believe that there is a conspiracy to confuse, but this is what social statistics often do unnecessarily. They are used selectively to support the policies and positions of those in power. The research staff and statisticians who collect and present those statistics often contribute to the confusion by claiming statistical correctness without justifying their claims in a way that can be understood by people who may be critical of the results. As we suggested above, you do not have to be exceptionally numerate to counter this power. What is useful is some scepticism, and the willingness to ask 'why?'.

The book is divided into eight parts to bring together chapters with a common theme. Because the same issues crop up so often, there are many connections between the various chapters, which we have tried to illustrate through cross-referencing. The index at the end of the book is also a useful route to finding where a problem may have been encountered in the different fields of social statistics. The book starts by looking at Collecting Statistics: the censuses, surveys, data providers and funders who determine what official statistics are available. The Models and Theories behind much of social statistics are then described, including examples of how research does not have to rely on official social statistics or on statistics at all. The third part of the book shows how Classifying People is problematic. Examples are taken from the statistics concerning pensioners, gender, ethnicity, religion and migration among other topics. The next five parts of the book each bring together chapters which address contemporary problems in some of the traditional fields of social statistics: Counting Poverty, Valuing Health, Assessing Education, Measuring Employment, and Economics and Politics. The themes that crop up in all the parts of the book are then brought together in the final chapter.

The Radical Statistics group campaigns for better statistics and for a better understanding of statistics. We still do not have the kind of statistics we need to understand and improve life. For example, in Britain:

- There are few statistics on gender, and children are generally treated as properties of their households. This is a relic of an era when women and children were 'goods and chattels'.
- Most of the work done by people at home, including looking after children, shopping and cooking, is not measured by either economic or social statistics. Only paid work is measured in the System of National Accounts and until recently we had no official time budget statistics.
- Unemployment statistics do not measure the number of people who want jobs.
- Health statistics do not measure health or care needs.
- Housing statistics do not measure homelessness or housing need.
- Poverty statistics do not measure the number of poor people.
- Economic and transport statistics do not measure the environmental and social costs of new roads or industrial activity.
- Until the 1990s, one of the few sources of statistics on 'ethnic minorities' were the 'mugging' reports of the Metropolitan Police, but there were almost no statistics on racially motivated assault.

Social statistics are mainly made available through computer databases (although most people have access only to the published summaries of a few of these). The vast majority can be used only within government. Social scientists with government contracts have access to a few (such as selected tabulations from the national surveys), but the general public, charities, housing associations, public libraries, schools and other groups have a right to see hardly any of these data that are collected about them.

Origins of this book

This book builds on the ideas and approaches developed over the past twenty-five years by Radical Statistics and is a successor to the group's classic, *Demystifying Social Statistics* (Irvine *et al.*, 1979). The authors of the current book have worked on key issues of social statistics. Although the book is not exclusively about England or the countries of the United Kingdom, most of its writers are based there. This means that events in these countries shape what is seen as most vital and provide most of the examples that are given.

Demystifying Social Statistics was written during the last year of the 1974–79 Labour government. We were, possibly, fortunate to be writing during the first year of the 1997 Labour administration. A great deal had happened in between; many things were changing, or at least appeared to be changing, as the book was taking shape. As a result of work done by the group over 18 years ago, to illustrate severe problems with official statistics, a Green Paper on official statistics was published in early 1998 on improving these (Office for National Statistics, 1998). It takes time to change long-held views and establishment opinion, but it is certainly timely now to review the role of statistics across the range of social policy.

We are grateful to many people in helping us produce this book as speedily as possible. First, we thank the members of Radical Statistics who sanctioned its production and encouraged us throughout. Second, we are indebted to all the authors for their enthusiasm, speed and tolerance with the project, our deadlines and our comments and criticisms. Third, we are grateful to the publisher's editor and assistant, Nicki Dennis and Marjorie Durham; without their efficiency and understanding of the project the book would have been a huge burden. Alison Macfarlane and David Gordon kindly agreed to write the book's preface on behalf of the group and have helped immensely in commenting on chapters. Finally, we should thank Tim Hunkin, who is responsible for the cartoons illustrating each section of this book, which first appeared in *Radical Statistics*.

This book has been a collaborative effort, allowing us to record, reflect on and revise our understanding of statistics in society at the end of the twentieth century. We very much hope that you find it both entertaining and useful. Moreover, we hope to encourage you to join in the effort to improve on the creation, interpretation and presentation of statistics in society.

PART I
Collecting Statistics

'I know the census is supposed to be completely secret, but I am surprised you've only got one toilet in that big house of yours.'
Source: *Evening Standard*, 24 April 1991

2

The Census

Ian Diamond

The biggest and best data source in the UK?

A census is essential. It is the only time when data are collected nationally at a very local level. This means that they can be used to allocate resources to a wide variety of geographies. The census data also provide the base from which population numbers can be estimated for the 10 intercensal years.

The census is certainly the biggest and probably the most important data collection exercise carried out in the United Kingdom (UK), as it is in most countries around the world. The aim is to get a snapshot of the population by collecting on one day, at an individual level, data on the demographic and labour force characteristics, on socio-economic circumstances and on the housing stock of everyone in the country. The UK undertakes three censuses: in England and Wales, in Scotland and in Northern Ireland. Although the context and outputs of the three censuses differ a little they take place at the same time and the three census offices coordinate their efforts so as to create economies of scale.

The UK started to undertake a census relatively late by European standards. The Scandinavian countries started early in the eighteenth century but it was not until 1801 that Britain first had a census. Prior to this it was argued that collecting data on each individual in the country infringed personal liberties; confidentiality still remains a fundamental concept of the UK censuses (*see* Chapter 4). However, at the end of the eighteenth century two contrasting factors led to agreement in Parliament for a census. First, Thomas Malthus had recently written an influential essay which argued that population growth rates in Britain were too high. Second, the military threat from Napoleon led a number of commentators to argue that there were not enough people to fight a successful war. The solution was simply to count the number of people! Since 1801 there has been a census every 10 years with the exception of 1941, when plans were diverted to form registers for food rationing, and in 1966, when an additional quinquennial census was undertaken. It should be pointed out that the first four censuses were not true censuses as they were merely counts of the population. A true census requires data to be recorded for each individual and these were not collected until 1841.

The census is a very costly exercise. The 1991 census of England and Wales cost around £135 million, roughly 25p per person per year over the decade from

1981. It is expected that the 2001 census will cost around the same in real terms. However, when one considers the multitude of purposes to which the census is put, this is not really so large a figure. As mentioned previously, a census is essential for a number of reasons. First, data are reported at a very local level: in England and Wales the basic geographical building-block typically comprises less than 250 households. It is used by all levels of government for resource allocation. For example, the Department of Environment, Transport and the Regions in central government allocates resources to the local authority districts for different purposes on the basis of their population and on 'need', which is typically defined using census data (*see* Chapter 21 on deprivation indices for an example of this). At a more local level the local authority district allocates funds on the basis of local populations. The census provides a decennial base for the population which is then updated annually throughout the decade using estimation strategies.

Resource allocation is not the only use of the census. It is used as the basis from which rates of births and deaths are calculated (*see* Chapter 27), it is one of the major sources of internal migration data and it is used for most aspects of town and country planning (*see* Chapter 42). In recent years there has been an enormous increase in the use of census data in commerce and industry (*see* Chapter 6). As marketing experts have seen the need to target their products more specifically they have wanted to know more about their customers. Many companies have linked the home address of their customers with census data for the local area in which that address occurs to learn more about the sort of people who buy their products and hence to target their advertising at the appropriate market. A number of commercial companies have formed clusters of enumeration districts – categorised into, for example, older people living in large houses or multi-occupied housing populated by young single people. These classifications are widely used.

Finally, it is important to note that there have now been censuses in the UK for almost two hundred years. By maintaining a degree of comparability in the questions across censuses it is possible for scholars to chronicle the demographic and social history of the nation for relatively small geographic areas (*see* Chapter 40).

Conducting a census

If one is undertaking a sample survey it is common to obtain a list of all those people in the population and to sample from them. This is not possible in a census, given that the aim is to identify all the people in the population. Indeed, if there was a list then there would possibly be no need for a census! Therefore the strategy has been to divide the country up into geographical areas known as enumeration districts (EDs) (in 1991 there were 109 670 in England and Wales) and to ask an enumerator to contact all households within the ED. These EDs are designed to be a workload for one enumerator and are typically around 200 households on a standard estate but could be rather less in a large rural area or in an inner-city area because these tend to be rather difficult to enumerate. In addition to the enumeration of residential households throughout the country, a number of groups are considered to be difficult to enumerate and receive special attention. In 1991 there were 23 such groups including, for example, prisons, rock lighthouses, people sleeping rough, the armed forces and the Royal Family.

In 1991, as in most previous censuses, the enumerator was first asked to walk around his or her ED listing all the addresses. Then the enumerator delivered a questionnaire to all households within the ED and collected it a few days later, after census night. One major issue for the enumerator is to identify households correctly. In suburban estates this is relatively easy, but in multi-occupied housing in inner cities it can prove rather difficult. The basic definition of a household is a group of people living under the same roof sharing at least one meal a day together. An example of a difficulty is whether to define a group of six young people sharing a house as six single-person households or one six-person household. Enumerators are given a set of rules to help them decide and, hopefully, to ensure comparability across the country. After collection the enumerators send the forms for processing. Where the enumerator cannot make contact he or she is asked to judge whether the dwelling is an 'absent household' or is vacant.

In order to maximise the use of scarce resources the Office for National Statistics (ONS) considered, for 2001, using some form of mail – either posting out the census form or asking the household to post it back. Under a mailout strategy it is likely that ONS would make use of a product called Addresspoint which has been developed by the Ordnance Survey and purports to give a complete list of addresses in the country. It seems to me that this could be effective in standard suburban estates but would represent a particular risk in inner cities, particularly for addresses with multiple occupation. A further innovation discussed by ONS is the use of teams of enumerators; this could be advantageous on the grounds of both safety and morale.

Following the enumeration the data are processed and a number of reports produced. In 1991 around 2000 people were employed to process the data and this operation took over a year. The forms themselves occupy over 20 km of shelving. In 2001 greater use will be made of computer scanning to improve the efficiency of this process.

Census content

Perhaps the most important decisions to be made with respect to the census concern the content of the census questionnaire. Throughout the decade preceding the census, the Census Office conducts a consultation which asks those in the user community for their views on what should be included. It is important to note that this user community is smaller than one would wish as a number of potential users, for example some voluntary groups, are prevented from using the data by the costs of obtaining tables particular to their needs or because of technological restrictions on access. There are always a much larger number of questions proposed and it is very difficult to make the final choice. Census questions need to fulfil a number of criteria. They should be easy and clear to understand; not sensitive; relevant to a large proportion of the population; perhaps comparable with previous censuses; and needed by a high proportion of the user community. This section provides an overview of the content of recent census questionnaires and a critique of a number of census questions.

Table 2.1 provides a list of the topics asked in all previous censuses. It can be seen that these mainly cover demographic topics, occupation, housing, education and health. In 1991 the majority of these questions had been asked previously,

Table 2.1 Census topics 1801–1991

Subject	1801	1811	1821	1831	1841	1851	1861	1871	1881	1891	1901	1911	1921	1931	1951	1961	1971	1981	1991	Subject
Names	—	—	—	—	GB	GB	GB	GB	GB	GB	GB	GB	GB	GB	GB	GB	GB	GB	GB	**Names**
Sex	GB	GB	GB	GB	GB	GB	GB	GB	GB	GB	GB	GB	GB	GB	GB	GB	GB	GB	GB	**Sex**
Age																				**Age**
quinquinnial age groups	—	—	GB	—	GB	—	—	—	—	—	—	—	—	—	—	—	—	—	—	quinquinnial age groups
in years	—	—	—	—	GB	GB	GB	GB	GB	GB	GB	GB	GB	GB	GB	—	—	—	—	in years
in years and months	—	—	—	—	GB	—	—	—	—	—	—	—	—	—	—	—	—	—	—	in years and months
Date of birth day, month and year	—	—	—	—	—	—	—	—	—	—	—	—	—	—	—	GB	GB	GB	GB	**Date of birth** day, month and year
Marital status	—	—	—	—	—	GB	GB	GB	GB	GB	GB	GB	GB	GB	GB	GB	GB	GB	GB	**Marital status**
Position in household																				**Position in household**
relationship to head of household	—	—	—	—	—	GB	GB	GB	GB	GB	GB	GB	GB	GB	GB	GB	GB	GB	GB	relationship to head of household
Whereabouts on census night	—	—	—	—	GB	GB	GB	GB	GB	GB	GB	GB	GB	GB	GB	GB	GB	GB	GB	**Whereabouts** on census night
Usual address	—	—	—	—	—	—	—	—	—	—	—	—	—	GB	GB	GB	GB	GB	GB	**Usual address**
Migration																				**Migration**
address one year ago	—	—	—	—	—	—	—	—	—	—	—	—	—	—	—	GB	GB	GB	GB	address one year ago
address five years ago	—	—	—	—	—	—	—	—	—	—	—	—	—	—	—	—	GB	—	—	address five years ago
Country of birth/birthplace	—	—	—	—	GB	GB	GB	GB	GB	GB	GB	GB	GB	GB	GB	GB	GB	GB	GB	**Country of birth**/birthplace
Year of entry to UK	—	—	—	—	—	—	—	—	—	—	—	—	—	—	—	—	—	—	—	**Year of entry to UK**
Parents' countries of birth	—	—	—	—	—	—	—	—	—	—	—	—	—	—	—	—	GB	—	—	**Parents' countries of birth**
Nationality[1]	—	—	—	—	—	GB	GB	GB	GB	GB	GB	GB	GB	GB	GB	GB	—	—	—	**Nationality**[1]
Ethnic group	—	—	—	—	—	—	—	—	—	—	—	—	—	—	—	—	—	—	GB	**Ethnic group**
Education																				**Education**
whether scholar or student	—	—	—	—	—	GB	GB	GB	GB	S	S	S	—	—	—	GB	GB	—	—	whether scholar or student
age at which full-time education ceased	—	—	—	—	—	—	—	—	—	—	—	—	—	—	GB	GB	—	—	—	age at which full-time education ceased
school-level qualifications	—	—	—	—	—	—	—	—	—	—	—	—	—	—	—	GB	—	—	—	school-level qualifications
scientific and technical qualifications	—	—	—	—	—	—	—	—	—	—	—	—	—	—	GB	—	GB	GB	GB	scientific and technical qualifications
higher qualifications	—	—	—	—	—	—	—	—	—	—	—	—	—	—	—	GB	GB	GB	GB	higher qualifications
Employment																				**Employment**
Activity: whether in job, unemployed, retired etc.	—	—	—	—	—	GB	GB	GB	GB	GB	GB	GB	GB²	GB	GB²	GB	GB	GB	GB	**Activity:** whether in job, unemployed, retired etc.
Students of working age	—	—	—	—	—	—	—	—	—	—	—	—	—	—	—	GB	GB	GB	GB	Students of working age

Working full-time or part-time
Weekly hours worked
Employment status
whether employee, self-employed, etc.
Apprentice or trainee
Name and nature of business of employer ('industry')
Address of business
Occupation
Family occupation[6]
Occupation one year ago
Workplace
Transport to work

Marriage and fertility
year and month of birth of children born alive in marriage
number of children born alive in marriage
whether live born child in last 12 months
year and month of first marriage and of end, if ended
year and month of present marriage
duration of marriage

Social conditions
religion (separate voluntary enquiry)
dependency: number and ages of children under 16

Topic												
Working full-time or part-time	—	—	—	—	—	—	—	—	—	—	GB	—
Weekly hours worked	—	—	—	—	—	—	—	—	GB[3]	—	GB	—
Employment status												
whether employee, self-employed, etc.	—	—	GB[4]	GB[4]	GB	GB	GB	—	GB	GB	GB	GB
Apprentice or trainee	—	—	GB[4]	GB[4]	GB	—	—	—	GB	—	GB	GB
Name and nature of business of employer ('industry')	—	—	GB[4]	—	—	GB	GB	GB	GB	GB	GB	—
Address of business	GB[5]	—	—	GB	GB	GB	GB	—	GB	GB	GB	GB
Occupation	—	GB	GB[5]	GB	GB	GB	GB	GB	GB	GB	GB	—
Family occupation[6]	GB	GB	GB	GB	GB	—	—	—	—	—	—	—
Occupation one year ago	—	—	—	—	—	—	—	—	GB	—	GB	GB
Workplace	—	—	—	GB[7]	GB[7]	E, W	—	GB	GB	—	GB	GB
Transport to work	—	—	—	—	—	W	—	—	—	—	GB	GB
Marriage and fertility												
year and month of birth of children born alive in marriage	—	—	—	—	—	—	—	—	GB	—	—	—
number of children born alive in marriage	—	—	—	—	—	—	GB[8]	—	GB	GB	—	—
whether live born child in last 12 months	—	—	—	—	—	—	—	—	GB	GB	—	—
year and month of first marriage and of end, if ended	—	—	—	—	—	—	—	GB[9]	GB	—	—	—
year and month of present marriage	—	—	—	—	—	—	—	GB	GB	—	—	—
duration of marriage	—	—	—	—	—	—	GB	—	—	—	—	—
Social conditions												
religion (separate voluntary enquiry)	—	—	GB	—	—	—	—	—	—	—	—	—
dependency: number and ages of children under 16	—	—	—	—	GB	—	—	—	—	—	—	—

Table 2.1 *Continued*

Subject	1801	1811	1821	1831	1841	1851	1861	1871	1881	1891	1901	1911	1921	1931	1951	1961	1971	1981	1991	Subject
orphanhood: father, mother or both parents dead	—	—	—	—	—	—	—	—	—	—	—	—	GB	—	—	—	—	—	—	orphanhood: father, mother or both parents dead
infirmity: deaf, dumb, blind, etc.	—	—	—	—	—	GB	GB	GB	GB	GB	GB	GB	—	—	—	—	—	—	—	infirmity: deaf, dumb, blind, etc.
eligibility to medical benefit	—	—	—	—	—	—	—	—	—	—	—	—	S	S	—	—	—	—	—	eligibility to medical benefit
limiting long-term illness	—	—	—	—	—	—	—	—	—	—	—	—	—	—	—	—	—	—	GB	limiting long-term illness
Language spoken																				**Language spoken**
Welsh	—	—	—	—	—	—	—	—	—	W	W	W	W	W	W	W	W	W	W	Welsh
Gaelic	—	—	—	—	—	—	—	—	S	S	S	S	S	S	S	S	S	S	S	Gaelic
Absent persons whole or part returns	—	—	—	—	—	—	—	—	—	—	—	—	—	—	—	GB	GB	GB	GB	**Absent persons** whole or part returns
Households																				**Households**
number of rooms[10]	—	—	—	—	—	—	—	—	—	E, W	E, W	E, W	E, W	E, W	GB	GB	GB	GB	GB	number of rooms[10]
number of rooms with one or more windows	—	—	—	—	—	—	S	S	S	S	S	S	S	S	—	—	—	—	—	number of rooms with one or more windows
sharing accommodation	—	—	—	—	—	—	—	—	—	—	—	—	—	—	—	—	GB	GB	GB	sharing accommodation
tenure of accommodation	—	—	—	—	—	—	—	—	—	—	—	—	—	—	—	GB	GB	GB	GB	tenure of accommodation
Amenities whether exclusive use, shared use, or lacking																				**Amenities** whether exclusive use, shared use, or lacking
cooking stove	—	—	—	—	—	—	—	—	—	—	—	—	—	—	GB	GB	—	—	—	cooking stove
kitchen sink	—	—	—	—	—	—	—	—	—	—	—	—	—	—	GB	GB	GB	—	—	kitchen sink
piped water supply	—	—	—	—	—	—	—	—	—	—	—	—	—	—	GB	GB	—	—	—	piped water supply
hot water supply	—	—	—	—	—	—	—	—	—	—	—	—	—	—	—	GB	—	—	—	hot water supply
fixed bath or shower	—	—	—	—	—	—	—	—	—	—	—	—	—	—	GB	GB	GB	GB	GB	fixed bath or shower
inside WC	—	—	—	—	—	—	—	—	—	—	—	—	—	—	GB	GB	GB	GB	GB	inside WC
outside WC	—	—	—	—	—	—	—	—	—	—	—	—	—	—	—	—	GB	GB	GB	outside WC
central heating	—	—	—	—	—	—	—	—	—	—	—	—	—	—	—	—	—	—	GB	central heating
Cars or vans	—	—	—	—	—	—	—	—	—	—	—	—	—	—	—	—	GB	GB	GB	**Cars or vans**

														Principal returns made by enumerators
Number of houses	GB	GB	GB	GB	GB	GB	GB	GB	GB	GB	GB	—	—	GB
Families per house	GB	GB	GB	—	—	—	—	GB	GB	GB	GB	—	—	GB
Vacant houses or household spaces		GB	GB	GB	GB	GB	GB	GB	GB	GB	GB	GB	GB	GB
House or household spaces otherwise unoccupied	} GB {	GB	GB	GB	GB	GB	GB	GB	GB	GB	—	GB	GB	GB
Shared access to accommodation	—	—	—	—	—	—	—	—	—	—	GB	GB	GB	GB
Non-permanent structures	—	—	—	—	—	—	—	—	—	GB	GB	GB	GB	GB

Note: GB = Great Britain = England, Wales and Scotland; E = England; W = Wales; S = Scotland

1 1841: only for persons born in Scotland or Ireland; 1851–91 whether British subject or not
2 also whether full-time or part-time
3 asked of part-time workers only
4 asked of farmers and tradesmen only
5 only distinguishing (a) agriculture, (b) trade, manufacture or handicraft, (c) others
6 1811–31: only distinguishing (a) agriculture, (b) trade, manufacture, or handicraft, (c) others
7 1901–11: restricted to whether those carrying on trade or industry worked at home
8 also number living and number dead
9 date first marriage ended not asked
10 1891–1901: only required if under 5 rooms; 1921–61: returned by the enumerator.

Source: Office for National Statistics

with only five topics appearing for the first time. Of these three are of particular note: ethnicity, long-term illness and central heating. Ethnicity was included in the 1991 census after being excluded from the 1981 census because pilot tests in the late 1970s suggested that the question was, at that time, too sensitive to ensure a good response. During the 1980s the representatives of different ethnic minority communities became convinced of the benefits of a question as a way of arguing for improved resources. As a result, a question was formulated which asked individuals to categorise themselves into one of a number of ethnic groups: White, Black Caribbean, Black African, Indian, Pakistani, Bangladeshi, Chinese, Other Asian, Other. This question was answered without undue problem and the results have been widely used by policy-makers. In addition, the then OPCS sponsored the production of four edited volumes which look at different aspects of Britain's ethnic groups. However, the question has also been criticised by many commentators (for example, *see* Chapter 16). Among the valid criticisms are that the question does not distinguish between race and culture; that it does not offer the option of Black British (a term which many second-generation British people use to describe themselves); and that by having only one 'White' category it does not reflect the wide heterogeneity of white ethnic groups – a vocal lobby in this latter case has been the Irish community.

The second new question in 1991 was to ask each individual whether he or she had a limiting long-term illness. There had not been a health question in the census since the early twentieth century (when there was a question asking whether anyone in the household had any of a number of infirmities including being an idiot!). Prior to 1991 there was a great demand for a question on health and a number of questions were trialled including one on disability. The question on limiting long-term illness correlated well with other health service indicators such as attending hospital or visiting a general practitioner and was deemed to be the best single proxy of health. Therefore despite the clear subjectivity in the question, it was asked. It has to be said that this question has been widely used and the results seem to reflect expected variations (*see* Chapters 3, 26, 27 and 29).

The third innovation was a question concerning central heating. There was a lot of demand, particularly from the commercial sector, for questions on standard of housing – for use in the market indicators described in the introduction. It should of course be recognised that this question is not a really good index of poverty as it does not measure whether occupants can afford to use their central heating. Another example was age of building. However, this latter question was answered poorly in the tests and it was deemed that central heating gave an indicator of (a) modernity and (b) gentrification.

In preparation for the 1991 census one question which was demanded almost universally was on income. This was required by local authorities, the commercial sector and the academic community (*see* Chapter 25 for evidence of the continued need). The ONS was very wary of this question, believing that it might be thought too sensitive by many people. (It should be pointed out that if one question is thought sensitive, people are also likely to answer other questions poorly. This is thought to have happened in 1951 with a question asking about marriage prior to the current one.) It remains to be seen whether a question on income will be asked.

Another question for which there was much demand concerned the relationship between any two members of a household. This question was needed to give information on the ever-changing patterns of household formation in Britain. However, it is very difficult to ask such a question in a simple way.

The final question worthy of mention here is that of religion. Following an extremely strong lobby from the different religious groups the ONS piloted a question on religion. There is a great difficulty in asking this question; one needs to decide whether one wants to know about religious belief, religious affiliation or religious practice. It remains to be seen whether a valid question on religion can be developed (Joanna Southworth considers this question in greater depth in Chapter 17).

To turn to the questions which have been asked, it is worth pointing out a number of interesting issues. First, it is clearly crucial to try to maximise coverage. For example, there is a great effort to ensure that very young babies are included and people are asked to write 'Baby' if they have not given a new child a name. However, despite this, comparisons with birth statistics reveal that there is always an underenumeration of young babies. It is not clear why this occurs.

Two questions which have been asked over the years but have often been criticised are those on number of rooms and on education. With regard to number of rooms people have been asked for the number of rooms in their house excluding kitchens under $6\frac{1}{2}$ ft wide. Clearly, this may have led to a large amount of measurement error, and continued lobbying has made it likely that a revised question will be asked in 2001. On the criteria described above this is an example of a 'difficult' question. With regard to education, the question has asked for educational qualifications at degree or diploma level. This excludes the majority of the country despite the fact that there is a lot of survey evidence that school qualifications such as General Certificate of Education (GCE) or General Certificate of Secondary Education (GCSE) passes can improve one's chances of success in many areas of life. This is an example of a question which is not relevant to a large proportion of the population. Again it seems likely that this question will be improved in 2001 (*see* Chapter 31).

Although this section has concentrated on some of the faults in the census questionnaire, it is important to recognise that in general the questionnaire is extremely sound and extracts a good deal of very useful information.

Census coverage

It would be good to report that everyone in the nation saw it as their civic duty to fill in their census form and that the census achieved a 100 per cent count of the population. Sadly this is not the case. A number of people are hard to enumerate and are missed by enumerators. This is a worldwide phenomenon. The 1991 census achieved a 98 per cent count, with the remaining 2 per cent of people being disproportionately grouped among those aged 18–30 (with men being more likely to be missed than women) and among the very old. In 1991 this level of underenumeration was higher than in previous censuses, and a number of innovative approaches were used to allocate the underenumeration by age, sex and geographical area.

For the 2001 census the aim at the time of writing is to develop a fully integrated census and underenumeration dataset at an individual level – known as a *one number census* – with those people who are imputed being flagged. ONS set up a large research project in this area with the aim of suggesting a strategy and consulting widely until a decision could be made in late 1998 to permit the procedure – which is based on statistical modelling of a coverage survey to be undertaken after the census together with the use of some administrative records – to be trialled in the 1999 dress rehearsal.

It should be added that the simplest way to do a one number census is to achieve a 100 per cent count. While this is extremely unlikely, much effort is always put into improving the publicity for the census and the friendliness of the census form. With regard to publicity it is essential to make the message clear and effective. In 1991 a television advertisement featuring a baby was thought by many commentators to be particularly abstract. However, it won a media prize for artwork!

Conclusion

The census is the key dataset for much of social statistics. Although expensive in terms of money it is needed by a very large user community and influences for a decade the lives of every person in the UK through its widespread use in resource allocation. If there were no census there would be enormous uncertainty in the allocation of resources. While it may be possible to get reasonable estimates of the population from other sources, such as administrative records, there is no other source which gives, at a local level, measures of deprivation and hence of need. I am convinced that it represents value for money, but only if the ONS continues its encouraging moves in the 1990s towards modernisation and consultation throughout its user community. Readers interested in learning more about the census or about its outputs are encouraged to consult one of the user guides which are produced at each census, or articles in the journal *Population Trends*.

3

Government Household Surveys

Charlie Owen

Using government household surveys in social research

There are five large-scale continuous household surveys conducted on behalf of the British government. Together these surveys each year conduct interviews with some 200 000 households, including data from over 400 000 adults. These surveys constitute a huge statistical resource. The published tables can be examined for alternative interpretations. Even more powerful is the secondary analysis of the raw data to address researchers' own questions. In either case, there are important considerations, both technical and conceptual, before the data can be used or understood. Part of the interpretation of the statistics must involve an understanding of the contexts in which they were produced. This chapter includes a brief description of each survey and some examples of how the data have been used for further analysis. It therefore allows insights into how it may be possible to make use of the government household surveys.

Introduction

Every year the government collects and publishes huge amounts of statistics. Many of these are catalogued in the Office for National Statistics (ONS) *Guide to Official Statistics* (Purdie, 1996), which is also available on a CD-ROM. Many are routine administrative statistics which get little publicity. Others are more likely to receive media attention, because of their political significance. Such statistics would include those concerning hospital waiting lists, recorded crime or unemployment. Barely concealed attempts at political manipulation of some of these statistics, dealt with elsewhere in this volume, have served to undermine confidence in official statistics more generally (*see* Chapters 29, 30 and 37).

This chapter concentrates on a particular set of statistics: the regular household surveys. First there is a description of the surveys in general, followed by a more detailed description of each one. There follows a consideration of some of the factors that need to be taken into account when using the data, together with some examples of such use.

The household surveys

There are five continuous national household surveys conducted on behalf of the government. These are listed, together with their sample sizes and dates of origin, in Table 3.1. As can be seen, taken together they collect data from over 150 000 households and almost 400 000 individuals each year. This is a pretty impressive volume of data, but quantity does not guarantee quality. If the samples are biased or if the questions are not appropriate, then the data may be unusable.

For each survey, tables of results are published in a printed volume, together with some analysis and commentary. In addition, the raw data for all the surveys are available, in a number of formats, from the Data Archive. (The Data Archive, University of Essex, Wivenhoe Park, Colchester, Essex CO4 3SQ; Telephone: 01206 872001; Fax: 01206 872003; email: archive@essex.ac.uk. Details of the datasets can be obtained from the Data Archive's online catalogue, BIRON, which can be accessed from the Data Archive web page, http://dawww.essex.ac.uk.) (*See* Chapter 6 for further information). Although it is possible to extract new information from the published tables, access to the raw data gives much more scope for new analyses.

A great benefit of all of the household surveys is that they collect data from (or about) all members of the household, and the data are coded hierarchically. People are coded within families, and families are coded within households, so that it is possible to link data between different members of the same family and/or household. This aspect of the data has, for example, been used to look at inter-ethnic marriage and cohabitation (Berrington, 1994). Examples of analyses which make use of the multi-level nature of the data are provided by Dale and Davies (1994).

It is, of course, important to remember when interpreting the data that the surveys do have a specific definition of household and family. A family is defined as:

(a) a married or opposite sex cohabiting couple on their own, or
(b) a married or opposite sex cohabiting couple/lone parent and their never married children, provided those children have no children of their own.

(McCrossan, 1991)

The definition of a household is:

a single person or a group of people who have the same address as their only or main residence and who either share one meal a day or share the living accommodation.

(McCrossan 1991)

Table 3.1 Approximate size of the samples in the household surveys

Survey	Started	Households	Individuals
Family Expenditure Survey	1957	7 000	18 000
Family Resources Survey	1993	25 000	60 000
General Household Survey	1971	10 000	20 000
Labour Force Survey	1973	96 000	250 000
National Food Survey	1940	7 000	22 000

National Food Survey

The oldest of the national household surveys is the National Food Survey (NFS), set up in 1940 by the then Ministry of Food. It was extended in 1950 to become a representative sample of households in Great Britain (Slater, 1991). The NFS is conducted by the Ministry of Agriculture, Fisheries and Food (MAFF). It is a survey of food and drink consumption and expenditure by households. Consumption data are also converted to nutritional intakes of energy, protein, fats, sugars and a number of vitamins. From 1995 results for eating out have been included. The survey covers Great Britain and the sample size is about 8000 households.

Results are published in an annual report called the *National Food Survey*, based on a calendar year, and as quarterly press releases. In addition a much more detailed report is published as the *National Food Survey Compendium of Results*; this volume contains detailed tables of household food consumption and expenditure and nutritional intakes from household food (Chapter 18 provides detailed criticisms of the survey).

Family Expenditure Survey

The Family Expenditure Survey (FES) has been carried out continuously since January 1957. It is conducted by ONS in Great Britain and by the Northern Ireland Statistics and Research Agency in Northern Ireland. The main government use of the FES is to calculate the weights for the Retail Price Index (Department of Employment, 1987) and as a source of data for estimates of consumer expenditure in the National Accounts (*See* Chapter 41), although it is widely used for other purposes by government departments. The survey collects both expenditure and income data from each member of the household aged 16 or over. Expenditure information is collected partly by interview and partly in diaries kept for 14 days by each person. There is also a detailed questionnaire on all sources of income for each adult.

The survey collects complete data from about 7000 households per year. However, this represents a response rate of only about 70 per cent. This loss of a large number of eligible households is the basis of much concern over the validity of data from the FES. Results are published in the annual volume *Family Spending*. From 1979 to 1993 data from the FES were used for the annual publication *Households below Average Income*; from 1994 onwards the Family Resources Survey has provided these data.

General Household Survey

Unlike the other household surveys, which have a single main purpose, the General Household Survey (GHS) is fundamentally a multi-purpose survey. It began in 1971 and is conducted in Great Britain; in Northern Ireland there is a similar survey called the Continuous Household Survey. Since 1988 the fieldwork has been conducted on the financial year. The sample is just under 20 000 households per year. The response rate has been between 80 and 85 per cent. Results are published in the annual volume *Living in Britain* (called *General Household Survey* prior to 1994).

Every year there are questions on fertility, housing, health and illness, car ownership, employment and education. Every two years questions are asked about smoking and drinking and about (since 1983) dental health (*see,* for instance, Chapter 26). Other topics are included from time to time. These have included bus travel, share ownership, childcare, sight and hearing, medical insurance and voluntary work. A great merit of the GHS is that it does cover a whole range of topics. This means that it is possible to look at the interactions between variables, e.g. between illness and unemployment (*see* Chapter 38). The long span of the dataset also makes it possible to examine time trends in some variables. Given the central importance of the GHS it was an unwelcome surprise that, prior to a review by ONS of the multi-purpose surveys, the GHS was cancelled for the year 1997/98. Fortunately, it was reinstated for 1998/99: there will be no survey in 1999/2000, but a redesigned GHS will begin in April 2000.

Labour Force Survey

All the countries in the European Union conduct a labour force survey, as a condition of membership, with a core set of questions. One of the roles of the LFS is to provide measures of employment and unemployment using the standard definitions from the International Labour Office (ILO) (*see* Chapters 36 and 37). The LFS was first conducted in 1973 and thereafter every two years until 1984, when it became annual. Since 1992 the LFS has become quarterly, with the same sample size (about 60 000 households) per quarter, as used to be interviewed per year. However, there are not 60 000 new households each quarter: each household is interviewed five times, for consecutive quarters. This means that successive quarters are not based on independent samples. Thus each year data are collected from around 96 000 different households.

The first interview is conducted face to face, but later interviews (where possible) are conducted by telephone. Some people who complete the first interview drop out for subsequent interviews. For example, for the spring 1997 LFS the response rate for the first interview was 81 per cent; the rate for the second interview dropped to 75 per cent and was 73 per cent for the fifth interview. The LFS includes households in Northern Ireland. For all aged 16 and over there are detailed questions on employment, including type of job and hours worked; unemployment; education and training; and health and disability.

The LFS was the first of the surveys to include a question on ethnic group, which was introduced in 1979. A question on ethnic group was included on the 1991 census for the first time: before then the LFS was the main source for estimating the size of Britain's ethnic minority populations (Office of Population Censuses and Surveys, 1986a). Indeed, because of the relatively small number of variables in the census and the difficulties of cross-tabulating them, the LFS continues to be a major source for data on ethnic minorities (Haskey, 1996).

Results used to be published in an annual volume, the *Labour Force Survey*, but since 1992 results have been published in the *Labour Force Survey Quarterly Bulletin*. Results are also presented in the ONS monthly journal *Labour Market Trends*, which includes a section called 'Labour Force Survey Help-Line'.

Family Resources Survey

The Family Resources Survey (FRS) is the most recent of the household surveys. It began in October 1992, with the first full sample beginning in January 1993. The FRS is conducted by the Department of Social Security (DSS) in order 'to provide detailed information about the characteristics and finances of households'. Its main purpose is to support monitoring of the social security programme; modelling and costing of policy changes to benefits, National Insurance contributions, and child support systems; and forecasting of benefit expenditure (Blackburn and Lincoln, 1992, p. 14). Before initiating the FRS the DSS had previously obtained this information from the FES. However, it was decided that the sample size of the FES was too small for DSS purposes. The DSS considered various ways of collecting the data that were required, including an expanded FES, but in the end decided on a new survey. The FRS collects data from around 25 000 households per year in Great Britain. The questionnaire covers primarily income, but also other areas of interest to the DSS such as informal care of the elderly and disabled, occupational pensions, childcare and savings. The fieldwork is conducted by ONS and Social and Community Planning Research (SCPR). The response rate to the FRS, like the FES, is around 70 per cent.

Using the data

This very brief summary of the structure and content of the five household surveys should have demonstrated what a wealth of data there is waiting to be exploited. Indeed, some landmark studies have drawn heavily on these datasets. The most obvious example is Townsend's massive study of poverty (Townsend, 1979), which made extensive use of the FES and also of the GHS. The Black Report on inequalities in health (Townsend and Davidson, 1982) made use of the GHS, and even had recommendations for improvements in its questions on income.

The published data are frequently used by campaigning groups, such as the Child Poverty Action Group (Walker and Walker, 1997) or Radical Statistics (Radical Statistics Health Group, 1987). However, much more can be gained by further analysis of the raw data.

There are a number of texts which set out methodological principles for the secondary analysis of statistics (Dale *et al.*, 1988; Stewart and Kamins, 1993). One factor that always has to be taken into account is the response rate. For a household survey this is not always straightforward to calculate. Each interviewer is issued with a set of addresses at which to conduct interviews. At some of those no contact will be made; at others the household will refuse to take part. For example, for the 1995 GHS, there were 11 914 households in the sample; 359 of these (3 per cent) were never contacted and 1810 (15 per cent) refused to take part (Rowlands *et al.*, 1997). This gives an overall response rate of 82 per cent. However, for some of the households some members were not contacted or would not take part; usually it was possible to obtain *proxy* data from another member of the household. In some cases, however, a respondent either would not answer a question themselves or could not answer for another person. For only 8431 households were there complete data from all members for all questions. This gives a response rate of 71 per cent. However, the response rate which is usually quoted is the one where partial data from some or all of the household members is acceptable: this is called the middle response rate and for the GHS in 1995 this was 80 per cent.

Even when an overall response rate has been established, it may be that non-response is uneven, so that respondents are different in important ways from non-respondents. The 1995 GHS reported its middle response rates by region: they varied between 76 per cent for Greater London and 85 per cent for East Anglia. For each of the surveys census data from responding households and from non-responding households have been compared. These comparisons have shown a number of important differences. For example, households comprising one adult aged between 16 and 59 or a couple with no dependent children were under-represented in the surveys, while households with dependent children were over-represented. So although the original, target samples were a random sample of the population, the achieved samples were not. This has to be borne in mind whenever the data are examined.

However, there are other considerations. Each of the surveys is conducted for a purpose, as summarised in Table 3.2. Any evaluation of the data not merely would have to be concerned with any technical errors or inadequacies, but also would need to consider the conceptual framework and presuppositions which lay behind the surveys (Hindess, 1973, pp. 44–5). The definition of family, given above, is one example: it makes sense in the context of the survey, but it does not necessarily coincide with other concepts of family. The LFS, concerned with paid work, can simply ignore domestic work or unpaid childcare. The FRS obtains detailed information about income into the household, but can say nothing about inequalities of consumption within the household. To say that these statistics are a social product that bear the inevitable traces of their origins is in no way to imply any misrepresentation, but simply to point out that part of the interpretation of the statistics must involve an understanding of the contexts in which they were produced. This does not make the data unusable, but it does mean that terms such as 'family', 'labour' and 'resources' have to be understood in their contexts.

Some examples of data use

The GHS includes a section on family formation and fertility. All adults aged between 16 and 59 are asked whether they are currently married or cohabiting and about their marital history; women are asked whether they have had children and if they expect to have any more. It is thus a unique source of data on family formation and intentions. The Family Policy Studies Centre has made use of these data in a number of reports. For example, Kiernan and Estaugh (1993) used the

Table 3.2 Main purpose for each household survey

Survey	Main purpose
Family Expenditure Survey	Retail Price Index
Family Resources Survey	Benefit levels
General Household Survey	Multi-purpose
Labour Force Survey	International Labour Office employment statistics
National Food Survey	Food and drink consumption; nutrition

GHS to study cohabitation and childbearing outside marriage. They found that cohabitation before marriage was becoming a norm, but whereas in the past cohabiting couples were very likely to marry around the time of the birth of a child, this was becoming less likely, along with a greater acceptability of child-bearing outside marriage. Burghes (Burghes, 1993; Burghes and Brown, 1995) studied one-parent families, the parent neither being married nor cohabiting. She found the familiar trend of a rapid increase in the number of one-parent families, with lone-mother families outnumbering lone-father families by almost ten to one. By using the GHS Burghes was able to look in much more detail not just at the numbers but at the circumstances of lone parents. She found that the most rapid growth had been in single (never married) lone mothers, and that it was these women who were in the most difficult circumstances: for example, they were the least likely to be employed, the least likely to have educational qualifications and the least likely to be receiving maintenance from the children's fathers.

Brannen *et al.* (1997) examined the relationship between parenting and employ-ment more broadly, using LFS data for the years 1984–94. They found a slight fall in the percentage of lone mothers in employment over the period, at the same time as a substantial growth in the employment of married and cohabiting mothers, especially in full-time employment. Furthermore, they found that this increase in employment for couple mothers was not uniform: it was almost entirely restricted to couples where the father was working full-time. Consequently, over the period, there was a growing polarisation between families with two earners – especially two full-time earners – and families with no earners, owing to the growth in the number of one-parent families. Families with one earner (with either one or two parents) dropped from 43 per cent of all families with children in 1984 to 33 per cent in 1994 (*see* Figure 3.1).

In a quite different vein, Bartley and Owen (1996a) investigated the association between employment, health and socio-economic status for men over the period 1973–93, during which there was a big decrease in the percentage of men in employment. Using GHS data they found a complex relationship (*see* Figure 3.2). All socio-economic groups experienced a fall in employment rates, but this fall was itself related to socio-economic group, being least in the professional and managerial group and most in the semi- and unskilled group. Men with a chronic illness were less likely to be in employment whatever their socio-economic group, but the effect was linked to socio-economic group in the same way. That is, men with a chronic illness were only slightly less likely to be employed if they were in the professional and managerial group but the difference was marked in the semi- and unskilled group. As employment rates fell, the differences between men with a chronic illness and those without were amplified, because the employment rate for men with a chronic illness fell more rapidly. This difference was also associ-ated with socio-economic group: the growing difference in rates of employment was most marked in the semi- and unskilled group and least in the professional and managerial group. Bartley and Owen concluded that 'Socioeconomic status makes a large difference to the impact of illness on the ability to remain in paid employment, and this impact increases as unemployment rises' (p. 445). (Chap-ters 28, 37, 38 and 40 also touch upon these issues.)

It has already been mentioned that the LFS has long included a question on eth-nic group. From 1992 onwards the LFS has used the same question as the 1991 census (*see* Chapters 16, 17 and 25). Prior to that the LFS question had included

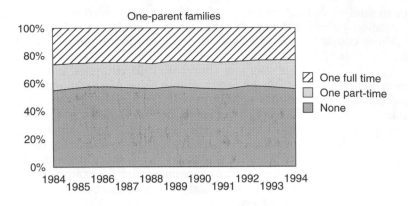

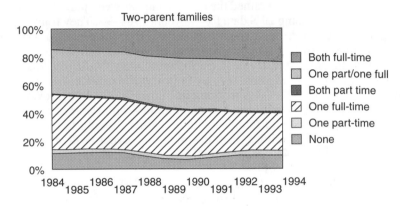

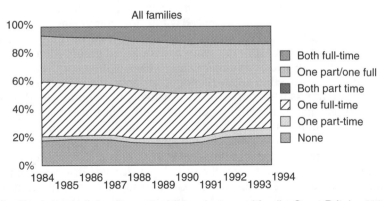

Figure 3.1 Employment in families with children by type of family, Great Britain, 1984–94
Source: Labour Force Surveys, 1984–94

a category of 'mixed'. Phoenix and Owen (1996) provided a context for a discussion of mixed relationships and mixed parentage using data from the LFS. They found that only about 0.5 per cent of the total population had described themselves as 'mixed', but that this constituted about 10 per cent of Britain's minority

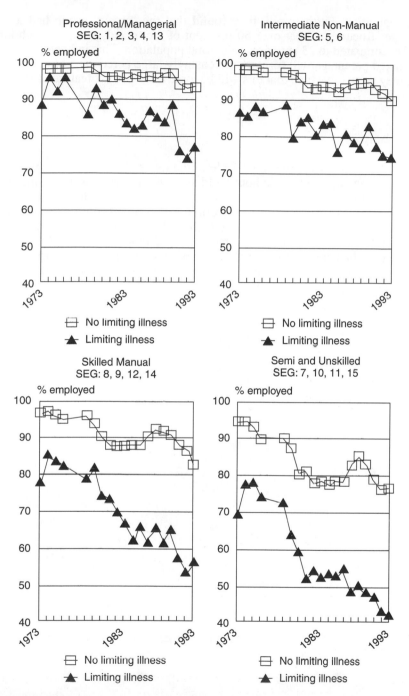

Figure 3.2 Percentages employed according to limiting long-standing illness and socio-economic group. SEG = socio-economic group as defined in the national censuses
Source: General Household Surveys, 1973–93

ethnic population. Furthermore, they found that the 'mixed' group had a very young age structure, so that over 50 per cent of the mixed group were children under 16, compared to 23 per cent of the total population. This indicates that people of mixed parentage will form an increasingly large percentage of the population as they grow older, and that this will become an important feature of Britain's population.

Further help

These few examples serve to show a little of the range of social issues that can be addressed using data from these household surveys. If anyone wants to make use of these data, there are many sources of help. The first place to look is in the published reports: these give a lot of background, including response rates. The Data Archive has copies of all the necessary technical documentation. There are also two ONS publications which contain more general information. *Statistical News* is a quarterly which contains non-technical news and articles on the work of ONS, including the household surveys; the *Survey Methodology Bulletin* is published twice a year and is a source of more technical articles. Both are recommended.

4

Confidentiality of Official Statistics: An Excuse for Secrecy

Angela Dale

Truly open government would reduce concerns about confidentiality of data

All governments collect data on the population they govern and such data form a fundamental tool in assessing the state of the nation, in the prediction of future requirements and in the allocation of resources. Many chapters in this volume refer to the availability or the quality of statistics collected by government and sometimes the need for more information is expressed. Any discussion of greater access to data invariably includes questions of confidentiality of the respondents who provide the data. This chapter aims to consider confidentiality in terms of the tension between the rights of the respondent and the public's 'right to know'. It begins by considering the dimensions of politics and power which may influence the collection and availability of official data.

The US census provides a vivid example of the political importance of official data. Political representation in the USA is directly linked to population size and, since its inception in 1790, the US Census of Population has provided the basis for apportionment. The speed of population growth – from 3.9 million in 1790 to 240 million in 1990 – has meant that apportionment has been a live issue through-out this period with frequent challenges from political parties (Anderson, 1988). Most recently, legal challenges have been mounted over adjustment to the 1990 census, with cities such as New York standing to gain considerably if census counts were adjusted for underenumeration (Choldin, 1994). Although this con-trasts with the much less politicised nature of the census in the UK, it nonetheless highlights the fact that these data make a powerful input to the allocation of resources. The availability of that information base for analysis outside govern-ment provides the opportunity to mount challenges – for example, in the UK by reassessing the methodology of the standard spending assessment that determines local government funding (*see* Chapter 21). Conversely, the unavailability of official data for public scrutiny removes the power of challenge.

Debate over UK unemployment statistics in the mid-1990s provides an example where the changing definition of unemployment was seen by many as the result of intervention by government for political purposes (Levitas, 1996a; *see also* Chapter 37). In order to address this perception, the Office for National Statistics (ONS) asked the Royal Statistical Society (RSS) to make recommendations on how unemployment should be measured and published a detailed guide to its measurement (Royal Statistical Society, 1995a; Office for National Statistics, 1997a). This example highlights the importance of open discussion over definitions, a clear and agreed basis for adopting a particular definition or methodology, and the timely publication of figures. Access to raw data in order to calculate alternative figures using different definitions provides an additional benefit, particularly welcomed within academia (*see* Chapter 36, for instance). One of the concerns of those who argue for greater availability of data is that issues of confidentiality may in some cases be used as a smokescreen to limit the release of data.

Context of publication

There is enormous variation over time and between countries in the way in which governments organise the collection of data, the purposes for which such data are used, whether they are available outside government and, if so, in what form. In general, one can argue that increasing the accessibility of official statistics to those outside government increases the ability to make an informed challenge, either to what is reported, how it is interpreted or how it is used. However, there is an important difference between access to a glossy report on, for example, change in household income over time and the ability to access and reanalyse the surveys conducted by government and used to compile that report. It is evident that any report by government has to be selective in coverage; it will use 'official' definitions and is unlikely to provide any depth of analysis. While many excellent critiques are based on such reports, the scope for reanalysis and reinterpretation is severely limited by the amount of data which is in the public domain. For example, the Family Expenditure Survey (FES) report provides descriptive statistics on the income and expenditure of households and families of different composition and at different stages of the life-course (*see* Chapter 3). However, the availability of the raw data from the FES has provided economists and social policy analysts with the scope for much more sophisticated analysis, using a range

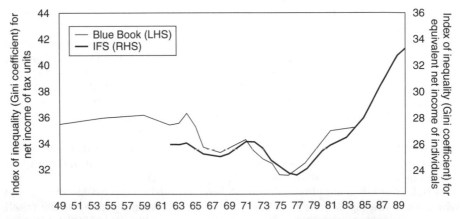

Figure 4.1   Income inequality in the UK, which grew very rapidly after 1977
Sources: Joseph Rowntree Foundation Income and Wealth Group (1995), Goodman and Webb (1995).

of different definitions, assumptions and models. Much of this analysis was used to compile the Joseph Rowntree Foundation's *Inquiry into Income and Wealth*, a two-volume report charting the growth of income inequality since the late 1970s (Joseph Rowntree Foundation Income and Wealth Group, 1995 and Figure 4.1).

The UK situation

ONS's dissemination policy explicitly aims to increase the accessibility of official statistics. The press release at the launch of ONS in 1996 quoted the director as saying:

> I want official statistics to be more widely available and more widely used not just by businesses and analysts but by ordinary people. ... Statistics that are collected or created and then never referred to or used again are of no value.

There is also a recognition that there is great potential for further analysis of the data collected by ONS and thus its dissemination policy includes a commitment to deposit data free of charge at the UK Data Archive for the purposes of academic research and teaching. However, the ONS confidentiality code also states, 'an overriding concern for ONS is the safeguarding of the information it holds about people and organisations, especially where that information has been given in confidence'. It goes on to say, 'ONS does not publish or release data or analyses unless it is satisfied that the risk of disclosure of information about individual data subjects, even indirectly, is effectively eliminated.' The question which occupies the rest of this chapter is how that balance is drawn, what are the relevant parameters and how they vary with the data source.

Range of official statistics

Official statistics take a wide range of forms; the diversity of sources and range of media on which they are supplied are increasing steadily. For the UK, the *Guide to Official Statistics*, published annually by ONS (and now available on the Internet), together with the ONS Catalogue, provide the most complete listing of sources. They include administrative records (e.g. births, deaths, abortions, National Insurance records); economic indicators (e.g. level of exports and imports); a census of population; official surveys, some repeated on an annual basis (e.g. the Labour Force Survey, the General Household Survey, *see* Chapter 3) and some *ad hoc*, conducted to meet a particular need (e.g. Women and Employment Survey; Living Standards and Unemployment). The form in which these data are available, and the amount of detail, vary according to the level of confidentiality attached to them, which, in turn, is related to the purpose for which they are collected and their legal status.

Some of the types of data listed above are collected under statute (e.g. registration data, National Insurance records and the census of population) whilst others are collected on a voluntary basis (e.g. social surveys). However, a further distinction is also important: that between data collected for administrative purposes and data collected for statistical purposes. This largely coincides with the former distinction except in the important case of the census of population, which is collected for statistical purposes but with a legal requirement for all households to complete a form (*see* Chapter 2).

These distinctions have implications for the level of confidentiality associated with the data. Data with the least confidentiality restrictions are those collected by social surveys where the respondent provides information on a voluntary basis and the use of the data for statistical purposes is clearly spelled out to the respondent. These surveys are routinely deposited as microdata (individual records) at the UK Data Archive and are available for further analysis with a basic set of safeguards which include removal of geographical identifiers and legally binding undertakings given by users.

By contrast, the Census of Population, although conducted for statistical purposes, is governed by the 1920 Census Act and the 1991 Census (Confidentiality) Act. These Acts place a statutory requirement on all households to complete a census schedule and also prohibit the census offices from releasing any identifiable records relating to households or individuals. Until recently the interpretation of this legislation resulted in data release being restricted to tabulations – either in report form or as electronic data. However, a reinterpretation of the Act in the late 1980s resulted in the judgement that samples of microdata could be defined as 'statistical abstracts', thereby meeting the publication criteria of the Census Act.

It is widely accepted that the use of data for purposes other than those for which they were collected requires special security, and this is particularly acute with respect to administrative data which are collected under statute (e.g. birth and death records, and National Insurance records). Administrative records are among the most restricted forms of data and are generally available only as aggregate data; for example, ONS publishes a series of statistics on births and deaths, available at local area level, but not as individual records. Some administrative data are, however, placed in the public domain: the Public Records Office allows public access to restricted versions of vital registration documents and the Electoral Register lists the names and addresses of everyone registered to vote and is available at local libraries. The Electoral Register is also available in electronic form and is widely used as a sampling frame for surveys and also in some geodemographic classifications.

The following sections use the census of population as a basis for a more detailed discussion of the various aspects of confidentiality.

The Census of Population

The census is the most widely used source of official statistics and is discussed in greater detail in Chapter 2. It provides essential information for central government (in particular the Department of Health and the Department of Environment, Transport and the Regions – both major spending departments); it is extensively used by local government in order to estimate demand for services and to allocate resources; it is used by the business sector, particularly in developing geodemographic indicators; and it is widely used in academia. As technology becomes more powerful and analysis techniques more sophisticated, demands increase for more detailed data. When this demand for data is set against the statutory basis under which the census is conducted – a legal requirement for all households to complete a census schedule and the prohibition on the release of any identifiable records relating to households or individuals – it is clear that the census provides an arena where confidentiality issues are most sharply focused.

The amount of data going into the public domain has been steadily increasing over time, not only as printed reports but as statistical abstracts, mainly supplied in machine-readable form. In 1991 two sets of local area statistics were released (the Local Base Statistics and the Small Area Statistics; Figure 4.3 provides an illustration of one SAS table layout) with a total of 82 tables and 8722 counts for each area; special workplace and migration statistics; and, new in 1991, two samples of anonymised records: a sample of 2 per cent of all individuals and a separate sample of 1 per cent of all households enumerated in the 1991 census (Dale and Marsh, 1993). Despite this seemingly large amount of data, the scope for analysis is invariably restricted by lack of detail or lack of sample size and there is an ongoing debate over the appropriate balance between ensuring the confidentiality of respondents and maximising the value of the data. These issues are discussed in more detail in the following sections.

Issues surrounding the release of census data

There are clearly important arguments why census data should be widely available. It is efficient use of public expenditure to ensure that the data are analysed as fully as possible and, in addition, the sale of census data provides a source of revenue. As discussed earlier, it allows those outside government to reanalyse the

Figure 4.2 'I know the census is supposed to be completely secret, but I am surprised you've only got one toilet in that big house of yours.'
Source: *Evening Standard*, 24 April 1991

1991 Census Small Area Statistics – 100%

Cowes Castle   **Medina**   Area Identifier – 29KYFA01   Grid reference – SZ48729646

PRODUCED USING SASPAC   ZONE 29KYFA01   **Isle of Wight**   *CROWN COPYRIGHT RESERVED*

Table 2 Age and marital status: Residents

Age	TOTAL PERSONS	Males			Females		
		Total	Single widowed or divorced	Married	Total	Single widowed or divorced	Married
ALL AGES	448	219	102	117	229	115	114
0–4	12	10	10	xxxx	2	2	xxxx
5–9	18	8	8	xxxx	10	10	xxxx
10–14	20	9	9	xxxx	11	11	xxxx
15	4	1	1	xxxx	3	3	xxxx
16–17	5	4	4	0	1	1	0
18–19	12	10	10	0	2	2	0
20–24	15	10	9	1	5	4	1
25–29	13	11	10	1	2	2	0
30–34	17	10	2	8	7	2	5
35–39	17	6	4	2	11	0	11
40–44	25	12	2	10	13	1	12
45–49	36	18	2	16	18	4	14
50–54	28	14	5	9	14	2	12
55–59	25	12	0	12	13	3	10
60–64	36	20	3	17	16	5	11
65–69	40	16	4	12	24	8	16
70–74	32	16	7	9	16	7	9
75–79	38	19	5	14	19	11	8
80–84	18	7	4	3	11	8	3
85–89	23	4	1	3	19	18	1
90 and over	14	2	2	0	12	11	1

Figure 4.3 SAS table for ED KYFA01

Source: 1991 census, Crown copyright

data used for resources decisions – for example, for standard spending assessments and, if appropriate, to challenge the decisions reached. It also provides a means by which the quality of data can be assessed. The analysis of data from the 1991 census has generated a vast amount of new and valuable information about society. What, then, are the constraints on the release of data?

First, it is important that the census offices ensure that the statutory requirements of the Census Acts are met and also that they keep the undertaking made to respondents at the time of the census. The 1991 census schedule contained an undertaking from the Registrar-General which read:

> Your answers will be treated in strict confidence and used only to produce statistics. Names and addresses will not be put into the computer; only the postcode will be entered. The forms will be kept securely within my office and treated as confidential for 100 years.

The importance of keeping trust with the public is seen by ONS as fundamental. The ability to collect official statistics depends upon the cooperation of the public. If this is lost, then whatever the legal requirement, it will become impossible to collect high-quality data. In the period immediately before the 1991 census in Britain, fears (totally unfounded) that census returns would be passed to local authorities to allow them to track down those who were not on the community charge (poll tax) register led to some local campaigns not to complete a census return. This highlights the importance not only of observing the undertakings given to the public but also of maintaining a climate of trust and public confidence.

This example also highlights a particular concern over the use of statistical data for administrative purposes. A proposal by the German government to use records from the 1981 census to update and correct local population registers caused such public disquiet that the proposal was abandoned and the census could not be conducted until 1987 (Redfern, 1987). There has been no census in Germany since this date and there are no plans to conduct a census in 2001. In other European countries, for example the Netherlands, a census has been abandoned because of a lack of public acceptability. There are, therefore, real public concerns over confidentiality which may rise unexpectedly in response to unpredictable events or to media attention. Although one may have much sympathy with the argument that the information collected by the census is so bland and innocuous that disclosure could not result in damage, that fails to address the fact that an undertaking has been given. It is very clear that, particularly around the time of the decennial census, concerns are often voiced by the public over confidentiality and, in particular, the uses to which the data may be put.

Balancing confidentiality against data availability

It seems clear that if major data collection exercises such as the census are to be successful, they have to retain public confidence. To ensure public confidence the data must be protected not just against a direct threat to confidentiality, but also against perceived and indirect threats (Cox *et al.*, 1986).

A *direct* threat to confidentiality occurs where a person can be identified and additional information thereby disclosed. Fellegi (1972) defined *disclosure* as requiring both identification of the subject (identity disclosure) and the disclosure of information additional to that used in identification (attribute disclosure). Marsh *et al.* (1991) argued, in the context of the release of samples of microdata, that the process of identification requires a match on a set of key variables between a subject in the dataset and in the population. However, to achieve this it is necessary, first, to be able to define an exact set of matching keys in the dataset and the population, and, second, for the individual to be unique on these key variables in both the sample dataset and the population. Finally, it is necessary to verify that the individual is, indeed, unique in the population. They made an attempt to quantify the risk of identification of a given individual in a 2 per cent sample of microdata drawn from a population census, coming up with an infinitesimal chance of about 1 in 4 million. These arguments formed the basis of the case submitted to the census offices to release samples of anonymised records from the 1991 census.

Most work has focused on direct threats to confidentiality, ways of estimating disclosure risk and finding methods of overcoming it; for example, by limiting the amount of detail contained in microdata files or, in the case of aggregate data, adding random noise to table cells with very few respondents. However, these risk assessment exercises are premised on the assumption that someone might actually wish to attempt to identify an individual or household. Little systematic analysis has been done on the circumstances in which this might occur. Elliot (1996), however, has set out an assessment of different scenarios under which an intruder might attempt identification in a microdata file. The conclusion from this is that there are very few circumstances in which any benefit could be achieved. The most likely occurrence would be an attempt to discredit the census or the census offices. However, it is also apparent that discredit could be achieved much more easily by making false claims – either of identification or claims over access to census schedules. This leads to the consideration of perceived threats to confidentiality.

Cox *et al.* (1986) point out that *perceived* threats to confidentiality may have as real an effect on public confidence as direct threats but, because they are not based on facts, they are much harder to counter. For example, fears about the passing of data to other government departments, or unverified claims to have identified a respondent, may be difficult to refute convincingly. Perceived threats can best be addressed by building up confidence in national statistical offices and by greater understanding by the media and the public of the data security procedures used. The level of detail on, for example, a microdata file has little relevance.

Indirect threats to confidentiality occur where data have been used in a way that invades the privacy of an individual without direct identification (Cox *et al.*, 1986). This could happen if several different tables are used to infer additional information about a respondent or set of respondents. However, the most widespread indirect threat comes from the use of census data to construct geodemographic profiles for small areas which are then used to target direct marketing mailshots to specific areas. In the US census the more sensitive questions such as income are asked only on the 'long form' which goes to about 20 per cent of households, and are not, therefore, available for very small areas. In Britain there is some concern over 'junk mailing', evidenced by the rise in recent decades in the number of parliamentary questions on census confidentiality which focus on the activities of the geodemographics industry (Marsh, 1993). However, geodemographic profiles cannot give precise targeting of addresses, and concerns may be premised on exaggerated assumptions of their efficiency.

Box 4.1 A typology of threats to confidentiality	
Direct threat	Respondent identified and personal information disclosed
Perceived threat	No identification but the public fear disclosure has occurred
Indirect threat	No identification but personal information inferred

If the public or the media fail to distinguish between these different kinds of threats then we need to be equally mindful of all three. To some extent perceived and indirect threats to confidentiality can be diminished by ensuring a good public understanding of how data are used, and by providing assurances of data protection. However, the volatility of public opinion and the difficulty of focusing media attention on factual information once an issue has hit the headlines suggest the need to be constantly aware of the importance of public confidence.

Conclusions

There will never be absolutes about confidentiality. What is acceptable will always be influenced by public opinion which, in turn, is influenced by history and culture. What seems to be essential is that a national statistical office can command public confidence. This will be facilitated if official statistics are readily and easily available to the public at large and are thus seen to be serving the public interest. To this end, the recent dissemination activities of ONS are greatly welcome. It is also important that the public believe that the responses which they provide will be used according to the undertaking given at the time of data collection. If these factors are in place, then perceived and indirect threats are much less likely to be damaging.

5

Who Pays for Research? The UK Statistical Picture

Ian Miles

Research and development produces knowledge, but is hidden in official statistics

Research and development (R&D) is the process of enquiry for new scientific and technological knowledge, including social scientific and engineering knowledge. Exactly what is and what is not included under this rubric has been extensively debated; most countries follow guidelines set out in the OECD's 1993 *Frascati Manual* – one of a family of manuals addressing statistical issues in science and technology. R&D is immensely important as the source of new knowledge of how to shape the world, and governments and business have much interest in it as a source of wealth. However, generating new knowledge is one thing, putting it into practice in commercial or socially useful innovations is another. This chapter looks at the UK's R&D profile.

The UK is at best an average performer of R&D among advanced industrialised countries, and has an unusually large orientation of its R&D towards military ends. The country's scientific reputation and outputs remain high, reflecting past leadership; many other countries are now catching up. The military orientation of much public R&D is compounded by a long-standing underperformance in industrial R&D. Much of the public sector effort is under severe strain.

Data sources

The *Guide to Official Statistics* (Office for National Statistics, 1996 edition) provides only half a page of information on R&D statistics, and indeed the volume is remarkably loath to deal with issues of science and technology. (The only substantive reference to 'science' will be of interest to animal liberationists: it points readers to the long-running annual series on *Statistics of Scientific Procedures Performed on Living Animals*.) Two major annual compilations of official data come from the Office of Science and Technology, now part of the Department of Trade and Industry. These are the *Forward Look at Government-Funded Science, Engineering and Technology* and *Science, Engineering and Technology Statistics*, which contain data on public- and private-sector R&D expenditures, the scientific labour force, etc.

The Office of Science and Technology funds R&D through the Research Councils, as well as commissioning its own studies. Notable among the latter has been the Foresight Programme, which has resulted in the publication of a series of reports dealing with UK performance and capabilities in many areas of technology. Government departments all also fund R&D to differing extents: the broad picture is outlined in the *Forward Look*, but for more detail on strategies and programmes it is necessary to go to individual departmental sources. The same is true for the programmes and detailed expenditures of Research Councils.

R&D in universities is funded by the so-called binary system. First, there are funds for specific projects from Research Councils, the European Union (EU), international and government agencies, charitable foundations and private business. Foundations are important sources of R&D funding in a few fields, with the Wellcome Trust's budget exceeding that of the Medical Research Council. Other large foundations funding R&D include Leverhulme, Nuffield, Rowntree, and various medical charities. Second, a share of university core funding from the Higher Education Funding Councils is intended to support R&D. The funding councils are responsible for the 4-yearly research assessment exercise which purports to assess the strength of different university departments and research groups, and which yields a considerable volume of data on their performance.

The EU itself publishes a number of important statistical compilations, notably Eurostat's *Research and Development Annual Statistics* and the very detailed *European Science and Technology Indicators 1994* (European Commission, 1997). Other useful sources of comparative data are the OECD's twice-yearly *Main Science and Technology Indicators*, and for a global view UNESCO's annual *Statistical Yearbook* and *World Science Report* (published every two years, with both data and commentary).

The big picture

Overall the share of national wealth devoted to R&D in the UK has remained roughly stable, at around 2 per cent of GDP, for many years – rather low in relation to most countries with which the UK is usually compared (see the trend line in Figure 5.1). Given the – perhaps not coincidental – tendency for these countries to have higher levels of GDP than the UK, the absolute gap is wider still.

What about the composition of UK R&D? Though the *Guide to Official Statistics* places R&D in the chapter entitled 'Public services', the bulk of UK R&D – about two-thirds – is actually performed in business enterprises. Such a dominant role for industrial R&D is common, and UK industry is actually below the OECD average – below Germany, Japan and the USA in particular (Figure 5.1). British companies, like the UK economy as a whole, tend to invest rather less in R&D than do overseas competitors. What is more unusual in the structure of funding is the high proportion of R&D that serves military ends (euphemistically described as 'defence'). Figure 5.1 demonstrates that the UK and USA stand out in this respect, with R&D systems highly oriented to military ends, compared to other OECD countries. (The countries lacking bars report zero or near-zero military R&D expenditure.)

A third important source of R&D funding is that which emanates from overseas. The UK is particularly successful in Europe in attracting such funding. Many

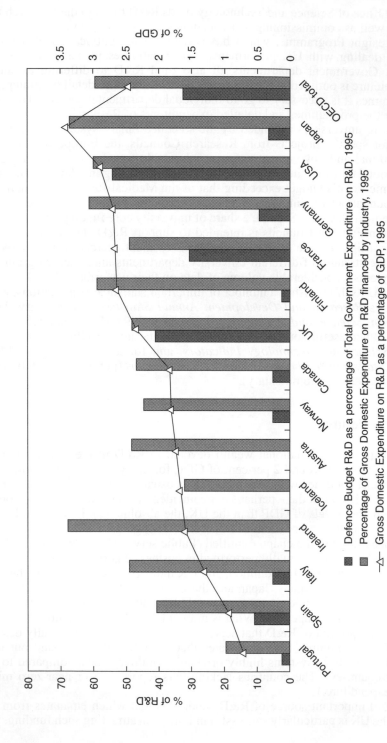

Figure 5.1 Comparative R&D data, 1995 or nearest year
Source: Data from OECD, Main Science and Technology Indicators, 1997 (tables 5, 13 and 64)

transnational companies choose the UK as a European base because of language, relatively low-cost but skilled workforce, and 'flexible' labour markets; some of these companies have graduated from using their bases simply as sales depots or assembly shops, to setting up regional headquarters which support region-specific R&D. Since UK public science capabilities remain relatively strong (and cheap), some transnationals locate R&D facilities near these sources of science and technology competence; and these UK science bases have been relatively successful in attracting EU R&D funding.

Public funding

Trends in public expenditure on science are displayed in Figure 5.2, based on data from *Science, Engineering and Technology Statistics 1997*. The graph in the original source displays only total defence and civil expenditures, giving the impression of a rough balance, with civil overhauling military budgets (thanks to Kieron Flanagan and Michael Keenan for pointing this out, and providing the reworked table used here; see their chapter on trends in UK science policy in Cunningham (forthcoming)). The table from which the data derive splits civil R&D into the two classes represented here, compared to each of which the military figure looks high. This tells us that, despite the end of the Cold War, military R&D still remains a leading area, showing only a slight decline. Public money is employing some of the best and best-developed talents in research activities that are often secretive, pursuing objectives that are infrequently subject to democratic scrutiny, and regularly contributing to the development of means of mass destruction that have at best limited defensive functions – and this from a country that is hardly a global superpower, and has long ceased to rule an empire. Admittedly there are stated aims to reduce the military component of UK R&D. While earlier

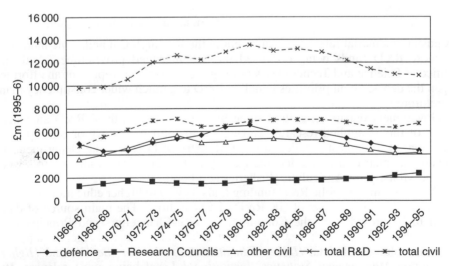

Figure 5.2 Trends in government R&D expenditure, 1966–95
Source: Office of Science and Technology, *Science, Engineering and Technology Statistics 1997*

ambitions to this end in the 1960s had a limited impact, at present the declining UK trend does run counter to that obtaining in some other countries. James and Gummett, in their chapter 'Defence' in Cunningham (forthcoming), note a fall in real terms of 25 per cent in UK government funding for military R&D over 1985–92, while France, Germany and the USA increased theirs by 36, 24 and 7 per cent respectively.

The sorts of knowledge developed from military R&D have limited spin-off to peaceful ends. The best case for economic benefits deriving from such expenditure has been for the creation of jobs, rather than that of commercial products or processes. It is, admittedly, also hard unequivocally to demonstrate large economic benefits from publicly funded basic research, at least in the short term (Martin and Salter, 1996). But to the extent that military R&D does yield knowledge useful elsewhere, this is likely to bring high-tech solutions to problems that might benefit more from other approaches – and from being addressed directly. Furthermore, there is also a noticeable skew in public R&D addressed to military ends as compared to civil ones: military R&D is much more oriented towards the 'development' pole, and away from pure research. The knowledge generated is thus less likely to be generic, more likely to be related to specific applications.

Numerous other issues in the allocation of public R&D expenditures could be examined critically. For example, there is the role of 'big science' (itself showing a military lineage – radar and nuclear bombs were influential in the development of radio astronomy and particle physics, respectively). As Flanagan and Keenan note,

> to a certain extent, the expensive sciences of the 1960s remain the expensive sciences of the 1990s, despite the rhetoric surrounding strategic, industrially-relevant research . . . . Nuclear physics and astronomy are still contentious because of the high costs of research in these fields, although . . . the proportion of research council funding directed to [them] declined markedly over the period 1990–1997, while support for engineering research in particular increased.
>
> (Flanagan and Keenan, unpublished)

A push to industrial relevance is apparent in the Research Councils: one indicator is more R&D funds being disbursed through managed programmes. And the Office of Science and Technology's Foresight exercise does represent an effort to open the debate about priorities in public R&D to a much wider constituency than heretofore.

Government departments fund some billion pounds worth of R&D annually. Most of this is applied research, carried out to fulfil regulatory or other policy functions. The biggest spenders are the Ministry of Defence, the Department of Health/National Health Service, the Department of Trade and Industry, followed by the Ministry for Agriculture, Fisheries and Food.

Of course, much public R&D funding finds its way to higher education (in 1995 £2695 million out of overall UK R&D of £14 billion). The main source of data used to be provided by the Committee of Vice-Chancellors and Principals in *Research in Universities*; now the Higher Education Statistics Agency supplies various volumes of data, *Higher Education Statistics for the UK* and *Higher Education Management Statistics, Institutional Level* being most relevant. Its website (www.hesa.ac.uk) allows users to download information on research funds by source of funding and university department. The Higher Education

Funding Council's research assessment exercise results (located on its websites) can also be readily processed into statistical form. Most university research is concentrated in a small group of elite universities. The Higher Education Funding Council strategies of rewarding universities on a sharply sliding scale, according to their departments' performance in the research assessment exercise, can only intensify this concentration. While there is a plausible argument that R&D should be concentrated in centres of excellence, this has been subject to little systematic scrutiny. The push towards a two-tier system of higher education, with elite research universities on one tier and a mass of teaching factories on the other, is more contentious. At the very least, it is arguable that teaching in many disciplines can only benefit from being provided by people with close contact with state-of-the-art R&D.

In addition to using sources already discussed to examine R&D in higher education, there are other avenues that can be pursued. One such is a detailed study by Georghiou *et al.* (1996, 1997) of research equipment and teaching equipment in universities. The first study shows 'a considerable concentration in the system' (p. 37) of expenditure on equipment in the 1990s: of 91 institutions, 13 account for half the expenditure and five for a quarter of it. 'Put another way, half of the population of institutions purchased only 10% of the equipment by value. This pattern of expenditure is . . . similar to that for the distribution of research funding in general' (p. 37).

Apart from demonstrating the concentration of R&D within the university system, this study also provided alarming data on the situation of UK laboratory equipment. Almost half of all departments surveyed considered themselves below the average in international terms, with only 18 per cent considering themselves better than average. Over three-quarters of departments were unable to perform critical experiments because of poor equipment; almost £500 million was reportedly required for priority items. While government response to such surveys has been to seek to match public and industrial funding, firms surveyed made it clear that they considered equipment purchase of this sort to be a government responsibility.

Business funding

As we have seen, UK firms' funding of R&D overall is low compared to that of many other industrialised countries. But some areas – in particular, chemicals and pharmaceuticals – are highly R&D-intensive in terms of, say, R&D being a high proportion of their turnover (similarly R&D per employee, and the share of R&D staff, are relatively high). Long-standing debates surround R&D in pharmaceuticals and other health-related sectors. The industry strongly argues against policies (such as the promotion of generic rather than brand-name drugs) which reduce high profits and discourage intellectual property protection, on the grounds that these are needed as continuing stimuli for investing in R&D. Critics claim that the search for such profits drives private R&D in pursuit of lines of knowledge which may not be optimal in health terms (e.g. treatments for diseases afflicting the wealthy rather than those affecting the global poor).

One problem with R&D statistics has been a neglect of the service sectors until recently; it was assumed that these were not really innovative. As efforts have been made in recent years to expand coverage of services, their share of reported business R&D has grown from only a few per cent to around 25 per cent of the total. We cannot be sure how much of this reflects real trends (for example, as services use more information technology they undertake more software R&D) and how much reflects the improved sampling.

The Department of Trade and Industry has long been concerned about the low level of research in UK industry. Thus, in an effort to raise awareness of R&D, and to jog individual firms into examining their performance, it publishes an annual *R&D Scoreboard*. This provides data on company expenditures, with international comparisons and sectoral and other groupings.

The standard definitions of R&D do not capture all the technological innovation effort undertaken in firms, which also involves efforts at technology transfer, training, even marketing, as well as technology development by end-users and production engineers. Accordingly, most EU countries have produced data on innovation efforts, based on a large sample of firms using the Community Innovation Survey. The results of a somewhat belated UK effort were due to be published by the Department of Trade and Industry in mid-1998. In compensation for the delay, the UK Community Innovation Survey does cover service industries, which were omitted in many earlier surveys. A source of annual data on R&D and related issues is produced by the Confederation of British Industry (CBI) and NatWest's Innovation and Growth Unit: the *CBI/NatWest Innovation Trends Survey*. This interesting annual survey investigates a wide range of topics, especially trends in expenditure on product and process R&D, market research, training, capital expenditure in support of innovation, etc. It suffers from various flaws, however. First, its sampling base is quite obscure (apparently it is weighted towards innovative firms, unlike the Community Innovation Survey, so giving little basis on which to extrapolate to the whole economy, or even to make reliable intersectoral comparisons). Second, the response rate is believed to be low. Third, the key questions about innovative activities are all couched in terms of 'To what extent are you doing/spending more/less than last year on this activity?' This tells us nothing about the baseline from which these trends manifest themselves, though there is some evidence that the data can be used to throw light on different firms' strategies and circumstances. For example, it appears that overseas companies in the UK are more likely to be pursuing radical innovation strategies than are UK firms.

Conclusions

Despite the cursory treatment in the *Guide to Official Statistics*, data on R&D and related issues are fairly rich in the UK. Though this chapter has only touched on a few key issues, it is clear that there are data resources for much more detailed exposition. Many dynamics of science and technology could be explored in far more depth – for example, the balance between pure, strategic and applied research in different areas, the programmes of Research Councils, the links between military R&D and regional and sectoral economies, and many more topics (but *see* Chapters 39 and 43). I have not here gone into the massive literature dealing with statistics of R&D outputs, such as publications and patents, let alone efforts to identify the economic benefits of R&D.

The statistics that we have considered do, however, also raise the fundamental question of the relations between R&D and democracy. The new scientific and technological knowledge generated by R&D does not just contribute to profits and careers. It is used to shape the world, and how we think about the world. But the origins of this knowledge remain opaque to most people – and the more applied forms of R&D, involving technology development, are in large part generated by private funds responding to perceived commercial 'imperatives' and technological opportunities. Further, public R&D funds are weighted towards military objectives, with their shrouds of secrecy.

There are signs that the public R&D culture is changing in many respects, with more emphasis on open evaluation of research programmes and Foresight processes. But at the same time there has been much effort to inject public R&D with 'industrial relevance' – with few opportunities for greater participation in business R&D processes. Of course, business decisions are often understandably subject to commercial confidentiality; and specialist technical knowledge is inevitably far from evenly distributed, even among that portion of the population with scientific and engineering training. Nevertheless, public policy does impact upon business R&D, from incentives for investment (in some countries tax breaks are used), intellectual property regimes (just what can be patented, and what rights should the originators of new knowledge have to dictate the uses of this knowledge) through to regulations governing animal testing and laboratory safety, to more general environmental regulations that affect both what can be done in the laboratory and the viability of the new products and processes at which innovation is aimed. Important steps forward include enhanced ethical systems and professional codes for scientists and engineers, improved regulation of research practices (ranging from animal experimentation and the use of human subjects to the treatment of contract research staff), and wider debate about the potential technological applications of developments in basic research (most of which are in the public domain). Political action on the openness of R&D is extraordinarily limited, given the importance of the subject.

6

Working with Government to Disseminate Official Statistics

Michael Blakemore

The contentious relationships in the dissemination of official statistics

This chapter concentrates on the motives, finance and resources for dissemination of official statistics. It uses examples from the UK to develop the arguments.

Dissemination of official statistics requires consumers (a term often freely interchangeable with 'users' or 'customers'). Such a simple statement can make the complex process of providing and using official statistics sound rather like selling ice-cream. Many national statistical institutes (NSIs) now have 'marketing' sections that do indeed mimic many of the functions of the retail sector, although the extent to which the activity truly is marketing is debatable. For example, the US Census Bureau (http://www.census.gov/) has a 'Marketing Services Office', although federal policy – *see below* – does not allow agencies to charge for data. Many NSIs and international statistical agencies distribute much information free of charge, and in the case of the World Health Organization (WHO) (http://www.who.ch/) the WHOSIS data extraction tool is provided as part of a 'public domain' policy for WHO data.

Contrasting policies

In a country where cost recovery is encouraged, the UK Office for National Statistics (http://www.ons.gov.uk/) has a marketing section which does specifically promote subscription services for data. The Australian Bureau of Statistics (http://www.statistics.gov.au/) promotes its own commercial product IRDB97 (Integrated Regional Database 97), and Statistics Canada (http://www.statcan.ca /start.html) provides information about its chargeable products. By contrast, even the home page of Statistics Singapore (http://www.ncb.gov.sg) has been password protected. The practices of many other NSIs can be explored using the ONS links page (http://www.ons.gov.uk/pages/links.htm).

Whatever the terms, and their implied meanings, dissemination is very different from a standard retail activity. Producers of official statistics collect the statistics for specific customers within government, and the usage is 'primary' – the statistics were intended for their needs. By contrast, the most vociferous customers tend to be those who use the statistics for secondary analysis; comments will range from the relevance of statistics for their use, accuracy and quality, timeliness and delivery schedules. Secondary users may have limited ability to influence the producers directly even though they pay for the collection process (indirectly) through taxation, and so they have to rely on liaison and lobbying groups, as well as a mutual self-interest on all sides in the production of quality-assured statistics.

At one end of the dissemination spectrum the US federal government position states that no direct data charges are to be made for the dissemination of statistics, and that only a nominal charge can be made for the indirect costs of the process of dissemination. In addition, no federal copyright exists, and this allows third parties to embed the statistics within their own products and to sell them without remitting royalties to the federal agency that produced the data. At the other end of the spectrum there are nations where considerable secrecy surrounds official statistics (e.g. North Korea), and those statistics that are for public consumption may be strictly filtered by the government.

In the USA there are powerful lobby groups (researchers, citizens' support groups, public libraries) that argue for the continued availability of free information. It is argued that information about the state, and which is gathered by the state for policy purposes, is essential to enable citizens to evaluate government activities. Furthermore, the citizens have already paid for those statistics through taxation, and it cannot be acceptable for them to pay again so they can use the statistics. Terms such as the 'information commons' hark back to the days when villagers could graze animals on common land which was a resource freely available to all. The 'information commons' are deemed to be analogous, all citizens having the right to access, process and analyse information. The 'information commons' are futher protected in the USA through the freedom of information policy, which in essence states that all government information is to be accessible by citizens unless the information is explicitly deemed to be secret, sensitive or confidential.

This idealised scenario of freely available information seems to be based on two macro-economic assumptions. First, there is the economic argument that assumes that freely available information stimulates more information use, and more businesses use that information in developing products or targeting markets more effectively. The businesses grow faster, employ more people and generate more tax revenue. The tax revenue accruing to the Treasury should then be more than the taxation costs spent generating the information. Second, there is the assumption that making information readily available directly enhances public participation in the democratic process. That in turn leads to a more informed citizenry, who are better educated and more able to evaluate their government.

At the opposite end of the spectrum are countries where there is no public right to access and process government information. Official Secrets Act legislation can be used in the reverse of the US argument by implying that all government information is secret unless it is explicitly deemed to be suitable for release to citizens. In Singapore there is no assumption that freely available information directly stimulates economic activity or that it produces extra tax revenues. There is no assumption either that readily available information is fundamental in developing a well-informed and well-educated citizenry. While the Singapore information release policy seems to be the reverse of the USA's, the country is renowned for high economic activity, high educational attainment and social stability. Conversely, given the US assumptions of business and social benefits, there are extremes of educational high attainment and social division to be found in that country, where less than 50 per cent of the voters participated in the 1996 presidential election. It is difficult to provide unequivocal economic proof of the macro-economic assumptions.

The debate about dissemination principles therefore is riven through with views that are almost at the level of religious zealotry. This chapter acknowledges them, but will focus instead on the practicalities facing most governments today. Dissemination is rarely a passive process. Even in the US situation dissemination is active in the sense that the federal government consciously makes available data collected for the purposes of the federal government. 'Active' also may imply the deliberate dissemination of misleading or deliberately focused information, involving everyone from spin doctors to propagandists. Deliberate focusing can imply the adoption of data definitions or methodologies which predispose outcomes favourable to the data owners, or lead to allegation of the 'massaging' of figures. Many examples of such practices are given in other chapters of this book.

This chapter is not the place to explore such concerns, even though one of the two dissemination systems used as examples, Nomis, is known best for the dissemination of the UK claimant count unemployment statistics. (Nomis is an online database of UK labour market statistical data covering employment, unemployment, job vacancies, demographics and earnings. It is run under contract to the Office for National Statistics by the University of Durham.) This statistical series was the focus of considerable contention during the period of Conservative government up to 1997. The series has been the subject of frequent revisions of the eligibility to be unemployed, and the definition of the workforce (the base rate denominator in the calculation of unemployment rates) also has changed, as described in Chapter 37. Each revision to the numerator or denominator of the equation led to a discontinuity in the data series, and so comparison over time may be affected by definitional changes that then mask any actual unemployment dynamics over time. But is this 'fiddling' or 'massaging' or 'manipulation'? Such accusations imply a deviousness, a sleight of hand and a desire to hide the reasons for change from the end-users of the data. If the data series are being repeatedly discredited, then usage would collapse and the user community would make a statement of no faith in the data.

The majority of users of data are sophisticated at making value judgements, and they acknowledge that most data they use are by-products from administrative systems, or that they are making secondary use of data intended primarily for the purposes of policy and governance. During the years of contentious debate of the nature of UK claimant count unemployment, usage of those statistics on Nomis increased, and the increase was from all sectors; the public, researchers, enterprise agencies and the private sector. Such a usage increase hardly implies that the users of the data are naive, or are prone to believing propaganda. Users of data generally are sensibly pragmatic, and they use and reprocess data in an implied partnership between the official producers of data, the reprocessors, brokers, value-adders of data, and the customer community (*see* Chapters 36, 37 and 40 for examples).

Producers of official statistics are primarily responding to the requirements of internal government customers. Claimant count unemployment statistics are not produced for the needs of academic researchers. They are a cost-effective by-product from an administrative computer system that pays unemployment benefit to those who are officially eligible to receive that benefit. The Labour Force Survey, while it enumerates unemployment to the internationally accepted International Labour Office definition, is not collected so that private-sector companies can monitor quarterly trends in occupational employment. It is collected to comply with EU legislative requirements and for the monitoring and policy needs of the UK government. Any external usage of the data is secondary and users need to be well aware that the data are not intended specifically for many of the secondary reuse purposes (*see* Chapter 3 for a discussion of some of the problems which arise).

Consequently, 'tools' are required by the secondary users so that they can make the best use of the data. Users want the data to be delivered in a timely and reliable form. Data need to be quality assured; accompanied by documentation describing collection methodologies, statistical techniques, geographical and thematic classifications; accompanied by metadata which highlight information relating to missing data, pointers to other sources of background information, and links to publications which are exemplars in how to use the data. Building the 'tools' is an expensive and resource-intensive process, and is not necessarily an important task for the producers of the data since generally it is expected that they will have the in-house knowledge and skills, accompanied by internal documentation, to process data effectively for government use.

Raw statistical data seldom have been made available by producers in a state that is directly suitable for secondary analysis. They will be in an 'as received' state with data formats and file structures set by internal practices, and documentation primarily written for internal users. Matching data to their end-use demands resources for the act of dissemination, documentation and metadata creation, and for providing advice and support to end-users. (Metadata are records, each of which describes another data source.) Such costs are of real concern to data producers. There is a justifiable fear that the more dissemination takes place, the greater the cost involved, and even greater is the cost of supporting the user community. Dissemination produces a conundrum: how to ensure unlimited dissemination using a finite resource, or, put another way, how to provide users with an unconstrained and flexible service for a finite cost.

Centralised research resources

In higher education one funding model that attempts to provide flexible and fixed-cost dissemination involves the top-slicing of finance to fund a centralized research resource, a key example for over 25 years being the ESRC Data Archive (http://dawww.essex.ac.uk/index.html) at the University of Essex. Here government data series and surveys are deposited by data owners in their original form. Considerable value-adding is carried out by the Data Archive in bibliographic documentation, with a key research product being the online BIRON data catalogue. Potential users of data can search intelligently on BIRON to identify data series and then place a request for data to be provided on one of a variety of media, or via FTP. The Archive catalogue is linked with that of other European Archives (http://www.nsd.uib.no/cessda/IDC/) to provide an integrated search facility. Many data can be supplied in formats suitable for the popular statistical packages, so some reworking of the data structures and formats is undertaken.

The Data Archive also partners with the MIDAS (http://midas.ac.uk/) service at the University of Manchester to deliver data online in popular formats, and MIDAS adds further value by allowing researchers to analyse data using super-computing facilities. The Data Archive has been a successful model of a research resource that has to cope with a thematically very broad range of data and to deliver it to a very wide range of researchers throughout UK higher education. In addition, dissemination partnerships such as the Data Archive have provided an important 'filtering' role on behalf of data suppliers, and have ensured that resource implications on data owners are minimised. Resources have been pro-vided by the Data Archive to license users where legislation demands it, to enforce confidentiality rules, again through licensing, and to build relationships between producers and users through workshops and seminars, or through a requirement for users to deposit copies of publications arising from use of the data.

These activities at the Data Archive indicate what is required to fill what can be termed a 'producer–consumer gap'. Organisations in the private sector acknowledge the need for an 'end-to-end quality process', and invest to ensure that this is delivered to their customers. However, to acknowledge that statement then presents another conundrum. If customers are best served by being near to the producers it may be logical to expect that the producers be encouraged to be the only official dissemination sources. In recent years the Citizen's Charter in the UK has promoted the awarding of Charter Marks to government agencies that focus more on end-to-end quality, and the Office for National Statistics has its own mission and goals as part of this initiative. But it is not realistic, even in the medium term, to expect Treasury funds to be released that enable ONS to address all its customers' needs. It is debatable whether that goal is achievable in an ever more complex global economy, though the emergence of the World Wide Web has given new impetus to NSIs that wish to control dissemination.

Economics and electronics

So far, examples have been provided to develop the core belief in this chapter that dissemination is often best achieved through partnerships that are mutually beneficial, rather than through legislative force which mandates that all govern-ment data must be available free of charge to end-users. A mandate invites further discussion of 'available in what form', and it encourages consideration of the pro-ducer–consumer gap as a resource as well as a quality issue. Put simply, the user community would like all relevant data to be made available by producers (or by some other agency such as the Data Archive, which is top-slice funded by the Economic and Social Research Council) in forms directly suitable for user needs. Producers, if they actually sign up to permit dissemination of their data, want to allow it without there being a resource impact on themselves. Where they do invest resources in dissemination, they may be expected to cover their costs by recovering the costs of the dissemination process – characteristically known as 'cost recovery'. The cost of dissemination model can then be extended to the cost recovery of an agency's operations where the maintenance costs are factored into end-user prices. This is the case, for example, with the dissemination of geospa-tial data by the Ordnance Survey (http://www.ordsvy.gov.uk/), and the Land Registry (http://www.open.gov.uk/landreg/home.htm) in Great Britain.

But dissemination may not 'cost' much at all nowadays. The World Wide Web has allowed us to consider abolishing the classic friction of distance argument. Even before the Web there were arguments that electronic media had reduced the frictional effects of reproduction of data for users. Conventional (the term 'traditional' is deliberately not used here, for to go back in history it could be argued that parchment and scribal copying is traditional) paper production has a cost overhead for each unit produced. There is a net price reduction through economies of scale, but with associated overheads of storage and distribution, and the problem that paper publications date quickly. The pressing of CDs and the reproduction of diskettes has a much lower unit cost and can be cost-efficient in small numbers, and so the economics of data dissemination look attractive.

Examining the economics alone, however, is to assume that the data 'product' is 'complete' at the time it is mastered onto the electronic media. This seldom is the case, which explains why there is a very healthy value-adding data industry which redocuments, enhances, packages, embeds data into desktop applications, provides help-desk and advice services, and uses data in consultancy services such as economic forecasting. The World Wide Web seems to be nearly a friction-free dissemination medium. It can be as easy to access a web site from New Zealand as it is from Greenland (subject to network traffic, etc.), so a web site located in Norway is as accessible to each, and the technical costs of reproduction of data are marginal. This is very attractive to those promoting free availability of data via the Web, but there are potential dangers. First, unless the data are well documented and flexibly available, the overhead costs of reworking and understanding the data can be passed on to end-users, so there still is a call for intermediaries to add layers of value unless and until data owners carry out all these tasks. The International Monetary Fund is imposing some standardisation through its Dissemination Standards Bulletin Board (http://www.imf.org/). Second, there is a real threat to the reputation of data owners. While it is very attractive to increase the number of people using data, it is just as easy to generate a cohort of disenchanted data users who find the data difficult to use, misuse or misunderstand the data, or at worst start publicly to criticise the data owners. The threat of adverse publicity is a very real fear in an era of unrestrained passive electronic dissemination.

The Web therefore can be a threat as well as an opportunity, and how it is viewed will be influenced by the initial dissemination policy of the data producers. In the USA the National Information Infrastructure includes the National Spatial Data Infrastructure in the federal geospatial sector, with Web 'clearinghouses' that make data accessible at no direct cost (http://nsdi.usgs.gov/nsdi/). More commercial approaches to the dissemination of official statistics are in the commercial sector, with (arbitrary) UK examples of the use of census of population and spatial data integration being produced by companies such as Business Geographics (http://www.geoweb.co.uk/), The Data Consultancy (http://www.geoweb.co.uk/) and WEFA (http://www.ibmpcug.co.uk/~wefaltd/). However, running through all these examples is the concept of value-adding and the resources needed to provide that value. Value adding activities can reduce the costs of documentation, restructuring, quality assuring, negotiating, licensing and integrating. End-users can then make sensible value judgements about whether they will invest local resources to undertake these tasks, or whether it is more cost-effective to purchase from a value-adder.

The last example illustrates hybrid partnerships and shared risk in value-adding. The Resource Centre for Access to Data on Europe, r•cade (http://www-rcade.dur.ac.uk/), is funded by the Economic and Social Research Council to provide integrated access to European statistical data. The core of the database service is statistical data from the Eurostat (http://europa.eu.int/en/comm/eurostat/) New Cronos database. Eurostat invests in the harmonisation of data from the 15 NSIs of the European Union. The activities of r.cade focus on delivering a quality-assured research resource, and it is acceptable policy therefore to invest directly in the creation of cohesive documentation, in building interfaces to complex data series, and in building quality assurance tools. The value to Eurostat in such a relationship is the direct feedback of the research activities from r•cade. Commercial resellers of data may then see r•cade as being a useful source of quality-controlled data suitable for use in value-added products. This will reduce the financial risks involved in negotiation, reprocessing, etc. In return, the commercial sector provides cost-effective access to marketing skills which in themselves are expensive for the academic sector to develop. In a triangular relationship all sectors may benefit by off-loading risks particular to one sector, and using sectoral strengths to mutual advantage.

Conclusions

It is highly likely in the future that dissemination of official statistics will remain a diverse and complex issue. The policies worldwide are diverse, as are the customer communities. While the US situation may remain the 'holy grail' for many, it is a situation unlikely to be repeated in many other nations. The technologies that allow data to be transmitted globally are both opportunities and threats, but perhaps this is no more than was the case when photocopiers were invented, or when the wood-block printing process threatened the dissemination powers, quality assurance and skills of the scribal copying guilds. A sense of historical analogy is important here, for any glib prediction of the panacea around the corner can easily become another of the examples used by people in the future, of how naive the future-watchers were in the past.

PART II
Models and Theory

PART II
Models and Theory

7

Eugenics and the Rise of Mathematical Statistics in Britain

Donald MacKenzie

Even the mathematical theory of statistics can be shaped by social circumstances and by political aims

The use of statistical techniques, as shown by other chapters in this book, is often interwoven with controversial social and political issues. But what of the techniques themselves? They are normally assumed to form a neutral mathematical realm, one that may be used or abused, but one that in itself stands apart from political questions. This chapter casts doubt on this assumption, examining a crucial episode in the development of modern mathematical statistics, the rise of a 'British school' of statistical theory between the 1870s and the 1920s. I shall argue that both the rise of this school (which was led by Francis Galton, Karl Pearson and, at the end of this period, RA Fisher) and the statistical theory it developed in part reflected features of the culture and politics of British society in this period. The point is not that the techniques developed by this school continue to reflect, in their modern use, the conditions of their origin; they do not. Rather, I want to establish, from this one instance, that there is nothing in statistical theory that makes it immune from this kind of analysis. Mathematical techniques are not necessarily neutral.

Eugenics

In the period 1870–1930 lie the historical origins of the debate about race, class and intelligence. Francis Galton (1869) suggested that there was an individual, definable, inherited and measurable entity called 'mental ability'. Karl Pearson (1903) designed the first major attempt to *prove* it to be inherited. RA Fisher (1918) provided the first estimate of 'heritability' (though of height, not intelligence).

This research was the work of people with clear social and political goals which were largely *prior* to their scientific investigations. For example, Galton was convinced that 'intelligence' was inherited and followed a normal distribution, and that the way to improve society was to increase the fertility of the 'able' and decrease that of the 'unfit', long before any method of measuring 'intelligence' was suggested (Galton, 1865, 1869).

The social and political programme embraced by Galton, Pearson and Fisher was eugenics. All three believed that the class structure of Britain reflected in large part a hierarchy of innate ability, at the top of which was the elite of the professional middle class and at the bottom of which were the poor, unemployed and 'criminal'. All three believed that the professional middle class should be encouraged to have more children, and that the 'unfit' should be discouraged (and perhaps ultimately prevented) from propagating. In this they were not alone. The eugenics movement, especially after 1900, was a prominent part of British intellectual life.

The eugenics movement was composed almost exclusively of members of the professional middle class, and its beliefs reflected certain crucial aspects of the situation of this group (Farrall, 1970; MacKenzie, 1976). It owed its social standing not to its ownership of capital or land, but to its possession of educational qualifications and supposedly uniquely valuable knowledge and mental skills. The eugenists' focus on individual mental ability as the determinant of social standing, and their assumption that this was unequally distributed and to be found in greatest quantity among the children of the successful of the previous generation, legitimated the social situation of the professional middle class. The social division between 'mental' and 'manual' labour was, according to the eugenists, the reflection of natural, genetic, differences between types of people (*see*, for example, Pearson, 1902).

At the same time, the eugenists had a solution for what was seen as a pressing social problem, the unemployed and semi-employed in the large urban centres, especially London (*see* Stedman-Jones, 1971).This group was perceived as politically volatile, liable to outbursts of rioting, mentally inferior and physically degenerate (e.g. Pearson, 1881). Britain's failures in international competition, difficulties in army recruitment and even British military defeats in the Boer War were attributed to this 'residuum' (e.g. White, 1901; Gilbert, 1966, pp. 85ff.). The eugenists' proposals to restrict the mentally defective, the alcoholic and the chronic poor to institutions so as to prevent them breeding did not, in such a climate, seem extreme. Even those on the left, such as the Fabian Sidney Ball, could argue for 'a process of conscious social selection by which the industrial residuum is naturally sifted and made manageable for some kind of restorative, disciplinary, or, it may be "surgical treatment"' (quoted in Stedman-Jones, 1971, p. 333).

Francis Galton (1822–1911) coined the term 'eugenics'. His work in statistics, genetics and the psychology of individual differences derived largely from his central, passionate concern for racial improvement (Cowan, 1972a, b, 1977). His key statistical innovations, the concepts of regression and correlation, arose from his eugenic research – directly in the case of regression, indirectly in the case of correlation. As a eugenist, he was interested in predicting the characteristics (say, the height) of offspring from those of their parents. The technique of regression enabled him to do this (Galton, 1877, 1886). Thus he could say that if a father had a height x inches above the average, then his son would have a height bx inches above the average, where b is a constant that can be evaluated from empirical data (the 'coefficient of regression'). Work on a scheme for the personal identification of criminals took him from the concept of regression to that of correlation (Galton, 1888).

Karl Pearson (1857–1936) developed Galton's notions of regression and correlation into a systematic theory. Galton had evaluated his coefficients graphically;

Pearson derived formulae for doing so. Galton had worked with only two variables; Pearson generalised the theory to any number of variables (thus making it theoretically possible to take account of the entire ancestry of an individual in predicting her or his characteristics). He established an institutional base for British statistics at University College London, where he was a professor; recruited and taught many students who were to become important statisticians; and for 35 years edited *Biometrika*, building it into the world's leading journal of statistical theory.

Pearson developed and championed a distinctive form of eugenic ideology, integrating it into a framework of social Darwinism, Fabianism and positivism. As a social Darwinist, he believed that the theory of evolution by natural selection applied to human societies: trading rivalry and war were the struggle for survival between nations (Pearson, 1901). As a Fabian (though never in fact a member of the Fabian Society), Pearson believed in gradual advance towards socialism by the use and extension of the existing state and the replacement of the bourgeoisie by technical experts and administrators (Pearson, 1888, chs 11 and 12; Pearson, 1902). As a positivist, he believed that only science was worthy of the name knowledge, and that science should seek only to describe appearances, not using theoretical concepts unless these can be shown to be reducible to sets of observational data (Pearson, 1892 and Ian Parker in Chapter 11 of this book). Pearson sought to construct a theory of evolution which would serve as the key social and political science (Pearson, 1900a, p. 468). It had to be properly scientific in his terms, which meant positivist and quantitative. Hence he wanted to give Darwin's rather loose concepts an exact mathematical form. Most of his statistical theory was developed in this attempt and is to be found in a long series of papers entitled 'Mathematical contributions to the theory of evolution' (e.g. Pearson, 1896).

Although RA Fisher (1890–1962) belongs to a later generation, eugenics played an important part in his earlier work. His interest in statistics and genetics developed largely as a result of his extracurricular undergraduate involvement with the Cambridge University Eugenics Society. Fisher was one of the driving forces behind it. He quickly read, mastered and went beyond Pearson's work. Thus a paper read to the society in 1911 ('Heredity: comparing the methods of biometry and Mendelism') already pointed in the direction that took him to heritability estimates and the statistical technique of the analysis of variance. His eugenic interests remained in his later work on population genetics: thus his classic *Genetical Theory of Natural Selection* (1930) concluded with several chapters on eugenics.

Not all British statisticians of this period were motivated by an interest in eugenics. Yule (of Yule's Q) and Gosset ('Student' of the t-test), for instance, were not, but Galton, Pearson and Fisher were the organisational leaders. Pearson in particular played the key role of 'intellectual entrepreneur' and builder of the discipline (teaching Yule and Gosset, for example). He was able to do this in large part because of the connection between statistics and eugenics. Though the department he founded in 1911 was called the Department of Applied Statistics, his chair had been established and funded as a chair of eugenics.

Up to this point my analysis has been in terms of individual motivation and institutional development. This, however, can only provide a background to the central question of the effect of social and cultural factors such as eugenics on the technical content of statistical theory. Might statistical theory not have developed in the way it did anyway, even had the eugenics movement not existed? Perhaps it was only incidental that some British statisticians happened to be eugenists, and they left this behind when they took up their pencils to write down formulae.

Hilts (1973) implicitly asks this question when he considers Galton's statistical work. He argues that Galton's concepts were a revolutionary breakthrough compared to the approach of such earlier statisticians as Quetelet and the mathematicians interested in the treatment of observational errors. There is no reason to suppose that workers in this earlier tradition would eventually have developed the concepts of regression and correlation. Some 'error theorists' did produce similar mathematical formalisms (Bravais, 1846, Schols, 1886), but they handled and interpreted these quite differently. For the error theorists, statistical variation was error; it was to be eliminated. For Galton, the eugenist, it was the source of racial progress. For Bravais and Schols statistical dependence (the value of one variable depending on that of another) was a nuisance. Good experimental technique should provide independent observations. For Galton, statistical dependence (offspring's height depending on parental height) was what made eugenics possible. Thus Galton, because he was a eugenist, worked in a different framework of meanings and assumptions from his predecessors; he developed the theory of statistics in ways that would have seemed pointless to them.

Pearson and Yule

Differences of approach *within* the British statistical community are of particular interest in throwing light on the relations between statistics and eugenics. The most bitter dispute (at least before 1914) was a controversy between Pearson and Yule over how best to measure the association between nominal variables (i.e. those variables for which no unit of measurement is available, which can at best be classified into different categories). The ordinary coefficient of correlation as developed by Galton and Pearson applies only to interval variables such as height, where a unit of measurement exists. Thus British statisticians needed a new approach to deal with problems such as, for example, that of measuring the relationship between being vaccinated and one's chances of surviving an epidemic. Here we have two variables, both nominal. Individuals are either vaccinated (X_1) or unvaccinated (X_2); and either survive an epidemic (Y_1) or die in it (Y_2). We arrange our data in the form of a fourfold table:

	X_1	X_2	
Y_1	a	b	$a + b$
Y_2	c	d	$c + d$
	$a + c$	$b + d$	N

Thus a individuals are vaccinated and survive, b are unvaccinated but still survive, c are vaccinated but die, d are unvaccinated and die. How does one measure the association between vaccination (X) and survival (Y)? The problem was not simply to produce an *ad hoc* measure for a particular problem such as this, but to develop a general 'coefficient of association' which could be used in all such four-fold tables, whatever they referred to.

Yule (1900) approached the problem by laying down three criteria that a coeffi-cient of association must meet. First, it should be zero if the table indicates no asso-ciation between X and Y. Second, it should be +1 if there is perfect positive association between X and Y (if, for example, $c = 0$ and all the vaccinated survive). Third, it should be -1 if there is perfect negative association between X and Y (if, for example, $a = 0$, and all the vaccinated die). He put forward his coefficient

$$Q = \frac{ad - bc}{ad + bc}$$

and showed that it fulfilled these three conditions. But he was aware that Q was by no means the only coefficient that would do this (for example, Q^3, Q^5 etc. also do so). Indeed, Yule (1912) later put forward other coefficients of a similar type to measure association; he was able to advance only loose pragmatic arguments as to which to use for a given table.

Pearson (1900b) argued differently. He proceeded not empirically, but theoret-ically. He proceeded *as if*, for example, X were height and Y were weight – both of which are normally distributed interval variables. X_1 might be 'tall' and X_2 'short', with a cut-off point at, say, 72 inches. Y_1 might be 'heavy' (weighing more than 160 lb), Y_2 'light' (less than 160 lb). Height and weight, being interval vari-ables, have an ordinary coefficient of correlation. Pearson showed that it was pos-sible, mathematically, to estimate the value of that coefficient knowing only a, b, c, d (the numbers of 'tall and heavy' individuals, and so on). The procedure for doing this is long and complicated, and does not give rise to a simple formula, such as that for Yule's Q. Now, if Pearson really had been dealing with interval variables like height and weight, the process would have been uncontentious (even though it would have been redundant, as much better ways of estimating the ordi-nary coefficient of correlation were known). His suggestion, however, was that this procedure should be used for *nominal* variables by *assuming* that the given fourfold table had in fact arisen by applying a cut-off to more basic interval vari-ables. In the majority of cases this could only be a theoretical assumption, and could not be proven. Yule argued that individuals were either dead or alive, and it was difficult to imagine a continuous variable of which these are subdivisions in the same way as tall and short are subdivisions of height. Pearson could reply that 'dead' and 'alive' corresponded to divisions of an underlying continuous variable 'severity of attack'. He had not, however, any way of proving that such underly-ing variables as severity of attack were normally distributed (and this was neces-sary for his coefficient to be valid). (In fact, an even stronger assumption, that the two underlying variables jointly follow a bivariate normal distribution, is neces-sary.) How did he (as a positivist, and opposed in general terms to theoretical assumptions of this kind) come to be in this situation?

His approach (which he called the tetrachoric method) makes sense if we look at it in the overall context of the relationship between statistics and eugenics. For interval characteristics such as height, Pearson had a measure of what he called 'the strength of heredity'. He would take a group of families, measure the heights of parents and those of offspring (or the heights of pairs of brothers or sisters), and work out the ordinary coefficient of correlation. The correlation of parental and offspring heights was, for him, the strength of heredity for height. It is interesting to note that this definition (Pearson, 1896, p. 259) assumes implicitly that resemblances between parental and offspring heights are entirely genetic and that, for example, similar nutrition does not play a part. But Pearson was already a convinced hereditarian and confident that the value of this coefficient of correlation reflected the genetic connection of parent and child, and not any similarities in environment. He had, however, a technical problem. Many characteristics of eugenic importance were not measurable in the way height is. The Binet scale of intelligence had yet to be invented; the best he had to go on was teachers' classifications of pupils into broad categories ('intelligent', 'slow', and so on). Similarly, eye colour (which was of theoretical importance because even environmentalists would agree that it was not influenced by environment) was not measurable, even though eyes could be classed into 'brown', 'blue', and so on. Pearson wanted to measure the 'strength of heredity' for such characteristics, and compare it with the values he had already obtained for characteristics such as height; to do this he developed the tetrachoric method.

A coefficient such as Yule's Q would not have served his purposes, because a value of Q cannot be compared to that of an ordinary coefficient of correlation, not even if we produce a fourfold table from the height data by dividing into 'tall', 'short', etc., because the value of Q varies considerably according to where we take the arbitrary cut-off point. A coefficient obtained by the tetrachoric method *can*, however, be compared with an ordinary coefficient, if one grants the assumption that the observed nominal data were in fact generated from underlying interval variables for which the ordinary coefficient of correlation was meaningful.

Pearson was thus able to use the tetrachoric method in the following way (Pearson, 1903). He had found the ordinary correlation of the height of pairs of brothers (say) to be about 0.5. Using teachers' judgements he would class each of a pair of brothers as either 'bright' or 'dull', and work out the tetrachoric coefficient of correlation for sets of brothers. Doing this for a wide range of mental and physical characteristics, measuring correlation by the ordinary method when the characteristics were interval and by the tetrachoric method when they were nominal, he found values of the coefficients closely clustering around 0.5. Thus he concluded that the 'strength of heredity' of all these characteristics was equal, and as the list included eye colour (on which presumably heredity was the only influence), he felt the eugenic argument to be proven. The way to improve the British race for survival in international competition was by eugenic measures, not environmental reform (Pearson, 1903, p. 207).

I would argue, then, that it was the needs of Pearson's research programme in heredity and eugenics that led him to choose to measure association in this way (and not, say, in the way developed by Yule, who was not a eugenist). Of course this was only one aspect of the development of statistical theory in this period (though it was the single statistical topic that occupied most of Pearson's energy in the period from 1900 to 1914). But this instance, together with such others as Galton's work on regression, demonstrates the influence of eugenics on the development of statistical theory as a system of knowledge.

Eugenics was not, however, to remain for long the dominant influence on British statistical theory. The needs of industrial and agricultural production came to play an increasing role, as seen for example in Fisher's work in the 1920s at the Rothamsted agricultural research station and the establishment by ES Pearson and others of an 'Industrial and Agricultural Research Section' of the Royal Statistical Society in 1933. These new needs led to new theoretical developments, but to describe them would take us far beyond the scope of this chapter.

Conclusion

The general conclusion to be drawn is, I think, that it is mistaken to see statistical theory as a field of knowledge developing simply by its own internal logic and giving rise to necessarily value-free techniques. Rather, statistical theory has evolved in historical interaction with conceptual change in other sciences, with the needs of production and with theological (Hacking, 1975) and political (Baker, 1975) developments. It is a social and historical product and not merely a collection of neutral techniques.

Note

This article is a revised version of a chapter from *Demystifying Social Statistics* (Irvine *et al.*, 1979); for an expanded account, *see* MacKenzie (1981). There has been much research since then on the history of statistics; two good sources, reflecting respectively a more 'cultural' and a more 'internalist' point of view, are Porter (1986) and Stigler (1986). The eugenics movement and its interaction with the development of scientific knowledge have, likewise, been the subject of considerable attention: a good source of up-to-date references is Paul (1995). For an excellent general introduction to a sociological approach to scientific knowledge, *see* Barnes *et al.* (1996).

8

Science, Statistics and Three 'Psychologies'

Daniel B. Wright

Scientific claims are too easily made and too easily believed in 'psychology'

When somebody mentions 'psychology', three different images come to mind. The first is of a greying figure in a white laboratory coat exploring some facet of cognition, physiology or behaviour. This coated figure might be demonstrating, for example, obedience to an authority figure, how to entice rats through a maze, what happens when you pluck at cortical areas with electrodes, or how undergraduate students learn nonsense syllables. The second image is of a couch and a Sigmund Freud-like character scribbling notes about the person lying there. This is what most people think 'psychologists' are. The third image is of therapists working out of, for example, a natural health shop that sells good-smelling oils (a large olfactory advantage over the rats-through-mazes psychologists!). This group are not always 'psychologists' *per se*, but they do provide many types of counselling for a variety of ailments and therefore are important to consider. Of course, these are caricatures and it is often difficult to draw boundaries differentiating these groups.

To describe the use (and misuse) of science and statistics among psychologists it is necessary to consider all three groups. For simplicity let's call the first group scientific psychologists, the second practising psychologists and the third fringe psychologists. In this chapter I will describe some concerns about each of these three groups. The progression from science to fringe moves from problems in the way that scientific research is conducted to the way in which beliefs about science are used in practice.

Statistics in scientific psychology

There are various criticisms about general methodology in psychology and how statistics are used (e.g. Cohen, 1990, 1994; Wright, 1998). I will concentrate on the dominant statistical approach. Over the past fifty years, scientific psychology has been plagued by an almost sole reliance on what is called null hypothesis statistical testing or NHST (Cohen (1994) suggested that *s*tatistical *h*ypothesis *i*nference *t*esting produced a more appropriate anagram). This over-reliance has occurred despite numerous critical papers. Meehl (1978, p. 817) summed up the situation: 'The almost universal reliance on merely refuting the null hypothesis is a terrible mistake, is basically unsound, poor scientific strategy, and one of the worst things that ever happened in the history of psychology.' Further, there have been several papers describing alternatives to NHST (e.g. Cohen, 1992; GR Loftus, 1993; Rosenthal and Rubin, 1994; Serlin and Lapsley, 1993). In null hypothesis testing, the researcher proposes a hypothesis that an experimental manipulation makes absolutely no difference on memory performance. The researcher collects data, then calculates the probability of observing data as divergent, or more divergent, assuming the hypothesis is correct. I will illustrate some concerns through a simple example: a psychologist interested in the effect of vitamins on IQ (an intelligence measure).

The psychologist would typically run a study in which some children were given vitamins and the others were given a placebo, a tablet with no chemical effect. The psychologist's statistical package will yield a p value. What should he or she do with it? Gigerenzer (1993) describes how our inner desires, what Freud called the id, want this p value to be the probability of the hypothesis under consideration. However, the p value is the probability of obtaining data more extreme than observed conditional on the null hypothesis being true. This misconception underpins most of the other concerns.

This first is that the null hypothesis is usually a point hypothesis. The psychologist makes the assumption that vitamins made no difference *at all*. This is a single point in a continuous space, and therefore the probability of its being true is zero. This makes NHST worthless if you have the id's misconception of the resulting p value, but NHST may be of value if the researcher does not have this misconception.

Another problem with the null hypothesis is that it is often not about the substantive model of interest. Consider two situations. In the first, some theory predicts that taking 50 mg of vitamin A each day will increase IQ by 5 points over 1 year. The researcher tests the validity of this hypothesis. Now, it is difficult to measure IQ, but as the scale improves, the attempts to falsify the theory become better and better (since measurement error is reduced). Similarly, if improved sampling procedures (or larger samples) are used, falsification again becomes simpler. If the theory is not refuted in these improved situations, then it remains and is thought of as a stronger theory. In the second situation, the theory still predicts a 5-point rise, but the hypothesis assumed is that vitamins make absolutely no difference. When our instruments improve we become more able to falsify a hypothesis that was not believed in the first place: as technology improves, we have weaker tests of our substantive theories.

Meehl (1967) discusses this and points out that while physics more often operates with the first situation, psychological research more often uses the second. In some textbooks, students are discouraged from ever operating in the first situation. Harris (1986), for example, has an entire section titled 'Why you shouldn't predict the null hypothesis' (p.139). In other words, Harris, and many others who teach statistics to psychologists, argue that you should *never* test your substantive model. It is not surprising then that, according to Meehl (1978, p. 807), 'psychology theories come and go more as a function of baffled boredom than anything else'.

One final concern is what the researcher does with the resulting p value. If p is less than 0.05, many psychologists simply describe the test as 'significant' because there is only 0.05 (a 1 in 20) chance of obtaining the data that were observed if the null hypothesis was really true. With such a small chance the psychologist rejects the null hypothesis. While all decent statistics books make clear that significance does not imply importance, many researchers still get excited when $p = 0.04$, but not when $p = 0.06$. There are several better options that provide more information. Most obvious is reporting the estimated confidence interval of the parameter. Another is not simply to say 'significant' or 'not significant', but to report the actual p value. I have had people argue with me on this. Suppose the vitamin researcher compared the control and vitamin A groups and found $t(47) = 3.00$. Many researchers believe this should be reported only as $t(47) = 3.00$, $p < 0.05$. Why? It is as if they want to keep it a secret that the p value associated with a t value of three with 47 degrees of freedom is very small, about 0.002. This can be found with any of the main statistics packages. If they are trying to hide the true value, why print the test statistic and degrees of freedom?

The obvious question is, does NHST still have a place in psychology? I think that it does, but without the many misconceptions people have of it and in combination with a variety of other procedures. This is the conclusion reached by a distinguished group of statistical psychologists who formed a task force of the American Psychological Association (APA). They found sole reliance on NHST damaging for the discipline, but that 'any procedure that appropriately sheds light on the phenomenon of interest to be included in the arsenal of the research scientist' (draft report).

Practitioners: effectiveness and belief systems

The second group of psychologists are those applying psychology theories to different areas, but are still able to register, for example, with the main psychology organisations. In the UK that would be the British Psychological Society (BPS). Psychology is one of the fastest-growing undergraduate disciplines in the UK and many different organisations employ psychology graduates. The biggest market, on the basis of the sizes of the BPS divisions and sections, is for those working in the mental health profession. They use a variety of techniques to treat many disorders and discomforts. Most practitioners do not spend much time conducting research or doing statistics. Their main contact with research is likely to be through reading some of the literature, attending meetings and taking courses on recent advancements. In this section I ask two questions. Given the multitude of therapeutic techniques, what does the research say about the relative effectiveness of these for therapy and for diagnosis? And how well informed are these individuals about the evidence for their beliefs?

In a powerful and disturbing book, *House of Cards: Psychology and Psychotherapy Built on Myth*, Robyn Dawes (1994) describes numerous studies examining the effectiveness of different therapeutic techniques and the ability of mental health professionals to predict behaviours and diagnosis disorders. Regarding effectiveness, he finds that psychotherapy, for example, works, but that what he calls paraprofessionals are just as effective as the more expensive psychologists with years of experience and many degrees. If the degrees and the experience are not predicting effectiveness, then surely this should be addressed in how and whether counsellors should be trained. It is not. In the UK there are

movements for licensing the title 'psychologist' on the basis of training and experience. This presumes that these are predictive characteristics for competence.

Predicting behaviour is something many mental health professionals do; for
example, to determine whether somebody is competent to stand trial, whether
somebody should be involuntarily hospitalised and whether a prisoner should be
released on parole. Dawes (1994) thoroughly debunks the myth that these professionals have some special ability to do this. To give an example, Goldberg (1959,
as cited in Dawes, 1994) asked psychologists and their secretaries to distinguish
schizophrenic and brain-injured individuals on the basis of a psychometric test
and found the psychologists were not reliably better. Further, simple statistical
models to predict outcomes are shown time and time again to be more accurate
than clinical judgements (Grove and Meehl, 1996). This raises the serious ethical
question of practitioners using clinical judgements to make decisions about people's lives when it is known that more effective procedures exist (Meehl, 1997). It
also raises disturbing ethical questions of the professional organisations that con-
done the use of clinical intuition when inexpensive alternative methods have been
shown to be more valid.

Dawes (1994) describes two worrying trends in the USA related to the lack of
differential effectiveness of therapists and their poor ability, on the basis of
clinical intuition, to predict behaviour. The first is licensing the title 'psychologist'. Many of the studies demonstrating poor clinical intuition and lack of
effectiveness have used as participants mental health professionals who are
either licensed or could be licensed. Dawes questions whether this is just vali-
dating invalid practice. He describes how licensed psychologists specialise in,
for example, helping people who have been abducted by space aliens, and points
out that one Harvard Medical School psychiatry professor (John Mack) estimated that 3.7 million Americans may have been abducted. Dawes's conclusion
is that techniques should be licensed. Treatments would need to be shown to be
safe and effective.

The second trend is the increasing costs in terms of public money (i.e. taxes)
and insurance premiums for what Dineen (1996) calls the psychology industry.
According to Dineen the psychology industry, as a collective organisation, 'man-
ufactures victims' in order to receive money to treat them. She uses the following
equations to describe the economics:

$$\text{Person} = \text{Victim} = \text{User/Patient} = \text{Profit}$$

She argues that the psychology industry operates like many others. There is a
potential client base, those either wealthy enough to pay for treatment or with
insurance policies (or the National Health Service) willing to cover the expenses.
Then pseudo-scientific theories are used to say that the person is a victim, explaining why he or she is not happy. Next, the idea is installed that the psychology
industry is ideally suited to cure the victims (which, given Dawes's (1994) comments, is suspect), equating victims with patients. Finally, the industry needs to get
somebody to pay for the 'service'. Many rely on state funding or insurance
premiums for this. In October 1997 the first criminal indictments against mental
health professionals were made in Texas. It was a 60-count indictment against five
people alleging that patients were diagnosed as having multiple personality disorder (now dissociative identity disorder) in order to collect insurance money. It
is alleged that they coerced nurses to write fraudulent reports in order to keep

patients in their unit. The cost for a patient on the dissociative disorders unit was $15 600 . . . a day (Pendergrast, 1996).

Dineen's (1996) book presents a frightening picture. She is not saying that most individual mental health professionals intentionally operate on the equation above. She is stating, however, that the equation exists because of the collective group, and argues that 'iatrogenic' disorders (those created in therapy) can be economically beneficial for the practitioners. The Royal College of Psychiatrists Working Group (1997) states, 'there seems little doubt that some cases of multiple personality are iatrogenically determined'. (The full report of this group was recently published as Brandon *et al.* (1998).)

The biggest debate in psychology this decade – the false/recovered memory debate (for example, EF Loftus, 1993) – has pitted practitioners against scientists (the division does not work out so cleanly, but these are often the labels given to the groups). The APA working party had to produce two separate reports because the members of these groups could not agree (American Psychological Association, 1996). Many mental health professionals argue that when people experience events such as war trauma, child sexual abuse and satanic ritual abuse (SRA) they repress these (or dissociate from them) and then recover them through techniques such as hypnosis. Others disagree. For example, according to a leaked copy of a report commissioned by the Royal College of Psychiatrists (*Guardian*, 12 January 1998, p. 1), 'despite widespread clinical and popular belief that memories can be "blocked out" by the mind, no empirical evidence exists to support either repression or dissociation'. The other group, often labelled the science group, argues that through extensive discussion, questioning, imagery, hypnosis and other 'memory recovery' techniques, false memories can be created. The dominant view now is that both of these occur (Lindsay and Read, 1995). This topic provides a good example to examine some UK practitioners' belief systems. Before we do so, it is worth making clear that people in both groups recognise that child sex abuse is both widespread and devastating.

With regard to the false/recovered memory debate, the BPS Working Party sent questionnaires to a large number of BPS-accredited practitioners, people who are likely to be able to be licensed if licensing occurs in the UK. Licensing is particularly worrying when one considers that approximately one-third of the sample felt that it was not possible for someone to have a false memory for repeated child sexual abuse (Andrews *et al.*, 1995). Is it possible? Consider Beth Rutherford (described in Loftus, 1997). A church counsellor 'helped' her to remember that her father repeatedly raped her from when she was 7 to when she was 14 years old (and that her mother assisted). She remembered her father twice forcing her to abort her foetus with a coat hanger. Her father was forced to leave his job as a clergyman. Fortunately for him, a medical examination showed Beth was still a virgin at 22 and had never been pregnant. She received a $1 million settlement against the therapist. What would have happened if she had not been a virgin? What then if she had one of the BPS-accredited practitioners who believes that false memories for this type of event are not possible? What if she then sued her father or pressed criminal charges?

In the BPS survey, 97 per cent of those questioned believed in the essential accuracy of some recovered memories for SRA (43 per cent 'usually' or 'always' believing). With such a high proportion, it is worth looking at this belief in more detail. There have been numerous media reports of widespread SRA. Patricia

Burges recovered memories of being 'royalty' in an inter-generational cult. The question is, could such horrific memories be created? Ethical and moral limitations mean that scientists cannot try to implant memories of this type. But are others implanting these memories? Patricia Burges left therapy and soon afterwards realized her memories were false. In November 1997, she reached a $10.6 million settlement with her former psychiatrist and the hospital where she was treated.

Being able to implant false memories does not mean that these events are not occurring. What evidence exists for SRA? The first large-scale study was by the US Federal Bureau of Investigation (FBI). After finding no evidence for the type of abuse being alleged, they concluded, 'now it is up to mental health professionals, not law enforcement, to explain why victims are alleging things that don't seem to have happened' (Lanning, 1992). A second large-scale US study investigated over 12 000 suspected cases of SRA (Clearing House on Child Abuse and Neglect Information, 1994). There was evidence for only one, and his case was unlike the typical events being described in therapy sessions. There was no infant killing, no cannibalism, no blood-drinking, etc. Gail Goodman, who led the study, said, 'If there is anyone out there with solid evidence of satanic cult abuse of children, we would like to know about it.'

The proportion of UK accredited practitioners who still believed SRA memories, even with the widely publicised FBI report, is worrying. Pendergrast, (1996) interviewed the chair of the BPS working party, John Morton, about this. He asked what the therapists would have said about past lives or alien abduction memories (p. 586). Morton replied, '"Oh, thank God we didn't ask them about that!"' (p. 586).

Fringe psychologists

Much pseudo-science is taken seriously. Consider astrology, where often complex algorithms are used to map star and planet positions onto people's lives. Millions of people read their horoscope regularly and many are just having a bit of fun. However, some people take the results seriously. What happens when the people writing these horoscopes have beliefs at odds with existing evidence, and want people to take their views seriously? Marjorie Orr writes horoscopes for the *Daily Express* newspaper. She is also a Jungian psychiatrist and founded a group called 'Accuracy about Abuse'. She has been vocal in the recovered memory debate described above. I know what you are thinking. Is there actually a person who is going to say they can tell by a person's star sign if she has been abused? '[A] personal birth chart will certainly show up clearly and in some detail the psychological dysfunctions which would indicate that someone was likely to abuse or had been abused' (Orr as cited in Sheaffer and Becker, 1995). Given the above discussion on predictive validity and the problems with unscientific beliefs, I do not need to explain the danger of this view.

The people registered with the large scientific organisations, such as the APA, American Psychological Society (APS), the Royal College of Psychiatrists and the BPS, are only a fraction of the people offering help to people. Fringe 'psychologists', as I have called them, are outside these groups. Many have their own groups and accreditation. I wanted to find out something about these people's practices in a way that a typical person might find out.

I am sitting, in Bristol, in front of my computer screen, trying to think how I can make various research deadlines, how to prepare tomorrow's lecture, and I look down at an unpaid electricity bill. I am feeling stressed. What could I do? I looked up hypnotherapy, psychotherapy, aromatherapy, Alexander technique, reflexology and just 'therapist' in the Yellow Pages. I wrote to all the places (56) with Bristol postcodes or telephone numbers and said:

> I am interested in finding some form of therapy to help with stress. I found your name in the Yellow Pages and am writing to you, and several others, to ask about the treatment you give and how it works for stress. If you have a pamphlet or something like that I would be very grateful. I am sorry for any inconvenience, it is just that there seemed so many different types of therapy in the Yellow Pages.

I received 32 replies (57 per cent response rate). I will concentrate on the hypnotherapists and psychotherapists, who made up the largest group (17, many doing both). Several people specialising in the other approaches also had qualifications in these. I chose the hypno- and psychotherapists because they are arguably the closest to the BPS-accredited psychologists referred to above. The original plan was to treat the letters and pamphlets from the different therapists as independent observations and to content-analyse them. However, examining these made it clear that the same phrases were being used. Given the size of Bristol, it is likely that many of these therapists had taken the same training courses and know each other. Therefore, I will simply report some of the beliefs as illustrative of the group, without making any strong claims for generality. I report just what is in the pamphlets which are available to the public. The names I list after a quotation refer to those on the pamphlets.

There were three beliefs that most concerned me: theories about memory; how hypnosis works in relation to memory; and the relationship between symptoms and 'causes'. There were two main views on memory that came out. The first is that every event that we experience is encoded and permanently stored. This is actually a common belief expressed by people in general population surveys. However, experimental research has shown since at least the end of the last century that memories for events decay; they are weakened and distorted by information presented after an event. Many of the therapists described how the 'subconscious' contained a memory for every event. The most vivid description was:

> The body holds memories of everything that has ever happened to it. . . . The body does not lie if listened to, it will tell stories as it has the memory of the physical grace and freedom that is its birthright. (S. Rathour, who went on to state, 'it allows us to reclaim our dreams and create realities we want')

Having events permanently stored allows the possibility of having accurate memories for everything. The question is how to access them. Many of the pamphlets came up with a solution: hypnosis.

> Hypnosis is used to aid the recall of the past emotional or distressing event or events. (D Angel)

The courts do not agree and are very reluctant to allow testimony based on memories obtained during or after hypnosis (Giannelli, 1995). This is because hypnosis can increase susceptibility to false memories while increasing confidence in them. In retractor cases, where people come to believe that their recovered memories were implanted by the therapist, hypnosis is often listed as a major contributing factor. The working group of the APA (1996) offered the following advice: 'Clients who seek hypnosis as a means of retrieving or confirming their recollections should be advised that it is not an appropriate procedure for this goal because of the serious risk that pseudomemories may be created' (p. 232, from the final conclusions agreed by the whole working group).

In reality, it is not always clear what is and what is not hypnosis (for example, despite 'neuro-linguistic programming' having a complex-sounding name, in many ways it is just a form of hypnosis). With that in mind, the Royal College of Psychiatrists Working Group (1997, p. 5) cautions against a number of techniques:

> drug-mediated interviews, hypnosis, regression therapies, guided imagery, 'body memories', literal dream interpretation and journaling. There is no evidence that the use of consciousness-altering techniques, such as drug-mediated interviews or hypnosis, can reveal or accurately elaborate factual information about any past experiences including childhood sexual abuse. Techniques of regression therapy including 'age regression' and hypnotic regression are of unproven effectiveness.

The final belief was expressed in almost all the pamphlets discussing psychoanalytical therapy. It is the doctrine of cause and effect: 'every symptom has a cause'. This simple line was repeated numerous times. It is used to justify the search for memories, which once revealed, the story goes, allow the symptoms to dissipate. One therapist sent some newspaper clippings that describe this:

> I consulted a local hypnotherapist, John Hudson, who said my terror was probably rooted in something that had happened when I was young. If I could remember it as an adult, he said, there would be no phobia. (*Daily Mirror*, 'How I beat my terror of birds', 21 October 1996)

A person arrives at therapy and to cure them, according to this story, you say the person becomes a victim of some past event. This matches Dineen's (1996) description of how the psychology industry keeps financially afloat.

At the beginning of this section I described a variety of stressors that could prompt somebody to seek help for stress. What would happen if I recited this list to a therapist and suggested *these* could cause my stress?

> 'Very often someone comes and tells me they know what the cause of their problem is, but it usually turns out to be something which happened much earlier in their life,' he [John Hudson] pointed out. (*The Gazette*, 21 August 1987, 'Therapy succeeds where others fail')

Conclusion

From descriptions of psychological science developing by 'baffled boredom', to alleged criminal fraud occurring in practice, it is clear that the increasing popularity in psychology coincides with some difficult teething problems. Blame of course can be directed at some individuals and particular practices, but in this chapter I wanted to stress that no branch of psychology can be satisfied with its procedures. All branches should take steps to improve their use of data, theories and science.

Acknowledgements

I am grateful to Chris Jarrold for comments on this chapter and to all the therapists who sent information on my request.

9

Using Statistics in Everyday Life: from Barefoot Statisticians to Critical Citizenship

Jeff Evans and Ivan Rappaport

Not every user of statistics has to be a professional statistician

Introduction

There is a crucial need for a wider distribution of statistical insights and skills in the general population. If we consider a representative set of contemporary concerns, we usually find many where statistical reasoning is central: for example, the rate of spread of AIDS; the risks of global warming; the likelihood of a comet's hitting the earth, the likely spread of Creutzfeldt–Jakob Disease (CJD) – and the list could go on: *see*, for example, Paulos (1988) and Matthews (1997). Further, the attempt to resolve these problems may well have economic implications, in the form of costs to society as well as to individuals; political implications, such as a result of government regulation and restrictions; and cultural implications, in that some citizens may be excluded in the debates that ensue.

In this chapter we address the following questions: what sorts of statistical skills are needed by adults in everyday life? and how might these skills be developed, in formal education and/or in more informal settings?

We consider two proposals for answers to our two questions. Evans (1989, 1992) developed the concept of a 'barefoot statistician', a user (and sometimes producer) of information, whose expertise lies in a balance of basic statistical skills and communication links with their community (*see* Chapter 38 for an example of such work). Roberts (1990), focusing on business settings, argues the need for 'parastatisticians'. Again, though not a professional statistician, a parastatistician needs a grasp of basic statistics. Roberts's broad conception suggests that the notion of a barefoot statistician needs extending to include citizens who can draw on basic statistical capabilities in living their lives. Our basic focus is therefore on the needs of adults in general in an industrial society like the UK, for statistical reasoning as part of a *critical citizenship*. At the same time, we refer to the additional skills and thinking necessary to function as barefoot statisticians or parastatisticians.

Statistical skills needed by adults in everyday life and in work

Consideration of 'the mathematical needs of adult life' led the Cockcroft Committee – in its report on mathematics teaching in the UK, *Mathematics Counts* (Department of Education and Science, 1982) – to focus on the notion of *numeracy*. This is a useful starting-point – since numeracy in some sense must form at least part of the basis for good statistical work.

What is numeracy?

The term 'numeracy' was coined to 'represent the mirror image of literacy' in the Crowther Report (Ministry of Education, 1959). The meanings used nowadays range from Crowther's broad conception – which includes familiarity with the scientific method, thinking quantitatively, avoiding statistical fallacies – to narrower ones found in some dictionaries, e.g. the ability 'to perform basic arithmetic operations' (*Collins Concise Dictionary*).

The Cockcroft Committee (Department of Education and Science, 1982) used the word *numerate* to mean the possession of two attributes:

1. an 'at-homeness' with numbers, and an ability to make use of mathematical skills which enables an individual to cope with the practical mathematical demands of everyday life; and
2. an appreciation and understanding of information which is presented in mathematical terms, for instance in graphs, charts or tables or by reference to percentage increase or decrease.

That is, they were concerned with 'the wider aspects of numeracy, and not . . . merely . . . the skills of computation' (p. 11).

Further on (pp. 135ff.), a 'foundation list of mathematical topics' for a basic course at school includes the following aims (among others):

- to encourage a critical attitude to statistics presented in the media;
- to appreciate basic ideas of randomness and variability, and to know the meaning of probability and odds in simple cases;
- to understand the difference between various measures of average and the purpose for which each is used.

Clearly, the Cockcroft Committee's specification of numeracy reflects a number of areas that could be considered statistical.

Numeracy and statistics

The Royal Statistical Society (RSS) and the Institute of Statisticians (IoS), in their evidence to the Cockcroft Committee, insisted on *statistical numeracy* as an important aspect of numeracy. They suggested, for example, that an employer might expect a school-leaver to have had 'some exposure to the notion of statistical variation, and some informal contact, through the study of data, with the processes of statistical reasoning' (1979, p. 6).

Moore (1990), in his presidential address to the RSS, argued that numeracy demands a statistical awareness, and further that statisticians are well placed to advance it. A deep knowledge of mathematics was not required for all in the future, but 'a feel for quantities, orders of magnitude and simple mathematical models combined with the natural consequences of the laws of probability' were essential for the well educated (p. 11). Similarly, the influential American writer Paulos defines *innumeracy* as 'an inability to deal comfortably with fundamental notions of number and chance' (1988, p. 3).

Thus a number of documents and commentators have seen statistical insights as a crucial element of numeracy. The RSS/IoS report also emphasised the need 'to *put statistics into context* by the detailed study of a real-life problem' (1979, p. 6; emphasis added). However, surprisingly and somewhat contradictorily, they also argued that it was unnecessary to distinguish between the statistical understanding that is required by the ordinary school leaver for employment and that required for daily life (p. 6). On the contrary, our discussion of adults' statistical needs emphasises the importance of the contexts of their activities.

The barefoot statistician and community research

In developing ideas about the statistical numeracy needed by adults in their everyday and working lives, we begin with the idea of *community research*. Community research can be seen as a way for members to participate in the construction and implementation of social policies affecting themselves, by producing facts and figures or accounts of experiences, so as to present a picture of needs, and of the adequacy (or otherwise) of resources to meet those needs (Cooper, 1986).

To think about the roles involved, we use the metaphor of a barefoot statistician, itself based on the 'barefoot doctor'. Barefoot doctors or village health workers were first trained in China in the 1930s. Since then, there have been attempts to spread the idea to other parts of Asia, and to Africa and Latin America. The 'ideal-typical' barefoot doctor provides health care based on disease prevention and health education (about nutrition and sanitation), depending on the expressed needs of the local community. In some settings, barefoot doctors also diagnose and treat the most common complaints, using a simplified set of ideas and treatments. However, barefoot doctors know how to recognise a set of symptoms that they should *not* attempt to treat – but should refer instead to the nearest clinic or hospital. They have minimal formal qualifications (except for literacy), and training normally takes place in the rural community itself (Doyal, 1979, pp. 288–90).

This suggests parallel aspects of what might be a 'barefoot statistician's' role. Such people will take a lead in producing and analysing information, normally using very basic techniques, and responding to the needs of their community (for example, as part of pressure groups in a neighbourhood, or of work groups). They need to be able to present information in terms that are both comprehensible to their community, and powerful in the discourses within which they may sometimes need to argue (often those of state bureaucracies). Barefoot statisticians will normally have at most very basic training in numeracy and perhaps statistics, and will probably select themselves into any further training.

Are there any 'barefoot statisticians' in existence? People may not recognise themselves in such a role, but the following actual cases provide good examples of community research done by individuals or groups that might be characterised as barefoot statisticians:

- a survey of local feelings about dog excrement on the footpaths, done to put pressure on the local authority to enforce the regulations;
- a survey of the effects of 'restructuring' experienced by employees of all types in an organisation;
- the monitoring of clients visiting the offices of a campaign for secure accommodation for single people, to produce information to counter local housing policy;
- an investigation of changes in women's employment patterns, and reasons, by a women's support unit for employment (local authority funded);
- a survey of the evolving 'psychological identity' of young members of an ethnic minority group, by a university student on placement.

The last three examples come from a network of community groups in north London in the mid-1980s. Libby Cooper (1986), coordinator of the Community Research Advisory Centre at the Polytechnic (now University) of North London, saw the development of statistical and other methodological skills as vital for the effectiveness of these community workers' efforts, and for some years offered both short courses and consultancy sessions. In the 1990s, she offers similar services to voluntary groups at the Charities Evaluation Services in London (*see* Chapter 10).

We consider people doing such work as likely to be barefoot statisticians in our sense. What they have in common is minimal training (if any) in mathematics or statistics, the use of very basic techniques, and a commitment to service to the community. Of course, professional statisticians can use basic techniques in service to the community, as exemplified by Ludi Simpson's neighbourhood surveys in Calderdale in the early 1990s (Simpson, 1991).

Other contexts: the parastatistician and critical citizenship

We see the barefoot statistician as a resource person, doing community research within his or her locality. A similar proposal has been made for employees in business organisations.

The parastatistician

Roberts (1990) has developed the idea of a *parastatistician*. In the context of post-war developments of management and quality enhancement in Japan and the USA, some statisticians have been prominent (e.g. Deming, 1986). But, in Roberts's view, the number of highly qualified statisticians will always be small compared to the potential applications in business. Though not a professional statistician, a parastatistician is able to carry out a range of basic statistical tasks, using a small core of well-understood techniques, and can spot areas where the professional statistician's expertise is needed. An important feature of the business organization is that it tends to generate regular series of data on the performance of the firm. Thus the parastatistician often has access to much ready-produced data. This is an advantage not likely to be readily available to the barefoot statistician.

Roberts broadens the scope of our discussion by including as parastatisticians not only resource persons within an organisation (like the barefoot statistician within his or her community), but also 'ordinary' managers with basic statistical capabilities. This dual sense suggests that the notion of a barefoot statistician needs supplementing with another role in the community: that of an 'ordinary' statistically aware citizen.

Critical citizenship

Most adults in society participate in a rich variety of practices which involve 'numerate' aspects, including work (paid or otherwise); consumption of food and other necessities; child-raising or child-care; acquiring and maintenance of housing; and engagement with debates on public policy, along with promotion of one's social and political interests. This last cluster of engagements might be called *critical citizenship*.

The practice of critical citizenship, in the UK currently, is likely to involve engaging with issues such as:

- What have been the increases in income for different groups in society in the past twenty years, and have these been fair?
- What would it mean for the increase in unemployment – or inflation – to be slowing down, would it matter much, and is it doing so?
- What would be the consequences of reforming the welfare system, by cutting the level of benefits for single parents or the disabled?

The skills and resources likely to be necessary for the practice of 'critical citizenship' are likely to be very similar to those needed by barefoot statisticians (Evans, 1989; Evans and Thorstad, 1995). An especially important resource in the struggle for critical citizenship is access to, and a grasp of, *official statistics*, for information about government spending, unemployment, inflation, poverty, and many more (e.g. Irvine *et al.*, 1979; Levitas and Guy, 1996).

But these statistics are not transparent, and hence critical citizenship needs insights and experience in their use. In 1990s Britain, access to these sources of information is mediated heavily by the mass media, in a context constituted by the following practices and tendencies on the part of government and the state:

- *mystification* of trends through publishing of claims without reference to full details of the assumptions which are necessary to interpret them, such as on the oft-repeated claims that there has been 'increased government spending' in certain areas (*see*, for example, Radical Statistics Health Group, 1987; Radical Statistics Education Group, 1987);
- frequent changing of crucial definitions, including unemployment (*see* Chapter 37), 'new hospital schemes' (a measure of the claimed expansion of the NHS; *see* Chapter 29);
- the cessation of publication of certain statistics, e.g. the number 'living in poverty' (*see* Chapter 20).

The discussion above shows that there may be people doing research in their community or workplace, or attending to social and policy developments, who may lack training in statistics but nevertheless be interested in undertaking it, though perhaps not through a traditional formal course. We discuss training below, and useful resources in Chapter 10.

Training for barefoot statisticians, parastatisticians and critical citizenship

In this section, we assume that what we are calling 'critical citizenship' forms a subset of the skills and insights needed for the broader, more active practices of barefoot statisticians and parastatisticians – and that the latter two overlap greatly for our purposes here; thus, we describe them both as 'lay statisticians' when appropriate.

In discussing both numeracy/everyday mathematics and applied statistics learning, many researchers and educators have emphasised certain *problem-solving strategies*. Indeed, there has been much overlap in the skills seen as important for applied statistics courses, ranging from the Open University's introductory course *MDST242: Statistics in Society* (Open University, 1983) to more specialised ones (e.g. Anderson and Loynes, 1987).

We have found Chatfield (1988) useful in setting down a set of stages for statistical problem-solving. These (with one exception) can be linked to areas of necessary training for lay statisticians, though the difference in context is important. These stages are also important for critical citizenship, though data production here may be less important than data accessing. Of course these stages are strongly inter-related: how data are produced depends on the objectives and the resources available – and, in turn, affects the type of analysis:

1. Understanding and formulating problems presented into a researchable form, in basic statistical terms.
2. Research design and data production. For critical citizenship, knowledge of the existence of, and ways to gain access to, government (national and local), community and institutional information sources; e.g. local economic, health and educational indicators. For lay statisticians the small-scale production of data also needs emphasis.
3. Scrutiny of data for accuracy: experience in investigating errors, oddities and outliers.
4. Initial data analysis. This involves organising and summarising the data, in ways meaningful to local communities. These will include simple techniques for one-variable data, cross-tabulation (and perhaps scatter plots), data reduction (Ehrenberg, 1975, 1982), exploratory data analysis (e.g. Open University, 1983) and graphical methods (e.g. Tufte, 1983) generally. This may often be all that is needed for data analysis.
5. Formal statistical analysis of the data. We see this area as the province of the professional statistician, with the scope of whose work lay statisticians need to develop familiarity. They should also have links with a 'consultant statistician' who can offer advice and competence-building in connection with statistical approaches which fall outside their training, but which may appear to be useful for investigating a specific issue; for example, the design of complex survey samples, or the use of modelling procedures.
6. Interpretation and communication of the results. This stage includes formulating the results in terms comprehensible to the lay statisticians' community, and to other relevant audiences (using multiple reports on the same research, if necessary), and comparisons with other results. They need to learn to report results in terms of the original substantive units, rather than using 'professional statistical metrics' such as significance levels and r-squared statistics (Evans, 1982, pp. 247–8).

Thus the statistical techniques focused on include mainly basic data analysis and presentation. Experience with probability (and tables) might be added, especially using conditional probability and Bayes' Theorem. The level of expertise we have in mind is demonstrated in the questions of the Institute of Statisticians (now the Royal Statistical Society) Ordinary Certificate examinations (which have run since 1978).

There are a number of further aspects of the style of the proposed training:

- encouragement of students to bring problems from their communities or organizations, so as to maintain the links of lay statisticians with those settings;
- support for developing the ability to *estimate and approximate*;
- exploration of the usefulness of *decision aids* (analogous to those used by the barefoot doctor) to help choose methods of analysis; for example, the specifying of levels of measurement of relevant variables to determine which summary statistics to use; *see*, for example, the inside front cover of Blalock (1979), the software *Statistics for the Terrified* (*see* Chapter 10), and some developing *expert systems* (e.g. Hand, 1987);
- experience with *computational devices* which are appropriate technology for the relevant work settings – which depends in turn on the resources available: for example, the computing technology available to most parastatisticians working in business organisations is likely to be more powerful and up to date than that available to barefoot statisticians or to critical citizens;
- an effort to build on, and to challenge, the intuitive conceptions (e.g. of 'probability') and informal methods already used by trainees; for example, in showing the calculation of compound probabilities, the betting experiences of many students, recently broadened by the UK's National Lottery, might be used; and
- the centrality, for training/teaching and assessment (as necessary), of investigations and projects, with an emphasis on interpretation of the results.

For further discussion of useful resources, *see* Chapter 10.

Conclusion

1. In planning teaching or training in statistics for adults – either in conventional taught courses or in less formal ways – it may be helpful to think of these adults as *barefoot statisticians* working in community research, as *parastatisticians* working in business, or as *critical citizens*.
2. A wide variety of resources exists for such training; examples are given in Chapter 10.
3. Barefoot statisticians and parastatisticians need access to a 'consultant statistician', who can offer advice on statistical techniques, and also competence-building. More broadly, the offering of 'citizens' advice' on numerate matters, analogous to disinterested services sometimes offered by other professions (e.g. legal advice), has been recommended (Gani and Lewis, 1990). The advantages to statisticians of offering such community consulting might include the bringing to light of novel and challenging case studies; the recruiting of additional students, including to short courses; and an improvement in relations between academic institutions and the public.

10

Resources for Lay Statisticians and Critical Citizens

Ivan Rappaport and Jeff Evans

Help is at hand to understand social statistics

Because of the lack of formal professional training, critical citizens need access to a variety of resources, some of which will be local to their community or workplace. Following our account of training needs in the previous chapter, we illustrate the variety of resources available. For training, these include books, statistical courses and computer learning packages. For the practice of statistics there are statistical software, support groups, and electronic and other resources.

Books

Some statistical textbooks give attention to basic topics and to real-life problems. Table 10.1 gives a selection. The numbers refer to the stages for problem-solving listed in the previous chapter, namely (1) problem formulation, (2) research design and data production, (3) data checking, (4) initial data analysis, (5) formal data analysis, and (6) communication of results.

Statistics courses

There are many basic statistics courses available nationally and locally, for example:

- 'Statistics in Society' (MDST242) of the Open University addresses topic areas of application – namely economics, education and health – considered to be at least as important as the 'statistical thread': statistical concepts and techniques. The materials for this Open University course are accessible throughout the UK in some public and college libraries and bookshops.
- 'Introduction to Statistics' (STX1000) at Middlesex University focuses on descriptive statistics of univariate and bivariate nature. It covers regression and exploratory data analysis without studying probability theory. The students are expected to use their developing statistical knowledge to produce reports of

Table 10.1 Statistics textbooks

Stages	Author	Title	Features
1, 2	Gilbert (1993)	*Researching Social Life*	Discussion of research design and data production
1, 2	Kane (1987)	*Doing Your Own Research*	Basic introduction to research design and data production
1, 2, 4, 6	Ritchie *et al.* (1994)	*Community Works*	Twenty-six case studies showing community operational research in action
2	Fink (1995)	*The Survey Kit*	Nine slim volumes give a comprehensive guide for carrying out surveys
2	Marchant *et al.* (1996)	*Guide to Surveys*	Short (24-page) book tells one how to carry out a survey
2, 4	Curwin *et al.* (1994)	*Numeracy Skills for Business*	Has floppy disk provided with data and simple software to analyse it
2, 4	Moore (1997)	*Statistics, Concepts and Controversies*	Particularly good on data production
3, 4	Freedman *et al.* (1991)	*Statistics*	Real-life observational studies treated critically
4	Marsh (1988)	*Exploring Data*	Applied approach to statistical techniques using exploratory data analysis
6	Chapman and Mahon (1988)	*Plain Figures*	Explains how to present data in a clear understandable way

small statistical investigations. There are many introductory statistics courses; approach a local college or university.
- 'Community Research' (SR334) at the University of North London aims to present the main features of social research in the community and in the voluntary sector.

Software for learning (computer-aided learning)

There are now computer teaching aids for statistics. 'STEPS' (Statistical Education through Problem Solving) is a project set up in 1992 by the Teaching and Learning Technology Programme. The result of this project includes 37 modules for PCs based on statistical problems arising in biology, business, geography and psychology. The modules, intended to support teaching, not replace it, are available freely to educational institutions. The project is now handled by the

Computers in Teaching Initiative (CTI) Centre for Statistics at the University of Glasgow Mathematics Building, Glasgow G12 8QW.

There are commercial learning packages such as 'ActivStats', available on CD-ROM, which uses multimedia and incorporates 'Data Desk', a statistical package. It is available from Interactive Learning Europe, 124 Cambridge Science Park, Milton Road, Cambridge CB4 4ZS. Another learning package is 'Statistics for the Terrified'(Morris *et al.*, 1996). This is a Windows-based package divided into nine modules including, for example, 'Simple data description' and 'Analysing 2 × 2 classification tables'. The student is given the opportunity to alter the data and note the effect (for example, to see how the mean and median are affected by extreme values). There is now a book with the same name (Kranzler and Moursund, 1995). 'Discovering Statistical Concepts Using Spreadsheets' is an interactive learning package with work cards, based on the Excel package. It was developed by Sydney Tyrrel and Neville Hunt and is available from Coventry University Enterprises, Coventry CV1 5FB.

Software for statistical practice

The computer has changed the practice of statistics as well as its teaching. With Windows (or the Mac system) the use of statistical packages is much simpler and one can move easily between packages. Packages such as the Windows versions of Minitab and SPSS are assistants for statistical practice.

It is possible to develop one's statistical competence by using books linked to the package. For example, McKenzie *et al.* (1995) includes a cut-down version of Minitab for Windows.

Many textbooks incorporate the use of a statistical package in the text. For example, Bryman and Cramer (1996), Ross (1996) and Ryan and Joiner (1994) are geared to Minitab for Windows; Bryman and Cramer (1997) and Howitt and Cramer (1997) are based on SPSS for Windows.

The availability of spreadsheet packages is more common than that of statistical packages. The Association of Statistics Specialists Using Microsoft Excel (ASSUME) aims to share ideas and common problems about the use of statistical facilities in Microsoft Excel. The web page for ASSUME is http://www.mailbase.ac.uk/lists/assume/.

Support groups

The critical citizen need not feel isolated. The Radical Statistics Group (10 Ruskin Avenue, Bradford BD9, and described in the Preface) is a national network of social scientists in the UK committed to the critique of statistics as used in the policy-making process. The group is committed to building the competence of critical citizens in areas such as health and education.

Charities Evaluation Services (4 Coldbath Square, London EC1R 5HL) offers consultancy sessions for voluntary groups.

The Association for Research in the Voluntary and Community Sector provides publications and other support (ARVAC, 60 Highbury Grove, London N5).

Table 10.2

Organisation	Details	Addresses Electronic mail	The World Wide Web
Data Archive	Contains data (now qualitative as well as quantitative) from many (government and academic) surveys and other studies. Funded by the Economic and Science Research Council (ESRC) and based at the University of Essex	archive@essex.ac.uk	http://dawww.essex.ac.uk
Social Science Information Gateway (SOSIG)	An online catalogue of many resources relevant to social science education and research. Funded by the ESRC, the Electronics Library Programme and the European Commission		http://sosig.ac.uk
Office for National Statistics (UK)	Government statistical agency		http://www.ons.gov.uk
Mailbase	See text		http://www.mailbase.ac.uk
Radical Statistics	See text	radstats@legend.co.uk (administrator) radstats@mailbase.ac.uk (discussion list, seen by all subscribers)	
British Official Publication Current Awareness Service (BOPCAS)	Includes information on official statistics. Jointly funded by the Hansard Trust, the Higher Education Funding Council and the University of Southampton	bopcas@soton.ac.uk	http://www.soton.ac.uk/~bopcas/

The Community Operations Research Network puts volunteers with an operations research background in touch with community groups to help with their research and decision-making processes (CORN, c/o Leroy White, Faculty of Health and Social Sciences, South Bank University, 103 Borough Road, London SE1 0AA).

Electronic resources: Mailbase and the Internet

The Internet is a worldwide network of computers allowing the dissemination of information including electronic mail, mailbases and so on. The World Wide Web (WWW) is a facility for displaying information on electronic 'pages' that is based on the Internet. To access a specific area of information that is provided by the network one needs an address; relevant addresses are given in Table 10.2.

Mailbase is an organisation that enables computer users to share information and communicate via electronic distribution lists. Through Mailbase any group of academics can set up a discussion group. For example, a critical citizen able to join the Radical Statistics Mailbase group would have access to a wealth of experience and expertise in statistics.

Other resources

Much statistical information is published by government through the Office for National Statistics (*see* Table 10.2) and is available in local and university libraries; alternatively it may be purchased, but UK official statistics have become relatively expensive in the past twenty years. The daily press, radio and television are also sources of statistics. *Stats Watch* is a monthly digest of statistical news from newspapers.

Acknowledgements

Thanks to Clare Morris, Dave Drew, Duncan Scott, Roy Carr Hill, Jonathan Rosenhead, Cathy Sharp, Ludi Simpson, Sydney Tyrrel, Libby Cooper and Paul Marchant.

11

Qualitative Data and the Subjectivity of 'Objective' Facts
Ian Parker

Research fails unless it engages with subjectivity

Traditional research directs its attention outwards, onto individuals who are not seen as doing research. They are often assumed to be different from us 'real researchers'. When we call them our 'subjects' in research studies we are often only using a codeword to cover up the fact that we treat them as if they were objects rather than human beings. Quantitative methods which rely on organising data statistically lead us to this way of looking at individuals and their problems. This is not to say that quantitative methods necessarily make researchers dehumanise people, but there is a powerful tendency for the systematic fracturing and measurement of human experience to work in this way. That approach also fits with the surveillance and calibration of individuals in society outside the laboratory. Of course, there are researchers who use statistical approaches to combat this, and they try to empower their 'subjects' (*see*, for instance, Chapters 9 and 10), but they then, of course, have to turn around and look at what the research itself is doing.

This is where qualitative research perspectives are helpful, for they can help us tackle what quantitative researchers say about objectivity and their attempt to see statistics as simply dealing with 'objective facts'. If we do that, then we will see that what research usually takes to be a problem – subjectivity – can actually be turned into part of the research process itself. This would have to be a research practice that studied and conceptualised how the inevitable messiness of social life worked itself through in our action and experience in the world, rather than attempting the rather hopeless task of trying to screen it all out to get a crystal-clear 'objective' picture of the 'facts' that are really there underneath.

Interpretation in qualitative research

Qualitative research is essentially an interpretative endeavour. This is why researchers working in this tradition are often uneasy about including numeric data in their studies or in using computer software to analyse material (*see*, for an example, Chapter 12). This queasiness about numbers is understandable, but there is no reason why qualitative research cannot work with figures, with records of observations or with statistics as long as it is able to keep in mind that such data do not speak directly to us about facts 'out there' that are separate from us. Every bit of 'data' in research is itself a representation of the world suffused with interpretative work, and when we read the data we produce another layer of interpretations, another web of preconceptions and theoretical assumptions. Numeric data can help us to structure a mass of otherwise incomprehensible and overwhelming material, and statistical techniques can be very useful here, but our interpretations are also part of the picture, and so these interpretations need to be attended to.

Most social research is still deeply affected by empiricism, in which it is believed that the only knowledge worth having in science is that obtained by observation through the five senses (and only the five). Laboratory-experimental models which are used to study social issues by predicting and controlling behaviour and measuring it against the behaviour of people in 'control groups', for example, is empiricist (Harré, 1981). A guiding fantasy of the researcher is that he or she is making 'neutral' observations. The conceptual apparatus of hypothesis testing and falsification in research developed by Karl Popper (1959, 1963) is often wielded by social researchers in defence of 'objective research' of this kind against any use of theory, and especially against theories they particularly dislike (such as psychoanalysis or Marxism). This is ironic because Popper actually argued for the importance of theory, not as a fixed and final form of complete knowledge but as necessary to enable us to structure our observations so that we might develop a better picture as to what the world is like.

What most quantitative research tries to forget when it pushes aside Popper's arguments about the role of theory is that there is always an *interpretative gap* between objects in the world and our representations of them, there is always a difference between things and the way we describe them (Woolgar, 1988). How we conceptualise that gap is a difficult issue, and there are a range of different positions in traditional philosophy and recent discourse theory to account for the way meaning is produced and structured, and how and where it is anchored (Bhaskar, 1989). This is not the place to go into that further now. The point is that research conventionally deals with the problem by wishing the gap away. This 'interpretative gap' returns to haunt research, though, and so we need to take it seriously rather than pretend it is not a problem. Definitions of qualitative research which have attempted to respect interpretation rather than wish it away have been cautious about providing a final finished account of what this alternative kind of research is. In one case in psychology (for another, see Chapter 8), then, three different overlapping definitions are offered, in which it is

> (i) An attempt to capture the sense that lies within, and which structures what we say about what we do; (ii) An exploration, elaboration and sytematisation of the significance of an identified phenomenon; (iii) The illuminative representation of the meaning of a delimited issue or problem. (Banister *et al.*, 1994, p. 3)

When we interpret and reinterpret a social issue, we are always bringing ourselves into the picture, and so this is where reflexivity becomes a crucial aspect of the research.

Reflexivity

An attention to reflexivity is sometimes the most difficult aspect of research to tackle because it seems to strike at the heart of the researcher's scientific self-image. That scientific image is often supported by appealing to a 'positivist' account of what real science is (Harré, 1981). Positivism is the dominant approach in much research, and this insists that what we must do is 'discover' things about the world, and treat these things as 'facts' that are independent of us. We are told that empirical observations will identify them and statistical techniques will arrange them in the right order. Positivists often seem to believe that one day we will have set all the facts in their place. This view of science is challenged by philosophers of science and many scientists themselves (Harré and Secord, 1972), but the positivist search for little hard bits of the 'real' still goes on in much mainstream research. Statisticians can too easily be recruited to this endeavour if they do not reflect on what they are doing (*see* Chapter 13).

Once again, Popper is recruited to this positivist image of research to defend it against what is often scornfully called 'speculation'. This too is ironic because Popper was himself hostile to positivism, and argued instead that although theoretical frameworks could approximate to the real, they could never finally arrive there. He challenged the idea of total knowledge as arising either from steady fact-gathering or from an all-encompassing theory. There is something very valuable in his account of theory generation and rational discussion. Good statistical research is a part of that process, and this should be at the heart of how we understand ourselves and how we develop a reflexive critical consciousness of our place in the world. This brings us to a concern with subjectivity and social change.

Subjectivity

Researchers coming across qualitative methods for the first time usually respond to the argument that subjectivity is important in research by saying that they would like to be 'subjective' in their research but that they still have, at the end of the day, to produce an 'objective' report. We need to take care, though, for this kind of response falls straight into the trap set by positivist research. The discourse of positivist research positions the researcher such that he or she experiences the issue as if it must entail an opposition between being 'objective' and being 'subjective'. Instead, we should insist that the contrast which concerns us is between '*objectivity*' and '*subjectivity*'. There is something specific about the nature of subjectivity which differentiates it from the 'merely subjective'. And to put subjectivity at the heart of research may actually, paradoxically, bring us closer to objectivity than most traditional research which prizes itself on being objective.

It is worth stopping for a moment to reflect on the way in which the discourse of positivist research stretches subjectivity and objectivity apart and polices the opposition to devalue interpretation and reflexivity (Parker, 1994). Let us look at two ways in which the opposition is policed.

Zero sums and 'neutral' positions

First of all, the opposition is treated as if it were a zero-sum game, as if the more you have of one the less you can have of the other. The more objective you want to be, so the story goes, the less intuition should be used, the less strongly you should allow yourself to feel about the material. Likewise, if you are making use of your subjective responses to the material, then it seems as if you must necessarily have lost some of the objective value of the research in the process. We are made to play some peculiar rhetorical tricks along the way here, and we call the objects of our research 'subjects' at the very same time as we operate as if we ourselves were objects with no feelings about what we are doing to others. What this process of splitting in research does is to cover over the way in which *our* position enters into research investigation whenever and wherever we do it. If you think about the effort and anxiety that being 'objective' involves, you will quickly realize that you are always doing a lot of emotional *subjective* work.

The very difficulty that some researchers have in maintaining a distance from their objects of study is testimony to the experiential entanglement that starts the minute a research question is posed. Distance and neutrality are themselves aspects of a particular, and often bizarre, subjective engagement with the material. This problem here is made all the worse when that engagement is denied, when we pretend that we think we must have no feelings about the issue we are researching. There is no escape from this, but it is possible to address it by turning around and reflecting upon the subjective position of the researcher. We could think of the paradox here in this way: that the more we strive towards objectivity the further away we drive ourselves, but when we go in the opposite direction and reflect upon our sense of distance we travel towards a more complete inclusive account. In this way objectivity, or, rather, something more closely approximating to it, is approached *through subjectivity* rather than by going against it. This way of addressing subjectivity might seem a little too much like an individual meditative answer to the problem, as if it were a weird paradox from Zen Buddhism. Let us turn to the second aspect of discursive policing that research engages in to keep subjectivity out. Then we can show how that attention to subjectivity is not simply a kind of delving into the individual self for some mysterious inner truth.

Embedded objectivity and reflexive positions

Positivist research discourse maps subjectivity and objectivity onto an opposition between the individual and the collective. This is the way the trick works. Subjectivity is assumed to be something which lies in the realm of the individual, while objectivity, in contrast, is seen fundamentally as a property of the social order. So, individuals are supposed to have intuitions and idiosyncratic beliefs about things, and they can try to bring these into order by positing hypotheses and testing them out. Meanwhile the collective, embodied in and exemplified by scientific institutions, absorbs knowledge into a statistically arranged system of truth. A fine balancing act maintains this mapping on both sides of the split between individual subjectivity and collective objectivity, and if either the individual or the collective departs from its assigned position and fails to show those expected characteristics it is quickly and efficiently pathologised. For example, if an individual is too certain about an opinion and starts to take the standpoint of someone with objective truth, then that is seen as some kind of madness. On the other hand, if a collection of people starts to act as if it were endowed with agency and seems to be expressing a will to act in certain ways, then it too is seen as having gone mad (Reicher, 1982; Billig, 1985). In protest movements, individuals who resist too firmly – where they are operating as if they were objective – and crowds that act with too much will – as if they had a subjectivity – are pathologised. This is, in part, because the opposition between the objective and the subjective is itself starting to break down. But even without this breakdown, we can see signs that the mapping of the objective only onto the social and of the subjective only onto the individual is a mistake. Conceptions of self, for example, that are so different across different cultures are formed out of *social* resources, and they are constructed in relation to others (Shotter, 1993). Investigations of language-learning, memory and cognition in psychology have long indicated that such apparently individual processes are impossible without a network of people around the subject (Middleton and Edwards, 1990). Many of the characteristics that we attribute to individuals, then, are in fact a function of social relationships, and, in turn, social institutions are often modelled upon images of the self. There is, then, an interplay between the two sides of the equation – the individual and the social, and the subjective and objective – that is difficult to disentangle.

Now, the point is that the attempt to approach an objective standpoint through an employment of subjectivity should not be seen as a journey into the private interior of the individual researcher. Rather, reflection upon the position of the researcher is a thoroughly *social* matter and it involves the recruitment and mobilisation of networks of people. There is a progressive demystifying dynamic in this reflection which leads towards an engagement with others as part of the research process, and we always need to formulate our research goals *with* those we are researching. Our research will often involve participation and empowerment of a collection of people who are drawn in to produce a type of knowledge that will be useful because it is *connected* to them. The limits to this involvement of others are set, of course, by the institutions in which we conduct research, and the groups involved may be restricted to other researchers. This is a political problem that we need to signal here, something a researcher should reflect upon in any kind of enquiry, but we will have to leave it at that for the moment; other chapters in this book take up this issue, including Chapters 7, 12 and 13.

'Objective' facts

What we can do is dispose of some of the obstacles that bedevil traditional researchers, and we can treat their problems as opportunities rather than as threats. The activity of the researcher is treated as a problem, for example, in the literature on 'experimenter effects' (Rosenthal, 1966). The neutrality of the investigator was thrown into question by a series of studies which showed that the hypotheses and presence of the experimenter could be so powerful as to shift the data in the desired direction. Techniques which try to solve this problem by increasing the distance between researcher and researched just make it worse, and they certainly make action research, which involves people in studies of their own activity, impossible. We need to say that of course the researcher affects participants, and is affected by them too in return. Rather than trying to prevent that happening, though, we need to look at *how* it happens and what clues that gives about the nature of the phenomenon under investigation.

We then need to address, as a matter of course, the moral position of the researcher, something that is usually set apart as a peculiar optional extra in traditional research. An ethics checklist is sometimes added onto the research plan as if it were something to be considered *after* the study had been designed. Now, rather than the researcher permitting themselves the luxury of qualms of conscience in an idle moment, as if ethical issues arise only as minor technical hitches, their subjective involvement is something that must be treated as part of the material under study as a moral question from the start.

Finally, we are able to take due account of the role of language in the research process. Empiricism, which leads the researcher to focus only on observable behaviour, and positivism, which leads the researcher to collect only small discrete chunks of data to be processed statistically, together make an engagement with language in research impossible. Many qualitative researchers would argue that since language is the stuff of human experience – that subjectivity is, in large measure, constituted in language – such empiricist and positivist assumptions lead us away from research reality. They are right, for research reality is, in many important respects, *discursive*, and the subjectivity of the researcher is implicated in the same language games as that of the researched (Parker, 1997). This is why I have referred to the work that positivist discourse plays in leading us into traps which try to make us suspicious of subjectivity. And bringing 'I' and 'we' and 'you' into the narrative of this chapter is an important part of the story. Qualitative research that takes subjectivity and interpretation seriously, then, also demands a new language, a different discourse and different kinds of subject position. Then facts are no longer 'objective' simply because they are in statistical form. Instead, they become things which we understand as embedded in a social world that we continually reproduce, and so they can be transformed as *we* and *you* reflexively connect the process of social research with the people who are represented within it.

12

Untouched by Statistics: Representing and Misrepresenting Other Cultures

David Sibley

Statistical methods on their own do not provide a way to understand the world

In 1978, the British government's Department of the Environment began to compile a statistical record of Gypsy caravans in England and Wales. The counting was done by local authorities and the figures were published as a caravan census by the government department. The numbers appearing in the tables were, first, caravans on public and private sites and, second, illegally parked caravans. This exercise had two immediate objectives. One was to assess the social needs of the Traveller population, particularly education and health. The other was to facilitate the control of movement and settlement of nomads using the instrument of 'designation', which was a ban on settlement in a district where the central government deemed that adequate provision had been made for those Travellers residing in or resorting to the area. A third, 'academic', use for the information, but not one used to justify the exercise, was to translate the numbers into patterns which would give a summary description of local and regional variations in the population of Gypsies and other Travellers (Sibley, 1984, 1990).

The use of these statistics to gauge the social needs of a population made up of several cultures, some very 'other' in relation to the dominant society – Romany Gypsies, Irish Travellers, Scottish Travellers, New Age Travellers – raises some interesting questions about useful and appropriate knowledge and about the ethics of this kind of enumeration. For control purposes, the information collected was probably sufficient. To ban Gypsies from certain localities, it was not necessary to know how many people were involved (numbers of caravans provided a crude but adequate surrogate) or to know anything about their cultures. The problem that I want to consider in this chapter is the role of statistics in distanciation, in the avoidance of close contact or engagement, thinking particularly about the cultural distance between groups like English Romany Gypsies and non-Gypsy bureaucrats in the then Department of the Environment, along with academic quantifiers. In this context we might ask, 'Is statistical analysis associated with a fear of touch?' I will then go on to examine claims made for qualitative, ethnographic knowledge as an alternative to statistical description, but will question the validity of ethnographies also, on ethical grounds.

Distanciation

The problem of distanciation originates in the elevation of one mode of argument, that based on 'reason' or 'rationality' over other ways of knowing (*see*, for instance, Chapter 11). This is deeply rooted in Western cultures. Thus, some ancient Greek philosophers insisted that disengagement from people, or distancing from experience, was a necessary condition for discovering truth. Paul Feyerabend (1987, p. 120) notes that Parmenides advocated a procedure for acquiring knowledge 'far from the footsteps of humans', as distinct from understanding gained from habit and experience. The former, he argued, was the only valid way. Modern science is more likely to find its foundations in Descartes than the ancient Greeks but the message is similar. Scientific investigation, based on the underlying premise of rationality, has been both authoritative and exclusionary.

In positive social science, statistical analysis, involving an essentially rational set of procedures, has played a key role in distancing from the subject, in keeping the observer and the observed well apart. Statistical methods are *powerful* because they have been constructed as such in relation to other forms of knowledge. In Western scientific culture, a clear hierarchy of knowledges was established with quantifiable, statistically testable knowledge positioned above other forms, such as folk knowledge. Writing in 1975, two mathematical geographers, Alan Wilson and Mike Kirkby (Wilson and Kirkby, 1975, p. 3), implied that explanation was a kind of progressive purification process, culminating in highly generalised mathematical statements, and the route to mathematical models, via statistics, was a necessary route to understanding: 'Indeed, it can be argued . . . that such deepening of understanding is associated *first* with statistical analysis and *second* with mathematical analysis as yet more progress is made.' Through the process of mathematical modelling, the messy bits and pieces of social and economic life are filtered and refined and their presentation becomes progressively more remote from the everyday, 'far from the footsteps of humans'.

The quantitative spatial analysis which I indulged in when mapping the Gypsy census data was the kind of exercise which valorised not just numbers but also the *visual*. It was presumably acceptable as an instance of scientific practice because it was seeking order and pattern, which might, conceivably, have provided clues to underlying processes. However, I would argue that the discovery of interesting, and even beautiful, patterns, is often an end in itself (but this kind of indulgence has to be legitimated by making token reference to processes (*see* Chapter 46 for examples of beautiful patterns). Spatial statistics provide their own aesthetic gratifications and keep the analyst at a safe distance from the world. As Kevin Robins (1996, p. 29) has argued, the dominance of the visual sense in the modern world has resulted in 'a drive to disembodiment and [a] retreat from experience'. Robins maintains that intellectual life has become powerfully associated with vision, while the sense of touch has been 'repressed and devalued'. The visual, because it is detached, has been equated with the real and objective, while the tactile, metaphorically at least, has been associated with engagement with or immersion in cultures with a consequent inability to articulate experience because to do so requires getting outside, viewing others from a distance. Robins suggests that the visual has been further emphasised with the growth of cyber-technologies (and here I would include computer applications of fractal analysis and geographical information systems). As he puts it, 'the cyber-world is utopian because it is a world of order, and it is a world of order because it is pre-eminently a visual world' (Robins, 1996, p. 30).

Statistics generally, and spatial statistics in particular, have been used to confirm the existence of an ordered world. Lurking behind these practices, however, are anxieties about disorder, about people and things which disturb the stable world-view of the observer. These people and things I would term *abject* in the sense in which the psychoanalytical theorist Julia Kristeva uses the term: that is, they cause discomfort, unease, because they 'threaten apparent unities and stabilities with disruption and possible dissolution' (Gross, 1990, p. 198). The abject is a negative, something potentially polluting, but it is always a source of discomfort because it can never be finally removed. Thus, distanciation, a retreat from experience, is for some a strong urge and the more a person seeks refuge in an imagined cosmic order, the more abject things beyond the boundary become. Order, spatial or otherwise, homogeneity and purity, if these are considered to be real or desired characteristics of the social world, accentuate the threat posed by difference. Difference becomes abject because it threatens boundaries.

Getting close to others

Social anthropologists seem to me to have a better chance of understanding cultural difference than do statisticians because of their tradition of engagement with the subject, and it is notable that social anthropology failed to share the enthusiasm of other social sciences for quantification during the 1960s and 1970s. There have been anthropologists who looked on their project rather differently, notably Claude Lévi-Strauss, who defined ethnology's ultimate goal as 'to arrive at certain universal forms of thought and morality' (cited by Todorov, 1993, p. 60) and thus advocated a degree of detachment and distanciation. More typically, however, anthropologists have been aware of the need to explore difference and have been sensitive to the problems involved in articulating the world-views of others. Their concerns seem to me to throw doubt on the general applicability when describing population characteristics. Statistics may mask important aspects of cultural difference.

Mutedness

Writing in 1975, Shirley Ardener maintained that

> a society may be dominated or overdetermined by the model (or models) generated by one dominant group within the system. The dominant model may impede the free expression of alternative models which subordinate groups may possess and, perhaps, may even inhibit the generation of such models. (Ardener, 1975, p. xii)

This observation on the power of knowledge led to the assertion by Ardener that some groups are *muted*; that is, because they communicate in an idiom different from that of the dominant groups in society, they are subject to distorted representations or rendered invisible in 'authoritative' constructions of the social world. Ardener was writing about the representation of women but the mutedness of other groups is equally or more evident. Cultural and social differences are never transparent. To women, in relation to men, we could add children, in relation to adults; nomads, in relation to sedentary populations; and so on.

To bring the problem of muteness back to my initial concern about counting Gypsy caravans, I will describe an incident which demonstrates the very small overlap between the world-views of Gypsies and *gajes* (non-Gypsies). In the early 1970s, a friend was asked by an elderly Gypsy woman to take a record-player which had belonged to her late husband to a shop in the city to be repaired. The friend later returned the record-player to the woman but it still did not work. On discovering this, she swore and waved an axe round her head rather theatrically, insisting that the record-player be returned to the repairers. When the machine was returned a second time, it was in working order and the Gypsy woman, looking pleased, threw it on the fire. This behaviour is clearly perplexing without knowledge of a Gypsy world-view. The record-player was *mochadi*, unclean, because it had belonged to a dead person. It had to be destroyed but it would have been an insult to the dead if it had been burned when it was not in working order.

Here we have a different cosmology which can be sensed and represented to a sympathetic readership only after long-term involvement with the group. It is necessary, as Linstead (1994, p. 1323) has argued, 'to stand apart from everyday categories', to acknowledge that there are systems of classification different from those used by the majority, and, at the same time, to 'stand close to the phenomenon as given, as experienced'. I would argue that long-term engagement with other groups is desirable – first, for the obvious reason that it takes time to gain acceptance, and second, because ambiguity and contradiction are characteristic of most people's attempts to articulate their experience, both of the researcher attempting to produce a personal account of other people's experience, and of the subject who has to cope with a researcher as well as the demands of everyday life. If one is trying to find out whether an English Gypsy family prefers living on a council site to travelling, for example, it is likely that at one moment the comforts of the site – hot water, electricity and concrete pitch – will seem preferable to the insecurity and material privations of a nomadic existence. At another moment, the site may seem boring and oppressive, and the travelling life will be fondly remembered. It is from this kind of mixed and muddled information that the researcher has to produce a persuasive narrative, and engaging with the subject for only a short period may well lead to firmer, less ambiguous conclusions than are warranted.

This theoretical and methodological reflection has implications for practice. It is particularly evident from past experiments in planning that a lack of engagement with clients has had serious negative consequences. From the French architect Le Corbusier through to urban renewal schemes in the 1960s and 1970s, and the provision of Gypsy sites in the 1970s and 1980s, it is evident that transparency has been assumed, that social and spatial categorisations were uniform and stable, and that needs could be gauged through scientific analysis. During this period, statistical and mathematical modelling were enthusiastically adopted by planners whose Corbusier-inspired schemes such as high-rise local authority estates can be seen, in retrospect, to have had damaging effects on the lives of many people. Dissenting voices in the 1960s and 1970s, like Paul Davidoff (1965), who argued the case for advocacy in city planning, or anarchist critics like John Turner and Ruth Fichter (1972) and Colin Ward (1991) were all insisting that people, 'clients', should be involved in planning rather than kept at a distance and treated as objects in mathematical modelling exercises. The language of planning, including statistical analysis, effectively insulated the professionals from the people for whom they were producing plans. Planning with people, or facilitating planning by people, as John Turner advocated, involves the dissolution of boundaries between the professional and the client. The planner, in a position of power in the 1960s, probably had a fear of merging or of 'touch', a fear of argument and making way for other people. My own impression, based on experience as a planning assistant with a mathematical bent in County Durham in the 1960s, was that the language of statistical modelling was used to keep people who were objecting to plans (particularly those people objecting to the reorganisation of the rural settlement pattern of the county) at a distance, to render them inarticulate. Plans had a geometrical purity which the preferences of the (mostly working-class) population might have disturbed if they had been acknowledged and acted upon. People had strong ties to places that were deemed redundant and upset the spatial order of the new settlement plan. Expert arguments about the need for a reorganisation of settlement were difficult to counter, however, if they drew on the language of statistical modelling which gave them a spurious authority. Solicitors representing the county council recognised this and emphasised the scientific and, therefore, authoritative nature of the planners' arguments.

Statistics, ethnography and ethics

Statistical analysis encourages remoteness and is unlikely to contribute to an understanding of cultural difference. Statistics can, of course, describe difference – for example, in household size or unemployment rates – but such differences may be misleading if abstracted from culture. For example, expert witnesses representing oil companies at judicial hearings about pipeline construction projects in northern Canada during the 1970s emphasised the high unemployment rates of the Inuit and other native peoples, arguing that pipelines would be to their benefit because they would provide work on construction projects. However, anthropologists supporting native peoples emphasised the importance of hunting in the 'hidden' native economy, something not registered in official statistics. Pipelines would disrupt caribou migrations and thus have a detrimental affect on hunting. To support their argument, they put a cash value on the products of the hunter–gatherer economy. They produced statistics in an appropriate cultural context but this would not have been possible without ethnographic knowledge (Sibley, 1981, ch. 11).

Ethnography should encourage both an understanding of and a respect for difference. This means, first, giving others a voice (*see* Chapters 9 and 10 of this book). Otherwise, there is a danger that received stereotypes will persist and will be projected onto others, maintaining distance and negative conceptions of difference. Other cultures, particularly vulnerable minorities, should also have some control over knowledge. It may be the case, for example, that an indigenous population wants to keep its traditional economic activities out of sight to avoid regulation by the state. My point is that it is only through the kind of engagement which is a necessary feature of ethnography that questions of power and authorial authority surface as ethical issues. In relation to other cultures, any kind of distanced, quantitative analysis will get nowhere near this kind of problem, but this does not mean that quantification is never appropriate. It does mean, however, that it has to be firmly embedded in culture.

Conclusion

I have argued that distanciation is a likely, although not inevitable, consequence of the statistical description of other cultures. If statistical order is desired, then disordered, marginal states represented, for example, by the cultures of peripheral minorities are more likely to appear abject. The anxieties felt about the abject then provide an incentive for removing abject things or people, through planning, for example. But this kind of knowledge fails to reflect other world-views, and I have suggested that engagement with others is necessary for the production of authentic knowledge. Only through engagement can it be appreciated that the intimate knowledge of other cultures can also be dangerous. We should ask: Is it invariably appropriate to represent others, even when respecting their difference and distinctive world-view? A postmodern acknowledgement of the importance of difference does not necessarily empower the 'other'. It may only add to the intellectual capital of the dominant groups in academia and the control regimes of the state. Statistical analysis, if it is isolated from other ways of knowing, does not touch on this issue. The ethics of knowledge production are more likely to surface in the process of ethnographic research but the experience may suggest that, in some circumstances, we should not touch the other either.

13

Models Are Stories Are Not Real Life

Jane Elliott

Statistical models are the means by which researchers construct meanings for variables

Introduction

It may seem obvious to suggest that models are stories and are not real life. When we talk about how well a statistical model fits the data we are making it clear that the model is not the same as the data. There will always be residuals, indications of how far each individual deviates from the norms suggested by the model. As Cathie Marsh (1988, p. 17) has written, 'The term model has many different meanings: all suggest attempts to summarise, formalise, and generalise some aspect of the world. The notion of fitting a model – some kind of provisional description – to a set of data values is fundamental.' The aim is usually to construct the simplest or 'most parsimonious' model, which fits the data and minimises the residuals (this is the term used to mean the difference between observed values of the outcome variable and the values predicted by the model). A model may therefore be thought of as a story because it is an attempt to make sense of the data by extracting the salient or important elements and presenting them in a form which is easy to understand. In the same way, when we return from work in the evening we may tell our family stories about what happened during the day. We will spare them the irrelevant details and select the most interesting and significant events and conversations to recount and explain.

In this chapter I want to explore some of the implications of thinking about models as stories. In particular I will draw on sociological work on auto/biography and consider how far it can usefully be applied to thinking about event history models such as Cox proportional hazards models (which are explained below). Many of the issues raised about how models might be understood as the means by which we construct, rather than simply describe, individuals' experiences within the social world can be applied to a whole spectrum of statistical methods and not just event history analysis.

Life histories and auto/biography in sociology

The 1980s and 1990s saw a resurgence of sociological interest in life documents and particularly in biography and autobiography. Among others, Plummer (1983), Denzin (1989) and Stanley (1992) have traced sociological interest in the in-depth analysis of individual lives back to its origins in the Chicago school and the classic work *The Polish Peasant* by Thomas and Znaniecki (1918). (*See also* Chapter 7.) The methodology then fell out of favour during the 1950s and 1960s but is now enjoying something of a revival. For example, Plummer argues that life history documents are essential to redress the balance within a discipline which has become focused on social structures and grand theories at the expense of understanding how these operate in individual lives. He argues that the narrative form of the material is also of importance here: 'stories . . . [give] flesh and bones to the injustices and indignities of the world where so frequently in social science there is only bland and horrible jargon that serves to overdistance from the issues' (Plummer, 1983, p. 149; *see also* David Sibley's arguments in this book (Chapter 12)).

Other sociologists with an interest in individual biographies approach them from a rather different perspective. In particular, the contribution of literary criticism and feminist theory has been recognised by many authors who would now question the notion that life documents can be assumed to be referential of a life in any straightforward way (e.g. Derrida, 1976; Brodzki and Schenck, 1988; Denzin, 1989; Stanley, 1992; and Chapter 14). For example Denzin (1989) makes it clear that his interest lies in the 'construction or on the doing of biography' as much as in its manifest content. He argues that 'the central assumption of the biographical method, that a life can be captured and represented in a text, is now open to question' (p. 9). In contrast to the modernist conception of the autobiography as a window through which it is possible to see the life of a person, writers such as James Olney (1980) and Liz Stanley (1992) have argued tha the self (autos) and the life (bios) do not exist in their finished form when the text (graphos) begins to be written. Indeed, it is through the very act of writing that the life and self take on a certain form, and the text takes on a life of its own.

In this context, the term *auto/biography* has come to be used to encompass all the different types of writing about lives. Conventional distinctions between different types of biographical and autobiographical documents are rejected, and even the difference between writing about one's own life and writing about the life of another becomes blurred (Stanley 1992). Given this interest in demolishing conventional boundaries between different types of biographical accounts within sociology, it is interesting to see how far ideas about how lives are structured through auto/biography may be extended to apply to the statistical analysis and construction of lives.

Once we question the ability of auto/biography directly to represent a life, and become interested in the form of narratives as well as their content, this opens up the possibility of considering the narratives that are produced through the collection and analysis of *quantitative* data. This is not to suggest that there is no difference between qualitative and quantitative research methods, but rather to allow for a new and different reading of quantitative research which uses some of the analytical devices which have been developed by those interested in auto/biographical texts.

Life histories in survey research

During the 1980s and 1990s large-scale surveys increasingly included questionnaires or interview schedules designed to collect work and life history information. In 1980 the Women in Employment Survey collected life history data on over 5000 women aged 16 to 59. Then in 1986 the large-scale Social Change and Economic Life Initiative, funded by the Economic and Social Research Council, included the collection of work histories. In 1981 and 1991 Sweep 4 and Sweep 5 of the 1958 cohort study, known as the National Child Development Survey, also collected life history data, while in 1992 and 1993 the British Household Panel Survey also included retrospective life and work histories as part of its interview schedule. These life histories all consist of recording the dates at which particular biographical events occurred.

To gain an understanding of the type of data that are being collected, it is perhaps helpful to focus on one example in more detail. As part of the fifth (1991) sweep of the National Child Development Survey, the 11 000 respondents were asked to fill in a self-completion questionnaire asking about various aspects of their lives since 1974, when they were aged 16. Each section of the questionnaire uses the same basic format, and asks the individual to record the dates of particular key events such as the birth of children, the beginning and end of relationships, the start and end dates of jobs, etc. As would be expected, the amount of information recorded by each respondent varies enormously; some have done the same job since they left school, have never had children and have had no cohabiting relationships, while others have changed jobs and cohabiting partners many times while parenting several children. Figure 13.1 illustrates a hypothetical, and simplified, example of a woman's life history from the National Child Development Survey.

Clearly, because this information is collected retrospectively, it relies on individuals' memories, honesty and interest for its accuracy. Indeed, for example, in nearly a tenth of cases, women could remember the years but not the months of all their job changes. Given this reliance on memory, respondents were not asked for detailed information about the events recorded, but nevertheless the parallel data about children, relationships, jobs and housing already provides the basis for detailed chronologies of the lives of a large sample of people.

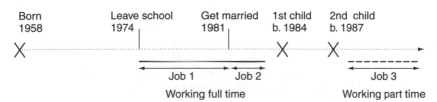

Figure 13.1 Hypothetical life history of a woman from the National Child Development Survey

Cox proportional hazard models

The collection of this type of life history data has partly been motivated by the development of new statistical techniques. Survival analysis or event history analysis, with its roots in engineering and biomedical science, has been developed for application to social data. There are two problems which mean that duration data (the time until an event occurs) is not amenable to analysis using standard statistical techniques such as ordinary least-squares regression. First is the problem of censoring, namely that not all members of the sample will have experienced the event in question before the end of the observation period. For example, if the analysis focuses on duration until re-employment following redundancy, or the birth of a child, there is obviously a problem of what duration to assign individuals who have not yet found a job or had a child, or perhaps never will. The second problem is how to deal with explanatory variables which change over time (commonly referred to as 'time-varying covariates'). Over the past twenty-five years, a number of different approaches have been developed to cope with these two problematic features of event history data. One approach is Cox's proportional hazard models. First discussed in the Journal of the Royal Statistical Society in 1972 (Cox, 1972), this procedure is now available for use as part of the most popular and widely used statistical package in the social sciences, namely SPSS (Norusis, 1994).

In essence, Cox proportional hazard models allow us to evaluate the relative importance of a number of different variables, or 'covariates', for predicting the hazard of an event's occurring. The hazard is a key concept in event history analysis, sometimes also referred to as the hazard rate or hazard function. It can be interpreted as the probability that an event will occur at a particular point in time, given that the individual is at risk at that time. (The terms *hazard* and *risk* both originate from early applications of the methods in biomedical research when the models were constructed to predict the relative likelihoods of death for patients with different characteristics.)

Both continuous and categorical variables can be included in the model. Each will be assigned a coefficient which indicates whether a particular change in the covariate increases or decreases the hazard of the event's occurring. For example, Table 13.1 shows the coefficients of variables in a Cox proportional hazards model where the event of interest was the return to work for mothers who left the labour market at the birth of their first child. The data are taken from sweep 5 of the National Child Development Survey described above. There were 2747 women who had left the labour market at the birth of their first child and 1020 of these returned to work before the birth of their second child and before the survey was carried out. In other words, there was a total of 1727 censored cases in this example.

The variables in this model include a dummy variable to indicate the absence of a cohabiting or marital partner when the child was born, two dummy variables to indicate the level of qualifications obtained by the cohort member, and a continuous variable to indicate the age in years of the cohort member when she had her first child. The general form of the Cox regression model is:

$$h(t) = [h_0(t)]e^{(B1 \times 1 + B2 \times 2 + \ldots + Bp \times p)}$$

where $h_0(t)$ is the baseline hazard and x_1 to x_p are the covariates. Therefore the

Table 13.1 Coefficient estimates for Cox proportional hazards model of mother's return to work after the birth of a first child

	B	SE B	sig.	exp (B)
Partner absent	0.1600	0.0833	0.0549	1.1735
Post-16 qualifications	0.1751	0.0825	0.0337	1.1914
O levels	reference category	NA	NA	NA
No qualifications	−0.2617	0.0738	0.0004	0.7697
Age at motherhood (yrs)	0.0097	0.0093	0.2931	1.0098

Note: In addition to examining the parameters for the individual variables in the model it is also important to test the overall hypothesis that all parameters are 0 (in other words, whether knowledge about the value of the variables for any individual will improve prediction of the duration being modelled). This is done by examining the value of minus twice the log likelihood before and after fitting the model. In this case the change in −2LL was 36.197. This statistic has a chi-square distribution and is significant given the four degrees of freedom (four parameters) in the model, so the null hypothesis that all the parameters are 0 can be rejected.

column labelled B in Table 13.1 gives the coefficient for each of the variables in the model. If this coefficient is positive the hazard is increased (in this case the hazard of returning to work); conversely, if the coefficient is negative, the hazard is decreased. The standard errors of the coefficients are shown in the column labelled SE B. From the column labelled 'sig.' it can be seen that the coefficients for the variables 'partner absent' and qualifications are significant, while the coefficient for age at motherhood is not. The column labelled Exp (B) in Table 13.1 gives the percentage change in the hazard rate for a unit increase in the covariate. For example, having no partner raises the baseline hazard rate by approximately 17 per cent.

It can therefore be seen, from Table 13.1, that the absence of a partner slightly increases the hazard of a mother's returning to work following the birth of her first child, and this effect is on the borders of significance at the 5 per cent level. Qualifications have a significant effect such that those with post-16 qualifications have an increased hazard of returning to work while those with no qualifications have a diminished hazard of returning to work, each compared with the reference category of women with O levels or equivalent. Finally, the age at which the cohort member became a mother apparently has no significant effect on the hazard of returning to work. In other words, a mother with a husband or cohabiting partner and with no qualifications is predicted to spend longer out of the labour market following the birth of a first child than a single mother with a degree or similar high-level qualifications. (For more detailed discussion of Cox proportional hazard models *see* Norusis (1994), and Allison (1984).)

Event history analysis represents an important advance within social statistics. The advantages of using this type of longitudinal method are twofold. First, longitudinal data allow us to model processes over time and to focus explicitly on duration effects. Second, the variation within individuals which is captured by longitudinal data resources allows control for the effects of unobserved variables (Davies *et al.*, 1992). However, it is important not to be seduced by the apparent sophistication of these techniques. In particular, it is important to pay attention to the ways in which the data which are analysed are not simply collected but may be thought of as being actively constructed according to the cultural and social norms that surround the research.

'Fictive devices' in life history data

By studying the content of a questionnaire such as that used to collect the NCDS work histories data, a number of what might be termed 'fictive devices' immediately become apparent. Although respondents are instructed to record the dates marking the beginning and end of *all* jobs they have held since leaving full-time education, they are simultaneously given a number of caveats regarding jobs which should *not* be recorded. First, jobs held while in full-time education are not to be recorded; second, all jobs lasting less than a month are to be omitted; and third, breaks from a job due to maternity leave are to be treated as continuous spells of employment.

These instructions are clearly intended to simplify the tasks of recording information for the respondent and processing information for the researcher. In the majority of cases omitting jobs which have lasted only for a few weeks would not be expected to make a substantial difference to analyses. However, there are some subgroups of the population for whom spells of short-term employment may be particularly salient. For example, those in cultural industries such as the performing arts may have work histories characterised by alternate jobs such that they spend a few weeks working as an actor followed by several months as a drama therapist or teacher. The instruction to omit short jobs is therefore likely to obscure and distort the work histories of certain minority subgroups within the population.

It is clear that if these instructions are adhered to, what will be produced is a specific type of job history which fits in with current cultural expectations about a 'career': namely, that it begins at the point when an individual leaves full-time education and consists of a number of spells in different jobs typically lasting for several years rather than only a few weeks or months. In other words, respondents are requested to provide a 'fictionalised' job history. This is the case not merely in the work history section of the questionnaire. In the relationship histories too, respondents are asked to record all cohabiting relationships, but to omit those which lasted for less than six months.

The data which are collected by this method are therefore already more than raw data. They are information which has been filtered through the instructions provided on the questionnaire to provide a particular version of an individual's work history or relationship history. At the point of data collection, then, what is elicited from the respondent might be considered a type of story or perhaps a partial story. In the case of the work history, the beginning is clearly specified as the end of full-time education; relevant episodes are jobs that last for more than a month; and the end-point is the respondent's current employment status at the time of the survey. However, a story is not merely a chronicle of successive events; it must have a 'point' and must conclude rather than simply ending (Mischler, 1986). As Polanyi (1985) has argued, for example, within conversational story-telling the narrator must guard against a response of 'so what?' from the audience. In this respect, a quantitative life history is not a fully formed story, but might be better thought of as an ongoing chronicle (White, 1987). It is only through aggregating these life histories and using them as the basis of statistical analysis that *the researcher* transforms the chronicles into a story.

Indeed, the account of an event history analysis may be understood to work as a story on two levels. First, the research process itself is narrated (*see* Aldridge,

1993, for a detailed discussion of this). If we use the now widely accepted terminology to describe the different parts of a narrative (Labov and Waletzky, 1967), the structure of a research paper may be understood as follows. The starting-point, or orientation, of the story is the research on the same topic that has been carried out in the past. The complicating action is described in the method section and consists of the data collection and the analysis of the data. The resolution is provided by the results, which are followed by the discussion and conclusion which form the evaluative components of the story. As Aldridge (1993) has discussed, the form of the research paper often gives a distorted view of the research process. Many researchers collect their data and analyse them before carrying out a thorough review of previous research in the same subject area. Although research papers are presented and understood as 'factual' – that is, as faithfully documenting the work carried out by the researcher – their structure frequently belies the rather more messy and iterative experience of actually doing research.

Event history analysis such as Cox's proportional hazard models itself takes the form of a story on a second level. In order to be able to carry out the analysis a clear starting-point must be defined, usually termed the *initial condition*. It is from this starting-point that the duration can be measured. In some cases this is relatively easy to define (for example, when analysing the length of time from redundancy to re-employment). However, in other examples this can be more problematic, such as modelling time until return to work after motherhood: the key question is, when does a woman enter the 'risk set' for returning to work? Is it immediately after she has given birth, or when she begins to look for another job? Some women may decide to complete their family before returning to work.

An end-point is also defined as part of the analysis. This is the time at which the event occurs or the individual case is censored. If there are time-varying covariates included in the model these might be interpreted as events which happened during the course of the story, namely the complicating action. This starting-point, complicating action and resolution are defined *by the researcher* as part of the process of specifying the data to be included in the model. As such they will be noted in the method section of the journal article describing the research. However, as in any story, the researcher/narrator's real interest is not just *what* happened but also *why* it happened. If we read event history analysis as a story, the point of the story is contained in the table of results which specify the parameter estimates for each of the variables in the model. Our interest, as readers, is therefore shifted from the individual respondents, who completed quantitative life histories as part of a survey, to the variables which provide an explanatory framework for why events happened as they did at the aggregate level.

If auto/biographies are stories about people, then perhaps statistical models are best understood as stories about variables. As Temple (1992) highlighted when discussing Shirley Dex's 'individual-level analysis' of work histories from the Women and Employment Survey (Dex, 1984), there is a sense in which these analyses obscure the individual. 'In none of the reports is a single total life history or even work history provided – there are no case studies of no matter how partial a kind. The analysis is entirely concerned with aggregations' (p. 147). This is illustrated in Table 13.1, where it is the covariates which are listed and not the individuals who formed the sample on which the model is based. In discussing the results of such analyses it is the covariates which take their place as the subjects of the statistician's narratives.

Taking this analogy a step further, while auto/biographies may be understood as textual means of establishing identities for individuals, quantitative analysis might be read as establishing an identity for a social group defined by variables such as gender and class. In other words, although variables are treated as individual attributes during the data collection phase of survey research, analyses and texts will subsequently be produced by the researcher which offer insights about the determining power of those variables as a social and narrative construction. Just as the self in the auto/biographical text can never be identical to a self which exists outside the text, the meanings which are constructed for variables through the process of statistical analyses represent only one possible set of meanings.

What, then, are the implications of this for those who are endeavouring to produce and/or interpret statistical models? First, it is important not to lose sight of the individuals whose lives provide the data for the models. Although variables rather than individual people may become the subjects of the statistician's narrative, it is individuals rather than variables who have the capacity to act and reflect on society. One possible mechanism for ensuring that these individuals do not become obscured is to combine qualitative and quantitative research methods. This should make it possible to provide case studies alongside statistical models which illustrate how the processes represented by variables operate at an individual level.

Second, it is important to maintain an awareness of the implications of any 'fictive devices' which may be operating when the data are collected. Although it will inevitably be necessary to simplify the complexity and diversity of individuals' lives when collecting data about a large sample, the possible distortions which may result for specific groups within that sample should not be ignored.

There is an increasing awareness among those who adopt a qualitative approach to the study of society that any research account is only one possible description of reality. By the same token it is important to be aware that statistical models are underpinned by a complex set of decisions, made by the researcher, regarding both the collection of data and the analysis of those data. A statistical model can therefore never be more than a 'provisional description' used to make sense of a set of data values.

PART III
Classifying People

'NOTHING BUT THE BEST FOR OUR BOY'

14

Missing Subjects? Searching for Gender in Official Statistics

Diane Perrons

How gender issues are emerging within official statistics

Gender issues in social research

With the mainstreaming of gender issues in the European Union (EU) and the United Nations arising from worldwide feminist pressure, all official statistics are required to be differentiated by gender and government-sponsored projects and programmes asssessed for their gender impact. The UK government has committed itself to ensuring that statistics are presented by sex and age following the Global Plan for Action arising from the UN Conference on Women in Beijing in 1995. More specifically, the Government Statistical Service (GSS) has agreed the following (somewhat qualified) policy statement: 'the GSS aims always to collect and make available, for example in publications, statistics disaggregated by gender, *except where considerations of practicality or cost outweigh the identified need*' (my emphasis) (Government Statistical Service, 1996a). It is important therefore for those interested in ending women's oppression to monitor progress towards differentiating statistics by gender and furthermore to ensure that the categories and concepts used do not incorporate gender bias. This chapter discusses and evaluates official sources of information relevant to some of the critical actions called for in the Beijing declaration. What becomes apparent is that some progress has been made in relation to public issues such as paid work, but less in the 'private' sphere, such as in relation to male violence (*see* Central Statistical Office, 1995a).

Women in the economy

The past two decades have been characterised by an almost universal feminisation of employment (Perrons, 1998; Rubery *et al.*, 1996), but statistical measurements have been criticised for failing to keep pace with the nature and scale of women's involvement and for being grounded in male experience (Women and Geography Study Group, 1997). Categories are, however, revised periodically to take account of employment restructuring – for example, the growing importance of the service sector – and some attempts have been made to devise more gender-aware categorisations. These changes improve the veracity of the statistics but clearly make it more difficult to identify long-run trends.

The most recent change took place in 1990 to harmonise UK statistics with International Labour Office (ILO) categories. The nine major Standard Occupational Classifications used in the Labour Force Survey (LFS, discussed further in Chapter 3) provide some indication of the extent of horizontal and vertical segregation. Women are concentrated in clerical and secretarial occupations while men outnumber women 9 to 1 in the construction industry. Women are under-represented in senior posts in all spheres (Equal Opportunities Commission, 1997). At the minor occupational level 'men's jobs' are highly differentiated, which facilitates men's social mobility and pay progression. In many supermarkets, for example, there has been a general reduction in the number of grades, yet in the butchery and bakery, where men predominate, grades remain. Thus women's progression and pay can be blocked by their limited formal occupational structure. Even so, men tend to colonise the more highly paid niches within minor occupational categories (Allen and Henry, 1995). Furthermore, in the contemporary labour market, characterised by employment deregulation and flexible working practices, the nature and security of the employment contract has become more critical. Data on different forms of flexibility are recorded in the LFS and these show that women are disproportionately represented among the flexibly employed (Dex and McCulloch, 1995; Casey *et al.*, 1997; and *see* Table 14.1).

Incomes and poverty

The New Earnings Survey (NES) collects gender-differentiated data by major and minor occupational groups for full- and part-time workers. Gender inequality in earnings is understated as women are over-represented in smaller organisations whereas the survey is biased towards large employers (Bruegel and Perrons, 1998). Furthermore not all bonuses and fringe benefits, from which women are disproportionately excluded either by informal practices or by time constraints,

Table 14.1   Gender differences in flexible employment, percentage of employees

	Women	Men	All employees
Full time			
Flexible working hours	13.2	8.6	10.2
Annualised working hours	4.2	3.7	3.9
Four and a half day week	2.4	2.7	2.6
Term-time working	4.7	1	2.3
any flexible pattern	24.7	16.6	19.4
Part time			
Flexible working hours	7.4	5.9	7.1
Annualised working hours	3.2	2.6	3.1
Term-time working	9.7	4.6	8.9
Job sharing	2.4		2.2
any flexible pattern	22.9	14.2	21.5

Source: *Social Trends* (Central Statistical Office, 1997) (from the Labour Force Survey)

are reported (Bruegel and Kean, 1995; Rubery, 1992). Some of these additional earnings can be picked up by surveys based on individuals rather than employers such as the Family Expenditure Survey (FES), the General Household Survey (GHS) and the more recently established and larger-scale Family Resources Survey (FRS).

The FRS focuses on income and benefits and includes earnings from a wide range of sources, even children's paper rounds. Computerisation has enabled interviewers to collect responses from all household members individually, which represents a significant advance. Nevertheless a number of problems remain. Data reliability rests on the respondent's ability to recall accurately his or her sources of income and on the respondent's truthfulness. Differences have been found when people are interviewed separately and then as a couple (Pahl, 1989). Furthermore, although data are collected from individuals, the household is often used as the primary unit of analysis in subsequent reports. Indeed, most sources of income data relate to household, family or benefit units and it is difficult to estimate women's welfare from these data because resources are rarely distributed evenly within the household (Pahl, 1989; Sen, 1989; McDowell, 1997). Many female-headed households find that control over a smaller budget provides greater well-being than a larger budget in a dual household (Chant, 1997). Thus, treating the household as a single unit can misrepresent the situation of individuals in both directions: that is, it can suggest that poverty exists where it does not and vice versa.

The original data from the FES and the FRS as well as other sources such as the LFS can be accessed directly and, for academics, at virtually no financial cost, via the MIDAS system at Manchester University. Help is available online and through special courses but the learning curve is quite steep. To utilise these resources it is necessary either to acquire the necessary computing skills and resist exclusion from contemporary technologies (Cockburn, 1985) or to campaign for more user-friendly access to this public information. The Equal Opportunities Commission (various), however, provides some summary information in its briefings.

Unemployment and 'inactivity'

Unemployment statistics are gender differentiated but even in the ILO standardised statistics women's unemployment is understated, as women often withdraw from the labour market in times of high unemployment (Metcalf and Leighton, 1989; *see also* Chapter 36). To assess the scale of differences in employment opportunities, 'inactivity', also differentiated by gender, needs to be explored. Formally, inactivity denotes absence from the labour market but the connotations are negative and fail to encapsulate the range of socially useful activities people might be doing outside of paid work or indeed that absence may be due to social exclusion rather than personal choice.

Although there is a wealth of detail on the different reasons for inactivity within the LFS these tend not to be reproduced in the aggregate statistics or summary reports and so it is again important to explore the original data sources. During the 1980s and 1990s the male inactivity rate increased while the female rate fell, reflecting the trend towards feminisation referred to above, although in absolute terms the male rate is lower (*see* Figure 14.1), reflecting the traditional division of labour between women and men. Clearly this term needs to be revised in order to

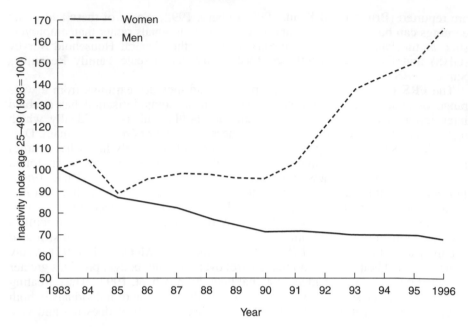

Figure 14.1   Gender differences in labour market inactivity
Source: Calculated from Eurostat, 1995 (REGIO database) and LFS, 1996 (MIDAS dataset)

provide appropriate social recognition for the wide range of activities outside of paid work necessary for social reproduction.

Domestic work

The unequal division of domestic work and childcare between women and men is widely believed to be the main, although not the only, foundation for women's subordinate position in the labour market. Some writers consider this unequal division of labour to reflect individual choices (Hakim, 1996). Others attribute it to social constraints such as women's low wages and the lack of affordable child-care, to conventions about gender roles, or to patriarchal oppression (Hartmann, 1986). The intermeshing of all of these influences could also be important, together with individual characteristics such as qualifications and the number and age of children (Glover and Arber, 1995). Irrespective of the reason, however, this unequal access to paid employment gives rise to women's over-representation among the poor, especially in old age as a consequence of having only a very small (if any) independent pension on retirement or divorce (Ginn and Arber, 1993; *see also* Chapter 15). So far in the UK there has been little formal data-gathering on the domestic division of labour.

In 1997, however, the Office for National Statistics (ONS) produced a set of household accounts in which domestic labour is recorded. The accounts were based on the Omnibus survey carried out by the Office of Population Censuses and Surveys (OPCS) in May 1995, which covered 2000 adults. Valuing domestic work would increase gross domestic product (GDP) by between 40 and 120 per cent depending on how it was measured (Murgatroyd and Neuburger, 1997). On the lowest valuation, domestic production is still greater than manufacturing sector output. There is no indication, however, that this official valuation will enable 'domestic' workers to claim social benefits, such as pensions.

The survey also investigated the division of domestic labour and childcare between women and men. Respondents were asked to keep a pre-coded diary for the previous day in 15-minute blocks, and the results show continuing inequality (Murgatroyd and Neuburger, 1997). Women spend about 150 per cent as much time on unpaid work as men but less time in paid work (*see* Figure 14.2). Thus the 'new man' is more a figment of optimistic imagination than a social reality. The small-scale nature of the survey restricts further disaggregation to employment and family status and it is not possible to explore variations by age or sexual orientation. Yet a more balanced division of labour between domestic and paid work has been found in lesbian households with children, indicating that it is not simply the presence of children that enforces a subordinate role for mothers in the labour market (Dunne, 1997).

Including domestic work within official statistics contributes to its social recognition and it is important to support the development of this work. (In the late 1990s the GSS was seeking sponsors from a variety of public- and private-sector agencies to carry out time use surveys and so bring the UK into line with other EU states.) Gender bias can, however, be introduced in the identification, measurement and recording of domestic work in time use surveys. Multi-activities – that is, when domestic work is combined with child or elderly care – which are

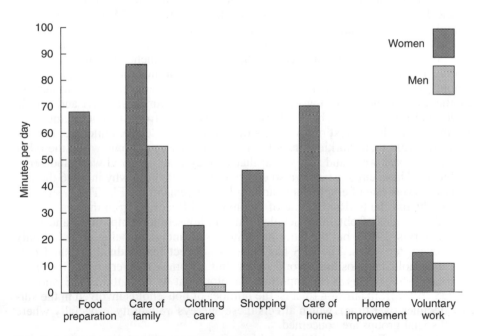

Figure 14.2 Gender division of domestic labour
Source: Drawn from Murgatroyd and Neuburger (1977)

thought to be practised more by women than men, are particularly difficult to measure. Similarly, tangible activities such as the production of goods, home repair and maintenance may be more likely to be recorded than less visible but arguably equally important tasks such as caring for the educational or the emotional needs of children (*see* Himmelweit, 1995).

Linking paid and domestic work

The British Household Panel Survey (BHPS) is a longitudinal survey carried out on an annual basis. It covers a wide range of social issues including the changing household composition and employment status of individuals (Gershuny *et al.*, 1996; *see also* Chapter 3). It is possible to trace individual movements into and out of poverty and employment and thus it allows analysis of changes associated with becoming a parent, or becoming separated or divorced, and thus overcomes some of the limitations of income figures from cross-sectional surveys based on households. Similarly, by using these data on changing household composition it is possible to show that lone parenthood is of temporary duration for many people (4.7 years on average for half of all lone parents) as they form new partnerships and because their children cease to be dependent (Gershuny *et al.*, 1996). The usefulness of this source will increase with time.

The National Child Development Study (NCDS) provides longitudinal data on a wide variety of issues. It is a cohort survey of individuals born in 1958 and as a consequence ethnic minorities are under-represented. There is a similar survey covering ethnic minorities from 1970, the National Survey of Ethnic Minorities, but to obtain statistics disaggregated by gender in this survey the original data at the ESRC data archive have to be used. The latest NCDS survey, by self-completion questionnaires and interviews, was carried out in 1991 when the individuals were 33 years old. Individuals were questioned about their household composition, work situation, income, childcare arrangements and domestic division of labour. Analysis based on the survey demonstrates the rise of paid work among mothers, the long hours worked by many fathers, and the numbers of mothers and fathers carrying out work in unsocial hours (*see* Ferri and Smith, 1996, and Figure 14.3). These patterns arise primarily from the continuing need for income in the context of low levels of affordable collective childcare. Parents have to juggle their working times to ensure that one is at home while the other engages in paid work and vice versa; that is, they practise serial work and serial childcare. These circumstances also contribute to explaining why in 1996 the UK had the lowest rate of employment among lone parents in the EU. The Longitudinal Study and the SARs (Sample of Anonymised Records) from the census also provide a means of obtaining detailed data on household composition and work status (Jarvis, 1997), but the latter have the disadvantage of being provided only once every ten years. The LFS has begun to collect longitudinal or panel data on household composition, work status and incomes (Government Statistical Service, 1996b). This is an important development as both the BHPS and the NCDS suffer from attrition (that is, the number of households involved in the survey declines over time), which affects these surveys' reliability, especially where small social groups are concerned.

In 1997 the Family and Working Lives Survey was carried out to update the Women and Employment Survey of 1980. Men were also included in this survey

Total hours worked by cohort members

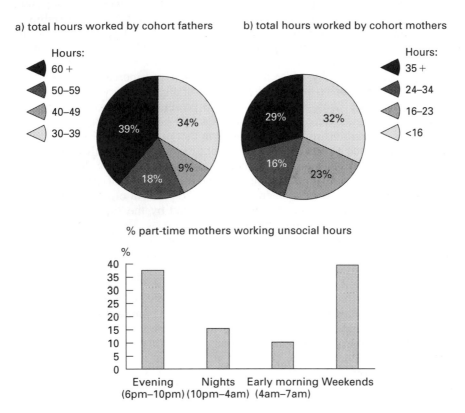

a) total hours worked by cohort fathers   b) total hours worked by cohort mothers

Hours:
60 +
50–59
40–49
30–39

34%
39%
9%
18%

Hours:
35 +
24–34
16–23
<16

29%   32%
16%
23%

% part-time mothers working unsocial hours

%
40
35
30
25
20
15
10
5
0

Evening   Nights   Early morning Weekends
(6pm–10pm) (10pm–4am) (4am–7am)

Figure 14.3 Working hours and working patterns of parents
Source: Ferri and Smith (1996)

so that changes in working and living experiences of women and men could be compared. The Equal Opportunities Commission (various) publishes briefings on Work and Parenting which provide some summary longitudinal data on different work–family combinations. So overall there is a growing volume of gender-differentiated longitudinal data which would permit analysis of the working and living experiences of women in different types of households and as their family circumstances change.

Women in power and decision-making

The election of the Labour government in the UK in 1997 led to a historically unprecedented number of women 'in power' (overall 120 women MPs), and this will enhance Britain's position on the United Nations Gender Empowerment Index. However, just at the moment when women are moving into formal positions of power, which is relatively easy to monitor, there has been a shift from government to governance, especially at the local level. 'Governance' refers to the growth of non-elected quasi-state bodies with executive powers (Jessop, 1995). Although women form a higher proportion of the local quango membership (32 per cent) and leadership positions (20 per cent) than in local government (19 and 10 per cent respectively), their presence is concentrated in gender-stereotyped areas such as the caring services (Peck and Tickell, 1996). The more powerful quangos are dominated by white middle-aged men and increasingly reflect and legitimate the voice of business and entrepreneurship in policy formation (Peck and Tickell, 1996). Thus, while women are present in the new structures of governance, the new quangos are very different in politics and orientation from the women's committees in local authorities in the 1980s. In public appointments more generally, such as heads of public corporations and nationalised industries, men continue to dominate. A similar situation is found in the judiciary and the police. Only 5 per cent of circuit judges and 21 per cent of the metropolitan stipendiary magistrates were women in 1996 and 6 per cent of police sergeants in 1995. Statistics on these latter themes are collated by the Equal Opportunities Commission (1997) on an annual basis.

Women and the media

Although women's representation in the media has increased, they remain similarly under-represented in terms both of the control of the main TV companies and of their representation on screen (Baehr, cited by Bakewell, 1996) and on the radio (Gill, 1993). Monitoring employment and pay has become more complex with the deregulation of the industry, the growth of independents and the increasing use of individualised short-term contracts. Individual instances of inequality, however, are not hard to find. Ironically, in a situation comedy called *Men Behaving Badly*, the men in real life were found to be doing exactly that as the pay of the two female co-stars was only two-thirds that of the men (Bakewell, 1996). However, there is little systematic reporting of gender inequality and yet the media and cultural activities more generally play an important role in socio-cultural normalisation and in shaping people's expectations of themselves and others (Gunter, 1995).

Violence against women

One of the central problems is that much of the violence daily endured by women fails to come to public attention as it does not fit in with established categories of crime. There are two main sources of official statistics: the Home Office and the British Crime Survey. (Chapter 24 also discusses these sources of data.) First, the Home Office statistics report on notifiable crimes, and only the more serious forms of violence come to public attention. Second, the British Crime Survey (biennial) of victims is likely to under-record domestic violence simply because it is carried out within the household where both perpetrators and victims may be simultaneously present. *Social Trends* shows that the number of reported rape cases has increased significantly since the 1980s (Central Statistical Office, 1997). This increase may be due to more rapes taking place or because of higher levels of reporting, which in turn have been attributed to more sensitive police treatment. However, some cases which are initially reported as rape become 'no crime' if the Crown Prosecution Service (or the woman) decides not to proceed with the prosecution (Lees, 1996). Thus, even though more cases are being reported, it is known that these understate the extent of rape and the prevalence of male violence more generally (Radford and Stanko, 1996).

On an international level, estimates of the extent of under-recording indicate that one in three women living with men have experienced some form of violence at some time during their lifetime (Römkens, 1997). To obtain more accurate data for the UK a national prevalence study is required that has official status, reports for a specific policy purpose and is designed in such a way as to provide the respondents with a degree of control over their involvement. A combination of structured and semi-structured interviews, with the respondent being able to write and reflect on their responses or responding to pre-recorded taped questions, as in the BHPS, may be a way forward. Qualitative work based on in-depth questionnaires, especially on questions of violence, can be intrusive and may simply lead to the informants having to relive the pain of their experience (Valiente, 1997).

Conclusion

Overall there has been some improvement in the collection and presentation of gender-differentiated statistics, and in some spheres women are no longer missing subjects. The Government Statistical Service plans in the 1990s to harmonise existing surveys and the Office for National Statistics has published a *Guide to Gender Statistics* in cooperation with the Equal Opportunities Commission (Office for National Statistics, 1998a). However, other aspects of life that are central to women's status and welfare are socially under-recorded and under-explored. More funding is required to improve the frequency and depth of surveys and in order to make the data collected more accessible.

As the providers of statistics become more responsive to potential user demands it is rather ironic that there has been a drift away from analyses based on large-scale numerical surveys towards more qualitative, intensive work. This shift is partly a consequence of the development of postmodernism and the interest in diversity and individual identities which in some instances denies the usefulness of the category 'woman'. Clearly, diversities between women need to be recognised in official statistics. However, data can only be collected on and social policy can only respond to outcomes, such as age, social class, ethnicity and gender, not to what we think about ourselves, which is a truly private matter (Williamson, 1997).

For women to have a voice in the formulation of national policies so that the continuing structural constraints that remain and that affect all women, albeit in different ways, can be addressed, valid and reliable data are required. The combination of large-scale surveys and qualitative analysis provides a way forward in terms of providing accurate and reliable knowledge of women's lives, including those aspects that may take place in private and yet have a profound effect on women's lives and well-being in the public world.

Acknowledgements

I would like to acknowledge help and advice from a number of people including Armando Barrientos, Irene Bruegel, Sylvia Chant, Ros Gill, Jay Ginn, Sue Himmelweit, Brian Linneker, Linda Murgatroyd and Jill Radford. I would also like to thank Mina Moshkeri for the graphics.

15

Playing Politics with Pensions: Legitimating Privatisation

Jay Ginn

Misleading statistics have helped to reinforce gender inequality in later life

The function of official statistics in the United Kingdom was defined by the Government Statistical Service as being 'to serve the needs of government' (Macfarlane, 1990). This definition neatly combines a publicly acceptable function (to guide policy-making) with a more dubious one: to manipulate public opinion in support of ideologically preferred policy options. The second function of statistics becomes particularly important when distributional issues involving substantial public resources are concerned, as with pensions policy.

This chapter shows how official representations of statistics can play a role in legitimating policy stemming from an ideological preference of the governing party. Pension policy provides an example, in which statistics were used to make the shift from state to private welfare appear necessary and fair. The chapter shows how independent research can provide an alternative account with different policy implications. First, the context in which pension privatisation has been promoted is outlined. Second, the way gendered assumptions in the production and reporting of statistics have contributed to a misleading impression of pensioner incomes, helping to justify cuts in state pensions, is considered. Third, the statistical basis for using different actuarial tables for men and women is questioned. Finally, the impact of pension privatisation on gender inequality of income in later life is summarised.

Apocalyptic demography and pension privatization

Pensions policy has become an increasingly controversial issue internationally over the past two decades. Alarmist media accounts have warned of a 'grey time-bomb' and a 'rising tide' of pensioners, with frequent references to older people as a 'burden' on society. Projections of the rise in the number of older people and in the proportion of the adult population over state pension age have been used to fuel concern about the sustainability of state welfare.

In the 1980s, warnings in the USA that the growth in the older population was creating inter-generational inequity (Preston, 1984) lent impetus to the complaint, spearheaded by Americans for Generational Equity (AGE), that older people received too large a share of federal resources (Minkler, 1991). AGE prophesied an 'age war' over resources unless social security pensions were cut. Significantly, this pressure group purveying 'apocalyptic demography' was financed by banks, insurance companies and health care corporations whose aim was to expand private insurance (Quadagno, 1990). An academic version of the inter-generational inequity theme warned of 'a large, growing and possibly unsustainable fiscal burden on the productive populations in developed nations' (Johnson *et al.*, 1989, p. 9), threatening collapse of the generational contract in which current contributions to social security are used to pay pensions.

In the UK, New Right think-tanks such as the Institute of Economic Affairs, the Adam Smith Institute and the Centre for Policy Studies emerged as powerful influences on policy, proposing pension reforms that were implemented by the Conservatives (Nesbitt, 1995). Cuts in state welfare were backed by projections of growing numbers of older people and rising social security spending (Department of Health and Social Security, 1981, 1985; Department of Social Security, 1990, 1993a). Apocalyptic demography has been less strident in the UK, yet here too the pensions, insurance and finance industry constitutes a powerful lobby for expanding private pension coverage.

Pressure for pension reform has continued in the 1990s, with economic reports of a 'looming old age crisis' (World Bank, 1994, p. iii) in developed societies. Such reports have contributed to an 'ascendant creed' (Downs, 1997, p. 9) that macro-economic performance was impaired by state pensions and that an expanded role for private-funded pensions would boost investment, benefiting the economy (World Bank, 1994; Poortvliet and Lane, 1994; Kotlikoff and Sachs, 1997).

Although apocalyptic demography and the associated economic arguments have been widely challenged by social scientists (Townsend and Walker, 1995; Hills, 1993; Lloyd-Sherlock and Johnson, 1996; Downs, 1997; Mabbett, 1997), a common response by European governments to the perceived crisis has been to seek cuts in state provision and to promote the private sector of pensions. In the UK, reforms have been far more radical than elsewhere, with a substantial diversion of resources from the public to the private sector of pension provision.

Three major reductions in state pensions were set in motion by the Conservative government after 1979. First, indexing the basic state pension according to prices instead of national average earnings has eroded its value since 1980 from 20 per cent of average male earnings to 14 per cent by 1993 (Commission on Social Justice, 1994). By 1997, the full basic pension for a single pensioner was £23 per week less than it would have been under earnings indexing; it is also over £6 per week below the level of Income Support. Whereas in 1948 the full basic pension for a single person was approximately 10 per cent above the level of means-tested benefits, by 1992 it was about 10 per cent below (Hancock and Weir, 1994). Second, the value of the state earnings-related pension scheme (SERPS) has been substantially reduced from its original 1975 formulation by basing the pension on revalued earnings over the whole working life instead of the best 20 years, by cutting the rate at which pension entitlement is built up relative to earnings and by halving the survivors' pension from the year 2000. Third, for women born after 1950, raising the state pension age will further reduce the amount of both basic and SERPS pensions by lengthening the required contribution period. Only those able to obtain full-time employment until age 65, currently only 6 per cent of women, will escape this reduction (Hutton *et al.*, 1995; Ginn and Arber, 1995).

While state pensions were cut, financial incentives were offered to those opting out of SERPS into a personal pension, following the Social Security Act 1986. The net cost to the National Insurance fund of the incentives from 1988 to 1993 was estimated as £6000 million (at 1988 prices) (National Audit Office, 1990). Tax relief for private (occupational and personal) pensions grew from £1200 million in 1979 to £8200 million in 1991 (Wilkinson, 1993) and has grown further since that year. In 1995 the 70 per cent of the National Insurance Fund spent on pensions amounted to £35 billion, but over a fifth of this amount went to the private sector in rebates and incentives (calculated from *Hansard*, 21 November 1996, col. 597).

In sum, there has been a substantial shift in resources from state to private pensions in the UK since 1980, in accordance with New Right ideology, pressure from the finance industry and ageist rhetoric in the media. The next section discusses how statistics were used to give a misleading picture of pensioner incomes, serving the government's need to justify reductions in state pensions.

Statistics used to legitimate pension reforms

State welfare was persistently denigrated by the Conservative government in the 1980s, accompanied by claims that pensioners were no longer poor and had a diminishing need for state pensions (Moore, 1987; Falkingham and Victor, 1991). The media amplified the message with provocative claims of pensioner affluence. Turning the traditional stereotype of needy pensioners on its head, the British media invented a new form of ageism in which older people were portrayed as affluent, enjoying a jet-setting lifestyle on their occupational pensions. For example, '[Pensioners] frequently enjoy a standard of living better than they had during their working lives . . . the over 60s are our nouveaux riches' (Paterson, 1988).

Older people's mean income, which was substantially higher than the median, was usually quoted, giving undue weight to the minority of well-off younger pensioners. Reports of the end of pensioner poverty were common, based on displacement of pensioners from the lowest income quintile by the growing numbers of young unemployed. However, media commentators stopped short of the US practice of applying a lower poverty threshold for those aged over 65 in support of such claims (Minkler, 1986).

In debates over pension policy, assertions that two-thirds of older people had an occupational pension were frequently repeated in the media to back the argument that the importance of the state pension was diminishing. The assertions were presented as uncontentious 'facts', neither attributed nor explained, let alone questioned. They are most likely to have stemmed from official statements similar to the following:

Now almost two thirds of those who retire can do so with an occupational pension. (Social Security Advisory Committee Report, July 1992, p. 12)

the additional element of occupational provision comes into play. Nearly 70 per cent of those now retiring in the UK will have supplemented their income in this way. (Peter Lilley, September 1992, p. 8)

57 per cent of all pensioners and 69 per cent of recently retired pensioners had occupational pensions in 1988. (DSS official, 5 September 1992, in a letter to British Pensioners and Trade Union Action Association)

The context of these official statements shows that they were all produced in support of the contention that for most elderly people the state pension is of minor importance as a source of income (*see* Ginn, 1993), legitimating the Conservative government's policy of reducing the value of state pensions and promoting private provision, mainly in the form of personal pensions, instead.

The statements were very misleading since the percentages quoted applied either to men only or to pensioner units (couples and lone pensioners). For example, Peter Lilley's (1992) statement was true for men but excluded women; among those aged 55 to 69, 68 per cent of men had entitlement to an occupational pension in 1988 but only 29 per cent of women did so (Bone *et al.*, 1992, table 6.2). By excluding women and failing to indicate this, the official statements gave the impression that only a minority of older people rely entirely on state pensions, whereas over half do so. The substantial gender gap in occupational pension entitlement was concealed.

Equally misleading is the implication that the value of the state pension is less important for those with an occupational pension. The distribution of income from private pensions (mainly occupational pensions) is highly dispersed, with only a minority receiving a substantial amount. For example, in 1993–94, among the 67 per cent of men aged over 65 with a private pension, a tenth received over £200 per week from this source and a quarter over £113; but half received less than £46 per week, a quarter less than £20 and a tenth less than £9 (*see* Figure 15.1b). This compares with older men's median income in 1993–94 from state benefits, £66 per week (Figure 15.1a).

For most older people with a private pension, the amount received is much less than from state benefits and is often too small to transcend the pensions poverty trap; because the basic pension is well below the level at which means-tested Income Support is payable, additional income may merely disqualify the recipient from Income Support and associated benefits. For example, it has been calculated that a lone pensioner in rented accommodation would need an additional income of £55 per week in order to be £6 per week better off (Collins, 1993). A small additional pension may bring no financial gain at all.

In spite of the declining value of the basic pension, state benefits still comprised 65 per cent of older men's and nearly 80 per cent of older women's income in 1993–94 (Ginn and Arber, 1997). Thus state provision is the vital mainstay of older people's income. Cuts in state pensions represent a serious loss of income, especially for older women.

In the UK the case for such reforms was markedly weaker than in other countries. Hutton *et al.* (1995, p. 15) note that 'the UK pension system is currently one of the "cheapest" of all OECD [Organization for Economic Co-operation and Development] countries', costing less than 7 per cent of GDP, and that state pensions provide the lowest earnings replacement rate in the European Union, with the exception of Ireland (*ibid.*, table 3.4). Looking to the future, Britain has one of the smallest projected rises in the elderly dependent ratio (population over state pension age as a percentage of the working-age population) among 20 OECD countries (Roseveare *et al.*, 1996, figure 1). The projected increase in the elderly

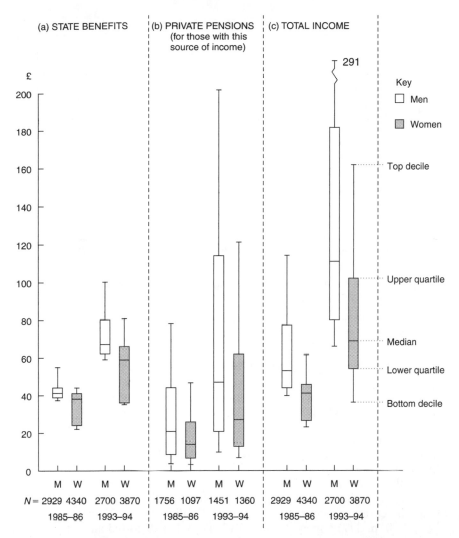

Figure 15.1 Distribution of personal income in the UK from (a) state benefits, (b) private pensions and (c) total, per week, men and women aged 65+, 1985–86 and 1993–94
Sources: Ginn and Arber (1991); General Household Surveys 1993–94 (author's analysis)

dependent ratio is also modest compared with past increases which have proved manageable (Hills, 1993; Vincent, 1996). Between 1960 and 1991, the elderly dependent ratio increased from 24.5 to 30.2 per cent, an average annual rise of 0.18 percentage points (*see* Figure 15.2). The government actuary's projections based on the 1991 census, which take account of the future increase in women's state pension age, show an expected rise in the elderly dependent ratio between 1991 and 2060 to 40.6 per cent. This represents an average annual expected rise in the elderly dependent ratio of 0.15 percentage points, somewhat less than in

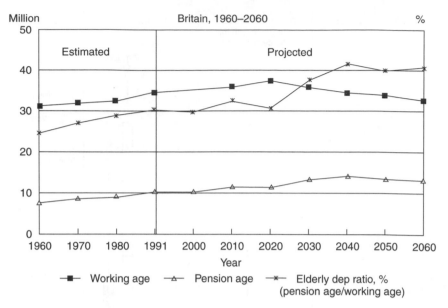

Figure 15.2   Population estimates and projections, Britain, 1960–2060
Source: Dilnot *et al.* (1994)

the three earlier decades. Only in the relatively short period 2020–40 is the rise expected to be faster.

The projected rise in the elderly dependent ratio is uncertain, owing to the unpredictability of future fertility rates and hence the size of the future working-age population. It is also an unsatisfactory measure of the economic viability of the National Insurance pension schemes because it takes no account of variation over time in the economic activity rate (at all ages) and in productivity (Falkingham, 1989).

The weakness of the demographic and economic arguments for cutting state pensions suggests that a crisis has been socially constructed in order to present a political choice as an economic imperative (Walker, 1990; Vincent, 1996). As Walker (1990, p. 377) argues, 'Political ideology has distorted and amplified the macroeconomic consequences of population ageing in order to legitimate anti-welfare polices.'

Actuarial tables: sex matters but not class

Government statistics on life expectancy might appear relatively uncontentious but the way they are used is not. British women's greater life expectancy at birth – 79 years compared with men's 74 (Tickle, 1996) – is taken into account in calculating annuity pensions (Keighley, 1992). For the same purchase price a woman's annuity is about 90 per cent of a man's, a discriminatory practice which is legal because of an exception in section 45 of the 1975 Sex Discrimination Act.

The gender differential in annuity rates may seem to have an unassailably objective basis in the life expectancy statistics: women live longer on average so the annuity must cost more. Yet white people live longer on average than those belonging to ethnic minorities, non-smokers than smokers, Cambridge residents than those in Manchester and non-manual workers than manual. For example, among British men aged 65–74, the standardised mortality rate (SMR) for those in the Registrar-General's social class I is 70 and for those in class V is 120, relative to the mean for men in this age group, defined as 100. Thus for class I men, their SMR is 30 per cent lower than the mean and for class V men their SMR is 20 per cent higher (calculated from Harding, 1995). The mortality difference between men in class I and class V is comparable to that between women and men. Yet people who are middle class (or white or non-smokers) do not suffer reduced annuities. Why is gender used as a criterion while class and ethnicity are not?

This awkward question was raised in the USA, where it was pointed out that if men have better annuity rates than women, blacks should enjoy a similar advantage over whites (Scott-Heide, 1984). Following a Supreme Court decision in 1983, company-sponsored pension schemes must now use sex-neutral annuity rates or risk legal action by women employees.

Thus a population group's average life expectancy is not the only criterion for determining annuity rates. Statistics on life expectancy may be selectively used, so that socially powerful groups obtain better rates than those with lower social status, unless the practice is successfully challenged in the courts.

With the growth of personal pensions in the UK and the trend towards the defined contribution type of occupational pension schemes, increasing numbers of older women will be affected by sex-based annuity rates. In the next section, the effect of pension reforms on women in the future is summarised.

Privatisation of pensions: the gender impact

In debates over pension policy, little attention has been paid to the effect on older women of privatisation of pensions. This omission is particularly serious since women constitute the majority of older people and are more vulnerable to poverty in later life than men. Women's lesser access to occupational pensions and their smaller incomes from this source have been well established and the reasons discussed fully elsewhere (Sinfield, 1978; Groves, 1987; Ginn and Arber, 1991, 1993, 1996; Arber and Ginn, 1991, 1994). The cuts in state pensions, combined with the maturation of occupational pensions, can be expected to reinforce class and gender inequality of income in later life.

Secondary analysis of the General Household Survey (GHS) has been used to examine the effect of the changing pension mix since the mid-1980s. Published results using the 1985 and 1986 GHS combined (reported in Ginn and Arber, 1991) were compared with results from the GHS in 1993 and 1994 combined (Ginn and Arber, 1997). In the analysis of private (occupational and personal) pension income, pension benefits 'inherited' from a deceased spouse were combined with the individual's income from his or her own private pensions. The analysis showed widening structural inequalities in private pension income. Over the 8-year period, the contribution of private pensions to the total income of those aged over 65 had increased from 17 to 25 per cent for men but only from 7 to 11 per cent for older women. A higher proportion of older people had a private pension income in the 1990s, but substantial gender and class differences remained. For example, in 1993–94, 67 per cent of older men and 35 per cent of older women had some private pension income, a slightly narrower gender gap than in 1985–86; the class gradient in private pension receipt persisted for both men and women (*see* Figure 15.3).

Gender inequality in the amount of private pension received had widened over the 8 years. Among older people with some income from a private pension, women's median amount had fallen as a proportion of men's from 65 per cent to 56 per cent; the median weekly amounts in 1993–94 were £46 for men and £26 for women (*see* Figure 15.1b). The gender gap in total income had also increased, older women's income as a proportion of men's falling from 77 to 62 per cent (*see* Figure 15.1c) (Ginn and Arber, 1997).

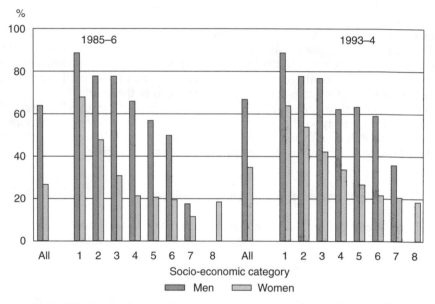

Figure 15.3   Private pension receipt, men and women aged over 65 by socio-economic category, 1985–86 and 1993–94
Sources: Ginn and Arber (1991); General Household Surveys 1993–94 (author's analysis)

Key
1 Large managers/professionals (2, 5, 6)
2 Small managers/intermediate non-manual (4, 7, 8)
3 Junior non-manual (9)
4 Skilled manual/supervisory (11, 12)
5 Semi-skilled manual/personal service (10, 13, 18)
6 Unskilled manual (14)
7 Employers/self-employed (1, 3, 15, 16, 17)
8 Never worked (21)

Excluded: Armed forces/full-time students/no answer (19, 20, 22)

The numbers in parentheses are Office for National Statistics socio-economic groups, and refer to the respondent's last or main occupation.

Women's disadvantage in pension income is likely to increase still further in the future as the pensions mix shifts further towards private provision. Personal pensions will generally provide a poorer return on contributions for women than men, because of women's lower earnings and interrupted employment, combined with high flat-rate administrative charges and sex-based annuity prices (Davies and Ward, 1992).

In sum, privatisation of pensions has tended to benefit those who were already better off. Those with high incomes, who tend to be working-age middle-class men, are the main beneficiaries of tax reliefs whereas working-class older women, because they rely most heavily on state welfare, bear the brunt of cuts in state pensions. Those whose income the Conservative government was to attack were first rendered invisible in official statements.

Conclusions

Pension reforms since 1980 have increased the public subsidy to private pensions while reducing the NI fund, a process of 'reverse targeting' (Sinfield, 1993, p. 39). Although driven by New Right political ideology and primarily serving the interests of the private finance industry, reforms were portrayed as necessary and just by the construction of increasing longevity as a fiscal problem and of older people as a burden. Official statements, repeated in the media, played a part in this process of legitimation by exaggerating occupational pension coverage among older people and implying that only a minority were solely reliant on state benefits.

The privatisation of pensions has exacerbated class and gender inequality of income in later life. The gender gap in older people's income, already substantial in the 1980s, has widened in the 1990s. As one economist has pointed out, 'Only the state can systematically, and with democratic legitimacy, redistribute resources to reduce inequality' (Downs, 1997, p. 12). Private pension funds will pay more tax as a result of the abolition of advance corporation tax by the Labour government in 1997; yet there is no sign that the government intends to restore the cuts made by the Conservatives in state pensions. If an increasing concentration of poverty among older women and others with low lifetime earnings is to be avoided, the policy of shifting resources from state to private pensions must be fully reversed.

Acknowledgements

I am grateful to the ONS for permission to use the General Household Survey data, to the ESRC Data Archive and Manchester Computing for access to the data and to the Leverhulme Trust for funding the research project on which the chapter draws.

16

Ethnic Statistics: Better than Nothing or Worse than Nothing?

Waqar Ahmad

Standard categories of ethnicity are often not appropriate

Introduction

Ethnicity is an important facet of self and group identity. However, its operationalisation into a concise, valid and reliable question has proved difficult. Further, the politics and pragmatics of ethnic or 'race' statistics have generated considerable debate. Those in favour include both those who wish to use statistical evidence to affect progressive change for minority ethnic groups and those who wish to propagate fears about the loss of British character because of the increasing number of non-white Britons. Those against have argued that the collection of statistical data gives the illusion of progress; that we already have ample evidence to promote positive change rather than constantly demanding more, finer and bigger datasets; and that statistics can be and often are used against the minorities rather than in their support. This chapter explores some of these problems. To begin with, however, I address issues in defining ethnicity.

Ethnic statistics: what are we measuring?

Among the concepts invoked by 'ethnicity' are culture, heritage, community, language, religion, lineage, geographical origin and shared symbolic elements ranging from food to clothes and rituals. Like many other identity claims, it is also flexible and situational, and, as Bulmer (1996, p. 35) notes, 'Members of an ethnic group are conscious of belonging to the group'. Such identity claims are not infinitely flexible; nor are they impervious to external relations. Others' reactions and responses to our identity claims confirm or question our notions of selfhood and may lead to the repackaging of identities. Resistance to external impositions of identity is also an important consideration; it may lead to a reinterpretation of values and symbols as in 'black is beautiful'. Ethnicity, then, is an important means of identification; and particular aspects of ethnicity become important in particular relationships, historical periods and geographical locations. In Pakistan, for example, language is an important symbol of ethnic affiliation; conflicts in

Karachi between the *mohajirs* ('immigrants' from India during the 1947 partition) and others were termed *lissani jhagrdey*; literally, language battles. In Kashmir, struggles around self-determination are focused on a discourse of ethnicity which places religion and nationhood centre stage. Berthoud (1998) notes the differences in key distinguishing factors between different places:

> If you look around the world, the lines of cleavage will vary from place to place. In Northern Ireland it is religious denomination. In Quebec it is language. In Belgium it is both of those in combination. In Bosnia, the gap is between those of Muslim and of Christian background; but in Croatia there are equally clear ethnic boundaries where religion is not relevant.

This flexibility and interconnectedness of ethnicity poses problems when one attempts to operationalise it into questions about ethnic identity. In relation to ethnicity, social policy has been closely associated with considerations of colour and culture, although the relative emphasis of the two has shifted. The multiculturalism of the late 1960s and 1970s conceptualised the 'race problem' in terms of cultural difference. The more radical anti-racism of the 1980s reconceptualised it in terms of racism and marginalisation. Attempts at ethnic categorisation encompass this duality of conceptualising non-white minority groups: white people are just that, 'white'; ethnicity resides in those who are different in culture and colour. As Berthoud (1998) notes, in Britain, 'the primary line of cleavage is based on colour. This is a characteristic which has undoubtedly been derived from one's family; and it has geographical associations common in ethnic differentiation'. However, as we will note, the assumption that colour (above religion or culture) is the non-negotiable part of ethnicity, and uniquely associated with discrimination, is challenged by many.

One anomalous variable in relation to ethnicity is religion – both in its status in law, and its relation to ethnicity (*see also* Chapter 17). British law, peculiarly, affords privileges to certain religions and 'religions/ethnicities'. Blasphemy laws cover only Christianity. Religious discrimination is outlawed in Northern Ireland. In Britain, the adherents of Judaism and Sikhism are formally defined as 'ethnicities' and thus protected under the race relations legislation, while those professing Islam, Hinduism and other religions are not. Further, Muslims contest the claims that colour is the major line of cleavage between groups; there is mounting evidence that Muslims experience forms of oppression and discrimination which are associated with their religion; colour discrimination is secondary for many. Following the Rushdie affair of the late 1980s, numerous cases of open discrimination against Muslims have been reported, and religious discrimination remains legal.

If we accept ethnicity to be a fuzzy, flexible and contingent concept then it follows that people's self-definitions of ethnic affiliation will be situational, will vary between people and time, and will be dependent on circumstances. I may regard myself as Pakistani, Asian or Punjabi; indeed, I do carry these as important aspects of my identity. Nor are these the only ethnic affiliations I allow myself. Within Punjabis there are cleavages by regional dialects and caste-like divisions. In terms of administrative statistics, then, ethnic self-definition, because of its variability and flexibility, is fraught with problems. Operationalisation of ethnicity in a standard and intelligent manner therefore is important if statistics are to have validity and reliability.

Early attempts at operationalising ethnicity, however, have been crude: defining non-white minorities, for example, at the level of 'born in the New Commonwealth' in some health statistics. (For a fervent criticism of past work on ethnicity and health *see* Chapter 25.) Table 16.1 shows more recent conceptualisations with similarities and differences betwen them. Significantly, CCETSW, an organisation centrally involved in anti-racist practice in the 1980s and early 1990s, gives importance to 'racial origin' in addition to 'ethnic origin'.

One casualty of any operationalisation of 'ethnicity' is that the resultant questions will be constraining and rigid. Most attempts also show other confusion. This confusion is also shared by the 1991 census question (Figure 16.1); *see* Coleman and Salt (1996) for accounts of other statistical series relating to minority ethnic communities. Two questions in the census related to ethnicity; the first (on country of birth) indirectly, the second (on ethnic group) directly. Country of birth cannot be directly related to ethnicity because of the substantial number of 'white' Britons born in pre-independence India, and an increasing number of minority ethnic people born in the UK. In the 1991 census, out of a total of 1.51 million non-white people, 0.7 million (46 per cent) were born in the UK (Salt, 1996). Question 11 asks about belonging to a particular 'ethnic group'. Those answering 'Black – Other' are invited to describe this; those ticking the final category 'Any other ethnic group' are asked for description with the additional instruction: 'If the person is descended from more than one ethnic or racial group, please tick the group to which the person considers he/she belongs, or tick the "Any other ethnic group" box and describe the person's ancestry in the space provided.' The ethnic question in the census is an amalgam of categories based on colour (white, black), notions of national background (Pakistani, Bangladeshi, Indian), and geographical origin. The difficulties in formulating a satisfactory

Table 16.1   Some non-census ethnic and racial classifications

Local government (Application for planning permission, Islington, north London)	Quango (Application for Diploma in Social Work, CCETSW)	University (Oxford University survey questionnaire)	Official survey (National Dwelling Housing Survey, 1978)
African	Racial origin:	White	White
Caribbean	Black	Black (Caribbean)	West Indian
Other black	Black (African)		Indian
Indian	Other	Black (other)	Pakistani
Pakistani	Ethnic origin:	Indian	Bangladeshi
Bangladeshi	African	Bangladeshi	Chinese
Chinese	Caribbean	Chinese	Turkish
Asian (other)	Indian	Asian (other)	African
Greek/Cypriot	Pakistani	Other	Arab
Turkish/Cypriot	Bangladeshi		Other
Irish	Chinese		
White	European (UK)		
Mixed/other	European (other)		
	Other		
	Welsh-speaking		
	Religion (NI only)		

Source: Coleman and Salt (1996, p. 12)

10 Country of birth	England ☐ 1
Please tick the appropriate box	Scotland ☐ 2
	Wales ☐ 3
	Northern Ireland ☐ 4
	Irish Republic ☐ 5
If the 'Elsewhere' box is ticked, please write in the present name of the country in which the birthplace is now situated	Elsewhere ☐

If elsewhere, please write in the present name of the country

11 Ethnic group	White ☐ 0
Please tick the appropriate box	Black – Caribbean ☐ 1
	Black – African ☐ 2
	Black – Other ☐
	please describe

	Indian ☐ 3
	Pakistani ☐ 4
	Bangladeshi ☐ 5
	Chinese ☐ 6
If the person is descended from more than one ethnic or racial group, please tick the group to which the person considers he/she belongs, or tick the 'Any other ethnic group' box and describe the person's ancestry in the space provided.	Any other ethnic group ☐
	please describe

Figure 16.1 Questions on country of birth and ethnic groups in the 1991 census of population
Source: 1991 census of population, enumeration form for private households, reproduced in Bulmer (1996), Crown copyright

question must not be underestimated; *see* Bulmer (1996) for accounts of some of these difficulties. However, the current categories appear conceptually haphazard and the current formulation suffers from some particular problems; we return to this later.

Statistics and social policy: why collect ethnic data?

Leech (1989), Bhrolchain (1990), Bulmer (1996), and others have summarised arguments for and against ethnic data collection. Only a brief and selective summary of these arguments is presented here.

A number of arguments can be put forward for ethnic data collection. One is that ethnic data collection is not different, in principle, from other forms of data collection in that such data represent a requirement of policy formulation and implementation, Ahmad and Sheldon (1993) summarise six main benefits of ethnic data collection. First, some sources of government funding have been available for work on or with minority ethnic groups; to apply for such funding, local authorities need figures for the numbers and the make-up of minority ethnic groups in their area. Second, with detailed and appropriate statistics, authorities and other agencies can tailor their services appropriately. Third, ethnic data can help in siting services in appropriate localities. Fourth, they can provide evidence of discrimination both generally, and at different levels within an organisation, which can then be tackled. Fifth, they provide baseline or targeted data for policy formulation and implementation. Sixth, minority ethnic groups can use and have themselves used such research for campaigning purposes.

Recent work by the Policy Studies Institute (PSI) (Modood *et al.*, 1997) gives evidence of diversity within the minority ethnic groups, as well as disadvantage experienced by many groups. Without ethnic data, the differences in the trajectories of different minority ethnic groups would be difficult to identify. Many powerful organisations, including the Commission for Racial Equality and the Runnymede Trust, have supported ethnic data collection in the census and ethnic monitoring to ensure equity of employment and access to services in statutory, independent and voluntary-sector organisations. Indeed, evidence on discrimination, from *ad hoc* studies as well as the repeated Policy Studies Institute Surveys, has contributed to Britain's having anti-discrimination legislation. However, not all regard ethnic data collection as having benign intentions or positive outcomes.

Arguments against ethnic data collection are of varying kinds. The greatest fear is perhaps about their use against the minority ethnic groups. From the days of Queen Elizabeth I's repatriation of non-white people to safeguard 'public relief' for white citizens to current times, measurement of difference has been associated with containing and controlling diversity. Statistics about immigration and about crimes, for example, have been used as justification of punitive and selective immigration control and oppressive policing, respectively. The distrust of official data collection by some minority groups and many white critics is therefore not mere paranoia. An organised campaign of non-cooperation with census tests before the 1981 census led to the abandonment of the ethnic question in that census. Similar fears of abuse have been expressed in other European countries and in North America, as Leech (1989) notes; *see also* Coleman and Salt (1996). This distrust was partially responsible for the considerable under-counting of certain minority ethnic groups in the 1991 census.

A second fear is that ethnic data collection becomes an end in itself rather than a vehicle for formulating, implementing or evaluating policy. Further, critics argue that there is no further need to prove racial discrimination and that the situation of minority ethnic groups remains poor. This rests on the lack of political will to effect change; having more, or finer, data will not change this. And, although having off-the-shelf categories allows for standardised data collection, their utility for social policy remains questionable; standardisation without utility is of little value. As Ahmad and Sheldon (1993) note:

> Let us take the example of a health authority that wishes to improve its employment practices and its service delivery with regard to minority ethnic groups. For employment monitoring the type of data used in the Census may have some validity, though we suspect that some of the categories are too

broad in order to be of particular use. To offer appropriate diet these categories become meaningless. 'Indian', 'Black Caribbean', 'Black African', for example, tell nothing about diet habits. An 'Indian' may be a Punjabi, Bengali or Gujarati; Muslim, Hindu, Sikh or Christian; vegetarian or meat eater, and amongst meat eaters requiring (or wanting) halal meat or non-halal meat; rice eater or chapati eater. If the same authority wishes to improve its interpreting services then the category 'Indian' tells it nothing about the mix of languages spoken (for example, Punjabi, Urdu, Gujarati, Hindi, Bengali).

One fear is that use of standard categories dissuades data collectors from thinking through the information they need for their purposes and collecting relevant policy-oriented data. General-purpose, routine data collection may not be sufficiently targeted to identify or address many social policy issues.

Third, state definitions of needs and financial support for these needs have resulted in minority groups' packaging and repackaging themselves into relatively arbitrary collectivities; identities become created in response to how the state defines groups and communities. For example, the anti-racist era witnessed some initiatives at the level of 'black', encouraging collective organisation between local non-white populations. Within this discourse, the non-white colour and shared experience of racism became the most salient symbol of self and group identification, above notions of shared history, language, religion and area of origin. More generally, the now defunct Section 11 funding encouraged repackaging in terms of 'culturally specific' need or 'integration' into the wider society.

Finally, ethnic statistics hide as much as they reveal. Internal diversity becomes hidden under what is regarded as the primary identity. This is what has been referred to as 'racialisation' (Ahmad, 1993; *see also* Chapter 25). Nazroo, for example, in Chapter 25, shows how ethnic inequalities in health become substantially diminished when controls for socio-economic positions are employed in comparisons. Over-reliance on ethnic data, accompanied by a neglect of socio-economic disadvantage, is in danger of pathologising the very minority ethnic groups which these data purport to support.

Some problems with the ethnic question in the census

Problems of ethnic categorisation generally apply equally to the census. Three issues are worth highlighting in particular. First, the category 'White' is problematic for a number of reasons. It posits white people as the 'norm' against whom the 'ethnic difference' of others is measured. Further, it privileges white colour and broad European heritage over other identity claims; thus the linguistic and other service needs of Eastern Europeans, racialised discrimination experienced by the Irish, the high rates of coronary heart disease experienced by the Scots, go uncharted. Nor is the 'Country of birth' question any defence against this, with increasing numbers of people of non-UK heritage being born here. Importantly, many 'white' people find it offensive to have their identities reduced to a colour.

The two other major considerations apply to the 'Black other' category, and to identification of mixed ethnic group, as noted by Berthoud (1998), among others. The difficulty in the case of the 'Black other' category resides in its ambiguity, which has encouraged hundreds of thousands of people, who in the census terms would be classed 'Black Caribbeans', to record their ethnicity as 'Black other' (Ballard and Kalra, 1994) – a confusion confirmed in the PSI's Fourth National Survey, which tested the link between 'family origin' and 'ethnic group' (Berthoud, 1998). The vast majority of those born abroad gave both their 'family origin' (92 per cent) and their 'ethnic group' (94 per cent) as 'Black Caribbean'. In contrast, for those Caribbeans born in Britain, whereas 80 per cent gave their 'family origin' as 'Black Caribbean', only 62 per cent gave this as their 'ethnic group'; a further 23 per cent described themselves as 'Black British'. This ambiguity needs to be addressed in the next census and in routine data collection using the census categories (*see* Chapter 2).

There is a greater problem in categorising those of mixed ethnic origin. In the PSI survey 39 per cent of Black Caribbean and 15 per cent of Chinese had one white parent (Berthoud, 1998). The 1991 census showed that 39.5 per cent of Black Caribbean and 15.8 per cent of the Chinese had a white partner (Coleman and Salt, 1996, p. 200). If one considers mixed unions by generation, more younger Black Caribbean (aged 16–34) people had a white partner than did older people (aged 35–59); the proportion increased from under 4 per cent (older) to over 6 per cent (younger) for Pakistani men and 0.6 per cent to over 2 per cent for women; and increased from 12 to 16 per cent for Chinese men. Indian men and women showed similar rates between the two generations (around 7 per cent and 4 per cent for men and women, respectively); Chinese women showed a drop from 26 per cent among older women to 22 per cent (still a high proportion) of younger women being in mixed ethnic relationships. Mixed ethnicity is a fact of life of modern-day Britain; it looks set to increase overall, even in populations which until recently have experienced relatively few ethnically mixed unions. This creates problems for ethnic data collection. How does one accurately or meaningfully capture the ethnic identity of those of mixed-ethnicity parentage? And whereas the suggestion in the census question for people of mixed ethnicity to assign themselves to that ethnic group to which they consider they belong, is based on assumptions of self-assignation of ethnicity, such self-assignation is not entirely without problems. Self-assignation also needs to be accepted by the selected group for the individual to have meaningful ethnic group membership. Further, the census also invites people to place themselves in the 'Any other ethnic group' box and give a description. This is a recipe for considerable confusion, with some 'mixed ethnicity' people assigning themselves to one specific ethnic group and others being described as of 'mixed ethnicity'.

Berthoud's (1998) proposed solution is to have a question about the ethnicity of both the mother's and father's family (Table 16.2). While this will be an improvement on the current question, it is, at best, a partial solution (it does not address the issue of 'White' ethnicity at all). With considerable numbers of respondents not just being of mixed ethnicity themselves, but having one or both parents of mixed ethnicity, this formulation has a short shelf-life.

Solutions to these difficulties are not easy to find. However, without credible responses, the ethnicity data for some may continue to lack credibility, will not provide maximum utility and will need to be used with caution.

Table 16.2 What are your family's ethnic origins?

	Your mother's family	Your father's family
White	1	1
Black Caribbean	2	2
Black African	3	3
Black other	4	4
Indian	5	5
Pakistani	6	6
Bangladeshi	7	7
Chinese	8	8
Asian other	9	9
Other	10	10

Source: Berthoud (1998)

Ethnic data: better than nothing or worse than nothing?

Debates on the ethics and utility of ethnicity statistics will continue. Operationalising what is essentially a flexible and interconnected concept into a small number of categories which are meaningful to respondents, are valid and reliable, and have social policy utility is a tough task. The varied attempts at doing this have only partially succeeded. The changing demography means that classifications have to keep pace with social change – a difficult task in itself and one which leads to difficulties in historical comparisons. Nor should the objections to routine ethnic data collection be brushed aside as paranoia or scaremongering. The populations who have objected have legitimate fears (Leech, 1989). Equally, ethnic data collection can facilitate policy development, implementation, monitoring and evaluation. But with one proviso: the presence of a political will to effect positive change.

Acknowledgement

This chapter is written as part of the ESRC project 'Comparative methods for studying socio-economic position and health in different ethnic communities' (L128251007). The grant is jointly held with George Davey Smith, Helen Lambert and Steve Fenton.

17

The Religious Question: Representing Reality or Compounding Confusion?

Joanna Southworth

There should not be a question on religion in the census

Arguments for and against a question on religion in the census

A religious question was proposed for the UK's 2001 census. This chapter suggests that there is some value to be gained from data following such a question. However, the various groups involved have not thoroughly addressed the implications. There is obvious overlap with the existing ethnic question, and the arguments for a religious question are as yet not strong enough to warrant its inclusion in addition to the ethnic question.

During the mid-1990s groups representing various opinions and interests discussed the content and form of the next census, to take place in the year 2001. These discussions supported the Office for National Statistics in its review of existing categories and their wording, for example the question on ethnic group, and tests of new questions in trial censuses, such as a question on religion.

A new question warrants careful examination and consideration, not only in wording, but also whether the need for it is adequately justified. Once a question is included it is likely to remain in the census for many decades and so can have long-term implications, some of which may not be immediately obvious. The religious question, like the ethnic question, deals with a sensitive and personal area, and so especial care must be taken. Statistics, once they are collected, are open both to use and to misuse; the minimisation of the latter must always be a priority. Once the data are collected their uses are not just those initially proposed or anticipated. To be of use, a question must be clear, be fair and fulfil a need. None of these is always simple to achieve.

At the time of writing, test censuses are still being devised and conducted, hence this chapter has a fairly broad and speculative approach. However, the question format of Figure 17.1 was tested in 1997, and it is these categories which will be used as the basis for discussion. One of the major areas of debate is whether or not to divide the Christian category into its various denominations. Other religions, Islam for example, also have different factions, which can be as diverse as those within the broad term 'Christian'.

> Do you consider you belong to a religious group?
>
> No
>
> Christian
>
> Buddhist
>
> Hindu
>
> Islam/Muslim
>
> Jewish
>
> Sikh
>
> Any other religion (please write in)

Figure 17.1   Census question on religion tested in 1997

There are two initial points of interest regarding the question as tested. First, it refers to a sense of belonging rather than belief or even attendance at religious events. There is, however, a great difference between the 'belonging' and a deep personal commitment, especially as regards the extent to which these views affect behaviour and actions, which are important for purposes of policy. Second, the categories are put in alphabetical order, with the notable exception of 'Christian', which is placed first. Be it covertly, in its present form the question will serve to re-emphasise the idea of a 'white Christian' Britain in majority, with the rest being 'others', minorities, such as has been criticised in relation to the ethnic question.

A preliminary report from the Academic Business Case for the census (Rees and Dale, 1997) proposes that the value of a religious question would be twofold. First, it would be used to further differentiate cultural groups. The uses of the data will be similar to those of the ethnic question, for example to ensure there are fair practices in education, employment, housing, etc., as well as to identify culturally and religiously specific welfare, health and service provision needs. This is based on assumptions such as that religious families will look after their elderly to a greater extent than non-religious families, or inferences about likely health at old age from attitudes to drinking, smoking, etc. However, here is where the differentiation between a sense of belonging and action following from a commitment is important. Studies like those of Modood *et al.* (1994, 1997) show that many members of minorities still 'belong' to their family religion, e.g. Sikhism or Hinduism, yet have adopted more 'Western' attitudes on careers, marriages and social behaviour such as smoking and drinking.

The second value of the religious question is for establishing cases for state support for schools administered by different religious groups, or for ensuring suitable religious education for the area the school is in. Issues of multiculturalism in British schools are prominent, but these usually relate to religious culture, rather than to ethnic culture.

A senior academic theologian and leaders from all religious faiths in Britain have lobbied for a census question on religion. Having failed to gain enough support from local or central government to be included in the shortlist of questions to be tested for the 2001 census (Office for National Statistics, 1996), a question on religion was nonetheless included in the census test of 1997.

In terms of personal identity, religion may be more significant than ethnicity, but the census is not intended to capture personal identity. Religious beliefs no doubt do affect behaviour, and hence may have an impact on policies and service needs. But these need to be thoroughly thought through, and the assumptions examined for their validity. The question should not be included simply because of pressure from certain support groups.

So who is campaigning for the religious question on the next census? There is wide support from religious establishments for this question; however, it is unlikely that this stems solely from desire for data. Indeed, at present data are already collected on attendance, membership and people's religious affiliation. For example, *Social Trends* in 1996 included data on church membership (adult active members or, in the case of Roman Catholics, attendance at mass). Figures 17.2 and 17.3 show that membership of Trinitarian churches (Roman Catholic, Anglican, Presbyterian, Methodist, Baptist, other free churches and Orthodox) has declined by 2.3 million (26 per cent) between 1970 and 1994, while the Muslim membership has increased from 0.1 million to 0.6 million (500 per cent) between the same years (though it is still much smaller). Census data would cover the entire population, but they also have an official status that statistics collected by other means do not. Present discussions surrounding religion have focused on falling church numbers and the rise of Islam. It is probably felt by all that official recognition on a census may improve the image and profile. What is not fully realised is that the nature of census data allows numerous cross-tabulations to be performed; it is not a case of just collecting data on religion. It is not easy to predict what other outcomes may be, but with present discussions surrounding the Royal Family and its continuing role, the issue of disestablishment is likely to emerge (*see later*, 'Religion and the state'), as occurred in Northern Ireland, which is examined next. Official use of religious categories would also pave the way for legislation on the religious aspects of services other than religion. The cons as well as the pros of having a question must be considered, and Northern Ireland provides us with the best example to date.

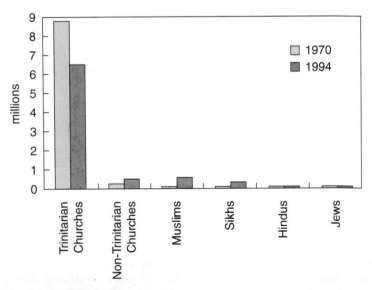

Figure 17.2 Church membership 1970 and 1994
Source: Christian Research, in *Social Trends*

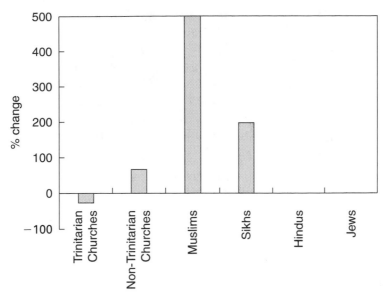

Figure 17.3 Percentage change in church membership, 1970–94
Source: Christian Research, in *Social Trends*

The religious question in Northern Ireland

Northern Ireland includes a voluntary question on religion on its census, which was included for the first time in 1861. The debate that surrounded this at the time is reported to be similar in nature to that which occurred over the ethnic question in the 1970s and 1980s (Macourt, 1995). The question is different from that proposed for the rest of the UK in that it focuses primarily on Christian denominational comparisons. One subsequent use has been to explore whether employment or other inequalities exist between the Catholic and Protestant groups. However, the question has been confronted with problems such as differences between the level of active participation compared with the profession of belonging, which vary from denomination to denomination, religion to religion, and over time. This could also apply to the UK situation. The census tends to be completed by the head of household, who could answer erroneously on behalf of the other members. The question in Northern Ireland still remains voluntary, and variation in response (as shown below) reflects various opposition campaigns and political and civil unease associated with religious connections (*see* Macourt, 1995, for details). Test censuses should reveal whether there is sensitivity or suspicion regarding the asking of a religious question in the UK, and it is to be hoped attention is paid to responses from all groups and not, as was the case for the ethnic question, just from minorities.

In Figure 17.4, the category 'non-Catholic' includes Presbyterian, Church of Ireland, Methodist, Protestant, Christian, interdenominational, undenominational and other Christian denominations. 'Other' encompasses those who gave other religions (including Jewish) as answers, those who expressed an absence of religion, or an indefinite answer. In 1991 the category 'None' was introduced.

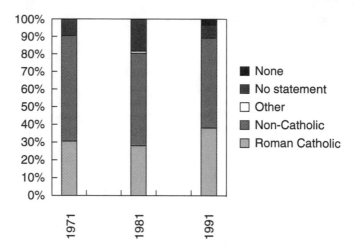

Figure 17.4 Percentage responses to the religious question in Northern Ireland
Source: Macourt (1995)

Religion and the state

> All human beings are religious if religion is broadly defined as the impulse
> for coherence and meaning. The strength of the impulse varies enormously
> from culture to culture, and from person to person. A non religious culture or
> person is described as secular. What does secular mean? It means the reli-
> gious impulse reduced to a minimum. (Tuan, 1976, pp. 271–2)

The 'religious impulse' in Britain may vary in strength and form from person to
person, community to community, and the society in which we live may appear
for the most part secular with few links between the church and the state. However, in an institutional capacity there are many legal remnants. The Queen is still
the head of the Church of England, this being the established church. The Bible is
still used in court for swearing on, and many mainstream schools still have periods of 'Christian' worship, carol concerts, etc., and to some extent are required to
by law. Thus Christianity is still covertly embedded in much of society today. Relatively little focus has been placed on this aspect. When a question, albeit voluntary, on religion was introduced in Ireland in 1861, disestablishment of the Church
of Ireland was not specifically mentioned but 'it was widely understood to be the
"hidden agenda"' (Macourt, 1995, p. 595), and this was indeed the end result. In
the present context disestablishment again has not been mentioned. The focus has
instead been on cultural and policy implications. This is a shift from previous
studies, which for the most part concentrated on the effect of religion on the landscape, rather than the religion itself and its cultural manifestations (previous studies have also been mainly concerned with the different Christian denominations,
e.g. Gay, 1971). Thus it may be that the religious question could open up a whole
new arena of debate, but the context is far from simple or uncontentious.

Aside from the need to consider uses which are not proposed, it is argued that
the question will further differentiate cultural groups. Religion's overlap with
ethnicity needs to be considered therefore to see whether additional information

is indeed gained. Can changes be made to this existing question to gain the extra information or is another question needed? Will the information gained add clarity or confusion? Which is most appropriate for gathering the information needed?

Religion/ethnicity, religion/community

It has become increasingly clear that in any mapping of the ethnic minority populations in Britain and Europe religion is an important differentiating factor. It divides groups with shared geographical, national, linguistic and ethnic origins, and sometimes unites groups with different origins of these other sorts. (Rex, 1993, p. 17)

Waqar Ahmad in Chapter 16 discusses how ethnicity takes on a meaning that is specific to the place and time in which it is used. Where a commonality of ethnicity *and* religion exists then the bonds are likely to be strongest. Examples include Jews and Rastafarians. Because of this variation in importance, religion could end up causing more confusion than clarity. It is not possible to talk of 'religion' as a coherent entity, as the different religions are very diverse in their nature. For example, some Christian and Muslim denominations take a very evangelistic approach, and in their pursuit of converts tend to span a wide range of cultural and ethnic groups. Judaism, on the other hand, is non-evangelical and therefore tends to remain confined to the Jewish ethnic population. The degree of organisational structure, taboos, etc. also varies, hence the impact on behaviour, and thus the extent to which it can provide a basis for policies, is also variable.

[W]hen assessing what religion adds to ethnic identification and inter-ethnic relations, it is essential to note that there are critical differences among religions which bear directly on how ethnicity is expressed and maintained collectively. It is not simply a matter of 'religion' being a part of the boundary setting package, but *which* religion. (Enloe, 1996, pp. 198–9; emphasis in original)

Barot (1993) argues that a common religion may be used by ethnic minorities. If minorities are defined with reference to visible difference (which may or may not correspond to cultural differences) with stigmatisation ensuing from this, then religious unity and practices may be used as arguments for more legitimate groups and communities. However, within each religion different factions and sects exist (such as division by caste, language, origin, etc.), so on what level does the community exist or specific needs occur? At present this question has not been addressed.

The overlap between religion and ethnicity therefore varies according to which religion and ethnic group we are concerned with. Space does not permit a full discussion of each, but some issues raised by the Christian category, and the religions of people of South Asian descent, will be discussed briefly.

Christian and Irish categories

Though the Irish are less visible than the traditionally considered ethnic minorities, their inclusion in the 'white' category does not in turn exclude them from prejudice or discrimination. 'The fact that the arrival of hundreds of thousands of Irish people in the mid-20th century was barely officially acknowledged did not mean that a racial discourse or racist practices against the Irish no longer persisted' (Hickman, 1996, p. 30), and many would argue the situation still has not changed and that the Irish presence in Britain should be identified (Commission for Racial Equality, 1997). However, if a suppression of Irishness has occurred, as the study by Hickman suggests, second/third/fourth-generation Irish people may now express themselves as English, Irish and English, or of Irish descent. Therefore their response to an ethnic question including the category 'Irish' may not produce the desired results.

In religious terms the identification of the Irish might be achieved by dividing up the Christian category into its various denominations, including Roman Catholic. This in itself would greatly increase the number of categories on the census form, and many Christians reject affiliation to a particular denomination. But being a Catholic does not automatically imply being Irish as well, nor is Irishness confined to Catholics, so an attempt to identify Irish people through religion would create more confusion than it would remove. Of ethnicity and religion, religion is the less appropriate means to identify Irish people in the UK.

There are other problems with the Christian category, which includes Mormons and Jehovah's Witnesses although some would argue that this is inappropriate.

South Asians

It is somewhat of a generalisation, but Indians tend to be Hindus, Sikhs, Muslims or Christians, whereas the majority of Pakistanis and Bangladeshis are Muslims. Hindus and Sikhs, again to generalise, tend to be non-evangelistic about their religion, and religion is not closely tied to their ethnic identity. However, for Muslims, religion is a very important defining element of their identity, separate from and often taking precedence over their ethnicity. As with Christianity, there are different groups within the general heading of 'Muslim'. The main division in Britain exists between what could be termed traditional Muslims and the Jamaat-i-Islami and its various organisations. The number of these latter Islamic organisations in Britain and Europe is growing, but, as Barot (1993, p. 10) says, '[h]ow far such bodies represent views and feelings of ordinary Muslims is a question which needs further investigation'. The Jamaat-i-Islami is a religious and political organisation which seeks to establish an Islamic state and seek the political transformation of society, considering its teachings to be the basis for social conduct (its slogan is 'Capitalism has failed, so has Communism. Islam is the answer' (Rex, 1993, p. 24)). The parallels with the church–state links discussed earlier cannot be denied as Muslims' numbers increase and those of the Church of England decrease.

The other need for a religious question that was identified was for establishing cases for state support for schools. A campaign exists for state-funded Muslim schools. The argument is that the Muslim population is growing in Britain, and as the government recognises Christian and Jewish schools, the same should apply to Islam. The Jamaat-i-Islami 'has decided to propagate its ideology within the host community by targeting religious and educational establishments in order to affect the host communities' perception of Islam, and to establish its version of teaching within religious studies' (Andrews, 1993, p. 77). Whether this idea is also backed by 'traditionalists' as well is less clear (just as we also need to establish who from the Christian community is supporting the census religion question). Lewis (1994) investigates these issues in more detail than is permitted here, and should be referred to. He reaches a pertinent conclusion:

> While few young Muslims in Bradford would identify themselves as other than Muslim, the precise nature of that identification in both the present and the future is still a very open question. Unless Islam in Britain gains the capacity to present itself in an intellectually intelligible and socially relevant way, it could become little more than a vehicle for the feelings of anger and bitterness generated by the exposure of the young to the chill winds of racial and ethnocentric exclusion. (p. 87)

The lobby for any particular census categories could be seen as a bid to define each religion, and perhaps also to differentiate each religion from those of other or no religion.

Conclusion

It was argued earlier that a question must be clear and fair, and fulfil a need to be included in the census. This chapter has demonstrated that clarity is lacking both in terms of what constitutes belonging and how it would be interpreted, and over the need it would fulfil. When needs have been identified, these are not adequately addressed by the question in its present form. There is an overlap between this question and that on ethnicity. Not enough is known at present about the ethnic–religious interface, so while adding detail to the ethnic dimension, it would also provoke more confusion.

For the broad geographical comparisons which the census allows, existing ethnic group categories largely satisfy the proposed few purposes of a question on religion. Identification of Jews and Irish within the 'white' category has some justification, but also serves to emphasise the otherness of minorities and consolidate the white Anglo-Saxon Protestant majority. We should take seriously the resistance in the rest of Europe to using categories of ethnicity and religion in official enquiries, which originated in the Fascist persecution of minorities.

While the major religious leaders appear to be in favour at present, the consequences of including a religious question on the census have not been adequately thought through. More questions than answers have arisen, and until these are properly addressed it would be premature to include the question before 2011.

18

Measuring Eating Habits: Some Problems with the National Food Survey

Mary Shaw

Is there such a thing as an average diet?

Introduction

As a topic for discussion, both among academics and in the lay arena, food has received a growing amount of attention in the 1980s and 1990s. There has been much public discussion, for example, surrounding issues of food and health, food scares and safety, eating disorders and the body, as well as increased interest in vegetarianism, and cooking and eating out as leisure activities. Over this period the sociology of food has developed (although other disciplines have a longer history of looking at social aspects of food and eating, most notably history, social anthropology and psychology). As our interest in food and related issues grows, aspects of methodology and the statistical data available must be critically considered.

Eating is not a matter merely of ingesting vital nutrients to support the body physiologically; the process of eating is also imbued with symbolic meaning. Food is produced, prepared and consumed in social, cultural and economic contexts. Research has established that eating habits exhibit the social dimensions and divisions of class, gender, age and ethnicity. Much of the emerging work on social aspects of food also considers culture; for example, examining the symbolic meanings of meat (Fiddes, 1991) or how we think about health, beauty and food (Lupton, 1996). However, solid empirical evidence on the nature of contemporary eating habits is also needed if we are to form a rounded picture of food and society. This chapter examines how some of the existing statistical data can be useful for understanding eating habits, if a critical perspective is taken.

Measuring eating habits

Most of the statistical data on eating habits which are available have been collected for the purposes of assessing nutritional status. There are basically three main types of methods which are used in nutritional research: the recall method, food diaries and food frequencies. These have to various degrees been used as a source and also adopted by researchers, though rather uncritically. Even within the discipline of nutrition itself the discussion of research issues is relatively scarce (Darnton-Hill, 1988).

The recall method, which usually refers to a time period of 24 hours, asks the respondent to recall everything that they have eaten as well as the amount of each food (Block, 1982). It is assumed that while the food eaten on the previous day can be fairly accurately recalled this may be an unstable measurement technique, as an individual's diet varies day to day (Samet *et al.*, 1984). Thus the day of the week as well as the time of year need to be considered. Block (1982) also mentions that 7-day recall can be used, but that inaccuracies and biases in memory become a major problem. The method is used in large-scale studies because, while it may be unstable in terms of individual measurements, it is considered to be adequate for looking at the intakes of a population as a whole (individual variations are averaged over a sample) and is thus useful for epidemiological purposes.

The second method commonly used in nutritional studies is that of food diaries, or food records as they are sometimes termed, which are usually collected over 1, 3, 7, 14 or occasionally 16 days. These require the respondent to record all the food he or she consumes during the allotted time and often involve weighing, hence subject training and sustained cooperation are required (Samet *et al.*, 1984). In a bid to improve reliability, this recording method is sometimes backed up by an interview with a nutritionist. However, this method usually incurs a high refusal rate because of the time and application needed by the respondent (Nelson and Naismith, 1979). Issues of literacy also arise owing to the need to keep written records. An additional problem is that accuracy tends to diminish towards the end of the time period (Block, 1982); understandably, people become bored with filling in forms on every morsel they ingest.

The third method commonly used in nutritional studies is that of the food frequency questionnaire, a method which was developed in the 1960s for use in large-scale studies investigating the link between diet and disease (Horwath, 1990a). Here the respondent answers questions (usually in a self-administered questionnaire) on how often certain foods are eaten, e.g. times per day, or per week, and sometimes questions on amount are included (e.g. serving size) (Horwath, 1990a). The central advantage of food frequency questionnaires is that they can be administered quickly and to large numbers of people; they can even be administered through the post (Horwath, 1990a). Because less is demanded of the respondent they also gain a high response rate (at least when compared to food diaries, which ask for a high degree of cooperation). They are also less unstable as they assess usual intake rather than that at one particular point in time. They are relatively inexpensive and the results are much more easily standardised than weighed records. However, while there are advantages to this method, in that it is less demanding of the respondents, it is a less precise form of measurement. Most people's habits vary from day to day and week to week, so can it be said with any certainty how often we 'usually' eat a particular food?

In endeavouring to design a sound measure of food intake, nutritionists have been concerned with the validity and reliability of the above measures. However, as all of these methods rely on self-reports, it can never be taken as certain that a true measure of food consumption has been attained. Respondents may wish to present themselves as conforming to behaviour patterns which they perceive to be socially desirable; there might be particular reasons for thinking this given the current public emphasis on healthy food (shall I admit to that third slice of chocolate cake?). This point is particularly pertinent when we consider that in 1994 the market for reduced-calorie foods amounted to £1.5 billion (Economist Intelligence Unit, 1994). Having interviewed 200 women on the subject of food, Charles and Kerr (1986) assert that virtually all these women had a problematic relationship with food; only 11.5 per cent of this sample said they had never been on a diet. Indeed, direct observation might be the only way to measure exactly what people consume, but this also brings a number of inherent problems, most notably ethical issues and the potential intrusion of an observer effect (Axelson and Brinberg, 1989).

There is thus no 'gold standard' by which to judge these various methods. However, Block (1982) sees the concern over validity as somewhat misguided and as the biggest impediment to nutritional research; it is not always necessary to be exact. She classifies two dimensions of dietary research: information on the group level – for example, mean cholesterol of a population, which can be adequate for epidemiological research; and the classification and ranking of individuals. Thus it seems that it is only when studying the individual in isolation that the absolute validity of the measurement of intake should be a concern. As Block (1982) points out, depending on our research question, we may not need such precision, and the concept of relative or accepted validity will have to suffice; frequencies can provide rankings of respondents and enough data for the comparison of groups of individuals. The methodological point is that different methods give different data, have their own advantages and disadvantages, and are useful for different research questions.

The National Food Survey

The National Food Survey (NFS) is the main source of statistical data about eating habits in the UK. The NFS is thought to be the longest-running continuous survey of household food consumption in the world. It was originally devised as a tool for monitoring the nation's diet during the Second World War, with the aim of devising food policy which would ensure that people were consuming a diet of a minimum nutritional standard (Ritson and Hutchins, 1991). It is now used as a source of information, by the government as well as other bodies, for a number of purposes, such as planning the production of food and the construction of the Retail Price Index. The NFS is also used to monitor the changing diets of British households and the consumption of food at the household level as compared to dietary recommendations.

The NFS uses a version of the food record data collection technique, as the amount of food brought into the household (whether bought or home-grown) is measured, rather than the actual amounts consumed. Data have been collected continuously for over fifty years, and the annual sample size is about 7000 households. On behalf of all household members a diary keeper records details of the description, quantity and costs of foods entering the home over a period of 7 days and there is also a questionnaire component. Some data on eating out are now also collected.

While this survey represents a valuable source of information about the nation's diet, and in particular about changes over the past five decades, it has a number of weaknesses which must be acknowledged. First of all, the survey measures foods taken into the household and not those actually eaten, so food that is wasted, fed to guests or fed to pets will lead to some distortion. There is also the issue of food bought in bulk and not used during the survey week, or which was bought during a previous week, although these variations are assumed to average themselves out across households. Another problem is that the response rate is relatively low (for example, of the 18 major datasets referred to in *Social Trends* 1994, the NFS has the second lowest response rate, at 59 per cent). It is also known (through comparisons with census data) that the NFS sample under-represents single people, housewives with full-time employment, married couples without children and the elderly (Ritson and Hutchins, 1991). This issue of sampling is significant, as there is commonly found to be a class pattern in eating habits, and if certain social groups are underestimated, then socio-economic differences in eating habits will likewise be underestimated. There are thus a number of methodological problems with the survey, as Fine *et al.* point out:

> In short, the survey is almost certainly biased by language, literacy, education, domestic stability, social class and availability of spare time. The intrusion of the survey is itself liable to bias behaviour, and the questionnaire is dependent upon subjective interpretation by interviewers. There are differences between purchase and ingest, not only in waste but also through the intra-household distribution of consumption. (Fine *et al.*, 1996, pp. 184–5)

As well as there being issues with the way these data are collected, the NFS statistics also need critical attention in terms of the interpretation of results. Perhaps the most crucial point (as Fine *et al.* suggest) is that data are aggregated at the household level. While Chesher says that this household level creates a 'veil of anonymity' (1997, p. 390) in that people will be less likely to under-report (as they will not feel that they as an individual are being surveyed), this aggregation means that the distribution of food within the household is not known. How problematic this use of aggregation is can be seen by looking at the example of meat consumption.

Table 18.1 shows consumption and expenditure for meat, fish and eggs per person per week in both grams and pence spent, and Figure 18.1 shows consumption of meat and fish per person per week in ounces from 1961 to 1992 (source: *Social Trends*, 1994). With the use of averages such as these there is an implicit assumption that meat-eating is equally distributed throughout the population. However, there is growing evidence that this is not so. Not only are increasing numbers of people becoming vegetarians, but men eat more red meat than women, as in Western culture it is associated with power and masculinity whereas poultry and chicken are seen as more appropriate for women: 'The macho steak is perhaps the most visible manifestation of an idea that permeates the entire western food system: that meat (and especially red meat) is a quintessentially masculine food' (Fiddes, 1991, p. 146). However, there is no mention in the NFS of the growth of vegetarianism. There are very few national or comprehensive data available on the actual numbers of people who are vegetarians (Beardsworth and Keil, 1997). However, a survey conducted in 1995 (Realeat Survey Office, 1995) suggests that as many as 12 per cent of the adult population regard themselves as non-meat-eaters, and that women are twice as likely as men not to eat meat. There is also an age gradient, with those in the 16–24 age group being most likely not to eat meat:

Table 18.1   Consumption and expenditure per person per week for meat, fish and eggs

	Consumption (grams)[a]			Expenditure (pence)		
	1985	1994	1995	1985	1994	1995
Meat						
Beef and veal	185	131	121	65.3	62.3	59.0
Mutton and lamb	93	54	54	25.4	22.6	22.5
Pork	98	77	71	25.8	25.6	25.0
Total carcase meat	375	262	247	116.5	110.5	106.4
Bacon and ham, uncooked	105	77	76	29.8	30.3	32.5
Poultry, uncooked	186	208	215	33.1	52.4	53.7
Other meat and meat products	377	395	407	94.1	161.7	168.3
Total meat	1 043	943	945	273.5	355.0	360.9
Fish						
Fresh	35	29	30	10.1	15.3	15.8
Processed and shell	15	15	16	6.1	10.1	11.3
Prepared, including fish products	41	56	52	15.5	28.1	27.3
Frozen, including fish products	47	45	47	12.8	18.0	18.5
Total fish	139	145	144	44.5	71.5	72.8
Eggs (no.)	3.15	1.86	1.85	21.1	17.3	17.4

[a] Except where otherwise stated
Source: National Food Survey 1995: Annual Report on Food Expenditure, Consumption and Nutrient Intakes (Ministry of Agriculture, Fisheries and Food, 1996). London: The Stationery Office (p. 7)

as many as 25 per cent of women surveyed aged 16–24 did not eat meat. The use of averages, then, obscures intra-household distributions as Chapters 3 and 15 highlight too. The use of averages can also be problematic in that they hide the extent of variation in the sample. For example, in 1995, males spent an average of £7.44 per week on eating out whereas women spent an average of £4.36 (National Food Survey, 1996). But how many spent nothing at all, and how many more than, say, £10?

The notion of a 'household' is a concept which is widely used in the NFS. Table 18.2 shows household food consumption by household composition groups within income groups for selected items (National Food Survey, 1996). This indicates that as income rises, so does expenditure on food. It is also apparent that household composition also affects expenditure: more children in the household means less expenditure per capita. As Fine *et al.* say,

> There is an implicit understanding that these socio-economic variables influence preferences, although why and how is rarely ever discussed (nor why these types of households should be considered rather than those partitioned by other criteria such as education, region, occupation, etc.). (1996, p. 153)

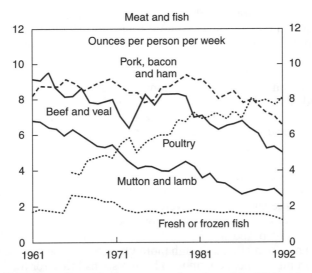

Figure 18.1 Changing patterns in the consumption of meat and fish at home
Source: Central Statistical Office, *Social Trends 24* (1994, p. 9)

Table 18.2 Household food expenditure by household composition groups within income groups, 1995 (includes food, but not soft drinks, alcoholic drinks or confectionery)

Income group[a]	Adults only	1 adult + 1 or more children	2 adults + 1 child	2 adults + 2 children	2 adults + 3 children	2 adults + 4 or more children	3 or more adults, 1 or more children
A	£19.22	data n.a.	£16.60	£14.77	£11.19	£10.45	£13.68
B	£17.03	£14.73	£14.28	£12.01	£10.95	£9.89	£12.55
C	£15.72	£12.83	£12.83	£10.61	£9.01	£7.48	£10.81
D	£14.87	£9.70	£9.68	£9.03	£7.45	£8.44	£10.40

Source: Adapted from *National Food Survey 1995: Annual Report on Food Expenditure, Consumption and Nutrient Intakes* (Ministry of Agriculture, Fisheries and Food, 1996). London: The Stationery Office.
[a] Income categories refer to gross weekly income of head of household. A £570 or more, B £300–£569.99, C £140–£299.99, D less than £140

However, in order to begin to understand the workings of the household and the distribution of food within the household these statistical data need to be augmented with data derived from other sources and other methods.

Sociological studies indicate that there are hierarchies of age and gender at play within the household. For example, Charles and Kerr (1988) note that in British culture there are certain foods that are deemed to be children's foods and that these are the foods located at the lowest part of what they term the 'food hierarchy' as they hold less social status. Examples of such foods are sweets, crisps, fish fingers, baked beans and breakfast cereal. Charles and Kerr found that younger children ate more of these low-status foods (using food frequency measures) than older children, whereas adults ate more of the high-status foods: 'The food eaten every day at home, therefore, marks social status and reflects social relations and divisions within families' (1988, p. 108).

Foods can also symbolise gender, as has already been noted, with red meat being seen as a masculine food, and white meat and fish as appropriate for women. However, it is not only the appropriateness of individual foods which is gendered; there is a much broader cultural framework determining the amount of food eaten and the most desirable body shape, especially for women. The ideal of slimness is one of the most dominant cultural values in Western societies and its counterpart, obesity, is devalued and stigmatised. Whereas slimness is the ideal body image for women, for males the dominant cultural preference is for muscularity. These ideals lead to very different relationships with food.

Not only the symbolism of food eaten within the household, but the organisation of food and the division of labour are gendered/hierarchical. Charles and Kerr (1988) and also Mennell *et al.* (1992) point out that the organisation of food within the family/household reflects familial ideology regarding the roles of men and women. In this dominant ideology, men are seen as providers, women as carers and homemakers. Research conducted by Warde and Hetherington (1994) found that food preparation tasks were still mainly performed by women: 'A woman was seven times more likely to have cooked the last main meal, ten times more likely to have baked a cake' (1994, p. 764). Research by Davidson and Cooper (1984) has found that although more women are now members of the paid workforce, working mothers continue to have responsibility for food and cooking at home. Moreover, it is the role of women to perform household tasks, including cooking, which are for the benefit of others, symbolic of care and nurturing, the good wife and mother. In short, the provision of food for the family is seen as a central part of a woman's role; 'as part of her role as the provider of food, it is the mother's responsibility to ensure that her husband and her children are well fed' (Graham, 1984, p. 130). This behaviour may lead to an unequal distribution of food within the household, especially when food is not plentiful. The gender division of labour within the home is only hinted at by the NFS in that for most of its fifty-year history the 'diary keeper' was referred to as the 'housewife' (whether that person was male or female), reflecting the social assumption that managing the food of a household is the responsibility of the woman in the household. Thus the statistical data may tell us how many grams of a food item are eaten in a household of a certain composition per day, but it does not give us any indication of intra-household divisions or the social and cultural processes which affect that outcome.

Conclusion

The National Food Survey and other sources of data collected for primarily nutritional purposes provide a valuable source of information on the British diet. Food has in many ways been a taken-for-granted topic in society, but data on food should not be taken at face value; they should be examined as critically as any other source of socially produced knowledge. There are a number of methodological issues which need to be borne in mind when collecting and interpreting data about eating habits. In particular, eating behaviour may be an area where people are particularly likely to be sensitive about revealing their habits. Second, when data are collected and presented at the household level (and when they are averaged), this masks layers of social practices and divisions within the household (such as patriarchal relations) which are crucial to understanding contemporary eating habits. Eating habits, and food statistics, need to be understood in their social and cultural context. For a fuller understanding a critical awareness of the limitations of food statistics is needed and, moreover, these data should be combined with data collected by other methods.

19

Measuring International Migration: The Tools Aren't up to the Job

Ann Singleton

A case study of European cross-national comparisons

International migration and the control of immigration are high on the agenda of the governments of the United Kingdom and the member states of the European Union in the late 1990s. Political debate is increasingly dominated by concerns both over numbers of documented migrants and over the supposed numbers of undocumented migrants in Europe. This was particularly evident in the debates which preceded the signing of the Maastricht Treaty and subsequent inter-governmental agreements on the control of immigration and of asylum, a period spanning the late 1980s and early 1990s. More recently, discussion has focused on the success or failure of immigration control measures and on monitoring and demonstrating their effectiveness. The production of statistical data on these issues has mirrored the concerns of policy-makers and resulted in the tabulations of international migration data now available to the general user. Critical research although hampered by the factors which influence production of the data, may also be aided by an understanding of their genesis, by the ways in which knowledge about migration is produced and reproduced. Migration data drive and underpin much of the production of knowledge on migration and they provide an insight into the processes of categorisation of migrants and the development of the control of population movements. This applies to national datasets, as it does to the cross-national datasets of the European state, a state which has emerged since the Palma document of 1988 and which is expected to be in place legislatively and politically in 1999 when the provisions of the Amsterdam Treaty become fully applicable (Statewatch, 1997).

This chapter examines the systems and processes involved in the collection, processing and dissemination of international migration data. Four main types of international migration statistics are considered: those on the resident population by citizenship, flows of international migrants, asylum-seekers and labour migrants. The strengths and weaknesses of the available data are discussed and common problems of incompatibility of sources, concepts and definitions are identified. The problems of using international migration data in policy-related and academic research are discussed. It is suggested that an understanding of these problems is necessary to informed and critical use of the data, as much theoretical work as well as political decision-making in the field of international migration has been driven by data availability (Salt and Singleton, 1995). Although migration statistics remain at the core of discourses around international migration at national and international levels (Singleton and Barbesino, 1998), the parameters of political debate and academic enquiry have been limited and defined by the partiality of the sources, by the categorisation of migrant types in the data sources, and by the lacunae in cross-national datasets. The limitations of the data, combined with a legacy of positivism in migration studies and an often uncritical attitude on the part of researchers towards their own role in the production of knowledge about migration, present a challenge to the social scientist wishing to analyse any aspect of migration.

One should beware of the uncritical use of international migration data, which (of their nature) present only a limited and partial view of a complex and dynamic set of social phenomena. Imaginative enquiry on the part of researchers is urged, both in the identification of what data availability tells us about policy priorities and what the lack of availability tells us about those issues which are not government or state priorities.

Construction and production of migration statistics

A central feature of the development of the nation-state has been the establishment and control of geographical borders and the control of movement of people – citizens and non-citizens – across those borders. The consolidation of new political, social and economic relationships between national governments within the European Union (EU) has coincided with a different type of external border around the EU and, arguably, with the emergence of a European state. This has generated new demands for data on migration at a European level.

Eurostat, the Statistical Office of the European Communities, has responded to the needs of policy-makers in creating a database on international migration. The database comprises data provided by the member states of the EU and the European Free Trade Association (EFTA) and by central European countries in response to an annual questionnaire. This database is probably the most comprehensive cross-national dataset on international migration stocks and flows in Europe. Other international organisations with similar databases in Europe include the Organization for Economic Co-operation and Development (OECD) (data are collected through the SOPEMI network of national correspondents) and the United Nations Economic Commission for Europe (UNECE). All cross-national datasets are compiled from data from the same package of national sources, in different combinations. The Eurostat database uses the sources indicated in Table 19.1.

As Table 19.1 shows, migration statistics are collected, processed and disseminated from a wide range of sources, including population registers, censuses and surveys, work permit records, records of asylum applications and National Insurance records. Most sources were not designed for the collection of migration data *per se*. They are often the administrative records of government ministries whose primary concern is with employment, policing or National Insurance. The multiplicity of sources on all recorded migrants, of methodologies, variables, concepts and definitions within and between countries (*see* Tables 19.2 and 19.3) presents major difficulties. It is necessary to take into account the type of organisation, the

Table 19.1 Institutions responsible for statistical data production

	Stocks of immigrants	Flows of immigrants	Employment of immigrants	Asylum seekers and refugees	Acquisition of citizenship
B Belgium	INS (Population register) (Census)	INS (Population register)	Ministry of Labour and Employment	INS, Ministry of Justice Foreign Office, General Commissioner for Refugees and Stateless	Ministry of Justice
DK Denmark	Danmarks Statistik (Central Population Register)	Danmarks Statistik (Central Population Register)	Danmarks Statistik (Central Population Register)	Directorate of Foreigners Ministry of the Interior	Danmarks Statistik (Central Population Register)
D Germany	Statistisches Bundesamt (Ausländerzentralregister: Central Register of Foreigners) (Census and microcensus)	Statistisches Bundesamt (Ausländerzentralregister, Microcensus) (Labour Force Survey)	Statistisches Bundesamt Ministry of Labour Social Insurance (Microcensus, Labour Force Survey)	Statistisches Bundesamt Bundesamt für die Anerkennung ausländer Flüchtlinge Bundesministerium des Innern, Bundesausgleichsamt	Statistisches Bundesamt
EL Greece	National Statistical Service of Greece (Census and residence permits)	Ministry of Public Order (Residence permits)	National Statistical Service of Greece (Work permits)	National Security Service	National Statistical Service of Greece
E Spain	INE Ministry of the Interior	INE (Padrón)	Ministry of Employment and Social Security (Work permits)	Home Office	INE
F France	INSEE (Census)	INSEE (OMI, OFPRA)	INSEE (Labour Force Survey)	INSEE OFPRA	Ministry of Social Affairs and Integration
IRL Eire	Central Statistical Office (Labour Force Survey)	Central Statistical Office (Census)	Central Statistical Office (Work permits)	Department of Justice	Central Statistical Office
I Italy	ISTAT Ministry of the Interior (Census)	ISTAT (Anagrafe) (Residence permits)	ISTAT Ministry of Employment (Residence permits)	ISTAT Home Office (Residence permits)	Home Office
L Luxembourg	STATEC (Répertoire National des Personnes Physiques)	STATEC (Répertoire National des Personnes Physiques)	Social Security General Inspection	STATEC Ministry of Foreign Affairs	STATEC
NL Netherlands	NCBS: Statistics Netherlands (Municipal population register)	NCBS: Statistics Netherlands (Municipal population register)	NCBS: Statistics Netherlands (Labour Force Survey) (Work permits and others)	NCBS: Statistics Netherlands Ministry of Justice	NCBS: Statistics Netherlands (Population Register)
A Austria	Österreichisches Statistisches Zentralamt (Census)	—	Ministry of Employment (Work permits)	Ministry of Interior	Österreichisches Statistisches Zentralamt
P Portugal	INE (Census) (Labour Force Survey)	INE Foreign and border services (Survey, census, residence permits)	INE (Residence permits)	Home Office Divisão de Planeamento	INE

FIN Finland	Tilastokeskus: Statistics Finland (Central Population Register)	Tilastokeskus: Statistics Finland (Central Population Register)	(Work permits)	Tilastokeskus (Central Population Register) Ministry of Interior (Central Aliens Affairs)	Central Aliens Affairs Ministry of Interior
S Sweden	Statistiska Centralbyrån: Statistics Sweden (Total Population Register)	Statistika Centralbyrån: Statistics Sweden (Total Population Register) Swedish Immigration Board (Residence permits)	Statistika Centralbyrån: Statistics Sweden (Work permits)	Swedish Immigration Board (SIV) Ministry of Culture	Statistika Centralbyrån: Statistics Sweden
UK United Kingdom	Office for National Statistics (Labour Force Survey)	Office for National Statistics (International Passenger Survey)	Office for National Statistics (Labour Force Survey)	Home Office	Home Office
IS Iceland	Hagstofa Islands: Statistics Iceland (Population Register)	Hagstofa Islands (Population Register)	—	Hagstofa Islands Ministry of Foreign Affairs	Hagstofa Islands
FL Liechtenstein	Amt für Volkswirtschaft (Population Register)	—	—	Amt für Auswärtige Angelegenheiten	Amt für Volkswirtschaft
N Norway	Statistisk Sentralbyrå: Statistics Norway (Central Population Register) (FREMKON – Register of Foreigners)	Statistisk Sentralbyrå: Statistics Norway (Central Population Register) (FREMKON – Register of Foreigners)	Directorate of Immigration (Settlement permits) Directorate of Labour (Unemployment data)	Directorate of Immigration (Central Population Register) Ministry of Justice	Statistisk Sentralbyrå: Statistics Norway Directorate of Immigration
CH Switzerland	Bundesamt für Statistik (Zentrales Ausländerregister – ZAR)	Bundesamt für Statistik (ZAR) (AUPER – database)	Bundesamt für Ausländer (ZAR)	Bundesamt für Statistik Bundesamt für Flüchtlinge (ZAR, AUPER)	(ZAR)

Note: Information in parentheses indicates data collection systems and sources. A dash indicates that data are not available
Source: Eurostat: 'Migration Statistics 1996'

Table 19.2 Comparison of types of population register by member states

Computerised and centralised	Municipal register[a]	Register of foreigners	No register
Belgium (Registre National)	Belgium	Germany[b] (Ausländerzentral-register)	Greece[c]
Denmark (Central Population Register)	Denmark	Liechtenstein (Zentralpersonen Verwaltung)	France
Luxembourg (Répertoire général des personnes physiques)	Germany (Melderegister)	Switzerland (Ausländerzentral-register)	Ireland
Finland (Central Population Register)	Spain (Padrón)		Portugal
Sweden (Total Population Register)	Italy (Anagrafe)		United Kingdom
Iceland (National register of persons)	Luxembourg		
Liechtenstein (Zentralpersonen Verwaltung)	Netherlands (Persoonregister)		
Norway[d] (Central Population Register)	Austria (Melderegister)		
	Sweden		
	Liechtenstein		
	Switzerland (Melderegister)		

[a] Partially computerised in Germany, Luxembourg, Spain, Italy, Austria and Switzerland; fully computerised in the Netherlands
[b] Computerised and centralised register
[c] Records are kept at municipal level for Greek nationals only
[d] Since 1994 the municipal registers have been linked to the Central Population Register
Source: Eurostat: 'Migration Statistics 1996'

method of collection (e.g. surveys or registers) and the type of administrative form used to record the data (e.g., residence or work permit applications). In some countries there is one responsible organisation, the national statistical institute (or central statistical office). In others there may be several, including the institutions or ministries responsible for the collection of raw data and the institution tabulating and publishing the data. Methods of collection may be direct or indirect, computerised, centralised or not. At the European level a bewildering combination may be involved. This diversity may be considered both a strength and a weakness for the purposes of research. On the one hand it hinders attempts to obtain harmonised datasets; on the other it allows for a wider range of migrants and migrant characteristics to be identified.

Table 19.3   Sources of data on international migration

Computerised central population register	Entry/exit form	Sample survey	Indirect sources for non-nationals
Belgium	Germany[b]	Ireland[c]	Greece
Denmark	Spain	United Kingdom[d]	France[a]
Luxembourg	Italy		Portugal[a]
Finland	Netherlands		
Sweden	Austria[e]		
Iceland	Liechtenstein (for nationals)		
Liechtenstein (for non-nationals)	Switzerland (for nationals)		
Norway			
Switzerland (for non-nationals)			

[a]   No source available for nationals
[b]   A parallel source (the 'Ausländerzentralregister') exists for the non-national population
[c]   A specific question is asked during the Labour Force Survey
[d]   International Passenger Survey: a voluntary survey organised at frontiers by the Office for National Statistics. Of the 100 000 persons leaving the country each year, approximately 1000 can be regarded as emigrants, and of the 60 000 persons entering the country some 2500 can be regarded as immigrants
[e]   in force since 1 April 1995
Source: Eurostat: 'Migration Statistics 1996'

Who is an international migrant?

The use of 'objective facts' about individual migrants moving in time and space and of concepts of citizenship, country of origin and nationality to classify and define migrants is a fundamentally problematic process. In addition, international migration statistics, by definition, record only those migrants known to the civil, administrative or police authorities (Singleton and Barbesino, 1998).

The basic unit of analysis in migration studies is the international migrant. The United Nations definition of an international migrant is a person who moves across an international border with the intention of living in another country for more than a year, having previously lived for more than a year outside that country. In practice, rather than this 'ideal type' of migrant being recorded, the variety of data collection methods, legislative systems and definitions involved at national level results in different groups of migrants being included in the datasets. The qualifying time period varies between recording countries from 3 months to a year. In some countries the intention of the migrant is recorded (as in the UK), in others it is the *de facto* situation as recorded in a population register.

Foreign population of countries in the European Economic Area

In most countries census data provide information on the resident population by citizenship. Citizenship is generally that defined in the passport of an individual.

This may appear to be a straightforward definition, but it is not unproblematic, particularly when cross-national datasets are constructed. National legislation is often complex and reflects the historical development of nation-states which formerly controlled empires. The legacy of imperial administrations has produced complex citizenship and nationality laws in both France and the UK. The result is that the population defined as nationals or citizens of one country include different and overlapping groups.

One measure of the social effect of international migration flows is often taken to be the size and composition of the resulting foreign population of a country (*see* Figures 19.1 and 19.2). Eurostat (1998) estimates that most of the foreign population of the European Economic Area (EEA) are citizens of non-member countries. According to these data, 73.7 per cent of the foreign population lives in three countries: Germany (5.1 million), France (2.3 million) and the UK (1.2 million), where non-EU foreigners account for 6.3, 4.0 and 2.1 per cent respectively of the total population. The data also show that 70 per cent of the total number of foreigners from other member states live in these three countries.

The figures illustrate a complex picture. On the one hand, they reflect the historical economic and social factors which have influenced and determined the direction and size of the movement of migrants into these three countries. On the other hand, they disguise the differential access to residence rights in and to citizenship of these countries. An example of the first case may be found in the figures for Luxembourg, which reflect the country's dependence on foreign labour

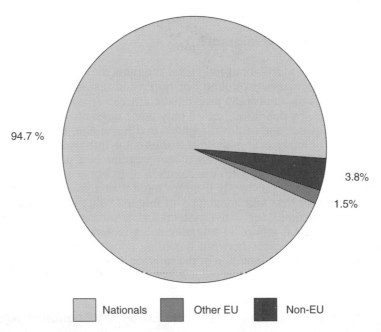

Figure 19.1   Population by citizenship group on 1 January 1995 in the 15 European Union countries
Source: Eurostat: 'Migration Statistics 1996'

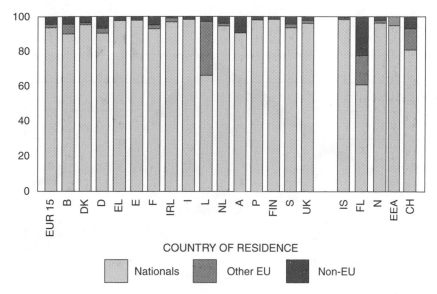

Figure 19.2 Population by citizenship group on 1 January 1995 by member state
Source: Eurostat: 'Migration Statistics 1996'

from within the EU. It has the largest foreign population in the EU (over 30 per cent), of whom most are citizens of other member states.

The significance of labour migration is most dramatically illustrated by the number of foreign residents from outside the EU who are from other countries in Europe. These are estimated to be over 50 per cent of the foreign population, the main citizenships being Turkey and former Yugoslavia. Over 18 per cent of resident non-nationals are from Morocco, Algeria and Tunisia. Conventional explanations for these numbers focus on the EU's need for labour, but the other side to the story is differential access to citizenship rights.

Asylum data

Asylum, alongside 'illegal migration', has become the big issue in Europe's immigration policy arena, with asylum-seekers becoming increasingly the subjects of state control. Control of entry to 'receiving states' and of access to application procedures is a central concern of policy-makers and of the immigration lawyers who act on behalf of asylum seekers. The available data starkly indicate policy priorities and the impact of legislative measures on access to asylum. Figure 19.3 shows the pattern of asylum-seeking in the EU and EFTA states and in the overseas members of the IGC (Inter-governmental Consultations on Asylum, Refugee and Migration Policies in Europe, North America and Australia). The leap in numbers between 1989 and 1992 is almost entirely explained by the numbers of applicants from Yugoslavia and former Yugoslavia. Inter-governmental policies at EU level subsequently controlled the numbers eligible to apply, and total applications fell.

The most sought-after variables are total numbers of applications, citizenship of the asylum-seeker, and 'sending' and 'receiving countries'. The media publicise

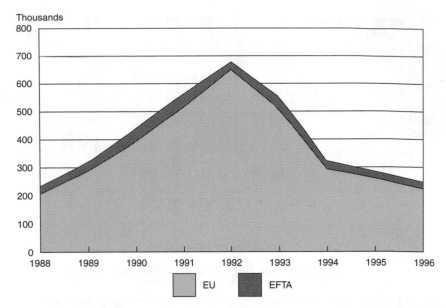

Figure 19.3 Asylum applications to the European Union and European Free Trade Association, 1988–96
Source: Eurostat: IGC 'Quarterly Bulletin on Asylum-Seekers 1, 1998'

government estimates of the numbers of unsuccessful asylum-seekers who subsequently evade deportation. Knowledge of asylum issues is driven, informed and validated by the availability of these data. Policy-makers rely on them and researchers use them to interpret changing patterns of asylum-seeking. In contrast, statistics on the length of time spent in the asylum process and of the numbers refused the opportunity to claim asylum are rarely collected or available. This impacts also on the crude approximation of recognition rates which governments allow to be published. These recognition rates are calculated using unrelated applications and recognitions which are also unrelated in any single year.

The asylum application data which are collected, tabulated and disseminated (with varying degrees of openness) are not unproblematic. Statistical and definitional problems connected with asylum data in the EU and EFTA countries have been thoroughly examined in two volumes prepared for Eurostat by van der Erf (Eurostat, 1994a, b).

Labour migration

The main sources of data on labour migration in the EU are the Labour Force Survey (Chapters 3, 36 and 37 discuss the UK version of this survey in this book) and Regulation 311/76. The two sources were examined, assessed and found wanting on labour migration in a report for Eurostat (Salt and Singleton, 1993). The main problem with the sources is that they do not cover the new types of labour migrants who are responding to economic, social and political change in the global economy and in particular in Central Europe. These are the new types of tourist workers, suitcase traders and those finding employment in many different forms of temporary work.

Categories of unrecorded, or inadequately recorded, labour migrants also include EU nationals such as posted workers, transit migrants of all nationalities, the self-employed, 'forced' migrants and hidden migration or clandestine movements. Additional statistical definitional problems arise from the many outstanding areas of unharmonised legislation on labour migration across Europe. 'Category-switching', the movement of an individual over time between different administrative categories and legal statuses, further complicates the question of obtaining up-to-date data.

International migration flows

As is the case for all types of data discussed above, statistics on immigration and emigration flows by country of previous residence and by citizenship should be used with caution. They are good indicators of the size and composition of population movements, for those migrants who are recorded. Data on emigration, however, notoriously under-record the numbers of people leaving a country and it should be remembered that they provide only a partial and limited picture of these movements.

The dynamics of international migration flows, particularly the new and emerging types of flows within Europe and between European countries and other countries are complex and generally not recorded in any administrative or statistical records. It is precisely these undocumented types of migration which may provide clues to help us gain a real understanding of population change in Europe.

Documenting the undocumented

The biggest challenge to the imagination of students of migration is posed by those migrants who are not recorded or who are deemed to be 'illegal'. The two categories (unrecorded and 'illegal') are not equivalent. Recorded migrants may be in an illegal situation of one kind or another, and unrecorded migrants may be legally present in a country. These migrants, often assumed to be 'illegal', represent complex and dynamic patterns of human mobility. They present an opportunity to understand and analyse realities about migration which are not simply those produced and reproduced by the availability of the data.

In order to analyse the changing dynamics and patterns of international migration, we need to move beyond dependence on statistical data which informs, shapes and limits our definition, questioning and understanding of the subject. The use of qualitative methods of research offers an opportunity to develop tools of greater potential than the ones on which much research to date has been reliant.

Acknowledgements

Eurostat produces regular publications and working papers which describe and analyse the key elements of the migration database. These may be obtained in the UK from The Stationery Office Ltd, International Sales Agency, 51 Nine Elms Lane, London SW8 5DR.

The author would like to thank:
Tony Bunyan, Gillian Hall and John Salt.
The Immigration Law Practitioners' Association.

PART IV
Counting Poverty

20

Poverty and Disabled Children

David Gordon and Pauline Heslop

Households with a disabled child are found to be among the 'poorest of the poor'

What is poverty?

Poverty, like evolution, is both a scientific and a moral concept. Many of the problems of measuring poverty arise because the moral and scientific concepts are often confused. In scientific terms, a person or household in Britain is 'poor' when they have both a low standard of living and a low income. They are not poor if they have a low income and a reasonable standard of living or if they have a low standard of living but a high income. Both low income and low standard of living can be accurately measured only relative to the norms of the person's or household's society. *See* Chapters 21, 28 and 40 for specific historical and contemporary examples.

A low standard of living is often measured by using a deprivation index (high deprivation equals a low standard of living) or by consumption expenditure (low consumption expenditure equals a low standard of living). Of these two methods, deprivation indices are more accurate since consumption expenditure is often measured only over a brief period and is obviously not independent of available income (as Mary Shaw shows in Chapter 18 of this book).

This 'scientific' concept of poverty can be made universally applicable by using the broader concept of resources instead of just monetary income. It can then be applied in developing countries where barter and 'income in kind' can be as important as cash income. Poverty can then be defined as the point at which resources are so seriously below those commanded by the average individual or family that the poor are, in effect, excluded from ordinary living patterns, customs and activities. As resources for any individual or family are diminished, there is a point at which there occurs a sudden withdrawal from participation in the customs and activities sanctioned by the culture. The point at which withdrawal escalates disproportionately to falling resources can be defined as the poverty line or threshold (Townsend, 1979, 1993).

Poverty and disability

There has never been a British study specifically designed to measure poverty among disabled people. Specific poverty studies have usually been carried out on relatively small representative samples of the British population. They include only a small number of disabled people and therefore cannot be used to produce reliable figures on the effects of disability on poverty. By contrast, studies of disabled people are generally not designed to measure poverty directly, therefore only indirect evidence is available at present.

There is a widely held belief that disability in childhood is an 'Act of God'; a 'misfortune' that is just as likely to befall the rich as the poor. Indeed, this view is often strengthened by the fact that many of the voluntary organisations that care for and campaign on behalf of disabled children are run by people from the middle classes or with reasonably wealthy backgrounds. The prevalence of childhood disability is not perceived to have a social class gradient in the same manner as diseases like childhood tuberculosis. (Social class is often used as a proxy (indirect) indicator of poverty since numerous studies have shown that social class IV and V households (e.g. partly skilled or unskilled occupations) have a much greater probability of being poor than social class I and II households (e.g. professional or intermediate occupation). (*See* Chapter 27, for instance.)

This perception is hard to understand given the crucial effects that maternal health and nutrition are known to have on the prevalence of congenital impairment (*see* Chapter 26). Numerous studies have shown that women of child-bearing age are much more likely to have poor health if they are in social classes IV or V than if they are in social classes I or II (Townsend and Davidson, 1988; Whitehead, 1988). The same social class gradient in women's health is observed if they are classified by their partner's social class. Given this known social class gradient in women's health, it would be expected that childhood disability would have a similar gradient. There are, however, a number of factors that might mask this effect. In particular, the risk of congenital impairment in children is known to increase with maternal age. Since middle-class couples tend to have children at an older age than working-class couples, this 'life-stage' effect may mask any social class gradient.

Table 20.1 shows the percentage of disabled children in the 1985 Office of Population Censuses and Surveys (OPCS) Disability Survey analysed by the social class of the head of household. (The OPCS Disability Surveys were the largest and most comprehensive studies of both adults and children with disabilities ever undertaken in Britain; *see* Martin and White, 1988a; Bone and Meltzer, 1989. Head of household is defined as the man in a couple or the woman where the family type is lone parent.) This distribution can be compared with that of all the children recorded in the 1991 census, analysed by head of household. The 1985 OPCS Disability Survey recorded that 4.5 per cent of all disabled children lived in households with a head in social class I, whereas the 1991 census found that 7.3 per cent of all children lived in social class I households. There are far fewer disabled children in social class I households than would be expected. By contrast, there are 1.7 times as many disabled children in social class V households as would be expected. This result must be interpreted with some caution since there was a 6-year gap between the OPCS Disability Survey and the 1991 census. There are also slight differences in the definition of head of household between these two surveys.

Table 20.1 Percentage of disabled children in Britain by social class of head of household from the 1985 Office of Population Censuses and Surveys Disability Survey compared with the percentage of children in the 1991 census by social class of head of household

Social class of head of household	Disabled children (1985 OPCS Disability Survey) (n = 1200) (%)	All children in households (1991 census 10% sample) (n = 856 520) (%)	Average number of children per household (1991 census 10% sample) (number)	Ratio of percentage of disabled children/ percentage of all children (ratio %)
I	4.5	7.3	0.71	0.6
II	18.4	31.3	0.66	0.6
III NM	12.3	11.5	0.55	1.1
III M	36.3	30.9	0.76	1.2
IV	20.1	14.2	0.68	1.4
V	8.2	4.7	0.65	1.7

Therefore, a second analysis was undertaken to compare the distribution of disabled children by socio-economic group of head of household with the distribution of all children by socio-economic group of head of household as recorded in the 1985 General Household Survey (GHS, described in more detail in Chapter 3) (Table 20.2). There is a clear gradient in the prevalence of childhood disability by socio-economic group of head of household. The children of unskilled manual workers are more than twice as likely to be disabled as would be expected. To express this another way, in 1985 a child was more than three times as likely to be disabled if their father was an unskilled manual worker than if he was a professional, despite the fact that professional fathers have, on average, twice as many children as unskilled manual workers. (This is because parents in 'professional' households tend to be older than those in 'unskilled manual worker' households.)

Table 20.2 Percentage of disabled children in Britain by socio-economic group of head of household from the 1985 Office of Population Censuses and Surveys Disability Survey compared with the percentage of children in the 1985 GHS by socio-economic group of head of household

Socio-economic group	Disabled children (1985 OPCS Disability Survey) (n = 1200) (%)	All children in households (1985 GHS) (n = 6454) (%)	Average number of children per household (1985 GHS) (n = 10 653) (number)	Ratio of percentage of disabled children/ percentage of all children (ratio %)
Professionals	4.5	7.2	0.82	0.63
Employers and managers	15.5	20.1	0.68	0.77
Intermediate non-manual	7.6	8.9	0.52	0.85
Junior non-manual	7.9	8.2	0.45	0.96
Skilled manual	37.6	35.9	0.72	1.04
Semi-skilled manual	19.2	15.9	0.52	1.21
Unskilled manual	7.7	3.8	0.40	2.03

The Sample of Anonymised Records (SARs) (Dale and Marsh, 1993; *see also* Chapter 4) from the 1991 census can be used as a further check on the association between parental social class and the likelihood of childhood disability. The 1991 census was the first to ask a question about limiting long-term illness (LLTI). Question 12 asked if any household member had 'any long-term illness, health problem or handicap' which limited work or daily activities. The question wording is somewhat different from that of the General Household Survey (GHS), which asks a broader question about 'long-standing illness, disability or infirmity' which limited activities 'in any way'. The census question, by contrast, asks an altogether narrower question using the present tense and qualifying the extent to which the long-term health problem, illness or handicap might pose difficulties (Heady *et al.*, 1996b; *see also* Chapter 2).

Although limiting long-term illness and disability are very different concepts, there is a considerable degree of overlap between these two groups (Pearce and Thomas, 1990; Charlton *et al.*, 1994), particularly in younger adults and children (Forrest and Gordon, 1993). The Individual Sample of Anonymised Records is a 2 per cent sample (just over 1.1 million individuals) from the 1991 census which allows tables to be produced that were not originally published from the census. Table 20.3 shows the prevalence rates of LLTI in children, analysed by social class of family head and compared with the distribution of children in households that do not have a LLTI, analysed by social class of family head. Once again, there is a clear gradient of increasing prevalence of childhood LLTI with decreasing social class of family head. There are 1.44 times as many children with a LLTI in families with a head in social class V than would be expected. This is an almost identical figure to that in Table 20.1 despite the many differences in definitions between these two tables.

It is not possible to determine from this kind of cross-sectional (one point in time) data whether it is 'poor' parental socio-economic status that is causing childhood disability, or childhood disability that is causing family poverty. However, since the publication of the Black Report, there has been considerable debate

Table 20.3 Percentage of children in Britain with a limiting long-term illness (LLTI) by social class of family head compared with the percentage of children without a LLTI by social class of family head from the 2 per cent Sample of Anonymised Records of the 1991 census

Social class of family head	Children with an LLTI (n = 3867) (%)	Children without an LLTI (n = 190 269) (%)	Ratio of percentage of children with an LLTI/ percentage of children without an LLTI (% ratio)
I	4.2	6.7	0.63
II	22.9	29.0	0.79
III NM	12.6	12.8	0.98
III M	29.4	29.0	1.01
IV	21.3	15.3	1.39
V	7.2	5.0	1.44
Armed forces	1.1	1.3	0.85
Inadequately described	0.4	0.5	0.80
Not stated	0.9	0.6	1.50

on the nature of the association between socio-economic status, ill health and disability (Illsley, 1986; Wadsworth, 1986). Research using longitudinal, case control and retrospective recall studies has shown that there is a strong association between adult poverty and adult disability, a lesser association between childhood poverty and disability in later life and only a relatively weak association between childhood disablement and adult poverty (Power *et al.*, 1990, 1991; Elford *et al.*, 1991; Ben-Shlomo and Davey Smith, 1991; Kuh *et al.*, 1994; Rahkonen *et al.*, 1997). Put simply, it appears that poverty is much more likely to make you 'sick' than being 'ill' is likely to make you poor (but *see* Chapters 23, 27, 28 and 30).

The datasets used to analyse the prevalence of childhood disability by social class and socio-economic group are large, comprehensive and reliable, and there is little doubt that 'working-class' children have a higher risk of experiencing disability than children from the 'middle and upper' classes. This factor is rarely taken into account in the allocation of resources for the provision of services for disabled children and this must compound the disadvantage of 'working-class' disabled children.

Deprivation and disabled children

The OPCS Disability Surveys did not set out to try to measure poverty in families with disabled children. Their main focus was on the additional costs of disability and, to a lesser extent, family income. They did, however, ask a limited subset of questions which had been used by Mack and Lansley in the Poor Britain survey in 1983. This study pioneered what has been termed the 'consensual' or 'perceived deprivation' approach to measuring poverty. The methodology has since been widely adopted by other studies both in the UK and other countries (Veit-Wilson, 1987).

It is possible to compare the results from the 1983 Poor Britain survey (Mack and Lansley, 1985) and the subsequent Breadline Britain in the 1990s survey (Gordon and Pantazis, 1997) with those from the OPCS Disability Surveys. The items shown in Table 20.4 are those common to both sets of surveys. Table 20.4 shows clearly that families with disabled children and families with children and disabled adults are all much more likely than disabled adult households to lack the necessities of life because they cannot afford them. In turn, households with disabled adults lack more necessities as a result of financial constraints than does the average British household. The much higher levels of deprivation suffered by families with children, where either the adults or children are disabled, is very marked. For example, 35 per cent of households with disabled children and 32 per cent of households with children and disabled adults could not afford two pairs of all-weather shoes in 1985, compared with only 9 per cent of British households in 1983 and 4 per cent of British households in 1990 that suffered from similar impoverishment.

It is possible to map the results from the limited subset of deprivation questions asked in the OPCS Disability Surveys onto the results from the 1983 Poor Britain survey to yield an estimate of the percentage of households with disabled children that are 'poor', using the same threshold levels as were used in the 1983 Poor Britain survey. Using a definition that would be commonly accepted by a large

Table 20.4   Percentage of households unable to afford a selection of consumer durables and certain items considered to be necessities by the majority of the British public in the 1983 Breadline Britain Survey (by household type)

	Breadline Britain Survey 1983 (all British h'holds)	OPCS Disability Survey 1985 (all disabled adults)	OPCS Disability Survey 1985 (disabled adults with children)	OPCS Disability Survey 1985 (adults with disabled children)	Breadline Britain Survey 1990 (all British h'holds)
	(n = 1174) (%)	(n = 8945) (%)	(n = 954) (%)	(n = 1200) (%)	(n = 1831) (%)
Warm winter coats	7	8	21	19	4
Two pairs of all-weather shoes	9	15	32	35	4
Presents for friends and family once a year	5	13	15	14	5
Celebrations on special occasions, e.g. Christmas	4	13	13	9	4
New, not second-hand, clothes	6	17	30	33	4
Meat or fish every other day	8	7	13	10	3
Roast joint once a week	7	12	15	14	1
Cooked meal every day	3	3	5	4	1
Toys for children	2	—	12	8	1
Money for school trips	9	—	17	10	4
Telephone[a]	11	14	13	23	7
Washing machine	6	9	9	6	4
Fridge	2	2	2	2	2
Video[b]	—	21	37	33	11

Note:
[a] the telephone was not considered to be a necessity by the majority of people in 1983 but it was by a small majority in 1990
[b] a video was thought to be a necessity by only 13 per cent of respondents in 1990

majority of people, this exercise shows that 55 per cent of households with disabled children were likely to have been living in poverty or on the margins of poverty in 1985.

This is an extraordinarily high level of poverty in comparison with the average British household. It seems that families with disabled children were four times more likely to be living in poverty than the average British household. This is a higher rate of poverty than is found within any other social group. Families with disabled children are more likely to be 'poor' than lone-parent households, unemployed households, households with heads in social class V, ethnic minority households, households with large families, etc. Families with disabled children

Table 20.5   Satisfaction with standard of living by household type

Satisfaction	All disabled adults 1985 (%)	Disabled adults with children 1985 (%)	Adults with disabled children 1985 (%)	All adults, Breadline Britain Survey 1983 (%)
Very satisfied	21	11	12	17
Fairly satisfied	50	42	49	58
Neither	14	17	14	8
Fairly dissatisfied	9	17	12	10
Very dissatisfied	6	13	13	7

are, arguably, 'the poorest of the poor' (*see* Gordon *et al.*, 1996). In many other countries, disabled people often suffer from extreme poverty. For example, in a statement presented to the United Nations World Summit on Social Development in 1995 on behalf of Disabled People's International, Liisa Kauppinen, General Secretary of the Deaf, said 'We are the poorest of the poor in most societies . . . Two thirds of disabled people are estimated to be without employment. Social exclusion and isolation are the day-to-day experiences of disabled people.' (Kauppinen, 1995, quoted in Beresford, 1996)

Supporting evidence for the extraordinarily high levels of poverty experienced by disabled people in Britain is provided by the work of Berthoud *et al.* (1993), who, using a completely different methodology, estimated that 45 per cent of all disabled adults were living in poverty. However, it must be noted that some of the respondents (whom this analysis would define as objectively living in poverty) expressed themselves 'fairly satisfied' with their standard of living. Table 20.5 shows that 61 per cent of respondents in households with disabled children were either 'very' or 'fairly satisfied' with their standard of living. This compares with 75 per cent of all British households who were satisfied with their way of life in 1983.

Income and disability

The presence of a disabled child in the family may affect the household financially in two ways: by limiting the earning power of the parent(s) and by altering their pattern of expenditure (Reid, 1975; Loach, 1976; Piachaud *et al.*, 1981). In order to assess the effect of the extra expenses that the presence of the disabled child incurs, the amount of money coming into the house needs to be known. Put simply, you can't spend what you haven't got. Just as all forms of consumption are subject to budgeting constraints, so too is the demand for disability-related items, and actual expenditure on the extra costs incurred by a family with a disabled adult or child will rise as income rises.

There is a history of research which indicates that families with a disabled child have lower incomes than equivalent families (Baldwin, 1977, 1985; Piachaud *et al.*, 1981; Smyth and Robus, 1989; Walker *et al.*, 1992; Beresford, 1995). With the exception of the work of Piachaud *et al.*, each of these studies has been based on selected samples of families with disabled children. The work of Piachaud *et al.* was seminal in that it presented evidence from an analysis of nationally representative data from the GHS in the 1970s. More recent data are now available from the GHS which provide new evidence on the income and expenditure effects of children and adults with a limiting long-term illness in the household.

Table 20.6 details the mean gross weekly income of households with children in 1993. Three groups are shown:

- households with no limiting long-term illness;
- households where there are one or more children (aged 16 or less) with a limiting long-term illness; and
- households where there are one or more adults (aged over 16) with a limiting long-term illness.

In all types of two-adults households with children, those which include a child with a LLTI have a lower average income than those where there is no LLTI. The difference is greatest in household types consisting of two adults and four or more children. Here, households which include a child with a LLTI receive £164 a week less gross income than those not containing a child or adult with a LLTI – a difference of £8528 each year, before the extra costs of the child's disability are taken into account. Such differences in income are largely accounted for by the difference in earnings between families with a disabled child and those without. This may be due to a number of factors. Baldwin (1985) found that mothers with a disabled child were less likely to be in employment than mothers in a control group and that, when they were employed, they worked fewer hours and earned less. Fathers' earnings were also affected, not only through lower labour force participation, but also through having their employment and promotion opportunities restricted.

Lone-parent households differ from those with two parents in terms of weekly gross household income. The difference in income between those which contain a child with a LLTI and those not containing a child or adult with a LLTI is comparatively small. Lone-parent households with one child with a LLTI receive a higher mean income than lone-parent households with one child without a LLTI. This effect is largely explained by the high proportion of lone parents who are dependent on income received from social security benefits and who are entitled to disability related benefits if they are the parents of disabled children.

Table 20.6 Mean weekly gross household income of households with children, by household type and the presence or not of a child or adult with a limiting long-term illness (1993)

Household type	No child or adult with LLTI	Child with LLTI	Adult with LLTI	Difference in income between no child or adult with LLTI and child with LLTI
n = 2315	(£)	(£)	(£)	(£)
2 adults and 1 child	423	350	390	−73
2 adults and 2 children	437	401	388	−36
2 adults and 3 children	470	344	359	−126
2 adults and 4 or more children	417	253	345	−164
1 adult and 1 child	120	151	122	+31
1 adult and 2 or more children	134	124	124	−10

Note: Generally, the same pattern of distribution is found if the median or 5 per cent trimmed mean figures are used instead of the arithmetic mean

The potential to spend money is partly constrained by the income available. Table 20.7 details findings from the 1993 GHS which reveals the proportion of all households with children who are without a variety of consumer durables, compared with the proportion of comparative households which contain a child or adult with a LLTI. In the case of almost every consumer durable considered, households containing a child or adult with a LLTI are more likely to lack an item. The difference is greatest when households with a child with a LLTI are compared with all households with children.

Table 20.7 shows that, in 1993, 23 per cent of all households with children were without a car. However, this figure rises to 27 per cent of households containing an adult with a LLTI and 35 per cent of households containing a child with a LLTI. Although a car was not judged to be an essential item by the majority of the British public in the Breadline Britain survey, for families with a disabled child a car is often a necessity rather than a luxury. Glendinning (1983), for example, quotes one mother who stressed that, if it was not for her disabled daughter, they would not have a car but that, so essential did she consider it, she would 'give everything else up' before she gave up the car (p. 67). Further, the OPCS Survey of Disability reported how the use of public transport decreased as the severity of the child's disability increased and that families with a severely disabled child were dependent on transport in a private car or taxi or that provided by voluntary organisations, education or health authorities or social services (Meltzer *et al.*, 1989).

Similarly, a washing machine or tumble drier are often regarded by parents of disabled children as indispensable necessities rather than extravagant luxuries (Glendinning, 1983), and some of the most frequent applications to the Family Fund are for washing machines, spin driers and tumble driers. Table 20.7 shows that the percentage of households without a washing machine is the same – 2 per cent – irrespective of whether the household contains a child or adult with a LLTI but that the percentage of households that do not own a tumble drier is less where the household contains a child with a LLTI than all other households with children.

Table 20.7 Percentage of all households with children in the General Household Survey who are without selected consumer durables compared with the percentage of comparative households with a child or adult with a limiting long-term illness (LLTI) (1993)

Consumer durables	All households with children in GHS (n = 2451) (%)	Households with children with LLTI (n = 284) (%)	Households with adults with LLTI (n = 581) (%)
Colour television	2	3	3
Washing machine	2	2	2
Deep freeze/fridge-freezer	5	7	8
Video	8	11	9
Telephone	13	15	14
Car	23	35	27
Microwave	25	28	27
Tumble drier	34	31	34
Compact disc player	51	59	53
Home computer	59	55	58
Dishwasher	78	84	81

The only other consumer durable which households with an adult or child with a LLTI more frequently report owning is a home computer. Although this may seem surprising, the relatively high computer ownership may be due to the number of disabled children and adults who rely on computers as items of essential equipment, to aid their communication or enhance their development (McWilliams, 1984; *Disability Now*, 1995).

The GHS does not provide details of the extra financial costs which disabled people incur because of their disability, such as the cost of special equipment, extra heating or clothing, individual transport, or for cleaning, cooking or personal care services. Attempts to gauge these extra costs are methodologically fraught with difficulty and the results are dependent not just on the type, nature and severity of the disabilities but also on cultural factors and the availability and cost of social, educational and health services (Horn, 1981; Chetwynd, 1985; Graham, 1987). Studies have tended to be small scale and limited to particular medical conditions or focus on the type of expense disabled people face and not the exact level. In the UK, direct questions about extra costs in the OPCS Survey of Disability estimated an average of £6.10 per week for disabled adults (Martin and White, 1988b). However, two smaller-scale studies by the Disablement Income Group found the extra costs of disablement for adults to be up to seven times this amount (Disablement Income Group, 1988, 1990). Reanalyses of the OPCS data by Berthoud *et al.* (1993), taking into account the severity of the disability, income and standard of living, concluded that the extra costs of disability for adults amounted to over three times the OPCS figures, at an average of £19.70 per week at 1985 prices.

In the case of disabled children, the OPCS Survey reported the extra costs of disablement to be £7.65 per week for families with a disabled child (1985 prices). Berthoud *et al.* did not reanalyse the OPCS data on families with a disabled child using the same approach as they did for adults.

Using expenditure diaries and information from parents, Baldwin (1985) found that families with a disabled child incurred extra expense for items bought regularly (such as food, clothing or transport) at a cost of approximately £16 per week (at 1985 prices) for those on average incomes. Baldwin's estimate is thus more than twice that of OPCS. In addition, Baldwin found that extra expense was incurred for aids and appliances or special equipment, for larger items, such as housing adaptations, and for hospital in- and out-patient attendances or stays. Gough and Wroblewska (1993), in their study of children with motor impairment in Scotland, found 63 per cent of families reporting a need for extra expenditure against a background of reduced parental employment and earnings.

Conclusions

None of the specifically designed poverty surveys in Britain have had a sufficiently large sample size to provide direct evidence of the levels of poverty experienced by households with disabled children. However, all the indirect evidence available indicates that as a group these households are among the 'poorest of the poor'.

This is important because it implies that major policy legislation is failing. Financial supplements for the families of disabled children appear to be insufficient to bring them to a par with the rest of the population, let alone cover the extra costs of disability. The Children Act 1989 has placed a statutory duty on local authorities to examine the position of children and to provide services to minimize these effects. This legislation needs to be interpreted more imaginatively by local authorities.

Similarly, the Disability Discrimination Act 1995, which addresses (to some extent) discrimination in the provision of goods, facilities and services to disabled people, does not directly address discrimination against disabled children and their parents. In the short term, as Beresford's (1995) study of the families of disabled children highlights, the greatest need is for more financial assistance through higher weekly disability benefits, so that childcare and development needs can be adequately met.

In the longer term, what is really needed is a more comprehensive, legally enforceable anti-discrimination policy which is of relevance to all disabled children and their families. Until this occurs, the situation will continue to look bleak.

Acknowledgements

This chapter is based on a paper presented at a meeting entitled 'Disabled Children and Their Needs: Present and Future', Centre for Social Policy, Warren House, Dartington, Totnes. We would like to thank our colleagues at the Statistical Monitoring Unit for their encouragement and help and, particularly, Christina Pantazis for her help with the General Household Survey.

21

Where Are the Deprived? Measuring Deprivation in Cities and Regions

Peter Lee

Deprivation is not easy to measure: it depends on what you want to find

Introduction

Where is the most deprived area in Britain? The answer may be intuitively easy but is difficult to measure scientifically. Having an objective view of where the poorest areas are located is important. The distribution of funds for public investment, such as the Single Regeneration Budget (England), relies in part upon levels of deprivation within areas. Moreover, the politics of poverty dictates that there are, inevitably, vested interests in being identified as the poorest area. As one conference delegate recently remarked at an anti-poverty seminar in London, 'The best indicator of poverty is the one that puts my authority at the top of the rankings!' This chapter therefore introduces some of the statistical and theoretical considerations when measuring deprivation at area level. The chapter also gives illustrations of how different approaches to the same problem result in different places being identified as 'poor'.

The Census of Population provides the only comprehensive data source for mapping deprivation within very small areas: electoral wards and enumeration districts (EDS). (*See* Chapter 2 for more information on the census.) At the time of the 1991 census there were 9509 wards in England and Wales and more than 100 000 EDs. These nest within local authorities which vary in size – the Isles of Scilly being the smallest local authority in terms of population and Birmingham the largest (Table 21.1). As there were 56 local authorities in Scotland with, below them, a different geography based on postcode sectors (of which there were 1002; Dale and Marsh, 1993), this chapter restricts its analysis to England and Wales. Since the publication of the last census, however, introduction of unitary authorities has meant that, in a limited number of cases, the census boundaries from 1991 are now out of date.

However, while the census is still the best source for spatial analysis, it is limited for the analysis of poverty or deprivation. This is because it contains only a narrow range of variables and has not included a question on income or wealth. David Gordon and Pauline Heslop consider more detailed surveys of poverty in Chapter 20, but having many more variables at small area level would not be

Table 21.1   The census geography and average household population

	Total households	Local authority districts		Electoral wards		Enumeration districts	
	No.	No.	Ave. h'hold pop.	No.	Ave. h'hold pop.	No.	Ave. h'hold pop.
Isles of Scilly	765	(1)	(765)	5	153	8	96
Birmingham	374 002	(1)	(374 002)	39	9591	1 942	193
Eng. & Wales	19 789 166	403	49 104	9509	2081	113 196	183

practical considering that the estimated cost of each additional question in the decennial census is £100 000. As a result the task is to select variables which provide satisfactory proxy measures of poverty or deprivation.

The issue of measurement, at area level, is fraught with methodological difficulties. The first difficulty is to make the concepts operational: what do we mean by poverty or deprivation? This is constrained by the problem of data availability: what data are available to operationalise the concept chosen and best reflect its measurement? Finally, there is the problem of geography: how are the constructs and measures represented spatially and what meaning do such results possess? Local authorities and government departments will continue to use area measures of deprivation for resource allocation. This chapter argues that techniques need to be validated and never taken for granted.

Conceptualising disadvantage

Whereas studies are variously concerned with 'poverty' (a lack of material resources, income and savings), 'social exclusion' (increasingly being defined in terms of exclusion from labour markets, but its early origins imply a wider notion of exclusion based on access to citizenship rights) or 'deprivation' (lack of access to services or absence of resources), it is clear that these concepts attempt to measure, estimate and describe different phenomena and that such labels are not always interchangeable. Understanding how concepts differ requires an understanding of the derivation of knowledge on disadvantage.

Peter Townsend points out that there has always been an interest and concern with concentrations of poverty and the interaction between place and deprivation arising from 'notions of association and contamination, congregation, inheritance and environmental influence' (Townsend, 1979, p. 543). Indeed, from the early nineteenth century there emerged an interest in understanding the associations between place and poverty. Edwin Chadwick's *Report on the Sanitary Conditions of the Labouring Population of Great Britain* of 1842 (Chadwick, 1965) demonstrated for the first time that physical environment was a determinant of 'social ills' (rather than the other way around, as previously perceived). The evolution of the debate on poverty in Britain took to this empiricist approach and lent itself to the spatial measurement of association. Later in the century Charles Booth's study of life and labour in London took this to its ultimate conclusion by mapping the city's criminal and feckless communities (Booth, 1899).

In the UK the poverty debate has, therefore, tended to emphasise empirical methods. The development of 'standard of living' approaches to poverty is testimony to this tradition (Townsend, 1979; Mack and Lansley, 1985). These approaches conceptualise poverty as a combination or series of deprivations, both material and social, such that resources are so seriously below those commanded by the average individual or household that they are, in effect, excluded from ordinary living patterns, customs and activities. As income is withdrawn from individuals or households, poverty lines or thresholds can be drawn at the point at which there occurs a sudden increase in material deprivation (Townsend, 1979). However, there are problems with this definition as not all households below the so-called 'threshold' will be experiencing poverty. In a study of resources, deprivation and poverty, Callan *et al.* (1993) took current income and deprivation into account in the measurement of poverty as they found that many households on low incomes were not suffering basic deprivation. Criticism has also been raised against poverty lines which are based on income alone (Ringen, 1988) as expenditure patterns suggest that the 'expenditure poor' and the 'income poor' are not the same group (Goodman and Webb, 1995).

Graham Room contrasts this 'Anglo-Saxon' approach to the study of disadvantage with the theoretical debates that have taken place on the Continent (Room, 1995). These debates revolve around notions of inclusion and the rights and obligations of citizens, and surface in the term 'social exclusion'. The term implies a wider approach than is often adopted in British debates about poverty and deprivation and is generally distinguished from poverty:

> The notion of poverty is primarily focused upon distributional issues: the lack of resources at the disposal of an individual or a household. In contrast, notions such as social exclusion focus primarily on relational issues: in other words, inadequate social participation, lack of social integration and lack of power. (Room, 1995, p. 105)

The concept of social exclusion is especially relevant to area measurement because of the explicit spatial references. Some accounts of social exclusion suggest that it has emerged in the context of transition within communities through the twin processes of immigration and spatial polarisation, with the city providing 'a visible spatial embodiment of the cleavages of a dual society' (Castillo, 1994, p. 625). However, in some definitions, social exclusion is narrowly defined by reference to employment, presenting a dichotomy whereby all those outside the labour market are perceived as *excluded* while those in work are seemingly *included*. Chapters 36–38 in this book show how such an assumption can be misplaced, and Ruth Levitas criticises the narrow definition of social exclusion and notes how it has become 'embedded as a crucial element in a new hegemonic discourse. Within this discourse, terms such as social cohesion and solidarity abound, and social exclusion is contrasted not with inclusion but integration, construed as integration into the labour market' (Levitas, 1996b, p. 5).

Proxy indicators of deprivation

The point to note is that unemployment will increasingly serve as a proxy for poverty and social exclusion. This perspective is increasingly being emphasised in the context of European Union debates on social exclusion and in the UK has been endorsed by Prime Minister Tony Blair's Labour government in policies designed to tackle exclusion through employment policies and the so-called 'New Deal' (Mandelson, 1997). Furthermore, when concepts of poverty or deprivation are employed in spatial analysis the tendency is to reduce complex processes and simplify their dimensions. Indeed, owing to the absence of household-level data (and income data) which accurately reflect levels of material deprivation, the unemployment rate has been used as one of the principal indicators of poverty in area analyses (Green, 1994).

The use of unemployment as a proxy for poverty has been justified for two reasons. First, it is argued that 'unemployment results in reduced incomes and straitened circumstances and imposes other pressures on individuals through loss of self-esteem' (Carstairs and Morris, 1991, p. 10). The second argument is more methodological. Unemployment is often said to be a 'direct' measure of deprivation (Thunhurst, 1985), as opposed to 'indirect' indicators of those at risk of disadvantage (e.g. the elderly or lone parents). However, there are two associated problems with an approach to area measures of poverty which use single proxy indicators. First, it is not possible to read off the whole range of possible circumstances for an individual when using a single indicator; that is, one cannot be certain that even apparently 'direct' census measures (such as unemployment) always measure deprivation, as not all of the unemployed are deprived. The recently unemployed, for example, differ in their expenditure and consumption patterns from the long-term unemployed; while deprivation in the workplace (low pay, poor conditions) is omitted when the focus is on the unemployed. Second, while high unemployment rates in a particular area correlate highly with other deprivation indicators such as low car ownership rates (Lee *et al.*, 1995), the relationship between any two indicators is far from dichotomous, as demonstrated by Figure 21.1. This figure ranks small areas for Great Britain by households with no car and the unemployment rate. The scales depict the decile points when ranking the 10 511 wards and postcode sectors (OPCS, 1991). It cannot be read off that an

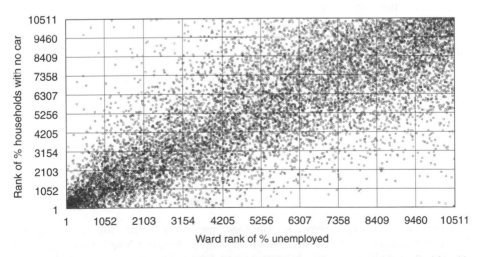

Figure 21.1 Plot of wards in Great Britain ranked by unemployment and households with no car

area with a high rank according to unemployment rates will also have a high rank according to the percentage of households without access to a car.

These two associated problems are related to a common problem in spatial analysis referred to as the 'ecological fallacy': 'the ecological fallacy is the assumption that when a relationship is found by correlating data for areas it reflects a relationship at the level of the individual or the household' (Dale and Marsh, 1993, p. 249). The problem associated with assigning deprived status to single variables leads to the conclusion that it is preferable to use several indicators in one index. Better than this, an indication of the likelihood that individuals or households are *actually* deprived when they are in a category designed to represent deprivation appears to be vital but is not attempted in many indexes of deprivation.

Indexes of deprivation

The past thirty years have seen a plethora of indexes of deprivation come into use. The reasons for this explosion in competing methods can be divided into at least four separate, but not mutually exclusive, categories: (a) *methodological*: the distinction between 'indirect' and 'direct' measures of deprivation (although somewhat artificial) has stimulated the development of competing methods; (b) *conceptual*: as poverty is relative to time and place, the indicators used will change in order to reflect changes in lifestyles and behaviour; (c) *technological*: data availability and the release of more detailed information creates demand for new and more 'sophisticated' measures; and (d) *policy*: the emergence of spatial urban policy in the 1970s and its re-emergence in the 1990s have driven the creation of new techniques for measuring area deprivation; at the same time different policy applications (e.g. payments to general practitioners (GPs), urban regeneration, etc.) require different indexes. Two indexes developed and used extensively during the 1980s were Professor Brian Jarman's Underprivilege Area Score (Jarman, 1983) and the Department of the Environment's 1981 Z score. They reflect the different policy uses of indexes.

The Jarman index was developed as a guide for the targeting and distribution of resources for primary health care. The method was innovative in that it provided an input from practitioners and resulted in an element of professional consensus built into the eventual index. Jarman surveyed over 1800 GPs in order to establish the types of *social* (e.g. the number of elderly) and *service* factors (such as the provision of ancillary services) which were felt to result in an increase in workloads or features envisaged to demand more primary health care resources. However, the index was weakened as Jarman dropped service factors from the final index and explained this omission on the basis that service factors were affected by administrative or political imperatives. Meanwhile, the Department of the Environment's Z score was used extensively as a guide to urban funding and to inform decisions concerning urban policy. The index was widely reproduced in the 1980s by government departments, local authorities and voluntary bodies. The index was heavily criticised for a number of reasons, not least Townsend *et al.*'s observation that 'what stands out from these measures is that "the most deprived areas of England" do not include districts drawn from the Northern Region and this flies in the face of most observation and experience' (Townsend *et al.*, 1988, p. 35).

More recently, following the release of 1991 census data, a new wave of indexes has emerged (Lee *et al.*, 1995). The 1981 Department of the Environment (DoE) Z score was superseded using a different set of indicators, and the resulting index (the Index of Local Conditions or ILC) has been used extensively in the allocation of urban policy resources (Department of the Environment, 1995a). Rather than attempt to summarise the differences between all competing indexes developed over the past thirty years, the next section focuses on the ILC and another of the more recent indexes developed, the Breadline Index (Gordon and Pantazis, 1997; *see also* Chapter 24), which illustrate contrasting statistical techniques and highlight differences in approaches to area measurements of deprivation.

The DoE Index of Local Conditions

The Index of Local Conditions (ILC) is constructed from census data held at enumeration district, ward and local authority district level as well as other non-census administrative data held at local authority district level. Included in the index are the following variables (census data): *unemployment* (unemployed persons); *poor children* (households with no earner or one parent in part-time employment); *overcrowding* (households with more than one person per room); *lack of amenities* (households lack or share bath/shower and/or WC, or live in non-permanent housing); *no car* (households without access to a car); *children in flats* (children living in flats, not self-contained or non-permanent housing); *education at 17* (17-year-olds no longer in full-time education); (non-census data): *long-term unemployed* (Employment Department, NOMIS); *income support* (Benefits Agency), *low educational qualifications (GCSE)* (Department for Education); *Standardized Mortality Ratios* (Office of Population Censuses and Surveys, vital statistics); *derelict land* (DoE, Derelict Land Survey); and *crime* (insurance company data).

The ILC uses the chi-square statistic to standardise for differences in population size (*see* Lee *et al.*, 1995; Dale and Marsh, 1993). The ILC also transforms all the indicator values by taking their natural logs. Logging the values has been justified on the basis of suppressing extreme values (such as the percentage of overcrowded households, which is an urban phenomenon, particularly in London) so that it makes for a more normal distribution.

The Breadline Index

The Breadline Index was developed from the results of the 1990 Breadline Britain survey in which 1174 adults were interviewed and a minimum standard of living, as defined by the general population, was established. Poverty was defined as the point at which withdrawal (measured by a material deprivation) increased disproportionately as income declined (*see* Townsend, 1979). Whereas the ILC contains no weights reflecting the degree of risk of poverty associated with each indicator, each Breadline variable was weighted in accordance with results from the Breadline Britain poverty survey. Characteristics of the poor from the results of that survey were matched with variables from the 1991 census and each component variable of Breadline has weights attached which reflect the degree of risk of being in poverty where each condition was present. Moreover, the interaction effects were also incorporated into the index by modelling the percentage that were poor on each component variable taking into account the other indicators.

The indicators used for the Breadline Index have been used in ward- and district-level analysis (Gordon and Forrest, 1995; Lee *et al.*, 1995), and are as follows: *unemployment* (economically active population unemployed); *no car* (households with no car); *rented housing* (households not owner-occupied); *lone parents* (lone-parent households as a proportion of all households); *long-term illness* (households containing a person with a limiting long-term illness); and *low social class* (persons in social class IV or V).

Reliability of indexes

Gordon and Forrest (1995, p. 5) observed that 'despite the fact that all indexes use different statistical methods and combinations of variables, the resulting maps are remarkably similar'. Figures 21.2a and 21.2b appear to uphold this at a local authority district level for England, as the quintile distribution of local authority scores on the ILC and the Breadline Index appear to highlight similar areas as deprived (deprived districts in the two maps are shaded more darkly). However, when the two indexes are ranked according to the 10 511 small areas in Great Britain, the association between the two indexes is not as great (Figure 21.3). Additionally, a regional breakdown of where the most deprived areas are located (the most deprived 10 per cent of wards referred to as a decile) shows that the ILC identifies 40 per cent as wards in Greater London and the South-East regions, whereas the Breadline Index identified only 24 per cent of the most deprived wards as being located in these two regions (Table 21.2). Figure 21.3 and Table 21.2 demonstrate that the choice of indicators and method of construction is, therefore, important when targeting small areas of deprivation.

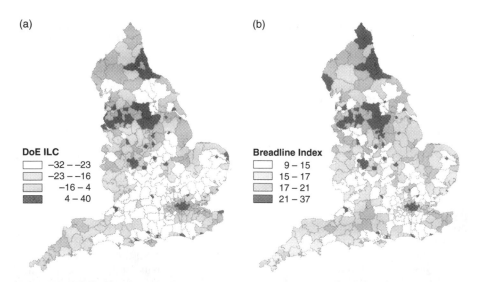

Figure 21.2   (a) Department of the Environment Index of Local Conditions for local authorities in England; (b) Breadline Index, also for local authorities in England
Source: From data in Office of Population Censuses and Surveys (1991). Boundary data by permission of OPCS, supplied by UKBORDERS, University of Edinburgh

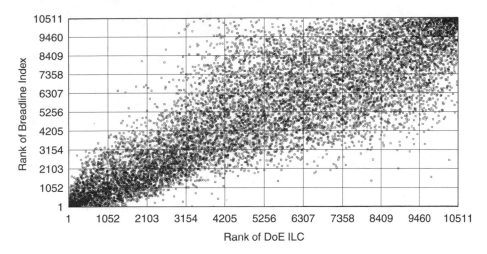

Figure 21.3   Scattergram for Breadline and Index of Local Conditions rankings, for wards in Great Britain
Source: As Figure 21.2

Table 21.2   Percentage of wards in most deprived decile by region by index

	ILC	Breadline
Greater London	32.8	20.6
North	4.5	14.7
North West	11.3	14.5
Yorkshire and Humberside	4.8	6.3
West Midlands	5.8	4.8
East Midlands	2.9	4.4
East Anglia	1.1	0.9
South West	3.2	1.7
South East	7.5	2.9
Wales	3.5	5.3
Scotland	22.5	24.1

ILC = Index of Local Conditions

The reason for the disparity in the identification of deprived areas is explained by the choice of indicators and the method of combining them into a single index. For example, standardising indicators in the ILC may also have made the index more sensitive to urban areas with larger than average ward sizes. This is partly acknowledged by the DoE, when the department commented that the method of standardisation (chi-square) is 'less appropriate for measuring precisely relatively low levels of deprivation' (Department of the Environment, 1994).

This sensitivity may be exaggerated owing to the auto-spatial correlation of building features associated with the choice of indicators in the ILC. Lee *et al.* (1995) found that the ILC was correlated most strongly with *children in flats* and *lacking amenities* and that this would show up in the pattern of poverty using such an index. In London, the proportion of households living in flats is greater than for other parts of the country, partly because of pressure on land values. To ascertain the environmental deprivation and developmental implications for children living in flats would require information on the floor level on which children are living. However, this information is not available from English census data. Without information on floor level combined with weights reflecting the degree of risk for children living in flats at different floor levels, the index is susceptible to this type of bias.

An associated problem with the ILC is that some indicators appear to count the target population twice: the ILC includes *unemployment* and *long-term unemployment* at local authority district level; it also includes *children in flats* and *poor children* – poor households with children (and little choice but to live in local authority flats) are, therefore, counted twice. Where there is overlap between categories in the Breadline Index (e.g. lone parents without a car), the method of weighting takes into account the interaction between categories so that a weight is calculated according to the additional explanation of the variation in poverty accounted for by each variable. This makes the index conceptually robust. However, the index is not standardised to take into account variations in ward size and this has meant that it is more sensitive to small areas.

Conclusions

Writing of socialist formulations to combat poverty, David Harvey wrote that a revolutionary approach adopting theory must be taken. For Harvey this does not entail

> yet another empirical investigation of the social conditions in the ghettos. In fact, mapping even more evidence of man's patent inhumanity to man is counter-revolutionary in the sense that it allows us to pretend we are contributing to a solution when in fact we are not. (Harvey, 1973 p. 144).

Harvey's agenda was and is perhaps more ambitious than that of a local authority policy officer charged with the responsibility of determining the allocation of resources according to need. It is for this reason that the identification of deprived areas and the mapping of deprivation will continue to attract attention from the worlds of policy and academe. What this chapter has demonstrated is that the measurement of deprivation at area level is a socially negotiated process which is also dependent upon the availability of indicators at sufficiently discrete spatial scales. What at first appears to be a straightforward exercise in the identification of deprived areas is far from simple in practice. Although concepts such as poverty, deprivation or social exclusion are used variously in area studies, little attempt has been made at thinking through these concepts and reflecting their epistemological traditions in the choice and weighting of available data. Methods of measuring area-based deprivation such as Breadline may be preferable if only because they attempt to validate their construction with reference to the traditions of empirical measurement and existing knowledge of what it means to be in poverty. Ultimately, however, the identification of deprived areas gives only a partial view of social disadvantage, and therefore needs to be combined with measurement of constraints, opportunities and powerlessness as well as flows into and out of poverty, which quantitative cross-sectional area measures cannot capture.

22

The Limitations of Official Homelessness Statistics

Rebekah Widdowfield

Homelessness counts are getting further from reality

This chapter examines the difficulties involved in trying to measure homelessness and the limitations of the official homelessness statistics, while Chapter 23 considers the statistical treatment of homeless people themselves. The quantification of any social problem, such as homelessness, logically begins with a definition, yet there is considerable difficulty in defining a social issue which can mean different things to different people at different times and in different contexts (Hutson and Liddiard, 1994, p. 26). Consequently, there is no universally accepted definition, and 'what constitutes "homelessness" and how many people are homeless is a debate which has been running for thirty years or more' (Greve and Currie, 1990, p. 28).

A number of commentators (*see*, for example, Watson, 1984; Blasi, 1990; Hutson and Liddiard, 1994) have conceived of homelessness as a continuum of housing situations ranging from life on the streets to people living in inadequate or insecure housing. However, while there is a general consensus that those without any form of shelter are homeless, as the definition is extended to encompass people with recognisable but less extreme housing problems – for example, those living in insecure or overcrowded accommodation – it becomes increasingly difficult to draw a distinction between homelessness and housing need. Thus, as Neale (1997, p. 48) contends, 'homelessness is a highly ambiguous and intangible phenomenon which lies at one end of a spectrum of housing need', and is 'inseparable from other aspects of housing need'.

Indeed, such are the difficulties involved in defining homelessness that some commentators have cast doubt upon the practical utility of the term. Sophie Watson (1984, p. 70), for example, writes, 'my own view is that the concept of homelessness is not a useful one . . . the range of meanings attributed to the home and homelessness is both too vast and too complicated to have any explanatory or prescriptive use', while more recently it has been suggested that 'homelessness as currently conceptualised and defined, may not be a social category of relevance either to those experiencing it or to those trying to help them' (Nord and Luloff, 1995, p. 464).

However, definitions are a paramount basis for action. As Neale (1997, p. 55) questions, 'if policy cannot even define homelessness how can it hope to respond to it?', while Chamberlain and Mackenzie (1992, pp. 274, 283) point out that it is 'difficult to urge governments to meet the needs of homeless people, if the parameters of the homeless population are unclear'. In order for a problem to be addressed by policy-makers there is a need to know how many people are affected, with size playing a pivotal role in determining whether something is a problem or not. Statistical data thus assume a fundamental importance in policy debates and are often crucial in any bid for funding. Indeed, Hutson and Liddiard (1994, p. 38) argue, 'if they [campaigning agencies] cannot quantify the problem they cannot hope for resources'.

Ultimately, therefore, there is a need to move beyond semantics and formulate a working definition of homelessness as the basis of attempts to assess the extent of the problem. Yet, as the following section highlights, this is far from being a straightforward task.

Difficulties in quantifying homelessness

Definitions affect both the type and (subsequently) the level of homelessness identified such that, as Hoch (1987, p. 34) notes, 'the nature and scale of homelessness may look differently depending on how tightly or loosely the definitional boundaries are drawn'. This in turn influences the kind of measures and amount of resources committed to tackle the problem. Thus, far from being an academic exercise, the way in which homelessness is defined and quantified is very much a political process which reflects both value judgements concerning who is, and who is not, deemed to be deserving of support and more material considerations in terms of the level of resources available to deal with the problem. Indeed, as Hutson and Liddiard (1994, p. 32) contend, 'because different professionals have different definitions of homelessness, so they also produce different statistics. In this way, statistics can tell us more about the organisation collecting them than about the phenomena that are being measured'.

Of fundamental importance are the aims and agenda of the defining body or organisation and the purpose for which homelessness is being identified. In particular, given the moral and/or statutory duty incumbent upon central and local government to tackle homelessness, it is not surprising that they adopt a fairly strict definition in order to minimise the problem with which they have to deal. Voluntary agencies, on the other hand, without the ultimate responsibility for housing homeless households, can afford to make a more generous assessment of the circumstances in which someone is deemed homeless. Indeed, given the importance of numbers in securing financial support, there may be something of an incentive for voluntary organisations competing for limited resources to adopt as wide a definition as possible in order to maximise the number of 'homeless' households identified and consequently the amount of funding they are likely to receive.

The 'official' homeless

Official homelessness statistics simply record the number of households statutorily accepted as homeless under the provisions of the 1985 and, latterly, the 1996 Housing Acts. These statistics are widely considered grossly to underestimate the extent of the problem. Not only do the figures only record those households who actually approach their local authority for assistance, but the way in which individual authorities interpret and implement the homelessness legislation is critical in determining the outcome of homeless applications and, related to this, the scale of the problem identified.

Under the 1996 Housing Act a person is homeless if he or she has no accommodation available in the UK or elsewhere which it is reasonable for him or her to continue to occupy. However, while legislation lays down certain criteria regarding the cricumstances in which a household is considered homeless and the nature of the local authority's duty towards them (whether the council is obliged to provide accommodation or simply housing advice and assistance to help them find their own accommodation), determining homelessness and the duty owed is far from being a straightforward objective process, depending as it does upon the interpretation of terms (such as reasonable, vulnerable and intentional) which are themselves ambiguous and contestable. For example, in determining whether it is, or would have been, reasonable for a person to continue to occupy accommodation, a local authority *may* choose to consider the criteria posited in the 1996 Department of the Environment (now Department of the Environment, Transport and the Regions DETR) Code of Guidance – namely physical conditions, overcrowding, type of accommodation, violence or threats of violence from persons not associated with the applicant and security of tenure – but is under no compulsion to do so.

Given such room for manoeuvre, it is perhaps not surprising that studies (e.g. Evans and Duncan, 1988; Lambert *et al*, 1992; Hoggart, 1995) have found considerable differences between local authorities in both the number and the proportion of applicants accepted as homeless. Thus, in 1995/96, while overall 40 per cent of households making a homelessness application were accepted as homeless, the proportion of applicants accepted ranged from fewer than a fifth in some authorities to more than three-fifths in others (Chartered Institute of Public Finance and Accountancy, 1997). Such variation highlights once again the contested nature of the concept of homelessness and suggests that whether or not a household is accepted as homeless is as dependent on where they apply from, as on their housing circumstances. In this way, the number and proportion of applicants accepted as homeless can be seen to be as much, if not more, a reflection of council resources, attitudes and procedures than an indication of the extent of the problem. For example, authorities with plenty of council housing available, particularly those experiencing difficulty in letting certain parts of their stock, can afford to take a more generous interpretation of the Housing Act and be less restrictive about which households are accepted as homeless than local authorities with long waiting lists and a very limited supply of accommodation available to let.

Whereas the figures cited in discussions of homelessness and featured in the national press tend to refer to the number of households *accepted* as homeless and in priority need – namely, families with children, pregnant women, or someone who is 'vulnerable' as a result of old age, mental or physical disability or other special reasons – local authority returns to the DETR also provide details of homeless *applications*. However, these statistics similarly provide only an inadequate measure of the scale of the homelessness problem. Lack of awareness of legal rights, media images which predominantly portray homelessness as living on the streets or in emergency or other forms of temporary accommodation and a reluctance to undergo the humiliation often associated with making a homeless enquiry may deter many people in unsatisfactory accommodation from presenting or even seeing themselves as homeless. In addition, since local authorities have a statutory duty to rehouse the homeless, councils have an incentive both to deny self-expressions of homelessness and to discourage homeless enquiries (Hoggart, 1995, p. 60) through, for example, lack of (or perhaps as importantly, *perceived* lack of) responsiveness to enquiries, restricted opening hours of homeless persons units and the rehousing of homeless households into temporary accommodation or in some of the least desirable parts of their housing stock. Furthermore, it seems unlikely that single people and other households who do not evidently fall into a priority need category will approach the local authority if they know that, even if they are accepted as homeless, the council is statutorily obliged only to offer advice and assistance and has no duty to provide them with accommodation.

As a result, official enquiry statistics – although perhaps a better indication of the number of households experiencing homelessness than acceptance statistics – still severely underestimate the scale of the problem. Indeed, it would appear that official homelessness figures – whether related to enquiries or acceptances – reveal more about the way housing departments define and record homelessness than they do about the extent of the problem. Thus, as Hoggart (1995, p. 67) contends, 'irrespective of "real" levels of homelessness, what official statistics record is the willingness of councils to investigate and respond to housing insufficiency, in a highly subjective decision environment'.

Quite clearly, then, homelessness 'goes beyond the legal definitions and stipulations' (Webb, 1994, p. 28), and a household can be homeless even if not officially accepted as such for statistical purposes. Yet following the 1996 Housing Act the official statistics seem set to provide an even less accurate indication of the extent of homelessness.

Homelessness statistics and the impact of the 1996 Housing Act

Since 1977 local housing authorities have been statutorily obliged to help homeless people, either directly – through the provision of accommodation – or more indirectly – by offering advice and assistance. Although at the time of its introduction, advocates of the 1977 Housing (Homeless Persons) Act – under which these requirements passed into law – hoped for a gradual extension of its remit to statutorily provide for all homeless people rather than only those in priority need, in practice the Act has been interpreted in the narrowest possible terms, and the subsequent twenty years have, if anything, seen a toughening up of the homelessness legislation. This has been demonstrated most recently and most clearly in the 1996 Housing Act, which abolished priority for homeless households on the waiting list for a council property (now more commonly referred to as the housing register) and removed the duty on local authorities to provide permanent accommodation for these households, in response to the unfounded belief that homeless people were 'jumping the housing queue'.

Although the Act primarily altered the way in which local authorities discharged their duties to homeless households rather than changing definitions of homelessness, in order to avoid being restricted to allocating only temporary, 2-year tenancies to homeless households, many authorities revised their allocations policies (in particular, altering points schemes to take greater account of factors commonly associated with homelessness such as insecurity of tenure) to enable these households to be rehoused from the housing register rather than via the homelessness route.

Consequently, regardless of whether there has been any change – and, in particular, a fall – in the number of households in 'homeless' situations, the Act is likely to result in a drop in the number of households *recorded* as 'homeless' in the official figures as more and more households – who formerly would have been rehoused as homeless – are rehoused directly from the housing register.

In addition, changes to the P1(E) form, completed quarterly by local authorities and the main source of the DETR's homelessness statistics, in the light of the 1996 Housing Act mean that authorities are no longer required to provide details of the number of homeless applications or enquiries (individual authorities may of course continue to keep such records for internal monitoring purposes but there is no requirement upon them to do so). This means that the official homelessness statistics now provide information relating only to those cases assessed under the homelessness provisions of the 1996 Act. It is therefore no longer possible to turn to enquiries and applications as a better, albeit still limited, indication of the scale of the homelessness problem. Furthermore, the enumeration of assessments rather than applications is likely to reinforce the tendency of official figures to underestimate the extent of homelessness with both academic research (e.g. Lidstone, 1994; O'Callaghan and Dominian, 1996) and anecdotal evidence suggesting that a significant proportion of applicants do not pursue their application to a conclusion, but drop out at some stage in the assessment process.

This highlights the impact of policy and practice on the number of households identified as homeless in the official homelessness statistics. Indeed, the latest figures available at the time of writing show a drop in both the number of decisions made and the number of households accepted as homeless. As Figure 22.1 highlights, compared with the first half of 1996, in the first 6 months of 1997 (following implementation of the 1996 Act in the January of that year) there were some 8800 fewer decisions on applications for housing under the homelessness provisions of the 1985 and 1996 Housing Acts and more than 6000 fewer acceptances of households eligible for assistance, unintentionally homeless and in priority need (representing falls of 7 and 10 per cent respectively) (Department of the Environment, Transport and the Regions, 1997a).

It is reassuring that the Housing Minister at the time, Hilary Armstrong, in comments following the publication of these statutory homelessness statistics, while welcoming the drop in the homelessness figures, expressed a concern that this fall 'may be in part due to changes in legislation introduced by the previous administration, deterring people in genuine need from looking for help from local authorities'. However, beyond a commitment to 'monitor the homelessness statistics closely', she offered no indication that the new Labour government intended to revise the information recorded in the official homelessness figures or to seek

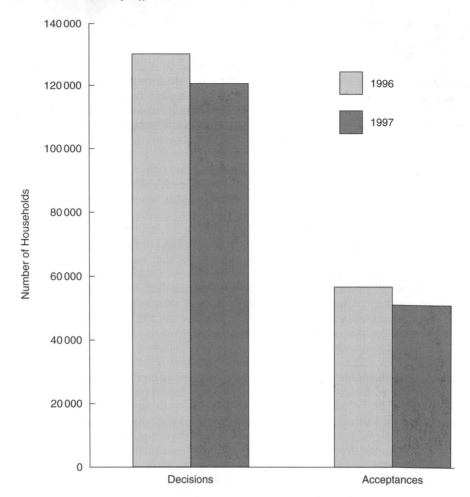

Figure 22.1   Homeless decisions and acceptances over a 6-month period
Source: Based on figures from the Department of the Environment, Transport and the Regions (1997a)

to establish the number of homeless households beyond those statutorily defined as such. As a result, it seems that under this government as under the last, there will be a continuing situation in which only some homeless count in the home-lessness count, with a large body of homeless people simply not appearing in the official statistics, namely:

- homeless households who do not approach the local authority for assistance;
- homeless households who do not pursue an application for rehousing;
- homeless households not considered eligible for assistance;
- homeless households deemed intentionally homeless;
- homeless households considered able to secure their own accommodation in the private sector; and
- homeless households not assessed as being in priority need.

The 'unofficial' homeless

A number of different agencies (including Shelter, Centrepoint and Crisis) have attempted to assess the extent of this 'unofficial' homelessness – namely those households who do not make a homeless application and/or fall outside the statutory definition of homelessness. In 1992 a Shelter report estimated there were some 1.7 million such 'unofficial' homeless households in England (Burrows and Walentowicz, 1992, p. 8), yet measuring the size of this population is a task fraught with difficulties not least because, as noted above, homelessness is an ambiguous and contested term. There is no universally accepted definition of homelessness and whichever definition is adopted is likely to result, albeit perhaps by default, in certain 'types' of homelessness being excluded from consideration, becoming or remaining, in effect, 'concealed'. For example, defining homelessness in terms of rooflessness excludes those in temporary or emergency accommodation, while extending the definition to include those in hostels still fails to consider those living in insecure and/or intolerable housing. Similarly, focusing on London and other major cities hides homelessness in non-metropolitan and rural areas, while a reliance on statutory definitions means that those not presenting as homeless or not accepted as being in priority need are in effect hidden (Webb, 1994, p. 28).

However, even assuming that the circumstances in which a household is deemed homeless can be agreed upon, a more fundamental problem presents itself in terms of whether, or how far, non-statutory homelessness *can* be measured empirically. The failure of some households to present or disclose their actual homelessness makes it difficult to gauge the extent of the problem in purely quantitative terms. Thus, even when a decision is reached on the form or forms of homelessness being measured, 'because the problem of homelessness is largely concealed, it is essentially unquantifiable and any estimates of its scale can be neither proved nor disproved' (Hutson and Liddiard, 1994, p. 41).

Measurement is further complicated by the essentially fluid nature of the homeless population, with movement into and out of homelessness over time meaning that the size of the problem identified is at least partly a function of the timescale over which it is measured. For example, the number of people sleeping on the streets on any one night is much lower than the number who experience such homelessness over the course of a month or a year. The area, as well as the time period over which homelessness is assessed, is also important, with spatial differences in 'type' of area (urban, non-metropolitan, rural) and circumstances (with regard to levels of unemployment, etc.) which are likely to affect the level and form of homelessness, making it difficult to establish the number of homeless people in a particular area or region by extrapolating from local figures.

While there are undoubtedly – given the contested nature of the concept – considerable difficulties in assessing the extent of non-statutory homelessness in purely quantitative terms, this lack of hard data should not provide an excuse for ignoring the issue. Instead, it requires that

> those in a position to achieve change examine the alternative forms of qualitative, and in some cases anecdotal, information presented to them and that they are wary of assuming a complete understanding of housing need based solely on the more visible indicators. (Webb, 1994, p. 109)

Conclusion

Statistical data continue to play a critical role in deciding whether something is a problem or not and what, if any, resources are required to tackle it. In terms of the current debate, not only does the reliance on numerical data to inform policy and determine the provision of resources overlook whether (or how far) it is possible to quantify homelessness, but it also accords such data with a degree of objectivity which belies the fact that homelessness statistics are social constructs rather than simply empirical measurements. (*see* Chapter 11, which extends this argument, or Chapter 12's example of Traveller surveys).

The type and extent of homelessness identified are heavily dependent upon the way in which the problem is defined, yet – as this chapter has emphasised – homelessness is an ambiguous and contested concept such that defining the term makes for a contentious debate. There are a large number of definitions which could be adopted, the scope of which reflects by whom and for what purpose the term is being employed. Defining, and subsequently quantifying, homelessness is consequently very much a political process. With statutory agencies keen to restrict demands on limited resources, perhaps seeking to depress figures, and voluntary agencies anxious to legitimise their existence at a time of increased competition for funds, perhaps seeking to inflate them, a substantial disparity exists between official and unofficial levels of homelessness.

Official figures are merely a record of the number of households accepted as homeless by local authorities. Not surprisingly, given the limited resources made available to deal with homelessness, the statutory definition excludes large sections of the population who are without a home (although not necessarily without a house). As a result, there are a substantial number of homeless households who simply do not appear in the official homelessness statistics. While attempts have been made to gauge the extent of this 'unofficial' homelessness, difficulties in defining the circumstances in which a household is deemed to be homeless and accessing a population which is, almost by definition, 'hidden' highlight the essentially problematic nature of trying to quantify homelessness. These problems suggest that, as with the official figures, unofficial homelessness statistics provide only an inadequate measure of the extent of homelessness in the UK today.

23

The Use and Abuse of Statistics on Homeless People

Walid Abdul-Hamid

We have a misconception of the mental health of the homeless because of flawed statistical studies in the past

Introduction

In the Victorian period homelessness in Britain was predominantly viewed as a policing problem. The Vagrancy Act of 1824 was used as a criminalising instrument with which to control vagrancy (Steedman, 1984). In 1912 Thomas Holmes, Secretary of the Howard Association and a retired police court officer, wrote a book entitled *London's Underworld*. In this book Holmes showed his concern for homeless mentally ill people. He used the 'epidemiological data' available at that time to 'prove' that homeless people were mentally ill. He used the statistics of the Royal Commission on the Care and Control of the Feeble-Minded. He concluded that as only a third of these people identified by the Commission were in workhouses, asylums, prisons, etc., the remaining two-thirds must constitute the homeless. He went on to suggest 'a national plan for the detention, segregation and control of all persons who are indisputably feeble-minded'. It is surprising to notice the similarity of Holmes's argument to those of some psychiatric researchers on the subject of homelessness today. This chapter contests the 'evidence' of very high rates of mental illness among the homeless by demonstrating many sampling and statistical flaws in previous studies. The homeless are particularly badly treated by official statistics as Rebekah Widdowfield's preceding chapter in this book also demonstrates.

Politicians tend to favour linking homelessness to mental illness and de-institutionalisation and community care because this diverts attention from other socio-ecomonic problems that contribute to homelessness. Two former presidents of the USA demonstrated this fact: George Bush said that mental illness was the 'principal cause of homelessness' while Ronald Reagan claimed that a 'large' percentage of the homeless were former mental patients (Hartman, 1984). Similarly, in the UK the Department of Health runs various initiatives to publicise and highlight the problems of 'the homeless mentally ill' (Department of Health, 1992b; Department of the Environment, Transport and the Regions, 1997b).

The perception of homelessness as being a mental health problem rather than a housing problem has arisen from publicity given to statistically flawed psychiatric studies that suggest a high prevalence of chronic psychiatric illness among the homeless and relate this to the closure of psychiatric institutions and community care policies. In fact, the main causes of homelessness lie in housing shortages, unemployment and under-spending on health and community services. Some psychiatrists also favour the mental illness hypothesis because they are still attached to old mental institutions and feel threatened by the principle of community care.

Also, the chronic mentally ill are a vulnerable and disadvantaged group that are affected by cuts in housing and other welfare benefits which have a major impact on homelessness.

Social origins of homelessness

Homelessness is only one manifestation of the most acute and visible form of class poverty and powerlessness (Brandon, 1974). However, society has regarded homeless people as undeserving poor and labelled them in different ways, representing them as spiritually weak and criminals, or Skid Row alcoholics. Since the 1980s, homelessness has been seen as a mental health service problem that has resulted from community care policies (Bassuk *et al.*, 1984; Weller and Weller, 1986; Weller *et al.*, 1987; Marshall, 1989; Connelly and Crown, 1994). This focus has shifted the emphasis from the need for decent housing for these people (Goering *et al.*, 1990), to the need for asylums and psychiatric interventions (Elpers, 1987; Third King's Fund Forum, 1987).

The Conservative government in the UK had cut back public expenditure on housing from £6.6 billion in 1979 to £2.1 billion in 1985 (Malpass, 1986). The net public capital spending on housing fell by 90 per cent between 1976/77 and 1988/89 and, even allowing for increases in construction costs, the fall in spending during that period was more than 37 per cent (Hills and Mullings, 1990). A report in 1985 by the British Association of Social Workers showed the depth of the problem. Fifty-three per cent of all repossession orders were where the owner had had a local authority mortgage. Between 1978 and 1985 there was a shortfall of 750 000 new homes due to inadequate government spending on housing programmes (Clode, 1985). The numbers of officially accepted homeless households in England placed in temporary accommodation rose from 4200 in 1982 to 44 490 in 1992 (Department of the Environment, 1992c).

The causes of homelessness are complex and multi-factorial. However, the policies of reducing spending on housing programmes, high unemployment and the cuts in housing and other financial benefits, in addition to changing family structures, have been shown to be major causes of increasing homelessness (Randall *et al.*, 1986; Fisher and Collins, 1993; Whitehead and Kleinmann, 1992).

Archard (1979a) noted that in spite of the obvious socio-economic origin of homelessness, studies on the homeless had concentrated on the 'biographic details of vagrants' moral careers' and avoided investigating wider socio-economic factors.

Families' homelessness has usually been seen by society as principally a housing problem, whereas homelessness among single people has been seen as a product of individual inadequacy or abnormality (Beresford, 1974; *see also* Chapter 22). This has caused the latter group to be exposed to control by law and correction and later to diagnoses and treatment by psychiatrists and social workers.

Sociologists define the problem of homelessness in terms of deviance and social control theory. This states that deviance is a behaviour that contravenes the norms of society, and that social control is a societal reaction to this deviance by labelling, sanction, treatment or control of such behaviour (Clinard and Meier, 1985). Cooper (1979, p. 138) stated that homeless people are 'waste – in the capitalist socio-economic machine – whether they are seen as waste product or as unusable is irrelevant, the point is that they form a human residue'. As the homeless rejected societal values, they were defined as deviant rather than merely eccentric. With the increase in the size of homeless populations, society started perceiving them as an 'urban guerrilla force' which threatened to disrupt society (Cook and Braithwaite, 1979; *see also* Chapters 7 and 40).

Archard conducted a historical review of the ideologies and the institutions that tried to control homeless alcoholics. He described the way in which social work and medical models of care replaced the moral and legal control methods (*see* Table 23.1). He noted how religious charities that work with the homeless have changed their philosophy of control from spiritual salvation to a more socio-medical approach (Archard, 1979b). It is interesting to note that some of these ideologies or combinations of them are still in action today. Loveland, in a study of two cities in California, found that homelessness was viewed as a policing rather than a housing issue. He noted that this ideology had temporarily changed after the October 1989 earthquake, when homelessness became 'normal'. In that period homelessness was viewed – even by the most right-wing politicians – as a housing need (Loveland, 1990).

Miller suggested that the visible subculture of drinking residents had shaped the public image of Skid Row residents as problem drinkers (Miller, 1982). Furthermore, Bahr and Caplow (1973) found that many of the Skid Row men were not drinkers at all and that many more were not problem drinkers. They compared a sample of Skid Row men with a sample of poor men in another area and another sample from a rich area. They found that the Skid Row men did not differ much from the poor-area men, but both differed greatly from the men from the rich area.

In spite of the importance of socio-economic factors as primary causes of homelessness in all its types (Fisher and Collins, 1993; Whitehead and Kleinmann, 1992), researchers have tended to study the characteristics of the homeless that are shaped by the public image of this group (Toro and McDonnell, 1992). Few studies have challenged this trend or used alternative survey methods (Bahr and Caplow, 1973; Koegel, 1992; Snow *et al.*, 1986).

Table 23.1 Social control matrix of the homeless people

Dominant ideological base	Definition of the problem of single homeless	Main strategy of treatment	Treatment institutions
Moral/religious	Spiritual weakness	Salvation	Missions
Penal	An offender	Correction	Police cells, courts, prisons
Medico-social	Alcoholic/social inadequate	Treatment/ rehabilitation	Alcoholism services/ shop fronts
Psychiatric	Mentally ill	Institutional treatment	Big shelters/mental hospitals

Source: After Archard (1979b)

Some psychiatric surveys on homeless people in the UK

Priest (1976) conducted a survey on Edinburgh hostels and their referral to psychiatric hospitals. He showed that the homeless people in contact with hospital-based psychiatric services were not representative of homeless people in general, their problems or their needs. A more representative profile of these people and their problems and needs could be deduced from community surveys of the same populations.

During the 1930s George Orwell lived as a homeless person and wrote his memoirs in *Down and Out in Paris and London*. He wrote:

> The Paris slums are a gathering-place for eccentric people – people who have fallen into solitary, half-mad grooves of life and given up trying to be normal or decent. Poverty frees them from ordinary standards of behaviour, just as money frees people from work. (Orwell, 1986, p. 7).

Researchers who have conducted interviews with homeless people have explained Orwell's observations in a different way (*see* Table 23.2). Edwards *et al.* (1966) studied 51 alcoholics from a soup kitchen, who had been sleeping rough. Twenty-one of them had drunk crude (methylated) spirits and most of them had experienced visual and auditory hallucinations. Edwards *et al.* suggested that drinking in this group may be a symptom of a central personality disorder.

Table 23.2   Psychiatric surveys of homeless people

Author (year)	No.	Schizophrenia/ psychosis	Affective disorder	Personality disorder	Other diagnosis	Substance abuse	No problem
Hostels and lodging-houses							
Lodge Patch (London) (1970)	122	15%	8%	51%	15%	—	11%
Timms and Fry (London) (1989)							
New arrivals	65	25%	6%	11%	—	1.5%	38%
Residents	58	31%	—	2%	—	—	53%
Marshall (Oxford) (1989)	43	67%	—	—	37%	—	12%
Camberwell reception centre (London)							
Edwards *et al.* (1968)	279	—	—	—	—	24%	—
Tidmarsh (1970)							
New users	130	16%	—	—	—	—	—
Casual	171	29%	—	—	—	—	—
People who sleep rough							
Edwards *et al.* (London) (1966)	51	—	—	—	—	92%	8%
Weller and Weller (London) (1986)	72	10%	—	—	55%	—	36%
Weller *et al.* (London) (1987)	100	42%	—	—	13%	—	45%
Reed *et al.* (London) (1992)	95	8%	—	—	—	37%	—

Excessive drinking also, of course, produces what look like symptoms of personality disorders. But many other studies have produced superficial evidence of high rates of mental illness, as the following summaries demonstrate.

In a survey of a London hostel, Lodge Patch (1970) stated that very few men (11 per cent of the residents) could be considered normal individuals. Fifteen per cent of them were diagnosed as schizophrenics and they had previously been in a mental hospital. Nineteen per cent of these men were alcoholics and 50 per cent had personality disorders. Although 23 per cent had been admitted to a mental hospital in the past, only 5 per cent were receiving any psychiatric treatment. The author suggested that this showed both a failure of community care and inappropriate early discharge (Lodge Patch, 1970).

Edwards *et al.* (1968) carried out a survey in the Camberwell Reception Centre. They found that the elderly residents of the Reception Centre were socially inadequate. The young dwellers were mostly 'immigrants' from outside London. They came to the reception centre either because of drinking problems or because of 'gross and positive psychopathy'. Forty-five per cent of residents had been arrested for drunkenness although only 24 per cent of the residents were chemically addicted to alcohol, which may suggest that a lot of the men were drinking excessively without being addicted. Twenty-four per cent had been in mental hospitals in addition to a further 10 per cent who had received hospital treatment for alcoholism, and 59 per cent had been in prison at one time or another (Edwards *et al.*, 1968).

Weller and Weller (1986) interviewed 42 residents from three hostels and 30 from Crisis at Christmas during Christmas Eve and Christmas Day in 1985. They found that 55 per cent of the people who completed the questionnaire reported mental illnesses leading to psychiatric treatment. Fourteen per cent of the people who agreed to discuss psychiatric history 'confessed' to hearing voices. The authors suggested that as wandering is common in psychosis, existing legislation may lead to destitution (Weller and Weller, 1986). A year later Weller *et al.* (1987) interviewed 100 men and women at Crisis at Christmas. They claimed to find that 35.8 per cent of the people who gave information had had in-patient treatment in the past. Thirty-one of the subjects were actively psychotic and 12 of those had never received treatment. Thirty per cent had a probable or definite diagnosis of schizophrenia. Seventy-eight per cent of the people with a psychiatric history had been imprisoned. The authors argued that the presence of a high proportion of mentally ill in their sample called into question the ability of the community services to deal with them (Weller *et al.*, 1987).

Both the two Weller studies had problems with the sampling methods which depended on the presence of homeless people at the site of their field work. Thus it was not a representative sample. In addition, they excluded people who were drunk on Christmas Day – as most of us are on that day! – giving no indication of the numbers excluded. The assessment procedures depended mainly on a short clinical assessment that was based on the subject's own reporting, which was shown to be inconsistent (Shanks, 1981) and might have overestimated psychiatric morbidity (Snow *et al.*, 1986).

Marshall studied two hostels for homeless people in Oxford in which he asked the staff to select the residents with mental disability. The staff selected 48 out of 146 residents. Of these subjects, 43 had had a previous psychiatric admission. The author used staff as informants to assess the level of social disability of the residents. He found that the mentally disabled were more likely to have stayed longer in the hostel. He then interviewed the staff about subjects' behaviour and symptoms. Twenty-nine had florid psychotic symptoms, 16 had clinically significant neurotic symptoms and only 5 had no clinically significant symptoms. The conclusion of the study was that hostels are having to care for increasing numbers of mentally ill people discharged into the community (Marshall, 1989). The main problems with this study, as in many of the others reported here, lie in the selection of the study sample. The exclusion of people with drinking problems might have biased the results in a similar way to that of Weller's studies. Also, there might have been a response bias in the disability assessment due to staff awareness of the research hypothesis (Abramson, 1984; Rumeau-Rouquette, 1978). The author asked the staff to report the degree of disability of the same residents that they had originally selected as disabled.

Reed *et al.* interviewed 96 single homeless people from a temporary shelter for homeless people who normally sleep rough. The researchers used a standardised psychiatric assessment method (the brief psychiatric rating scale), and were surprised that they found 'low' levels of psychosis (8 per cent) among these people (Reed *et al.*, 1992). In a second study, Marshall and Reed (1992) were less surprised to find that 45 out of the 70 residents interviewed in two hostels for homeless women fitted the DSMIII-R criteria for schizophrenia. The high morbidity in the second study could be due to the selection process of subjects to the survey, or to these hostels' admission criteria, or it might reflect a true high prevalence of psychosis among homeless women. Street homelessness is much rarer among women than among men.

Two more recent and methodologically more representative surveys have shown different pictures. An assessment of the disability and needs of a representative sample of homeless people in hostels showed that they are not chronically psychiatrically ill people as previous studies suggested (Abdul-Hamid *et al.*, 1995). Also, a recent survey of the health status of single homeless people in Sheffield suggested that psychiatric symptoms in these people were the result rather than the cause of their homelessness (Westlake and George 1994).

Limitations of previous psychiatric studies of the homeless

The above-mentioned surveys had aimed at assessing the mental health problems of homeless people (*see* Table 23.2). These studies are often quoted without taking into consideration the methodological problems and biases that limit their usefulness and generalisability. The methodological difficulties of these studies can be summarised as follows:

The problem of operational definition

Operational definition (or working definition) defines the characteristics that the study actually measures. It should be phrased in terms of objectively observable facts and be sufficiently clear and explicit to avoid ambiguity (Abramson, 1984).

Homeless people are composed of many non-homogenous subgroups, as we have seen from earlier discussions (and from Chapter 22), but researchers have used non-comparable operational definitions of the term 'homeless' (Bachrach,

1984). Most studies have defined the homeless in terms of the accommodation in which these people are living. These studies, however, did not give enough detail about the characteristics of the accommodation in which they took place. Different forms of accommodation differ in the type of people they accept, the people they exclude and the duration of stay. The lack of a comparable operational definition for studies makes it very difficult to compare their results (Bachrach, 1992).

Milburn and Watts (1986) noticed that most of the studies they reviewed used a 'theory-based definition' of homelessness. This is usually related to what the researcher is trying to prove rather than what is there in real life. In Marshall's study, hostel staff were asked to select residents according to his hypothesis (Marshall, 1989).

The other problem with these studies is that they assume that the homeless population is homogenous, which is far from the case (Milburn and Watts, 1986). For example, Weller *et al.* interviewed people who were sleeping rough and those from hostels without questioning the possible differences between the two populations (Susser *et al.*, 1989).

Sampling problems

The difficulties that face researchers in sampling relate partly to definition problems and partly to the lack of a sampling frame or a denominator. It is difficult to have an accurate estimate of the size of the homeless population (Fischer and Breakey, 1986). In their survey, Weller *et al.* selected 42 residents from three hostels and 30 people from Crisis at Christmas (Weller and Weller, 1986). The authors did not indicate why they chose their particular sample size and whether it reflected in any way the size or composition of the homeless population.

The sample size is another problem, as most of the studies of the homeless had a sample size smaller than 100 subjects (*see* Table 23.2). The confidence limits will be very wide, and this affects the accuracy of study results (Susser *et al.*, 1989).

The procedure of sampling is also of importance to the generalisability of the study. Very few studies have used random selection as the sampling method. Even studies that claimed to have a representative sample did not report the method of sampling (Timms and Fry, 1989; Priest, 1976). Out of 75 studies on the homeless reviewed by Milburn and Watts, 38 used the presence of study subjects at the data collection site as the sampling method; 27 used the first criteria in combination with other criteria such as the person's mental status, social support, etc. Nine studies used the duration of homelessness as a selection criteria.

Weller's two surveys depended mainly on the presence of the subjects in the research site (Weller and Weller, 1986; Weller *et al.*, 1987). Such sampling procedures will select only the visible group out of the homeless population, which might not represent the homeless generally in any meaningful way (Bachrach, 1992). Such studies tend to contribute to, rather than correct, public prejudices about the homeless (Toro and McDonnell, 1992).

The problem of detection of mental illness

Psychiatric interviews have been used as an instrument to detect the presence of mental illnesses in this population. Reports (particularly in the USA) have begun to throw doubt on the suitability of conventional psychiatric interviews for this purpose as they may measure homeless people's adaptation to their difficult circumstances rather than mental disorders (Snow *et al.*, 1986; Mossman and Perlin, 1992).

The validity of homeless persons' reporting to professional interviewers, whom they meet for the first time, has been found to be questionable. Shanks (1981) studied the quality of data obtained by a questionnaire from residents of a common lodging-house. He also compared the answers received from these residents on two or three occasions by different interviewers. He concluded that unless the resident knew the interviewer well, the constancy of the data would be poor (Shanks, 1981). This reporting bias may have a powerful effect in altering psychiatric diagnosis rates.

Priest found inconsistencies between his clinical assessment and the standardised psychological tests in assessing the presence of psychotic disorders in his sample. He claimed that the sample reporting the psychological tests was biased and discarded them (Priest, 1976). This bias could equally have affected his own clinical assessment. Very few of the studies mentioned in the previous section used a standardised method to reach a diagnosis and what they did use was not necessarily a reliable measure of disability or need for psychiatric services (Kendell, 1975; Wing, 1990), especially in community surveys (Bebbington, 1990). The only study that measured disability was Marshall's survey in Oxford (Marshall, 1989), but unfortunately, this study did not use a random sample.

Conclusion

In spite of the clear socio-economic roots of homelessness, society has continued through the ages to label homeless people in different ways, many of which characterise them as the undeserving poor. They are labelled as 'spiritually weak', 'criminals', 'alcoholics' and lately 'mentally ill'.

During the 1970s, the public image of homeless people was that of people with drinking problems, and many studies tended to confirm and maintain this image. Bahr and Caplow (1973), in a systematic study of the Skid Row homeless people in New York, found that the majority of these people were not problem drinkers. They suggested that the visibility of homeless people with drinking problems had shaped public opinion.

Since the beginning of the 1980s, and with the changes in the economic and demographic structure of urban areas, public opinion concerning homeless people has changed again. Homeless people started to be considered as 'wandering mad people' (Miller, 1982). Their problems were seen as a result of community care policies and the closure of mental hospitals. Some researchers – particularly those opposed to community care – have tried to reinforce this publicly held notion. Though most of these research projects have numerous methodological problems, they are often quoted and used in service implementation plans without taking into consideration their limitations. Public opinion has significant policy and service implications. In the USA, most of the services that work with homeless people are built on the assumption that the homeless are mentally ill (Abdul-Hamid, 1989; Mossman and Perlin, 1992). Homeless people whose only problem is the lack of affordable housing are being institutionalised on the basis of this assumption (Grunberg and Eagle, 1990).

In the UK, studies have tried to prove that homeless people are increasingly becoming chronically mentally ill. For example, Weller and his colleagues in their two surveys, though these were conducted within a 1-year period, produced two estimates of psychotic illness which varied from 10 per cent in 1986 to 40 per cent in 1987. The authors proposed that they showed real increases in the proportion of psychotic patients among the homeless due to community care policies (Weller and Weller, 1986; Weller *et al.*, 1987). The difference, however, is much more likely to be a manifestation of the unsystematic sampling methods used and the problems of the suitability and reliability of the diagnoses.

The question of whether homelessness is a mental health problem is important. It influences decisions about resource allocation in helping homeless people. It might make the difference between either giving these people only antipsychotic injections, or attending to their complex social needs and solving their health and mental health problems within the framework of meeting their overall needs (Abdul-Hamid and Cooney, 1997).

24

Are Crime and Fear of Crime More Likely to be Experienced by the 'Poor'?

Christina Pantazis and David Gordon

Crime has a greater impact on poorer people, although they lose less property

Introduction

Crime and fear of crime have emerged as major public and political issues in recent decades. This may, in part, be attributed to the enormous growth in recorded crime since the 1970s, when the average annual increase has been about 5 per cent. Five million crimes were recorded by the police in 1994 (Barclay *et al.*, 1995). However, the British Crime Survey estimated a total of 19 million crimes against individuals and their property in 1995 (Mirrlees-Black *et al.*, 1996). People's anxiety about crime has also risen in this period. Surveys have repeatedly shown that crime has surpassed unemployment and health as an issue of major public concern (Jacobs and Worcester, 1991). This has led some criminologists to conclude that fear of crime poses almost as large a threat to society as crime itself (Clemente and Kleinman, 1977).

Both official statistics and research show that there is an uneven distribution in criminal victimisation and in fear of crime (Barclay *et al.*, 1995; Mirrlees-Black *et al.*, 1996). Some individuals and some geographical areas are more vulnerable to crime than others and certain groups fear crime more. A consensus appears to exist between various government departments and some criminologists that there is a positive relationship between poverty and victimisation. Put bluntly, there is a strong belief that the 'poor' experience more crime than the 'rich'.

Home Office research has demonstrated high risks of crime in 'poor' areas, particularly in areas containing large council estates (Ramsay, 1983; Hope, 1986; Hope and Shaw, 1988). Evidence of high levels of crime on poor council estates provided the impetus for the Department of the Environment's Priority Estates Project in the late 1970s. This involved attempts to reduce crime on poor and disadvantaged council estates through improved management (Foster and Hope, 1993). Support for the view that there is a strong link between poverty and victimisation can also be found from within criminology (Lea and Young, 1984; Kinsey, 1984; Jones *et al.*, 1986; Kinsey *et al.*, 1986). One criminologist has even likened crime to a regressive tax which falls disproportionately on poor people (Downes, 1983).

This chapter explores the controversial relationship between poverty and victimisation, and between poverty and fear of crime. It is based on secondary analysis of the 1992 British Crime Survey (BCS) and the 1990 Breadline Britain survey (BBS). So far there have been six sweeps of the BCS in England and Wales (1982, 1984, 1988, 1992, 1994, 1996). In the 1992 sweep, a representative sample of 10 059 people aged 16 and over were questioned about the crimes committed against them in the previous year (Mayhew *et al.*, 1993). In the 1990 BBS, a nationally representative sample of 1831 households were interviewed about their standard of living. Households lacking three or more necessities were defined as 'multiply deprived' or 'poor'. The survey estimated that 20 per cent of British households (approximately 11 million people) were living in circumstances of multiple deprivation in 1990 (Gordon and Pantazis, 1997). In addition to measuring the extent and nature of poverty, the 1990 BBS also examined some of the problems such ill health, crime and fear of crime considered to be disproportionately experienced by poor households (*see*, for example, Chapter 20 and the spatial analysis of the BBS by Peter Lee in Chapter 21).

Measurement and definition of crime

The measurement of crime presents a major problem for criminologists. Many crimes are neither reported nor recorded, resulting in the police crime statistics underestimating the 'real' level of crime. Victimisation studies, or crime surveys, provide more reliable results. They provide a more accurate picture by assessing the crime that police statistics fail to include. The 1992 BCS found that only 43 per cent of crimes were reported to the police and that only 30 per cent of crimes were recorded by the police (Mayhew *et al.*, 1993). However, crime surveys cannot possibly claim to offer a complete account of the extent and nature of crime.

Surveys suffer from problems relating to measurement. People fail to report to interviewers all the relevant incidents which they have experienced within the so-called 'recall period', and they also record incidents which had in fact occurred earlier. Crimes are also underestimated when people conceal crimes committed against them (for example, as in some rape cases where the offender is a friend or family member). Thus, crime surveys may underestimate the extent of crime against women.

Crime surveys also have problems concerning response rates. For instance, although the BCS achieves a good response rate (77 per cent in 1992), non-respondents may include a disproportionately high number of victims. There is also evidence of systematic 'response bias': for example, better-educated respondents seem more adept at recalling relevant events at interview; and middle-class respondents seem more prepared than others to define certain classes of incidents as assaults (Sparks *et al.*, 1977). Furthermore, surveys are based on a sample of the population, and error may arise from this – particularly for rare crimes, such as robbery and rape.

There is also the problem that most victimisation surveys count only certain types of crimes. For instance, the BCS acknowledges that it excludes crimes against organisations, such as fraud, shoplifting, fare evasion and commercial burglary and robbery, and 'victimless' crimes, such as drug and alcohol misuse, consensual sexual offences and crimes where people may not be aware of having been victimised – as in many cases of fraud. Crime surveys thus rely on a narrow conception of crime. Critical criminologists have shown how the definition of crime is a social construction (Chambliss and Seidman, 1971; Box, 1983). It is argued that both legal and common-sense definitions of crime encompass the activities of poor people, but exclude many of the activities of the rich and powerful (Box, 1983, 1987; Cook, 1997).

The definition of poverty

The definition and measurement of poverty are also subject to much debate. The pioneering work of Townsend showed that poverty was best understood in terms of relative deprivation using a measure of standard of living (Townsend, 1979, 1993). However, others have argued for the use of both income and deprivation criteria in identifying poverty (Ringen, 1987, 1988). The BCS does not collect the detailed information on people's living standards which is needed to measure poverty directly. It does, however, contain detail on income and other indicators of poverty such as tenure, lack of car ownership, economic activity and social class. These indicators, when combined to form a deprivation index, can provide a useful alternative in situations where information on standard of living is lacking (Townsend *et al.*, 1986; Goldblatt, 1990; Phillimore, *et al.*, 1994a).

Poor areas and crime

A number of studies have established that 'poor' areas suffer from higher crime rates than 'better-off' areas (Morris, 1957; Ramsay, 1983; Hope, 1986). The 1990 criminal statistics recorded an overall crime rate of 12.8 per cent in the Urban Programme district authorities compared with only 9.6 per cent for the remaining English and Welsh district authorities. There was, however, wide variation within the 57 urban areas. The crime rate ranged from 5.8 per cent in Rotherham to 21.7 per cent in Kensington and Chelsea.

Findings from the combined 1984, 1988 and 1992 BCS reinforce the link between crime and some 'poor' areas (Mayhew and Maung, 1992). Figure 24.1 shows the relative crime rates (national average = 100) for burglary and robbery, for residents of different ACORN neighbourhood groups (*see* Appendix; CACI, 1992). The *mixed inner metropolitan areas* and the *less well-off* and *poorest council estates* suffer from relatively high crime rates. These ACORN neighbourhoods are characterised by low-income households. However, *high-status, non-family areas*, which are characterised by households with well above average incomes, also suffer from high crime rates. *Agricultural areas* and *older terraced housing*, which typically contain many low-income households, have respectively very low and average burglary and robbery rates.

This research demonstrates that relatively high rates of crimes can be found in both 'poor' and 'rich' areas. Indeed, some 'poor' areas suffer very little crime. Furthermore, despite linking high levels of crime with certain types of 'poor' areas, the research does not prove whether it is 'poor' people/households who suffer from high rates of crime. It may be that it is the 'better off' who live in 'poor' areas who are disproportionately the victims of crime. Thus, social

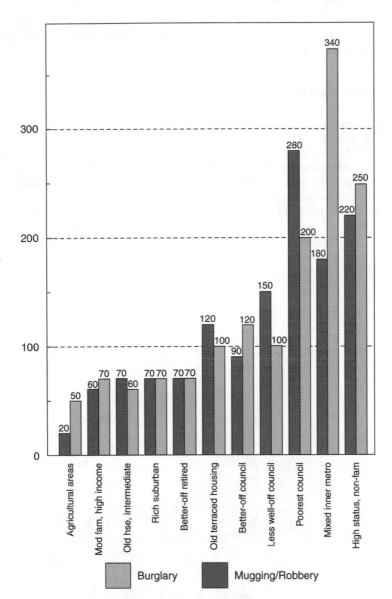

Figure 24.1 Indexed rate of crime for combined 1984, 1988 and 1992 British Crime Surveys by ACORN neighbourhoods. UK average = 100

scientists need to be aware of the pitfalls of the ecological fallacy (Robinson, 1950; Baldwin, 1979). The ecological fallacy refers to instances in which inappropriate inferences about relationships at the individual level are made on the basis of aggregate data obtained at the area level. In such situations, individuals are assumed to have the characteristics of the areas in which they live. Many of the early ecology studies of juvenile delinquency fell into the trap of the ecological fallacy (Polk, 1957; Willie, 1967). There is little discussion of the fallacy in the literature on victimisation, and there is a danger that it may influence some of our present assumptions about poverty and victimisation.

Results from the 1992 British Crime Survey

The BCS is now accepted to be the most accurate guide to measuring conventional crimes. The survey asks respondents about victimisation in the previous year. They are asked about their own experience and that of others in the household for household crimes: burglary, thefts of and from vehicles, vandalism and theft from around the home. They are asked only about their own experiences with respect to personal crimes: threats, assaults, robberies, thefts from the person, and other personal thefts. In 1991, 34 per cent of respondents experienced any one of these

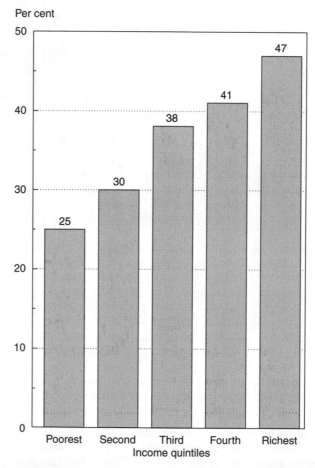

Figure 24.2 Percentage of people experiencing household crime by gross household income quintiles
Source: 1992 British Crime Survey

household crimes, while 12 per cent experienced any one of these personal crimes.

Figure 24.2 shows the relationship between the experience of household crime and gross household income quintiles. There is a clear linear relationship between victimisation and household income. Contrary to popular and academic opinion, the 'rich' experience the most crime whereas the 'poor' experience the least. One in two respondents in the richest 20 per cent of households experienced crime in 1991, compared with only one in four respondents in the poorest quintile.

A near linear relationship between personal crime and income can also be observed in Figure 24.3. In 1991, the 'rich' experienced marginally more personal crime than the 'poor'. Fourteen per cent of respondents in the richest 20 per cent of households experienced crime, compared with only 10 per cent in the poorest quintile.

In addition to income, victimisation may also be investigated according to three other indicators of poverty: lack of vehicle access, rented tenure and manual

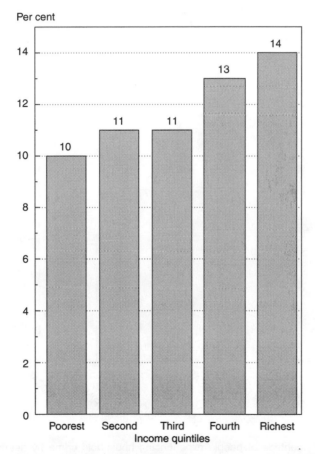

Figure 24.3 Percentage of people experiencing personal crime by gross household income quintiles

Source: 1992 British Crime Survey

employment. Many health studies interested in explaining health differences
between social groups have relied on deprivation indexes using combinations of a
number of socio-economic variables (Townsend *et al.*, 1986; Goldblatt, 1990;
Phillimore *et al.*, 1994a). Goldblatt (1990), for example, constructed an index
combining car access, housing tenure and social class in order to explain differences in mortality between socio-economic groups. These variables were chosen
for specific theoretical reasons. Car access was chosen because it was considered
the best surrogate for current income. For example, in the 1992 BCS, the average
gross annual household income for individuals with vehicle access was £20 832,
three times the income level of those individuals without vehicle access, while
non-owner occupation was thought to reflect both lack of wealth and lack of

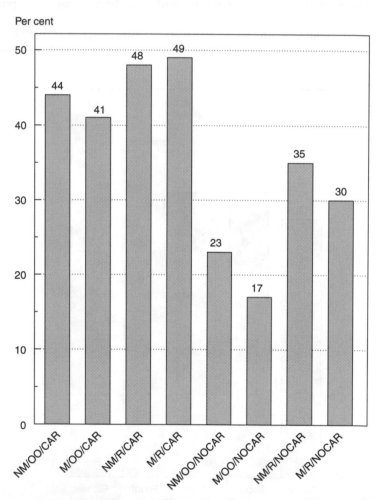

Figure 24.4 Percentage of people experiencing household crime by deprivation index.
NM = non-manual worker; M = manual worker; OO = owner-occupier; R = renter; CAR = car
owner; NOCAR = non-car owner
Source: 1992 British Crime Survey

income (Townsend *et al.*, 1986). The inclusion of social class (used in terms of manual or non-manual) had to do with historical reasons: the Registrar-General's Social Class Classification was developed specifically in order to measure the health differentials between different social groups.

Figure 24.4 demonstrates victimisation risks for individuals according to the deprivation index used by Goldblatt (1990). In contrast to health, there is no simple positive relationship between deprivation and victimisation. Figure 24.4 clearly shows that respondents with vehicle access have much higher levels of victimisation than those without. The victimisation rate of respondents with vehicle access ranges from 41 to 49 per cent. Individuals without vehicle access have a victimisation rate ranging from 17 to 35 per cent. Thus, vehicle ownership appears to be an important factor influencing levels of victimisation between different individuals. The figure also shows that there is little difference in rates of victimisation between manual and non-manual respondents. However, greater differences can be observed between renters and owner-occupiers. Respondents with a car who live in rented accommodation and who have a manual occupation experienced the highest levels of victimisation (49 per cent) in 1991.

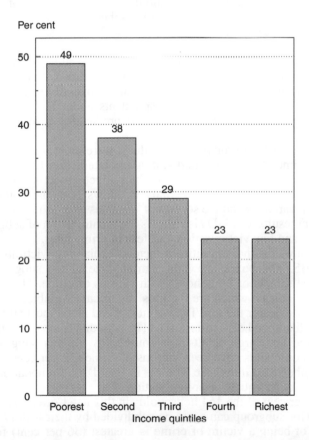

Figure 24.5 Percentage of people fearing crime by gross household income quintiles
Source: 1992 British Crime Survey

Respondents in the BCS, as well as being asked a range of questions about their own and that of their household's experience of victimisation, were asked how safe they felt walking alone in their neighbourhood after dark. This question is used in many surveys to measure fear of crime. Figure 24.5 demonstrates the relationship between household income and fear of crime. Respondents in the poorest household income quintile experienced the highest rates of fear (49 per cent), while respondents in the richest two income quintiles had a fear of crime of only 23 per cent. Thus the 1992 BCS demonstrates that despite experiencing higher levels of victimisation, respondents in the richest household income quintile are less fearful of crime.

Results from the 1990 Breadline Britain survey

In order to clarify the complex relationship between poverty and crime, the 1990 Breadline Britain Survey asked respondents whether they or members of their household had experienced burglary, assault, mugging or robbery, or any other crime in the last year. Fourteen per cent said that they or members of their household had experienced crime: 7.2 per cent of households had been burgled, 2.5 per cent mugged, 2.6 per cent had been assaulted and 2.9 per cent had been victims of other crimes.

Fear of crime was assessed by asking respondents whether they or members of their household felt unsafe in their local neighbourhood. Seventeen per cent said they feared crime. This is much lower than the overall 1992 BCS figure of 32 per cent. The BCS question asked respondents about feeling unsafe when walking alone at night, a question that would predictably give a much higher result.

In order to assess which groups of individuals were most likely to experience crime and fear crime, the Chi-Squared Automatic Interaction Detector (CHAID) was used to explore the most significant variables affecting victimisation rates and fear of crime (Kass, 1980). CHAID analysis allows both the combination of categories within variables and the sorting of variables to produce the most statistically significant results. CHAID also allows the identification of subgroups with particularly high and low victimisation and fear of crime rates.

Figure 24.6 shows the most significant factors 'explaining' victimisation levels in the 1990 BBS. The boxes show the sample size of the subgroup and the percentage of the sample of households with victims of crime. The stems of the CHAID diagram indicate which are the most significant variables, with those of greater significance nearer the top. Type of household is the most significant factor affecting the likelihood of victimisation. The type that is most victimised is the single non-retired and large, adult-only households. In this subgroup of 385 households, 85 households have been victims of crime (22 per cent). This group can be further subdivided into those who are in 'good' accommodation and those who are in 'poor' or 'adequate' accommodation.

Twenty-nine per cent of the 164 households in this latter group have been victims of crime. This subgroup can again be subdivided by their history of poverty. The likelihood of being a victim of crime is greatest (36 per cent) for the subgroup which has 'never' or 'rarely' been poor in the past. Student and ex-student households might fit this description (*Guardian*, 21 September 1993).

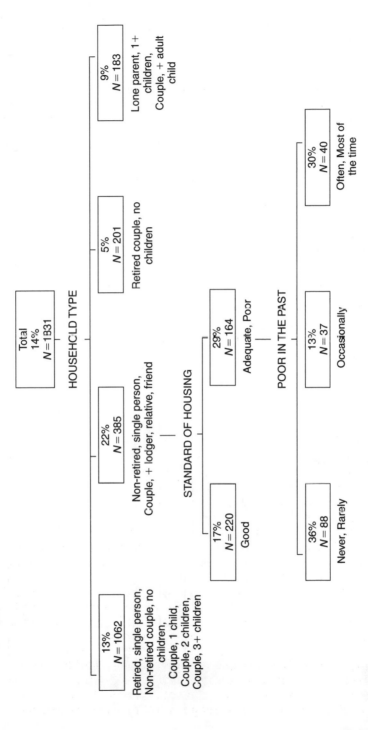

Figure 24.6 Victims of crime
Source: 1990 Breadline Britain Survey

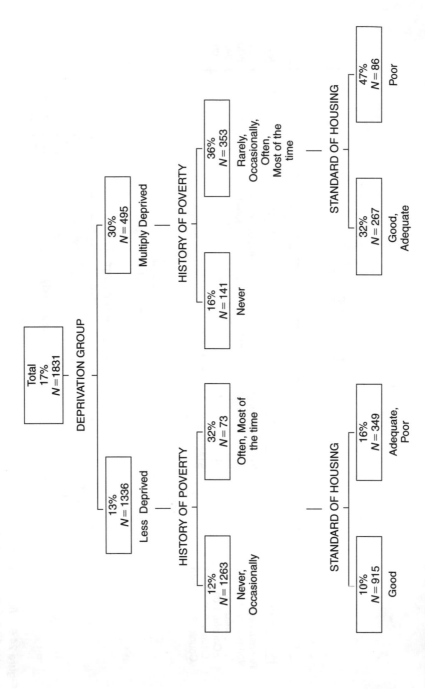

Figure 24.7 Fear of crime
Source: 1990 Breadline Britain survey

The CHAID analysis illustrates that poverty is not the most significant factor explaining victimisation. Type of household, followed by standard of housing and a history of poverty, were more significant than deprivation level, social class, household income, or the sex and age of the respondent in explaining experience of crime.

Figure 24.7 shows the most significant factors relating to fear of crime. In this case, deprivation is the most important factor for people fearing crime, affecting 30 per cent of the multiply deprived. Furthermore, fear of crime increases when deprivation is compounded with a long history of poverty and a 'poor' standard of housing. Of the multiply deprived, 36 per cent who had also experienced poverty in the past felt unsafe in their local neighbourhood and, of this group, 47 per cent of those in 'poor' housing fear crime. The age and sex of the respondent were not as significant in explaining fear of crime.

Fear of crime and lack of insurance

Although fear of crime is often seen as irrational by many criminologists, the effects of crime, particularly property crime, will be greatest for low-income households since many will simply be unable to afford to replace lost possessions. In these circumstances, fear of crime cannot be considered to be irrational. Respondents in the BBS were asked whether they had contents insurance and, if they did not, whether this was because they could not afford it. Figure 24.8 shows that households lacking insurance because of financial reasons suffer almost twice as much fear of crime as households with insurance, despite having only a 2 per cent higher crime rate.

Eighty-five per cent of the group with no insurance suffered from multiple deprivation (i.e. lacked three or more necessities). As Paula, who also appeared in the television series for Breadline Britain, made clear:

> 'We haven't got insurance simply for the fact that we can't afford it. . . . It's mainly for your personal possessions, if they break in, that's just gone. . . . It's really pot luck, take your chance. You go out and you lock the doors, if they're broken when you come back there's nothing you can do about it.'

Conclusion

This chapter has considered the impact of poverty on victimisation and on fear of crime. Contrary to the existing consensus, it has provided evidence from two national surveys (the 1992 British Crime Survey and the 1990 Breadline Britain survey) which have shown a complicated relationship between poverty and victimisation. Although victimisation may be concentrated in some poor areas, it is less clear how it impacts at the individual level. Evidence from the 1992 BCS, for instance, demonstrated that individuals in 'rich' households were significantly more likely to experience crime. Car ownership is an important aspect of this, since a large proportion of all crime is vehicle related. However, better-off individuals were also marginally more likely to experience personal crime. The analysis, on the other hand, showed a clear link between poverty and fear of

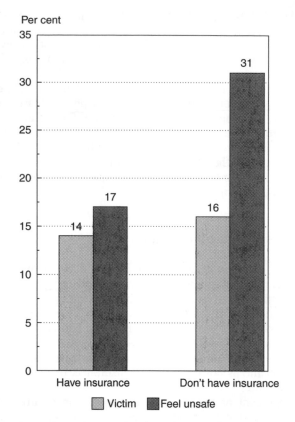

Figure 24.8   Percentage of people experiencing crime and fear of crime by possession of household insurance.
Note: 'Don't have insurance' excludes those who do not have insurance and do not want it
Source: 1990 Breadline Britain survey

crime. People living in poverty suffered from a disproportionately high level of fear of crime regardless of whether or not they had been victimised. This fear is not irrational, as some criminologists suggest, but results from the greater impact that crime has on poor people.

Acknowledgements

The findings presented in this chapter are based on a research project entitled Crime and Fear of Crime: A Socio-economic Approach, undertaken between 1995 and 1997 and supported by the Economic and Social Research Council under grant number R000221683.

Appendix: The ACORN system of neighbourhood classification

ACORN stands for 'A Classification of Residential Neighbourhoods'. It is a system of classifying households according to the demographic, employment and housing characteristics of their immediate neighbourhood. It was produced by CACI, a market and policy analysis consultancy, by applying cluster analysis to 40 variables from the 1981 census including age, class, tenure, dwelling type and car ownership.

There are 38 ACORN neighbourhood types, and these aggregate to 11 groups. Each of the 130 000 enumeration districts (EDs) in Great Britain (an average ED comprises about 150 households) has been assigned to an ACORN neighbourhood type on the basis of its scores on selected census variables. The principle of ACORN is that people who live in the same neighbourhood share the characteristics of class, income and lifestyle. There will be differences between individual EDs within the same ACORN classification – particularly in heterogeneous areas such as those in inner cities. Nevertheless, ACORN is a useful way of determining the immediate social environment of different households, and can provide better indicators for some purposes than individual characteristics such as income or class. For example, ACORN will show what types of targets for crime different neighbourhoods offer and what risks their residents face compared to those living close by (Mayhew *et al.*, 1993).

Appendix: The ACORN system of neighbourhood classification

PART V
Valuing Health

25

The Racialisation of Ethnic Inequalities in Health

James Y. Nazroo

Ethnicity does not reflect presumed genetic or cultural attributes

Introduction

Class inequalities have been at the centre of health research for some time, despite attempts to ignore their existence. Their importance was firmly established by the Black Report (Townsend and Davidson 1982), which came to the conclusion that they were primarily a consequence of material differences in living standards. Since then researchers have, on the whole, concentrated on providing additional evidence for a material explanation of class inequalities in health and unpacking the mechanisms that might link material disadvantage with a greater risk of poor health (e.g. Macintyre, 1997; Vågerö and Illsley, 1995; Davey Smith *et al.*, 1994; Lundeberg, 1991, and *see also* Chapters 20, 27 and 28).

There is also a burgeoning interest in the health of ethnic minority groups. Within the UK a large amount of research has been funded. Some examples include national surveys – the Health Education Authority's Black and Minority Ethnic Groups: Health and Lifestyles surveys (Rudat, 1994) and the Fourth National Survey of Ethnic Minorities (Nazroo, 1997a,b). This growing body of research reflects, at least in part, a public policy concern with the health of, and quality of health care provided for, ethnic minority groups. In theory, research should lead to policy developments that improve both of these, but in practice this may not be the case. This is not only because research can be poorly disseminated, or because it may provide unpalatable messages to those concerned with public finance, but also because the research itself may contribute to the racialisation of health issues. It does this by identifying the health disadvantage of ethnic minority groups as inherent to their ethnicity, a consequence of their cultural and genetic 'weaknesses', rather than a result of the disadvantages they face because of the ways in which their ethnicity or race is perceived by others.

Given the pattern of socio-economic deprivation faced by some ethnic minority groups in Britain (shown in Table 25.1), and the clearly established relationship between socio-economic factors and health described above, it would be expected that once their younger age profile had been taken into account, ethnic minority people would generally have poorer health as a result of their poorer class position.

Table 25.1   Ethnic differences in socio-economic position: percentages by class for each ethnic group

Registrar-General's class	White	Caribbean	Indian	Pakistani	Bangladeshi
I/II	35	22	32	20	11
IIIn	15	18	20	15	18
IIIm	31	30	22	32	32
IV/V	20	30	26	33	40
Unemployed	11	24	15	38	42

Source: Fourth National Survey of Ethnic Minorities (Nazroo, 1997a)
Note: 'Registrar-General's class' is based on the occupation, or most recent occupation, of the head of the household (where known)

However, in the UK, with one or two notable exceptions (Ahmad *et al.*, 1989; Fenton *et al.*, 1995; Smaje, 1995; Nazroo, 1997a, *see also* Chapter 16), class has disappeared from investigations into the relationship between ethnicity and health. Instead, work in this field has largely followed a trend set by the now classic work of Marmot *et al.* (1984) shortly after the Black Report was published. This used a combination of death certificate and census data to show how mortality rates varied by country of birth. The findings indicated that class and, consequently, material explanations were unrelated to mortality rates for most migrant groups, and made no contribution to the higher mortality rates found among those who had migrated to Britain (Marmot *et al.*, 1984). Indeed, for one group, those born in the 'Caribbean Commonwealth', the relationship between class and overall mortality rates was the opposite of that for the general population. The authors concluded:

> (a) that differences in social class distribution are not the explanation of the overall different mortality of migrants; and (b) the relation of social class (as usually defined) to mortality is different among immigrant groups from the England and Wales pattern (Marmot *et al.*, 1984, p. 21)

Rather than puzzling over why such an important explanation for inequalities in health among the general population did not apply to ethnic minorities, researchers have simply accepted that different sets of explanations for poor health applied to ethnic minority and majority populations and, of course, that for the minority population, explanations were related to cultural and genetic differences.

Demarcating 'race' and ethnic groups in health research

Part of the problem lies in the 'untheorised' ways in which ethnic groups are identified in the research process. There is, of course, a wider sociological literature on ethnicity and 'race', which can be broadly defined as concerned with understanding how ethnic and racial groups become social realities, and the relationships between them. Within this work there appears to be complete agreement that 'race' is a concept without scientific validity. For example, in a recent edited volume most authors started from the position that the notion that people can be divided into races on the basis of genetic differences had been shown to be false

(Barot, 1996). From this position, the term 'race' becomes an artificial construct, used to order groups of people hierarchically and to justify the exploitation of 'inferior races'. In contrast, many of these writers give credence to a notion of ethnicity. Crudely, ethnicity can be said to reflect self-identification with cultural traditions from which individuals can draw strength and meaning. Importantly, these cultural traditions are seen as historically located; that is, they are seen as occurring within particular contexts and as changing over time, place and person. And it is recognised that ethnicity is only one element of identity, whose significance would depend on the context within which the individual finds him/herself. For example, gender and class are also important, and in certain contexts may be more important.

Some elements of this message have been adopted by research on ethnicity and health. Within the UK 'race' is never explicitly measured and ethnicity is clearly the term favoured for describing these minority groups. However, the nature of research, with the need for easily used and repeatable measures, often results in the concepts of ethnicity and race being merged (despite the exclusive use of the term 'ethnicity') and the dynamic and contextual nature of ethnicity being ignored. So, the term 'ethnic' is frequently used to refer to genetic and cultural features that are undesirable (from a health perspective) and inherent to the minority groups under investigation.

Consequences of 'untheorised' research

Before we go on to explore why research on ethnicity and health tends to be reduced to genetic and cultural explanations, and how this can be avoided, it is worth illustrating this point with two examples. The first concerns the well-publicised greater risk of 'South Asians' in the UK of coronary heart disease (CHD). A *British Medical Journal* editorial (Gupta *et al.*, 1995) used research findings to attribute this problem to a combination of genetic (i.e. race) and cultural (i.e. ethnicity) factors that are apparently associated with being 'South Asian'. Concerning genetics, the suggestion was that 'South Asians' have a shared evolutionary history that involved adaptation 'to survive under conditions of periodic famine and low energy intake'. This resulted in the development of 'insulin resistance syndrome', which apparently underlies the greater risk of CHD affecting 'South Asians'. From this perspective, 'South Asians' can be viewed as a genetically distinct group with a unique evolutionary history – a race. In terms of cultural factors, the use of ghee in cooking, a lack of physical exercise and a reluctance to use health services were all mentioned – even though ghee is not used by all the ethnic groups that comprise 'South Asians', and evidence suggests that 'South Asians' do understand the importance of exercise (Beishon and Nazroo, 1997) and do use medical services (Nazroo, 1997a; Rudat 1994). It is important to note how the policy recommendations flowing from such an approach underline the extent to which the issue has become racialised. The authors of the editorial recommend that 'community leaders' and 'survivors' of heart attacks should spread the message among their communities and that 'South Asians' should be encouraged to undertake healthier lifestyles (Gupta *et al.*, 1995). The problem is, apparently, viewed as something inherent to being 'South Asian' – nothing to do with the context of the lives of 'South Asians' and as solvable only if 'South

Asians' are encouraged to modify their behaviours to address their genetic and cultural weaknesses.

The second example involves the high mortality rates from suicide among young women living in the UK who were born on the Indian subcontinent. Attempts to explain this have focused on cultural explanations, particularly on a notion of culture conflict, where the young woman is apparently in disagreement with her parents', or husband's, traditional or religious expectations. For example, despite an almost complete lack of evidence, Soni Raleigh and Balarajan (1992, p. 367) state:

> Most immigrant Asian communities have maintained their cultural identity and traditions even after generations of overseas residence. This tradition incorporates a premium on academic and economic success, a stigma attached to failure, the overriding authority of elders (especially parents and in-laws) and expected unquestioning compliance from younger family members. . . . These pressures are intensified in young Indian women, given their rigidly defined roles in Indian society. Submission and deference to males and elders, arranged marriages, the financial pressures imposed by dowries, and ensuing marital and family conflicts have been cited as contributory factors to suicide and attempted suicide in young Indian women in several of the studies reviewed here.

Again such research has led to the racialisation of the issue. The problem is located within a relatively permanent and pathological culture, and one that needs to adopt the 'freedoms' allowed to young white women.

Both examples provide us with notions of groups that are (from a health point of view) genetically and culturally inferior to the white population. There are a number of reasons why such racialisation has occurred and continues to be present in epidemiological work. Important here is the belief that comparative epidemiology can provide clues to aetiology and that ethnicity is a useful and easy tool for demarcating groups to provide the basis for comparison. There have been numerous critiques of the crudeness with which ethnicity is measured in such work (e.g. Sheldon and Parker, 1992; Senior and Bhopal, 1994; Ahmad, 1995), and work has shown that a more sensitive measurement of ethnicity can lead to quite different conclusions (Nazroo, 1997a). Figure 25.1 shows that although South Asians as a group had a greater risk of indicators of CHD than whites, once the group was broken down into constituent parts this applied only to Pakistanis and Bangladeshis; Indians and African Asians had the same rate as whites. In addition, using religion to provide a more detailed 'ethnic' breakdown in these data showed that within the Indian and African Asian groups Muslims had a high rate of CHD, while Hindus and Sikhs had low rates (Nazroo, 1997a). This approach was useful in uncovering the extent to which convenient assumptions of ethnic similarity within obviously heterogeneous groups could lead to the racialisation of health issues.

However, using more refined ethnic categorisations still allows differences in health to be racialised. For example, it could be suggested that the findings just described mean we can use the term 'Muslim heart disease', or 'Pakistani and Bangladeshi heart disease' – rather than 'South Asian heart disease' – to describe the situation, and explanations can be sought in assumptions about Muslim, Pakistani and Bangladeshi cultural practices or their shared evolutionary history. This

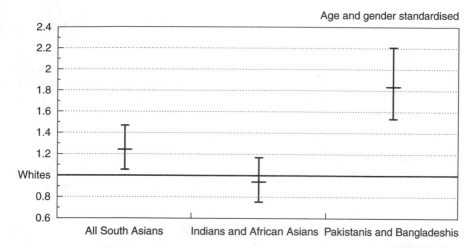

Figure 25.1   Relative risk of reporting indicators of heart disease for 'South Asians' in the UK as compared with whites. ('Relative risk' is simply the chance of being in a particular category compared with a reference group.) Each bar represents 95 per cent confidence limits: that is, the range within which there is a 95 per cent statistical probability of the true value lying. If this range does not cross the range representing whites, differences can be considered to be statistically significant
Source: Nazroo (1997a)

is because of another problem with work on ethnicity and health: the crudeness with which ethnicity is assessed in nearly all this work allows the status of ethnicity as an explanatory variable to be assumed. The view of ethnicity as a *natural* division between social groups allows the *description* of ethnic variations in health to become their *explanation* (Sheldon and Parker, 1992; *see also* Chapters 16 and 17). So, explanations are based on cultural stereotypes or suppositions about genetic differences, rather than attempting to assess directly the nature and importance of such factors. Here, it is worth emphasising the different status of explanations used for ethnic minority groups compared with the ethnic majority. While a lack of interest in exercise, or restrictive family practices, are a consequence of a pathological (minority) culture, high rates of smoking are not viewed as a problem arising from white ethnicity that should be ameliorated by a modification of white culture(s).

The need to measure hypothesised variables – and to be sensitive to class

This does not mean that cultural or genetic factors are of no use in explaining poor health. Rather, cultural practices and genetic differences need to be assessed directly, rather than assumed, and the contexts in which they operate and their association with health outcomes measured. Also, other explanatory factors have to be considered when exploring the relationship between ethnicity and health, in particular those related to socio-economic status. Although initial attempts to take socio-economic factors into account for migrant groups in the UK failed to show a relationship between class and health (Marmot *et al.*, 1984), there are a number of reasons why the relationship between class and mortality rates might have been suppressed in these data. Most important is that they used occupation as recorded on death certificates to define social class. It is likely that, because of the well-recognised process of inflating occupational status on death certificates (Townsend and Davidson, 1982), where relatives declare the most prestigious occupation held by the deceased, this will have failed to capture the downward social mobility of members of ethnic minority groups on migration to Britain (a process that both Smith (1977) and Heath and Ridge (1983) have documented). So, the occupation recorded on the death certificates of migrants may well be an inaccurate reflection of their experience in Britain prior to death. In addition, given the socio-economic profile of ethnic minority groups in Britain, this inflation of occupational status would need to happen only in relatively few cases for the figures representing the small population in higher classes to be distorted upwards.

More recent analyses of immigrant mortality data around the 1991 census have shown that socio-economic differentials might be an important explanation of inequalities in mortality rates *within* ethnic minority groups, but still have suggested that differences *between* ethnic groups remain unexplained by class effects (Harding and Maxwell, 1997). Again it is tempting to believe that the differences between groups must be a consequence of 'obvious' genetic and cultural factors. However, it is important to recognise that important shortcomings remain in the statistics used. As before, the analysis had to rely on occupation as recorded on death certificates, which has the problems outlined above. And, as the authors point out, this information is not present for all men and missing for a large number of women (who consequently were not included in the analysis). Most important, however, is that the attempt to control for socio-economic factors with an indicator such as occupational class, ignores how crude this measure is and how it may apply differently to different ethnic groups. In fact, there has been an increasing recognition of the limitations of traditional class groupings, which are far from internally homogeneous. A number of studies have drawn attention to variations in income levels and death rates among occupations that comprise particular occupational classes (e.g. Davey Smith *et al.*, 1990, *see also* Chapter 27). And within an occupational group, ethnic minorities may be more likely to be found in lower or less prestigious occupational grades, to have poorer job security, to endure more stressful working conditions and to work unsocial hours. For example, it has been shown that within particular class groups ethnic minority people have a lower income than white people; that among the unemployed, ethnic minority people have been out of work for longer than whites; and that some ethnic minority groups have poorer-quality housing than whites regardless of tenure (*see* table 5.2 in Nazroo, 1997a). The conclusion to be drawn is that, while standard indicators of socio-economic status have some use for making comparisons *within* ethnic groups, they are of little use for 'controlling out' the impact of socio-economic differences when attempting to reveal a pure 'ethnic/race' effect.

Figure 25.2 provides additional support for this conclusion. Using data from the Fourth National Survey (Nazroo, 1997a), it shows changes in the relative risk of reporting fair or poor health for Pakistanis and Bangladeshis (the groups who had the poorest health) compared with whites once the data had been standardised for a variety of indicators of socio-economic status. Comparing the first bar with

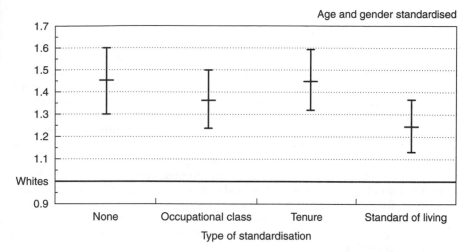

Figure 25.2 Relative risk of fair or poor health standardised for socio-economic factors: Pakistanis and Bangladeshis in the UK compared with whites. (For the meaning of 'relative risk' and the meaning of the bars, *see* the caption to Figure 25.1)
Source: Nazroo (1997a)

the second and third, shows that standardising for occupational class and tenure (the indicators commonly used in epidemiological research) makes no difference. However, taking account of an indicator of 'standard of living' (a more direct reflection of the material circumstances of respondents – *see* Nazroo (1997a) for full details) leads to a large reduction in the relative risk (compare the first and last bars). Given that this indicator is also not perfect for taking account of ethnic differences in socio-economic status (*see* table 5.4 in Nazroo, 1997a), such a finding suggests that socio-economic differences, in fact, make a large and key contribution to ethnic inequalities in health.

Another problem with using data that have been standardised for socio-economic status is worth highlighting. Such an approach to analysis and interpretation regards socio-economic status as a confounding factor that needs to be controlled out, in order to reveal the 'true' relationship between ethnicity and health. This results in the importance of socio-economic factors becoming obscured, and their explanatory role lost. The presentation of 'standardised' data leaves both the author and reader to assume that all that is left is an 'ethnic/race' effect, be that cultural or genetic. This also gives the impression that different types of explanation operate for ethnic minority groups compared with the general population and so leads to racialisation. While for the latter, factors relating to socio-economic status are shown to be crucial, for the former they are not visible. Differences are then assumed to be related to some unmeasured aspect of ethnicity or 'race', even though socio-economic factors are important determinants of health for all groups and cannot be assessed with sufficient accuracy in our statistics for us to be confident that they have really been controlled for (Kaufmann *et al.*, 1997).

Conclusion

If work on ethnicity and health is prone to such problems, how do we avoid them? Most important is to avoid reading off explanatory factors from the ethnic classification that we have imposed on our data without bothering to assess such factors directly. The ethnic classifications we use do not reflect unchangeable and natural divisions between groups. Also, ethnicity does not exist in isolation; it is within a social context that ethnicity achieves its significance, and part of that social context is the ways in which those seen as members of ethnic minority groups are racialised. Indeed, one of the most important purposes for undertaking work on ethnicity and health is to extend our understanding of the nature and extent of the social disadvantage faced by ethnic minority groups. Not only is poorer health potentially part of that disadvantage, it is also a consequence of it. Understanding ethnic inequalities in health raises the need to address them. And if such inequalities are rooted in the wider inequalities faced by ethnic minority groups in the UK, as some evidence suggests (Nazroo, 1997a), the impetus must be to address the wider inequalities.

Here it is also important to acknowledge the limitations of our assessments of socio-economic status and to remember that they do not account for the other forms of disadvantage faced by ethnic minority groups that might play some role in ethnic inequalities in health. For example, both the experience of racial harassment and discrimination, and the geographical location of ethnic minority people in particular locations, may have a direct impact on their health.

Acknowledgement

My current work is supported by a grant from the (UK) Economic and Social Research Council under the Health Variations Programme (L128251019). I am grateful to Karen Iley and Satnam Virdee for comments on an earlier version of this article that was, in an abridged format, published in the journal *Share*.

26

What Do Official Health Statistics Measure?

Alison Macfarlane and Jenny Head

Monitoring health administration is not the same as measuring health or health services

'Official health statistics' is a loosely defined term usually applied to a broad range of statistics collected by UK government bodies or the National Health Service about the health of the population and the care it receives. In this chapter, we start by outlining the main types of statistics collected, before going on to discuss what they measure, what is left out and the questions which should be asked when using them. A much fuller account is given in *Official Health Statistics: An Unofficial Guide* (Radical Statistics Health Group, 1999).

'Health statistics' can be subdivided into statistics collected about the health services and the people who use them, and statistics which are collected about the population as a whole. The main types of data are summarised in Table 26.1. Relatively few data are collected about private health care, an issue which is explored in Chapter 29. The first half of this chapter describes statistics collected by or through the National Health Service about the care it provides or commissions. At national level data are collated separately by the health ministry for each of the four countries of the UK listed in Table 26.2. The ways in which data are collected vary between the four countries, making it difficult to produce figures for the UK as a whole. The second half of the chapter describes statistics about the whole population and its health. Most of these data are collected through registrations of events or through population censuses and surveys. Again, the four countries of the UK differ in the methods they use to do this.

Statistics about the National Health Service

Data collection about health services has developed in response to the use of public funds to provide them. In the 1920s and 1930s, the Ministry of Health started to fund local authorities and voluntary organisations to provide services such as maternity and child health clinics. It collected data about the services it funded and published them in its annual reports.

When the National Health Service (NHS) started in 1948, a much fuller system of administrative statistical 'returns' was set up to enable the activities of the service to be monitored by the central government health departments responsible

Table 26.1 Types of official statistics relevant to health

Source of data	Examples	Characteristics
Civil registration	Births, stillbirths, marriages, deaths	Comprehensive coverage as documents required for legal purposes. Inflexible, as questions can only be changed by Act of Parliament
Statutory notifications	Births, communicable diseases	Coverage should be complete as notification is required by law, but under-reporting does occur, especially with infectious diseases, which cannot be notified unless the person consults a doctor
Voluntary notifications	Cancer registration, congenital abnormalities	More under-reporting, but more opportunity to collect data than with statutory notifications and registrations
Censuses of population	Census taken in 1961, 1966, 1971, 1981 and 1991	Statutory enumeration of whole population taken every ten years. Data become out of date, so demand for 'mid-term' census of a sample of the population arises, but the only one undertaken was in 1966
Claims for National Insurance and social security benefits	Sickness absence, industrial injury and accidents	Sickness absence statistics confined to those paying full National Insurance contributions. Industrial illness and accident benefits can be paid only if it can be readily established that the condition was occupational in origin. This is difficult for some occupational diseases
Administrative returns to central government health departments	Waiting list returns	Emphasis on use and availability of service and facilities rather than on characteristics of those who use them
Patients' contacts with the health service	Hospital Episode System, National Morbidity Survey	Data concentrate on hospital in-patients and very unrepresentative data from general practice. Because of the incompleteness of record linkage, data tend to deal with facilities and treatment rather than outcome
Special analyses and record linkage	Registrar General's Decennial Supplement, ONS Longitudinal Study	Combined analyses of data from more than one source. Much more powerful than data from a single source but problems may arise when discrepancies arise in data, e.g. when different occupations are given at census and death registration. The ONS (formerly OPCS) Longitudinal Study overcomes this, but has much smaller numbers in its 1 per cent sample
Surveys, including one-off surveys, surveys repeated at regular intervals and continuous surveys	Infant feeding, dental health, health surveys for England and Scotland, General Household Survey	Includes people who have not been in contact with the health services. Continuous surveys enable trends to be monitored over time. The General Household Survey is the only continuous government survey which relates people's perceptions of their health to a wide range of socio-economic factors. The health surveys for England and Scotland are more detailed but their content changes from year to year

Table 26.2 Official organisations concerned with health statistics for the four countries of the United Kingdom

Country	Health ministry	Vital statistics office	NHS register
England	Department of Health	Office for National Statistics (ONS)[a]	England and Wales
Wales	Welsh Office		
Scotland	Scottish Office Department of Health	General Register Office Scotland	Scotland
Northern Ireland	Department of Health and Social Services (NI)[b]	General Register Office Northern Ireland[b]	Northern Ireland

[a] On 1 April 1996, the Office of Population Censuses and Surveys merged with the Central Statistical Office to form the Office for National Statistics (ONS)
[b] The General Register Office for Northern Ireland and statisticians working in the Department of Health and Social Services are now part of the Northern Ireland Statistics and Research Agency

for England, Wales, Scotland and Northern Ireland. These included returns about the capacity and utilisation of hospitals, and about staffing and expenditure. As the Chief Medical Officer of Health for England put it in his report for 1949,

> The unification of the hospital and specialist services has made both possible and necessary the introduction of a uniform set of returns designed to keep the Minister and the hospital boards accurately informed of the amount and range of facilities at their disposal and the use being made of them. (Ministry of Health, 1950)

Systems grew up, rather more slowly, for collecting data about individual people's stays in hospital. Each regional health authority in England developed systems called Hospital Activity Analysis (HAA). For each in-patient stay, demographic, clinical and administrative data were collected when the person was discharged. Each region sent a 10 per cent sample to the Office of Population Censuses and Surveys to be analysed nationally as the Hospital In-patient Enquiry (HIPE). Information about people in mental illness and 'mental handicap' hospitals was collected through a separate system, the Mental Health Enquiry.

In the early 1980s, the Steering Group on Health Services Information was set up to make recommendations about the collection and use of data in the hospital and community health services. Most of the data collection systems came into operation on 1 April 1987 and the rest on 1 April 1988, except for maternity data, which were delayed by another 6 months. In practice, ten years later, some of the systems were still not fully implemented, or had some records with many items of data missing (House of Commons Health Committee, 1996b; Department of Health, 1997a). The greatest problems occurred with data for maternity and community services, and these systems have been under review for some years.

In its first report, the Steering Group said that its 'main concern is with information for health service management. Thus we have not tackled specifically the information needed by health professionals to evaluate the results of their care' (Steering Group on Health Services Information, 1982). It therefore concentrated on data needed for management of resources. Its approach in doing so was to define the 'episode of care' as a basis for data collection. Thus for hospital in-patient or day case care a 'finished consultant episode' was an episode of care in one department of the hospital. If a patient changes department during an in-patient stay, then he or she contributes more than once to the total count of 'finished consultant episodes'.

The Steering Group also defined 'minimum data sets' for each area of NHS activity. These consisted of a minimum set of items to be collected in each district, with a subset to be submitted to the Department of Health. Data about all episodes of in-patient or day case care were submitted centrally to form the Hospital Episode System (HES), the successor to HIPE. Data are now analysed and published by the Department of Health.

In April 1991, the Conservative government set up an internal market in the NHS. Under this, health authorities and general practitioner (GP) fundholders were allocated funds to purchase care from 'providers' of hospital and community services (UK Government, 1989). This meant that additional financial data were needed, for example to enable transfers of funds from purchasers to providers. Health authorities were asked to set up new district information systems to bring together data about their residents (Department of Health, 1990a). NHS trusts and any private concerns which provide services to NHS patients were required to return a minimum data set to the person's district of residence together with the invoice (Department of Health, 1990b, 1990c).

In preparation for the start of the internal market, the NHS Management Executive set up systems to collect 'fast-track' information, which is less detailed, but compiled more quickly than that in the statistical systems. Unfortunately, some 'fast-track' systems use different definitions and thus produce figures which are often different from those from the standard statistical systems. 'Fast-track' figures are not published formally, but were extensively used by the Conservative government, often fallaciously, in documents and press releases produced to claim that the internal market was a success (NHS Management Executive, 1992; Radical Statistics Health Group, 1992). In the run-up to the 1992 general election, the Conservative government put out increasingly frequent press releases in support of its claim that waiting times were decreasing.

In England, the reorganisation of the 14 regional health authorities into eight in 1994, followed by their eventual abolition in 1996, has been a major blow to NHS data collection. Regional databases, some built up over many years, have been abandoned, and most of the staff who knew their contents in detail have left the NHS. The Department of Health now has to collect data directly from a much larger number of individual health authorities.

In Wales and Northern Ireland, the data collected are fairly similar to those in England. Scotland has a far more highly developed system of health service statistics than the other countries of the UK and makes extensive use of record linkage.

The data collected in current systems are derived from information about three main aspects of the NHS: its activity, its workforce and its finances. Data are collected about services provided on hospital premises, including operating theatres, accident and emergency departments, radiotherapy departments, and diagnostic services. Less extensive data are collected about services provided on and off hospital premises, including out-patient clinics, day care facilities, paramedical services, family planning services and maternity services, and about services provided in or for the community including preventive services, community nursing and child health and school health services. Many systems produce counts of 'face-to-face contacts', with very little indication of the purpose of the contact or the clients' characteristics.

At the time of writing, it is unclear what implications the proposals to end the internal market, set out in the Labour government's White Paper *The New NHS*, will have for data collection (Department of Health, 1997a). It appears that greater reliance may be placed on data collected from general practice. These are fairly scanty despite the fact that the majority of patients' contacts with the NHS are with GPs. The fact that fewer data have been collected and published about their work may be a reflection of their status as independent contractors. Until recently, most data about care given by GPs have come from four national surveys of people consulting GPs in sets of volunteer practices in 1955–56, 1970–71, 1981–82 and 1991–92 (McCormick *et al.*, 1995). As well as background information about patients, diagnostic information is collected about each episode of illness and consultations within it. In the last three surveys, data have been linked to socio-economic data collected about the people in the population census, to allow analyses of social class differences in consultation rates.

Major developments have taken place in the 1990s now that most general practices have computer systems. Data from practices having the same systems have been pooled for research or monitoring purposes (Hollowell, 1997). More recently, the MIQUEST project has been attempting to aggregate anonymised data from a variety of systems (NHS Executive, 1996). A major problem with all these initiatives is the lack of consistency in the way different practices record clinical and other information.

Other data are collected through systems set up to monitor particular conditions or diseases. The oldest of these is the notification of communicable diseases. This dates back to the Public Health Acts of the 1870s, when these diseases were major causes of death (Thunhurst and Macfarlane, 1992). Over the years, cancer registries were established in each old NHS region, with information about cancers in residents of the region. They varied considerably in the way cancers were notified and the completeness of the registers (Leon, 1988). The registers were under threat when the regions were abolished but have now been reorganised. The Office for National Statistics combines data from all the registries to produce analyses of trends and variations in incidence and survival.

Other notification systems include those for sexually transmitted diseases, congenital anomalies, abortions, and industrial diseases and accidents. Local authority social services departments maintain registers of people with vision and hearing impairments, children with special needs and general 'handicap registers'. Some have registers of people with severe learning difficulties. Most of these registers are incomplete, for a variety of reasons.

All these data are collected through people's contact with health and social services. This means that people are counted only if a service exists and people are aware of it, want to use it and succeed in doing so. The introduction of market forces into both health and social services has led to increasing fragmentation of data collection and use of activity statistics as a measure of the success of policy changes. This is epitomised in the data in Figure 26.1, showing hospital activity in England. While it shows that the amount of activity recorded has increased, in the absence of record linkage there is no way of knowing how many people received care or what its outcome was (Radical Statistics Health Group, 1992, 1995). There is also no way of knowing how the changes in activity related to changes in the population. The 1997 Labour government's dismissal of this approach as 'bean counting' shows that it is aware of these problems, but it has

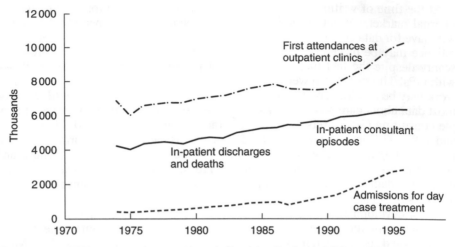

Figure 26.1   NHS acute and general hospital activity, England, 1974 to 1996–7
Source: Department of Health *Statistical Bulletins* 5/85, 1993/2 and 1997/20

yet to reveal what measures it will be taking to collect more meaningful statistics (Department of Health, 1997b). Meanwhile, there are other data which start from a population standpoint.

Statistics about the circumstances and health of the population

In England and Wales, most statistics about the population are compiled by the Office for National Statistics (ONS), formed in April 1996 from the merger of the former Office of Population Censuses and Surveys (OPCS) and the Central Statistical Office (CSO). The Population and Health Division of the ONS brings together data from records completed by local registrars of births, marriages and deaths, and conducts censuses, while its Social Survey Division conducts sample surveys of the population. As well as publishing national summaries of these data, it produces analyses for smaller areas. Some of these are published in its 'Monitors' and longer annual reference volumes, while others are fed back as printouts and on disk to local authorities and health authorities. Increasingly, these more detailed data are being sold in electronic form. In Scotland and Northern Ireland, similar data are produced by the General Register Offices for each country, as Table 26.2 shows.

Birth and death registration

Each year, ONS and the other two general register offices publish birth statistics tabulated by the parents' ages, area of residence, country of birth and marital status, the father's social class, and the baby's sex, place of birth and birthweight. Mortality statistics are published in considerably greater detail with tabulations by age, sex, area of residence and certified cause of death.

Every 10 years, data from death registration are compared with those from the census to produce 'decennial supplements', which include analyses by occupation and social class. In the 1990s, a much more wide-ranging approach has been taken to producing decennial supplements, which now include reviews and other related analyses (Britton, 1990; Drever and Whitehead, 1997). Nevertheless, as Chapter 27 shows, there is still a strong emphasis on the mortality of men. In tabulations by social class, married women's and babies' mortality tends to be linked to their husband's or father's social class respectively, because of gaps in the information about married women's and mothers' occupations.

Although birth and death statistics are very complete, there is little scope for changing the data which are recorded at birth and death registration, except by Act of Parliament. It is difficult to obtain parliamentary time for legislation on this subject. A White Paper published in 1990 (Office of Population Censuses and Surveys, 1990) set out proposals for improving the registration system, some of which related to data collection. Eight years later, it still had not become law, apart from two items which were the subject of private members' bills.

There are several approaches which can be taken to enhance the data collected at birth and death registration. One is record linkage. Since 1975 data collected when deaths of babies under the age of 1 year are registered have been linked to data collected at birth registration. This procedure has been extended to deaths of any children born from 1993 onwards. A more ambitious system is the Longitudinal Study, formerly known as the OPCS Longitudinal Study. In this, a 1 per cent sample of the population was selected from the 1971 census. Subsequent births, cancer registrations and deaths to members of this sample have been linked to their 1971, 1981 and 1991 census records. Chapter 27 describes some of the many analyses which have been published of the fertility, mortality and cancer incidence of members of the cohort, in relation to socio-economic data from the censuses. Even these, however, are restricted to the data already collected.

Another approach is 'confidential enquiries', in which additional data are collected from case notes, pathologists' reports and other sources about particular categories of deaths. These are then reviewed by panels made up of members of relevant professions, who attempt to assess whether deaths were 'avoidable' – or, in more recent years, involved 'substandard care' which may have contributed to the death.

The longest-running of these are the confidential enquiries into maternal deaths, which date back to 1952 in their present form. These enquiries bring together information from death registration with reports from relevant clinical staff. Later arrivals on the scene are confidential enquiries into perioperative deaths, which are enquiries into samples of deaths of people undergoing operations. Confidential enquiries into stillbirths and deaths in infancy (CESDI) started in January 1993 in England, Wales and Northern Ireland.

Despite their claim to make inferences about avoidability, these enquiries do not generally use 'controls', who did not die, to look at the extent of 'substandard care' among the survivors. Even when they do, the factors considered may be highly selective. For example, the 'Sudden Unexpected Death' study, which was part of CESDI, considered data which suggested that smoking by the parents and the baby's sleeping position play a major role in cot death, while playing down adverse social conditions, which could also have been harmful to the baby, on the grounds that these 'factors are not amenable to change' (Fleming *et al.*, 1996; Blair *et al.*, 1996).

Censuses and surveys

National censuses of the population have been held every 10 years since 1801, with the exception of 1941. Problems with the interpretation and use of census data are discussed in Chapters 2 and 3. Starting from 1851, attempts were made through the census to collect data about the prevalence of 'infirmities', in the form of seeing and hearing problems, mental illness and learning difficulties. A number of problems were encountered, particularly the stigma associated with these disabilities. In addition, with young children, the extent of the disabilities might not have yet become clear (Macfarlane and Mugford, 1984). The 'infirmities' question was abandoned in 1921 and it was not until 1991 that a question on 'long-term illness' was introduced (Charlton *et al.*, 1994).

The government surveys discussed in Chapter 2 are to a greater or lesser extent sources of data about health. Health is prominent among topics covered in the General Household Survey, a continuous survey by ONS of samples of the population of England, Wales and Scotland. People are asked whether they have any 'long-standing illness', whether they have recently had any short-term illness, and about their use of health services. In some years, questions are asked about smoking, drinking, use of contraception and sterilisation and use of dental services. Although it does not record clinical diagnoses, the strength of the General Household Survey is that the information about health can then be analysed in relation to the social and demographic variables.

As the General Household Survey has been conducted since 1972 in every year except 1997–98, when it was suspended in the wake of a funding crisis, it can be used for looking at time trends. Data from the General Household Survey, shown in Figure 26.2, give a rather different picture of the use of hospital services from that shown in Figure 26.1, with no increase in the proportion of people in the population having been admitted to hospital as in-patients.

When the Conservative government established its 'Health of the Nation' strategy, it wanted to set statistical targets to be achieved by the year 2000, at both

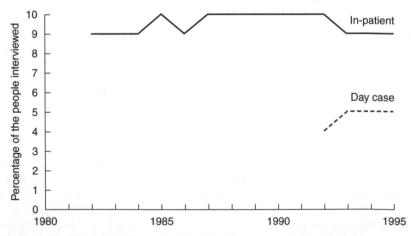

Figure 26.2   Percentage of people who had received in-patient or day case hospital care in the preceding 12 months, Great Britain, 1982–95
Source: Office for National Statistics, General Household Survey

local and national level (Department of Health, 1991, 1992a). Because of this approach, its strategy was driven by the availability of data at a time when it did not want to make a major investment in the collection of new data (Macfarlane, 1992). As a result, its indicators were predominantly measures of NHS activity or of mortality (Radical Statistics Health Group, 1991a, b).

Nevertheless, some attempt was made to collect data about health at national and regional level through new health surveys for England, Wales, Scotland and Northern Ireland. The Health Survey for England is an annual survey which began in 1991 and is commissioned by the Department of Health. The first three surveys were carried out by the former OPCS, but since 1994, as a result of market testing, the surveys are being done by a market research firm, Social and Community Planning Research (SCPR), in conjunction with University College London. The broad aims of the survey are to monitor health and the impact of health promotion on health and to collect data relevant to the planning and evaluation of health services. Core topics included every year include general health, smoking, drinking, obesity and blood pressure.

The first four surveys, from 1991 to 1994, focused on cardiovascular diseases and the risk factors associated with them, and were restricted to adults. The focus of the 1995, 1996 and 1997 surveys, which also included children, was on respiratory disease and accidents. Questions on disability were included in the 1995 survey and measures of general health status were included in the 1997 survey. The content of these surveys has so far reflected the target areas of the Health of the Nation strategy and has thus tended to concentrate on individual behaviours such as smoking (Prescott-Clarke and Primatesta, 1997).

In addition, a series of surveys was commissioned from ONS by the Health Education Authority to monitor trends in health-related behaviour, knowledge and attitudes. Data for 1995 and 1996 have been published by ONS (Office for National Statistics, 1997b).

There are also a number of *ad hoc* government surveys on a variety of topics, some related to health. Some surveys are repeated at regular intervals, enabling time trends to be followed. Thus, surveys of adult dental health are done every 10 years and surveys of 'infant feeding' every 5 years. Others are on a one-off basis. Topics covered in recent years include surveys of the prevalence of psychiatric morbidity (Meltzer *et al.*, 1995), smoking among secondary school children (Diamond and Goddard, 1995; Jarvis, 1998), teenage drinking (Goddard, 1996), and surveys of diet and nutrition (Finch *et al.*, 1997; Gregory *et al.*, 1995; and *see also* Chapter 18).

All these surveys provide data at national level and some provide data at regional level. None can provide data for areas below regional level, such as NHS districts. Some of the former NHS regions undertook 'health and lifestyle surveys' which provided data for districts within the region concerned, but each region did the survey in a different way, so the comparative data could not be collated at national level.

Interpreting official health statistics

As with other subjects, official health statistics have to be interpreted and used with an awareness of the process by which they came to be collected, analysed and published. It has long been recognised that many are collected as a by-product of other processes. For example, William Farr, who in the mid-nineteenth century set up the system for analysing and publishing data from birth and death registration to inform medical and public health practice, pointed out that

> The registration of births and deaths proves the connection of families, facilitates the legal distribution of property, and answers several other public purposes which sufficiently establish its utility; but in the performance of the duty with which you have been pleased to entrust me, I have to examine the registration under a different point of view, and with different object, which will ultimately prove of no less importance. (Farr, 1839)

It is therefore important to know which items of data have been collected and how. A major feature of NHS hospital data over the past years is the lack of data about the socio-economic characteristics of the people using them. When this issue was considered in the early 1990s, it was concluded that 'further work is needed . . . to establish the feasibility of introducing a social class indicator' (Department of Health, 1990b). In contrast, it was agreed to collect data about ethnic origin, using the census definition, despite all the problems and dilemmas described in Chapters 16 and 25.

Political sensitivity can affect data collection in other ways, particularly if the data item concerned is selected for use as a target. Waiting lists and waiting times are an area where this has happened. The Patient's Charter standard that people arriving in casualty departments must be seen within 5 minutes has largely been achieved, but there is no evidence that it has affected overall waiting times for actual treatment.

Reductions in numbers of people who have been waiting a long time for admission to hospital can reflect a number of different factors. Waiting lists have increasingly been tidied up, removing people who no longer require admission as they have died or been treated elsewhere. People who are offered an appointment that they cannot keep have their waiting times reduced to zero on the grounds that they have 'self-deferred' (Radical Statistics Health Group, 1995).

Another major problem is the fragmented nature of the data, with no linkage at national level to follow people through a series of encounters with the NHS. The exception to this is Scotland, where extensive record linkage is possible, making it possible to produce person-based statistics of the use of health services. Record linkage can also provide some information about the outcome of health care. For example, in Scotland, the outcome of hospital care has been monitored by using data about subsequent admissions to hospital, and death rates. This may be possible in the future in England, when the plan to put people's NHS number on all NHS records is implemented. Meanwhile, most NHS data will continue to measure, to some extent at least, the activity and resources of the service, but not the outcome of the care it provides.

Although surveys have a greater potential to ask questions of this sort, they are much more expensive, and this limits the extent to which survey data can be used to monitor differences between areas as small as NHS districts. Surveys which make clinical measurements are more expensive than those limited to asking people about their own views of their health without attempting to record diagnoses. On the other hand, surveys involving clinical measurements may well be biased towards the types of ill health it has been decided to measure.

These considerations can also affect the way data are classified. Diagnostic data are usually coded using classifications such as the International Classification of Diseases or Read codes which can be heavily conditioned by the medical views of the day.

The choice of data for publication and the way in which this is done can have a major influence on the way they are reported by the press. The Conservative government was particularly aware of this when using data selectively as 'good news' about the NHS. It relied, for example, on the way that most people are unaware of the difference between a 'finished consultant episode' and a person using the health service. The 1997 Labour government has announced its intention to publish data which measure the quality of health care and tell us 'what happened to individual patients; what their experience of care was; what the outcome of treatment was; how rapidly they got better or didn't' (Department of Health, 1997b).

Developing data collection systems which can answer these questions is not a trivial task and will need considerable investment. Some could be approached only through surveys, which tend to be more expensive than administrative statistics. On the other hand, for administrative statistics to be more informative, enhancements will be needed, notably for making greater use of record linkage. It remains to be seen whether the resources are made available to enable these aims to be brought to fruition.

27

Making Sense of Health Inequality Statistics

Mel Bartley, David Blane and George Davey Smith

Careful measurement shows that health inequalities are rising nationally

Most of the information we have on differences in health between people in different social groups is about men, and the social differences between groups are usually measured in terms of the sort of work they do, the income they earn and the status or prestige of their occupations. This type of social difference is referred to as 'socio-economic inequality', and more information is presently available about these inequalities than about many other sources of variation in health such as gender or ethnicity (*see* Chapters 25 and 28). Work on socio-economic inequalities has raised and addressed a number of issues about how we define and measure concepts and the methods we use, which are equally relevant to health differences between genders or ethnic groups.

Basic information on the socio-economic distribution of ill health is readily available from Decennial Supplements on Occupational Mortality (Office of Population Censuses and Surveys, 1986; Drever and Whitehead, 1997), the Office of National Statistics Longitudinal Study (Hattersley and Creeser, 1995), the General Household Surveys (Office of Population Censuses and Surveys, 1993) and Health Surveys for England (Department of Health, 1993). The first three of these continuous government surveys are carried out by the government office which conducts the census and collates the register of births, deaths and marriages: the Office for National Statistics (ONS, formerly called the Office of Population Censuses and Surveys, OPCS). The Health Surveys for England are carried out every 3 years by whichever organisation puts in the best 'tender': from 1994 to 1998 they were carried out by a consortium of the research organisation Social and Community Planing Research (SCPR) and the Department of Public Health at University College London; this consortium also carried out a similar Health Survey for Scotland.

Inequalities in mortality

The Decennial Supplements and the Longitudinal Study

Since 1911, information on mortality in different social classes (groups defined according to occupation) has been available from the Registrar-General's Decennial Supplements. These are called 'supplements' because they are supplementary volumes which accompany each UK census, which takes place every 10 years at present. The most recent of these (Drever and Whitehead, 1997) is written in relatively non-technical language and is strongly recommended to anyone interested in this topic.

In order to calculate death rates in each social class the numbers of deaths in each occupation, as recorded on death certificates, are collected for the 3 years around each census to form the numerator. The denominator is taken from the numbers enumerated in each occupation at the census. Occupational groups are combined into 'social classes' in a different way at each census, to try to take account of social changes (e.g. in the degree of skill needed for each type of job, the income earned and the prestige accorded; *see* discussion below). However, this is done not by means of any systematic process, but by informed guesswork undertaken by government researchers and officials involved in the process of designing and planning the census.

Traditionally, death rates in each social class have been reported some time after the census, based on the combination of death certificates with census data as described. This is known as the 'unlinked method' because there is no guarantee that the people enumerated in each occupation are the same ones as those who have that occupation on their death certificate. It was long suspected that people 'upgraded' their relatives on the occasion of a bereavement. Any such bias could easily produce distortion in the death rates. If the same person was more likely to be described as a 'miner' on their death certificates than at the time of filling out the census form, miners in general would seem to have higher deaths rates than they should. This kind of error, due to the mismatch of census data and death certificate data, is referred to as 'numerator–denominator bias'. In order to overcome this problem, in 1971 a new kind of study was set up within the OPCS, by linking 1 per cent of the census records for people in England and Wales to future censuses and records of their 'vital events' (births and deaths). In this way it became possible to follow the same 500 000 people from one census to the next, record any changes in their occupation, and be sure that this information could be used to produce more valid death rates in the different occupations (and thereby in the social classes). This study, the ONS Longitudinal Study (LS), uses what is known as the 'linked method', which is held to produce the most valid estimate we have of the real differences in life expectancy in the different social classes.

When the LS results on health inequality were first published, it did appear that the gradient in mortality between the Registrar-General's social classes had been an artefact and mostly due to numerator–denominator bias. There was little difference between classes, and a very high mortality was observed in men who had not been allocated an occupation at the 1971 census. However, on further investigation, this turned out to be a mistaken conclusion. It was due to the fact that men in the more privileged occupations can remain in employment when they have some degree of illness, while men in more arduous jobs may have had to withdraw from the labour force and describe themselves as 'permanently sick or disabled' (Bartley and Owen, 1996). The less privileged social classes, in other words, contain a high proportion of 'healthy workers'; they are 'selected' for good health. In 1971, the 'permanently sick and disabled' were not asked their previous job and not given an occupational code in the census. The result of this (which in fact tells us a lot about how social class works and is not just a point about statistical method) was that the presence of sicker men who *did* have an occupation in the higher classes produced an artificially inflated mortality rate early on in the follow-up period after the census (Fox and Goldblatt, 1982). After 5 years had passed, it was found that the mortality gradient reappeared. This was because of the initially high death rate of higher-social-class men who were ill at the time of the census, which removed them from the cohort. Because of this, when using such longitudinal studies to study differences in mortality between more and less privileged social groups, it is necessary to ignore the deaths which take place in the first 5 years, to give health selection time to 'wear off'.

Figures 27.1 and 27.2 give some examples of what it is possible to see using the LS: Figure 27.1 compares trends in heart disease mortality for men and for women, and Figure 27.2 compares mortality from all causes in men who were employed at the censuses of 1971 or 1981 with those who were not.

Inequalities in ill health

Data on health inequality are not confined to mortality, although we may think that death provides the strongest evidence. There are major government surveys carried out every year which can also be used to investigate inequalities in ill health and how they have changed over time.

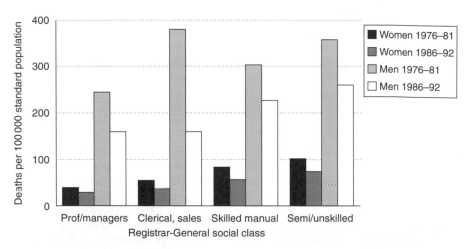

Figure 27.1 Ischaemic heart disease mortality in Britain, 1976–92: comparison of men and women
Source: Office for National Statistics, Longitudinal Study (Drever and Whitehead, 1997)

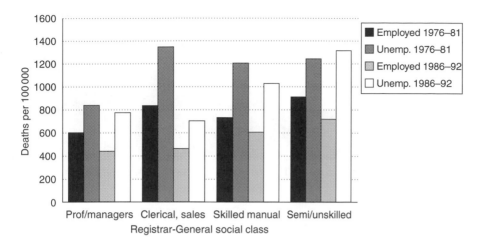

Figure 27.2 Class and employment status differences in mortality for British men aged 35–64
Source: Office for National Statistics, Longitudinal Study (Drever and Whitehead, 1997)

The General Household Surveys

The General Household Surveys are more fully described in Chapter 2 (and Chapter 26). Each year they ask people whether they have had to limit their activities because of illness in the past month; whether they have visited a doctor; what is the state of their health generally; and whether they suffer from any long-standing illness, and if so whether this limits them in any way. Figures are presented according to socio-economic group (SEG), which is not the same as the Registrar-General's social classes (discussed above). This classification is based on the prestige and shared lifestyle of the groups of occupations involved. However, like the Registrar-General's classes, these groups are made up and revised by officials with no reference to research. No one has ever tried to look at the extent to which the groups actually do share aspects of lifestyle.

The Health Surveys for England

The Health Surveys for England (HSE) are relatively new. They take a similar sort of sample to the General Household Survey (GHS) and collect detailed health measures (Prescott-Clark and Primatesta, 1997). As is the case for the GHS, households rather than individuals are sampled, and the sampling procedure is multi-stage. The first HSE was carried out in 1991, and successive surveys are repeated each year. Their aim is to monitor progress towards the government's Health for All targets, in particular the targets of reducing levels of cardiovascular disease and its risk factors. As a result, the 1993 and 1994 surveys concentrated on variables such as blood pressure, blood cholesterol, and others most relevant to heart disease. The sample size is now around 16 000 adults. Figure 27.3 shows social differences in the proportion reporting that they had fair or poor general health according to social class in 1993.

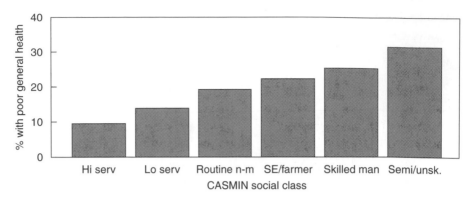

Figure 27.3   Self-reported general health by CASMIN class for men aged 30–55
Source: Health Survey for England

Other studies of health inequality

Data on health inequality have also been collected by special studies such as Whitehall I and II (Marmot *et al.*, 1991), Health and Lifestyle I and II (Cox *et al.*, 1993) and the British Regional Heart Study (Pocock *et al.*, 1987).

Measuring socio-economic position

A range of measures or indicators of socio-economic position have been used in social research. These include the Registrar-General's classification of occupational social classes, the Erikson–Goldthorpe or CASMIN schema (Erikson and Goldthorpe, 1993), the Cambridge scale (Stewart *et al.*, 1973) and Eric Olin Wright's class scheme (Wright, 1985). Housing tenure and car access can also be used as single-item measures of socio-economic position and, when used in this way, have proved strong predictors of mortality risk (Filakti and Fox, 1995).

There are a number of reasons for the existence of so many different measures and indicators of social position. In some cases new measures have been developed because of dissatisfaction with earlier methods. For example, the Registrar-General's classification lacks an explicit basis in sociological theory, and the perception of this as a serious weakness motivated the creation of more theoretically based and externally validated measures such as the CASMIN (Erikson and Goldthorpe, 1993) and Wright's (Wright, 1985). Internationally, education and income are the more widely used single-item measures (e.g. Kunst and Mackenbach, 1994; Hirdes and Forbes, 1992; Lynch *et al.*, 1996). Education, whether expressed as the number of years spent in formal education (National Center for Health Statistics, 1994) or the level of educational qualifications achieved (Morris *et al.*, 1996), tends to be associated with a steady 'stepwise' gradient in health; that is, people at each level of education have lower mortality or illness than those at the level below them, and higher than those at the next level above: those with degrees are not just healthier than those with no qualifications at all, but also healthier than those with the equivalent of A levels. Most studies using income, whether personal or household annual income (Sorlie *et al.*, 1995) or median income of residential locality (Davey Smith *et al.*, 1996; *see also* Chapter 28), have found the same relationship with health. All these associations between health and social position are in the same direction. The most socially advantaged tend to be the healthiest, the least socially advantaged tend to be the least healthy and in general there is a stepwise gradient between these extremes.

In preparing for the census of 2001 there has been a major change in the way occupations are grouped into social classes (*see* Chapter 2). The CASMIN schema reflects a belief that social groups based on the type of work people do and the sort of relationship they have to their employment (whether they are self-employed or an employee, whether they are a manager or supervisor, whether their job involves security and a career structure or not, etc.) are important influences on a wide range of life experiences and that these in turn affect attitudes, values and actions. The new government classification, called 'SEC' or socio-economic class, aims to combine occupations according to differences in conditions of employment in a similar way to the CASMIN schema. The composition of the categories in the SEC is the result of research commissioned by ONS. In order to make sure that occupations are placed in the appropriate classes, questions on methods of payment, job security and employment status have been asked as part of other surveys such as the Labour Force Survey, and the answers used to decide which class each occupation should go into (Rose and O'Reilly, 1997).

Measures of health

Social class differences in all-cause mortality remain the most certain evidence for the existence of socio-economic differences in health. Attempts to study the emergence of these differences, and the fact that health implies something more than the absence of premature death, create the need for measures of ill health. There are a number of these, many of which are as yet little used in health inequality research (Bowling, 1991; Fitzpatrick *et al.*, 1992; Jenkinson *et al.*, 1993), and each has potential shortcomings (Blaxter, 1989; Blane *et al.*, 1996). Most surveys contain a 'global self-rated health' question on whether the respondent considers their health to be generally excellent, good, fair or poor. Answers to this question are highly sensitive both to where it is placed in relation to other questions and to the social context in which it is asked. Some respondents saying their health is 'fair' or even 'good' also report quite serious diseases and impairments. Surveys often also ask about doctor-diagnosed diseases and recently experienced symptoms. If people are presented with a checklist of symptoms they remember far more than if they are asked to describe their symptoms without a checklist (Wadsworth *et al.*, 1971; Morrell and Wale, 1976; Hannay, 1979; Dunnell and Cartwright, 1979). It is difficult to decide the best way to combine the answers to such questions to arrive at an overall 'health measure' which would give us an accurate idea of social inequality in the burden of illness (Blane *et al.*, 1996). The relationship between 'objective' biological disease states, social disadvantage and subjective health is illustrated by the finding from one study that social status was related to illness scores on one widely used measure, the SF-36, as strongly as the presence of certain diseases such as arthritis (Hemingway *et al.*, 1997).

Methods for measuring inequality

The extent or size of social class differences in health can be expressed in different measures. Standardised mortality ratios (SMRs) reflect a class's mortality experience, once its age structure has been taken into account. Years of potential life lost (YPLL) reflect, in addition, the age at which death occurred (Blane *et al.*, 1990). Disability-free life expectancy (DFLE) combines data on both mortality and disabling morbidity (Bebbington, 1993). This measure tends to be highly sensitive to social inequality, and is the only one to show that the greater life expectancy of women is, on the whole, made up of time spent in a state of disability (van de Water *et al.*, 1996). The size of the class differences, as judged by the difference between the groups at the top and the bottom of the social hierarchy, appears to widen somewhat in line with the amount of information contained (i.e. SMR < YPLL < DFLE). There are obvious problems, however, in using only the top and bottom social groups to measure the degree of health inequality, and a number of measures have been proposed which use information on all groups in a social hierarchy. These include the index of dissimilarity, the relative index of inequality and the slope index of inequality (for a clear explanation of each of these, *see* Kunst and Mackenbach, 1994). They also allow comparison of the inequalities in health between countries in which social classifications divide the populations into different-sized groups.

Explaining health inequality

This chapter is intended to be about the measurement, rather than the explanation, of health inequality. However, measures always depend on the objectives of those who use them (MacKenzie, 1981; *see also* Chapter 8). So, a sketch of the major explanatory frameworks adopted in health inequality research may be helpful in understanding why certain data exist and others do not.

A major milestone in British health inequalities research was the publication of the Black Report (Department of Health and Social Security, 1980). The report summarised existing knowledge in the area and suggested an explanatory framework for investigating the causes of socio-economic variations in health. This explanatory framework proved most influential and it structured much of the thinking and research in the subsequent decade (Blane, 1985; Davey Smith *et al.*, 1990). The four types of explanation of health inequalities which the Black Report proposed are well known: (a) artefact; (b) social selection; (c) behavioural/cultural; (d) material/structural (for a longer introduction to these ideas for non-specialists, *see* Scambler, 1991, pp. 121–6; Anthony Staines also discusses their use in a local study in this book – *see* Chapter 28).

Are health inequalities an artefact of measurement?

Artefactual explanations put forward various types of measurement bias as the 'cause' of apparent health inequality (such as the numerator–denominator bias discussed above). One of the more striking examples of this argument was the claim that health inequality over time could be measured by calculating a Gini coefficient for the average age at death in different historical periods. This is a measure of the amount of variation around the mean age at which people die: the mean age at death is divided by the standard deviation of the mean. The argument is that if there are large inequalities in mortality between social groups, this must be reflected in a sizeable variability in the average age of death in the whole population.

Table 27.1 shows two hypothetical (very small) populations. In one the individuals die at very different ages and the Gini coefficient is high; in the other the opposite is the case, although the mean age at death is the same for both. Research using this measure found that, while comparing mortality rates (adjusted for age) in the most and least privileged classes suggested that mortality differences were widening over time, the Gini coefficient had narrowed considerably between 1921 and 1981. Comparing death rates in top and bottom classes, leaving out all the information on those in between, is indeed not a very satisfactory way to measure trends in health inequality. The problem is that measuring changes in the variation around the average age at death for the whole population does not necessarily give us a measure of changes in health inequality between social groups. In the extreme case where everyone died at more or less the same age, there would be little variation in life expectancy between social classes. But in any more realistic scenario where there is reduced but still considerable variation, the Gini coefficient alone cannot tell us whether health inequality has increased or decreased. Variation in the average age at death for the whole population may be partly or largely determined by variations in death rates within each social group, some of which may still be much higher than others. A Gini coefficient could fall, but it could still be the case that rich people were living a relatively long time and poor people a relatively short time.

Another version of the 'artefact' explanation made reference to the shrinking size of the bottom social group. Could the lack of any significant reduction in the

Table 27.1 The 'Gini coefficient for age at death'

Person	Age at death of individual person	Individual variations from mean	Gini coefficient
Large coefficient			
1	66	13.2	
2	77	24.2	
3	88	35.2	
4	22	30.8	2.75
5	11	41.8	
Mean age at death	52.8		
Sum of individual variations		145.2	
Small coefficient			
1	66	13.2	
2	55	2.2	
3	55	2.2	
4	39	13.8	0.666
5	49	3.8	
Mean age at death	52.8		
Sum of individual variations		35.2	

death rate in this group (Blane *et al.*, 1997; Drever and Whitehead, 1997) be due to a greater and greater concentration of unhealthy people at the bottom of society's hierarchy? But who is in this position? Those with the least well paid and hardest forms of work. This argument depends on acceptance of the idea that as the number of less skilled, routine, heavy jobs in a society decreases, those who do these jobs are increasingly confined to some kind of 'underclass' with poor health. However, there is no logical reason why this should be so. The argument smuggles in eugenic assumptions (*see* Chapter 7) and if these were justified, we would expect the health of those with professional and management jobs to get worse as these groups *increased* in size and became less of a highly selected 'elite'. In fact, as the proportion of the population able to get more privileged jobs has steadily increased over time, levels of health in these occupational groups have steadily improved (Drever and Whitehead, 1997).

Do sick people migrate into lower social classes?

Selection explanations put forward the idea that apparent health inequality is really due to the presence in less privileged social groups of people whose illness is the cause rather than the consequence of their social position (Stern, 1983). In particular, sick men may fall down the social scale into lower-status jobs, thus producing the impression that such jobs are bad for health when in fact it is the other way round. It is more likely, in fact, that sick men (this argument is seldom put forward in relation to women) leave the labour force altogether, and the spectre of an ailing stockbroker becoming a labourer does not seem plausible on common-sense grounds; nor is it supported by studies (Blane, 1985; Fox *et al.*, 1985; Goldblatt, 1988; Power *et al.*, 1991).

Health-related behaviour

The behavioural/cultural explanation holds that differences in health between social groups are due to their diet, smoking, sexual practices or other aspects of behaviour and 'culture'. Tobacco smoking, leisure-time physical exercise, dietary preferences and the informed use of medical services are the most frequently cited examples. Data on social variations in these factors can be obtained from the Health Surveys for England and Scotland and from some of the General Household Surveys (the latter asks 'health behaviour' questions only now and then; *see* Chapter 26). Information on diet can also be found in the National Food Surveys, which usefully show estimates of the amount spent 'per calorie' and 'per vitamin' by families in different social classes – this is how we know that poorer families make more cost-effective use of the money they do have to spend on food (Mary Shaw gives more details of these surveys in Chapter 18.) Each of these behaviours varies with social class (Department of Health and Social Security, 1980), and consequently they may be the intermediate pathway by which socio-economic position influences health.

Material factors in the environment and economy

The material explanation proposes that health differences originate in aspects of life experience which result from the nature of the social structure, and over which individuals have little control. Some studies of health inequalities using more accurate measures of social circumstances have been able to control for group differences in health-damaging or health-promoting behaviours. Typically their results show that group differences in behaviour explain some of the inequalities in health, but that clear inequalities remain after behaviour has been taken into account (Marmot *et al.*, 1991). The measures of social position used in most official data sources such as the GHS and HSE are less accurate, and it is hard to tell how much of the strong association found with behaviour after adjustment for social class is actually still there partly because these variables (such as admitting to being a smoker) are better measures of social position than is 'social class'.

Understanding health inequality: a life-course approach

By the mid-1990s, artefactual (Davey Smith *et al.*, 1994) and social selection (Blane *et al.*, 1993) processes had become less accepted as major contributors to socio-economic gradients in health. A more plausible idea is that there may be what has been called 'indirect health selection'. This concept refers to social mobility according to some individual characteristic which is not itself a disease, but which does influence the probability of developing disease in later life; family culture and stability and education have been suggested as candidates (West, 1991; Sweeting and West, 1995). People from difficult or disturbed family backgrounds are more likely both to end up in hazardous, low-status, badly paid jobs (or unemployment) and to engage in high-risk behaviour of various kinds (Montgomery *et al.*, 1996; Power and Matthews, 1997).

'Indirect selection' is an interesting concept. It can best be used by integrating it within the 'life-course' framework which has emerged from the 'post-Black' studies (Chief Medical Officer, 1995). This takes a longitudinal perspective on the development of disease and its socially unequal distribution, and attempts to understand their emergence in terms of the cross-sectional clustering and longitudinal accumulation of advantage and disadvantage.

The sorts of studies needed to progress further in understanding health inequality, therefore, are longitudinal ones, which either follow cohorts of people over long periods, or link data between studies. These allow selection explanations to be directly tested, and will eventually enable us to see how the chances of a good life and good health are distributed among populations over time.

Acknowledgement

Mel Bartley's work on this chapter was supported by Medical Research Council grant no. G8802774. Data from the Health Survey for England were provided by the ESRC Data Archive: those who carried out the original collection and analysis of these data bear no responsibility for its further analysis and interpretation. The Health Survey for England is Crown Copyright.

28

Poverty and Health

Anthony Staines

Local statistics can be used to show the widening gap between the health of rich and poor people in Britain

Poverty and ill-health

It is obvious that poverty is undesirable and, among many other consequences, that the poor have worse health, and die younger than the rich. Figures to show this are available as far back as seventeenth-century Geneva (Drever and White-head, 1997). Since the reports of the Poor Law commissioners in 1838 many studies have confirmed this observation (Wohl, 1983). It is very likely that poverty is the single largest cause of premature death and ill health in most developed countries. The Black Report (Townsend and Davidson, 1982) drew together much of the evidence for this link, and may be said to have started the current wave of interest in poverty and health. Despite this, the topic remains controversial, probably because the implication – that something ought to be done about poverty – has itself immense implications for how our societies should be organised.

Two arguments have been put as to why we should not really worry about links between poverty and ill health. The first of these is based on the distinction between 'absolute' and 'relative' poverty, where absolute poverty is defined as lacking the necessities for bare survival (Feuerstein, 1997). One argument against concern about gaps between the health of rich and poor people is to deny the existence of 'real' poverty. It is true that absolute poverty, as defined above, is rare in developed countries today. Thus it is said that poor people in these rich countries are only 'relatively' poor, and therefore that their poverty is irrelevant, and they should stop whining and 'get on their bikes'. Our idea of poverty needs to be wider than this, and should, on both moral and scientific grounds, include people who command inadequate resources to take part in our society (Feuerstein, 1997; Alcock, 1993; Townsend, 1993).

A cruder argument for not worrying about the gap between the health of rich and poor is that it's all due to the use and abuse of clever statistics which no one really understands. Of course plain people know very well that there isn't a major problem, the economy's on the up, all's right with the world, and all of this 'statistical jiggery-pokery' is just confusing the issue. This argument is nonsensical, but there is a dangerous half-truth in it. The statistical methods used in studies of health and wealth are complicated (*see* Chapter 27 for examples). In addition, the conventions of scientific writing often make the argument unnecessarily hard to follow. Sometimes the obscurity seems wilful, almost as if the authors did not intend that anyone else should be able to read their work.

Alternative explanations

Three major criticisms have been made of the work on links between poverty and ill health. These are serious alternative explanations for the differences in death and ill health observed between rich and poor in our societies. One is that the apparent differences are in fact spurious, and due to various statistical and technical problems – this is the 'artefact' explanation. The second is to argue that poor people have worse health than rich people because ill health leads to poverty, or because poor people are genetically inferior to rich people – this is the 'social selection' explanation. The third argument is to blame the poor for their ill health. On this argument poor people are ill because they behave badly: they drink too much, smoke too much and so on – this is the 'behaviourist' explanation, also known as 'victim blaming'. I think that the most credible explanation is that poor people have worse health and die younger than rich people because of the multiple ways in which our society mistreats them, and the ways in which they are excluded from full membership in it – this is the 'materialist explanation'.

Artefact

It has been seriously proposed that there are few substantial differences between the health of rich and poor, and that the apparent differences are due to a number of errors and statistical problems. It is best to summarise this lengthy debate by saying that there is no doubt at all that there are real differences between the health of rich and poor, most of which favour the rich. This is true using any sensible measure of poverty and affluence. For more details of this, read the Black Report (Townsend and Davidson, 1988), Margaret Whitehead's excellent book *The Health Divide* (1988) or the recent ONS report on inequalities in health (Drever and Whitehead, 1997). A series of more technical reports from the English Longitudinal Study have effectively excluded artefacts as a credible explanation (Fox and Goldblatt, 1982; Goldblatt, 1988, 1989; Fox *et al.*, 1990).

Social selection

The argument here is that people who are ill are more likely to move down in social classifications and those who are healthy are more likely to move up. Hence the apparent differences between social classes reflect changes in social class because of ill health and tell us nothing about the health of the different social classes themselves (Illsley, 1955, 1986; Klein, 1988).

Power *et al.* (1991) have analysed data from the National Child Development Study, a birth cohort study which started in 1958. Because of its design and because of the detailed data collected on health, social circumstances and attainments, this provides an exceptional opportunity to test these theories of social selection. Their finding, that social mobility is not an important explanation for social class differences in health, is convincing, and seems likely to be definitive. The work of Goldblatt (1988, 1989; Fox *et al.*, 1990) also supports the idea that social selection has at most a minor effect on social mobility. A recent study from the USA (Lynch *et al.*, 1997) confirms this finding.

Behaviour or victim blaming

'Blaming the victim' is the proposition that the difference between social classes is due to different types of personal behaviour; cigarette smoking and diet are often suggested (Le Fanu, 1993). The point of this argument is that the focus of criticism is directed to the wicked individuals showing the unhealthy behaviours, who are meant to change their behaviour – hence the term *victim blaming*.

There are two basic objections to this view. One is that one component of the social class effect is indeed different behaviour, but the place for action lies in the causes of this behaviour, which include limited opportunities, financial restrictions on diet, limited access to education and so on. The second is that studies which take diet and smoking into account still show strong social class effects (Davey Smith *et al.*, 1990). There are also studies showing that while health-related behaviour differs between social classes, these differences are not very large, and are unlikely to explain much of the difference in health (e.g. Smith and Baghurst, 1992).

Structural or materialist explanations

The preferred explanation (in the Black Report) for the bulk of the observed effects of social class on morbidity and mortality was described as 'materialist' (Feuerstein, 1997; Townsend and Davidson, 1982). This type of explanation emphasises that there are hazards inherent in society to which some people are exposed not by their own choice, but by their position in our society. The choice here, between behavioural and structural causes, has enormous implications for health policy.

Unfortunately, relatively little work has been done on 'materialist' explanations, although what has been done tends to emphasise the significance of this type of explanation (Feuerstein, 1997). Two good examples, which illustrate clearly what a materialist explanation looks like, come from Oakley (1989) and Calnan and Williams (1991), and *see* Chapters 20 and 27 of this book.

Ann Oakley (1989) studied the role of smoking in the life of pregnant women. Using data from a trial of social support in pregnancy carried out in England she examined the roles of cigarette smoking in the lives of women who smoked in pregnancy. Her conclusions strongly support materialist ideas on the origins of ill health. Specifically, she finds that smoking in pregnancy is an area of women's behaviour which is systematically linked with aspects of their material and social position. Smoking is used as a strategy for coping with a difficult life, and as a way of claiming some space and time for the woman herself.

Calnan and Williams (1991) carried out a study of the importance of health in the lives of a small group of families of differing social circumstances using in-depth interviewing techniques. They found that food, exercise and smoking were used in very different ways by working-class and middle-class respondents. Their analyses support materialist explanations of differences in behaviour between social groups. This study also revealed the extent to which opportunities and income determined diet and other aspects of lifestyle.

Poverty and death in Yorkshire

To make this discussion more concrete I will show the results of some work which my colleagues and I did in Yorkshire. We decided to ask one question: how did the gap between the death rates of rich and poor people change between the 1981 census and the 1991 census? We knew that the share of income held by poor people had fallen during this decade, and work by Richard Wilkinson (1990, 1992) suggested that this should lead to a wider gap in health and death between rich and poor people over the same 10 years.

Background

The Yorkshire Regional Health Authority covered the counties of West and North Yorkshire, and Humberside. Approximately 3.5 million people live in this area, mostly in big cities, although there is a large rural population.

Data

We used two sources of data, death registration data and census data, both prepared by the Office for National Statistics. To measure wealth we used information on where people lived at the time of death, and marked them as being poor if they lived in a poor area, and as being rich if they lived in a rich area. Peter Lee shows in Chapter 21 that this is very crude, but it is better than nothing. We used a score developed by Peter Townsend, one of the authors of the Black Report, for this purpose (Townsend *et al.*, 1988).

Graphics

I have chosen to show two pictures from our work which capture a lot of what we found out about the rates of death in rich and poor people. The first graph (Figure 28.1) shows death rates by age, for men and women separately, and for two time periods – the years from 1981 to 1983 and from 1990 to 1992. There is a big difference between death rates for men and women, in favour of women. This is well known, and has been seen many times before our work. The second thing which this picture shows is that death rates fell quite a lot between the years 1981 to 1983 and 1990 to 1992. In fact they fell by about 20 per cent, which is a very substantial drop. Death rates are shown on a log scale in this graph.

The second picture (Figure 28.2) shows something more complicated. It is possible to work out how long an average person will live, if present rates of death persist; this is known as life expectancy. It is also possible to work out how long someone who dies at a specific age could have expected to live. This is referred to as the 'years of potential life lost', and can be calculated for groups of people as well as individuals. What I have done here is to calculate this for five groups of people: those living in the richest fifth of the population down to those living in the poorest fifth of the population. Each of these groups is large, containing about 700 000 people.

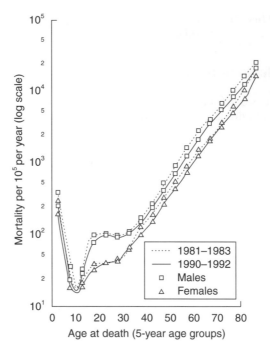

Figure 28.1   Death rates by age in 5-year age groups for males and females in 1981–83 and 1990–92, plotted on a logarithmic scale.

Figure 28.2 shows the number of years of life lost by a typical person in the four less affluent groups compared with a typical person in the most affluent group (shown as the straight line at zero years across the top). There are four graphs, for males and females, and for the years 1981 to 1983 and 1990 to 1992. This figure shows that for males and females the gap between the richest and poorest widened between the two censuses. Poorer people lost ground, compared with the richest fifth of the population, between the two censuses, with the poorest fifth losing nearly an extra year of life for females, and more for males over the decade.

Statistics

The other way to show these results is using statistics and tables of numbers. Pictures and graphs show the broad picture, and give a good general idea of what is going on. Tables provide the numbers. The statistically trained reader will know what it means to say that Poisson regression was used for these analyses, and all the data presented have been adjusted for age and sex. This is a common way of analysing this sort of data. What the tables show is the effect of a change of one standard deviation in the wealth (or poverty) score for the place where you lived, on your risk of dying. This takes account of difference in ages and proportions of males and females between the different groups compared.

Table 28.1 shows the effect of living in different types of area, using three different measures of deprivation: the Townsend index (Townsend *et al.*, 1988), the Carstairs index (Carstairs and Morris, 1989) and the Jarman score (Jarman,

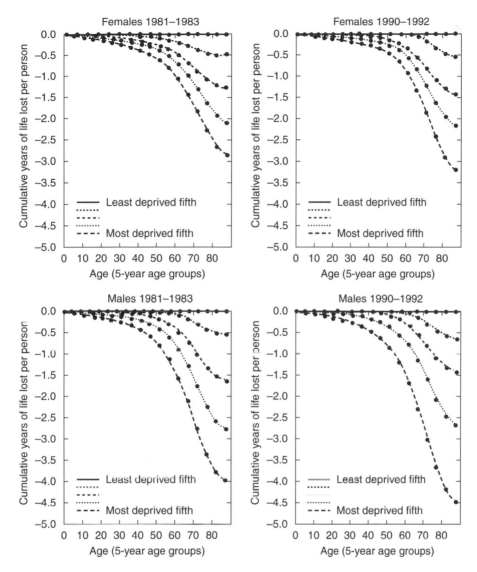

Figure 28.2 Potential years of life lost per person in the more deprived quintiles of the population, compared with those in the least deprived quintile.

1983, 1984). The figures in parentheses are 95 per cent confidence intervals for these estimates. In the first column of numbers, headed 'Townsend', the 1.08 in the first row means that the risk of death rose 8 per cent on average as the deprivation rose by 1 standard deviation. The 0.80 in the second row means that between the two time periods (1981–83 and 1990–92) the overall risk of death fell from 100 per cent in the first period to 80 per cent in the second, a fall of 20 per cent as discussed earlier. The 1.02 in the third row means that the risk of death for a change of 1 standard deviation in the deprivation score was 2 per cent higher in

Table 28.1   Change in mortality rates for a change of 1 standard deviation in the deprivation scores, and change over time from 1981–83 and 1990–92

Deprivation score	Townsend	Carstairs	Jarman
Score	1.08 (1.07–1.09)	1.07 (1.06–1.08)	1.08 (1.07–1.09)
Time period (1981–83 v. 1990–92)	0.80 (0.78–0.81)	0.80 (0.78–0.81)	0.79 (0.78–0.81)
Score. Time period	1.02 (1.01–1.04)	1.03 (1.01–1.04)	1.03 (1.02–1.05)

Table 28.2   Change in mortality rates for a change of 1 standard deviation in the Townsend score for 1981–83 and 1990–92 separately

Townsend score	1981–83	1990–92
Rate ratio	1.076	1.100
95% confidence interval	1.065–1.088	1.090–1.110

1990–92 than in 1981–83. In other words, the gap between rich and poor was wider in 1990–92 than in 1981–83. The results for the other two scores are very similar.

Table 28.2 tries to show this more clearly by separating data for the two time periods. In 1981–83 the risk of death rose 7.6 per cent for each change of 1 standard deviation in the deprivation score, while in 1990–92 for the same change in the deprivation score, the risk of death rose by 10 per cent, a change of 2.4 per cent in favour of the richer people living in Yorkshire.

Definitions of statistical terms

Poisson regression – this is an extension of ordinary regression used to analyse data on rates, such as death rates, birth rates and so on.

Standard deviation – This is a measure of how spread out values are. If one set of data, say a set of children's heights, has a standard deviation twice that of another set of measurements on children's heights, then the first set is about twice as spread out as the second set.

Confidence interval – This is a measure of how precisely something is known. To say that the mortality level in 1991–93 was 80 per cent of the mortality level in 1981–83 with a 95 per cent confidence interval of 78 to 81 per cent means that we are 95 per cent sure that the real change in mortality levels was between 78 and 81 per cent.

Discussion

These analyses suggest two things. First, the overall level of mortality in Yorkshire has fallen by almost 20 per cent in just over 10 years. The second is that this overall gain has been very unequally distributed. Compared with the top fifth of the population the other four-fifths live for significantly fewer years, and this gap has risen over the decade. These results are similar to those of Phillimore *et al.* (1994b), who did a comparable study at around the same time in the Northern Regional Health Authority. This is exactly what was predicted, given the changes in the distribution of incomes in England over the decade (Department of Social Security, 1993b).

Conclusions

There is really little doubt that poverty has an immense impact on health. What is more alarming is that those who are at the top of the economic ladder are leaving those at the bottom further behind as time goes by. This is not likely to improve the operation of society. This is a simple message, and needs to be spelled out clearly. Interest is now moving from describing differences in mortality between groups in our societies, to doing something about it. Two recent reports (Chief Medical Officer, 1995; Benzeval *et al.*, 1995) contain many useful policy prescriptions for state action to tackle inequalities in health. With the change of government in the UK in 1997 these and similar policies may start to be implemented. If this is done, the benefits to the health of the British population could be very substantial.

29

Statistics and the Privatisation of the National Health Service and Social Services

Alison Macfarlane and Allyson Pollock

Using statistics to investigate the hidden privatisation of health and social care services

Government policy has been to:

- reduce the size of the public sector through privatisation and contracting out;
- involve the private sector in providing existing and new services through the private finance initiative and challenge funding.

Financial Statement and Budget Report, HM Treasury, 1996

Background

While other parts of the public sector have been privatised with a single high-profile sell-off, privatisation has been taking place piecemeal in the UK's National Health Service (NHS) and social services. In some sectors, such as long-term care, the process is well advanced, while in others, such as acute hospital care, it is less well advanced but catching up fast. Many of the methods used to further private-sector involvement in the NHS are well known. They include tax relief on private health insurance for elderly people, compulsory competitive tendering for cleaning, catering and laundry services, the transformation of NHS beds into pay-beds, and charges to users of services. Others, such as capital charges, the private finance initiative and the subsidy of private nursing and residential care homes through the social security budget, are less well known.

These policies were brought together with the introduction of the NHS internal market in 1991 and the changes made in social services in 1993. Under these, health and local authorities have purchased care for their resident population from NHS trusts providing hospital and community services and also from private hospitals and private and voluntary organisations providing residential care and community services. This contrasted with the previous situation in which health and local authorities planned and provided the services themselves. The changes also offered scope for the private sector to use any spare capacity to sell services to the NHS.

This chapter uses the statistics which are available to describe the impact of these changes, while pointing to gaps in the data available. It is concerned first and foremost with the effects of the policies, rather than the mechanics by which they are implemented. It starts by describing changes made in long-term residential and community care before going on to discuss acute hospital care and charges to users of services.

Privatisation of long-term care

Before the 1980s, responsibility for funding and providing care was shared by two public-sector agencies. The NHS funded and provided care in long-stay hospitals, acute hospitals and community health services, while local authorities funded, and through their social services departments provided, residential care and community-based home care services such as 'meals on wheels' and home helps. All services provided by the NHS were free at the point of delivery but local authorities had discretionary powers to make charges for services and means-test their users. Until the 1980s few chose to do so (Association of Metropolitan Authorities, 1994).

Since the 1960s, the policies of community care and deinstitutionalisation had been leading to the rundown and closure of NHS and local authority institutions, as Figure 29.1 shows, with the transfer of people with learning difficulties and mental illness into what was described, often euphemistically, as 'the community'. For some people this may in reality mean homelessness, as Chapter 23 shows. These trends continued into the 1980s, and were extended to elderly people including those described as 'elderly mentally infirm'.

Up to 1981, the private and charitable sectors played a relatively minor though slowly increasing role in provision for long-term care. This changed when the Conservative government used an amendment of the Social Security Act to allow residents entering private-sector homes to claim board and lodging allowance

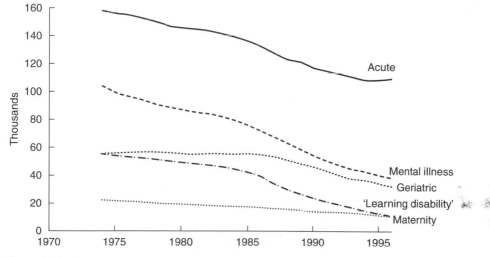

Figure 29.1   Average number of beds available daily, England 1974–96/97
Source: DHSS and DoH *Statistical Bulletins* 5/85, 1995/29 and 1997/20

from the social security budget to pay for their care. This option was not available for residents in local authority or NHS facilities or for community services for people in their own homes. It led local and health authorities to encourage their residents to opt for private care subsidised through social security payments. These changes had a dramatic effect on the pattern of residential and nursing home care provision, calling into question the stated policy of moving towards 'care in the community' (Audit Commission, 1986; Department of Health, 1988). In England, the overall number of places in homes for elderly and disabled people rose from 179 502 in 1979 to 233 587 in 1985 (Radical Statistics Health Group, 1987). The proportion of these places which were in private homes rose from 18 per cent in 1979 to 34 per cent in 1985.

These trends continued through the late 1980s and into the 1990s, with increasing numbers of places in private and voluntary homes for people with learning difficulties, and in private homes and hospitals for mentally ill people. Although the move to the private sector was the most outstanding feature, changes in methods of data collection in England made it difficult to measure changes precisely, let alone compare trends with other countries of the UK, each of which collected the data slightly differently. In 1987–88, a new category, 'elderly mentally infirm', was introduced to categorise people who had previously been described as either 'elderly' or 'mentally ill'.

Greater changes in data collection accompanied the much more widespread changes made in 1993 when the NHS and Community Care Act of 1990 was finally implemented. The key features of the Act were to devolve responsibility for funding long-term care to local authorities and to restrict the extent to which long-term care was funded through the social security budget (Pollock, 1995). The NHS' role in long-term care was reduced to caring only for the people who are most dependent on health care, thus requiring local authorities to provide care for increasingly dependent groups of people. Local authorities began to fund people in private nursing homes registered with the NHS as well as those in various types of residential care registered with local authorities. From 1994 onwards, the Department of Health began to collect data about places and residents in nursing homes. Small homes with four or fewer residents were required to register with local authorities from 1 April 1993 and they began to appear in the statistics from 1994 onwards.

The number of homes registering dually as both residential and nursing homes rose in the mid-1990s, from 602 in 1993 to 1537 in 1997 (Department of Health, 1997c). To counter the likelihood of double counting, from 1997 onwards data about residential and nursing care in these were explicitly subdivided. In the same year, data collection about places in unstaffed homes run by local authorities was stopped on the grounds that much of this accommodation was now being provided by housing associations, from which the Department of Health does not collect data. All these changes make it difficult to assess whether the rundown of NHS in patient facilities has been offset by facilities provided elsewhere. As a result of these difficulties, data for 1997 have not been added to the graphs in this chapter.

In England, the overall number of places in staffed residential homes of all types rose from 283 800 in 1987 to 307 896 in 1997 (Department of Health, 1996, 1997c). The proportion of these which were in local authority homes fell from 45 per cent in 1987 to 20 per cent in 1997 and most of the growth was in the private sector. In addition, the numbers of residential places in dual-registered homes providing both residential and nursing care rose from 11 308 in 1993, the first year for which data were collected, to 24 954 in 1997. The numbers of places in nursing homes rose from 124 369 in 1994 to 143 834 in 1995, then fell to 131 674 in 1997, while the numbers of nursing places in dual-registered homes rose from 13 831 in 1993 to 48 633 in 1997 (Department of Health, 1997c).

As can be seen from Figure 29.1, the rundown of 'mental handicap' hospitals has taken place over a long period. The process speeded up from the mid-1980s onwards, with the average number of beds available daily falling by 36 000 from 46 000 in 1983 to 10 000 in 1996–97 (Department of Health, 1997d). In contrast, the number of places for people with learning difficulties in staffed local authority, voluntary and private accommodation rose by only 28 896 from 17 328 in 1983 to 46 224 in 1997, as Figure 29.2 shows. The numbers of places available in local authority unstaffed homes apparently rose by only about 1250 from 1453 in 1983 to 2700 in 1995, but by 1997 there were 3389 people with learning difficulties living with local authority support in unstaffed homes, which could be owned by local authorities, housing associations or other organisations.

For people with mental illness, the average number of hospital beds available daily fell by 44 000 from 82 000 in 1983 to 38 000 in 1996 while the numbers of places in residential accommodation outside long-stay hospitals increased to a much lesser extent from 6500 in 1983 to 13 100 in 1996. The picture here is much less clear, as Figure 29.3 shows. People categorised as 'elderly mentally infirm' were not counted before 1987–88. Although the overall picture shows a major shift to the private sector, private mental hospitals are very different from the long-stay institutions which have been run down. They cater largely for people with less severe illness who are funded by private health insurance, although this may be changing.

The NHS and Community Care Act 1990 also requires local authorities to assess people's needs and financial eligibility for long-stay residential and nursing home care. While they continue to have discretionary powers to impose charges on

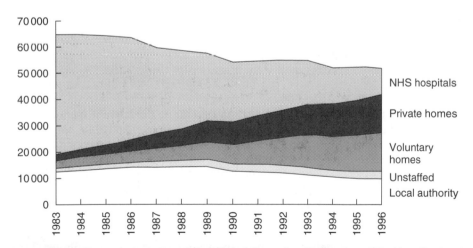

Figure 29.2 Places in homes and hospitals for people with learning difficulties, England, 1983–96
Source: Department of Health Local Authority Statistics A/F 91/11A and Statistical Bulletins 1994/13 and 1996/25

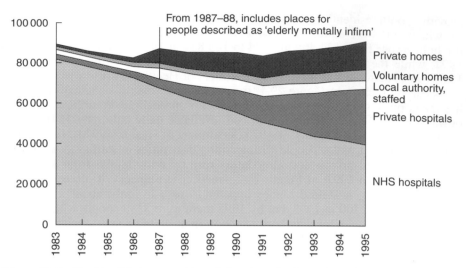

Figure 29.3   Residential places and hospital and nursing home beds for mentally ill people, England, 1983–95
Source: Department of Health Local Authority Statistics A/F 91/11A and Statistical Returns KN03, KO36, RAC5, RAC(5)S and RAU1

community-based services, means testing for residential and nursing home care follows national criteria. As a result, in the financial year 1994/95, £559 million was collected by local authorities from charges for residential and domiciliary care (House of Commons Health Committee, 1996a).

In 1996, the total expenditure on residential care for elderly, chronically ill and physically disabled people was just over £8.2 billion, of which 61 per cent was publicly financed, while the rest came from user charges, means testing and individual payments. Another £3.7 billion was spent on non-residential and domiciliary care including that provided by the NHS, social services and private and voluntary organisations (House of Commons Health Committee, 1996a). In 1995 over half of the long-term care expenditure in England, £4.9 billion, was spent on care in private residential and nursing homes and 49 per cent of the costs were met from public funds. There are no data available so far on the amount spent on community-based private care because this sector is fragmented and a national survey has yet to be done. In 1997, the average cost per person per year of nursing home care and residential care was estimated at £17 472 and £12 844 respectively in England (Laing, 1997). There are no national data on the number of people who have had to sell their homes to pay for care (Hamnett, 1992).

As a condition of central funding, social services departments increasingly use private organisations to provide community-based services such as 'meals on wheels' and home care. In 1996, the private sector provided 32 per cent of the total contract hours of home help and home care in England, compared with just 2 per cent in 1992 (Department of Health, 1997e). Although the numbers of hours of care had increased, the numbers of households receiving it had declined from 528 500 in 1992 to 491 000 in 1996.

These changes have yet to be thoroughly evaluated but the immediate impact of decentralisation has been to introduce wide differences in approaches to needs assessment, in both financial and other criteria for service eligibility (Leicester and Pollock, 1996; National Consumer Council, 1995). These differences create wide geographical and other inequities in access to long-term care services.

It is worth noting that many of the large health care corporations providing care in the USA have also set up operations in the UK. US experience shows little evidence that the shift to private-sector financing and ownership of long-term care will save money, especially if the corporations operating in the UK have similar patterns of spending on administration, similar capital and similar profit levels (Harrington and Pollock, 1998). At present, there are few data about the position in the United Kingdom.

Acute hospital care

The removal of restrictions on private practice led to a growth in the numbers of private acute hospitals in the early 1980s. The actual size of this growth is not so easy to assess, however. Data collected by the Independent Healthcare Association, shown in Figure 29.4, relate to the UK as a whole and do not tally exactly with the data collected separately by the health ministries for each of the four countries, because of the differences in definitions.

Data collected by the Independent Healthcare Association show that the numbers of private acute hospital beds in the United Kingdom rose from 6614 beds in 150 hospitals in 1979 to 10 155 beds in 201 hospitals in 1985 (Radical Statistics Health Group, 1987). After that, numbers of beds levelled off for a year or two before increasing more gradually to 11 681 beds in 227 hospitals in 1995 (Independent Healthcare Association, 1995). The percentage increases have been highest in the regions and countries, such as Scotland and the former Northern, North Western, East Anglian, Wessex and Oxford regions of England, which had the fewest private beds in the past. Nevertheless, private acute hospitals are very

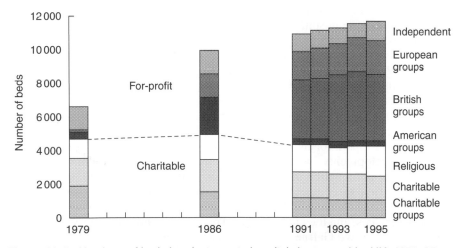

Figure 29.4 Numbers of beds in private acute hospitals by ownership, UK, 1979–95
Source: Independent Healthcare Association

unevenly distributed geographically. In 1995, 44 per cent of the beds in the UK were in the Thames regions and many of them were concentrated in London.

The increase has largely been in hospitals run by for-profit organisations, with a slight decline in hospitals owned by religious and voluntary groups. For-profit groups owned 63 per cent of private acute hospital beds in 1995, compared with 29 per cent in 1979. The early 1980s was a time of major expansion for British groups, such as the British United Provident Association (BUPA), and US companies expanded into the UK market. Since then European groups have entered the market and largely replaced the US companies.

The numbers of pay beds in NHS hospitals also rose in the early 1980s. In England, the authorised quota rose from 2405 in 1979 to 2919 in 1982, after which numbers levelled off to 2967 in 1985 and 2956 in 1991, the last year for which the data were collected in England. Revenues from NHS pay beds rose from £94 million in 1991 to £240 million in 1996. Over the same period, the NHS's share of the total private health care market rose from 10.9 per cent to 16.7 per cent and was predicted to rise to 20 per cent over the following three years (Fitzhugh Directory, 1997). Data from the Hospital Episode System for the financial year 1989/90 onwards show an increase in the use of private beds from 83 478 in-patient and day case episodes in 1988–89 to 99 399 in 1994–95 (Williams, 1997).

The increase in the provision of facilities in the private sector came at a time when facilities for acute in-patient care in the NHS were decreasing, in response to the move to day case surgery and shorter lengths of in-patient stay. In England, the average numbers of acute beds available daily decreased from 147 000 in 1979 to 108 008 in 1994/95 but have shown a slight increase in the past two years, reaching 108 895 in 1996/97 (Department of Health, 1997d). The decreases over earlier years were also associated with hospital closures, in many cases, meaning that people are having to travel longer distances for in-patient care. It is difficult to assess the extent of this inconvenience, as the Department of Health no longer records the location of hospitals in the published statistics and simply identifies trusts by the type of care they provide.

No data are collected routinely about the numbers of in-patient and day case admissions to private hospitals, but a series of three surveys estimated that numbers of admissions to short-stay private hospitals in England and Wales rose from approximately 275 752 in 1981 to over 655 350 in 1992/93 (Williams and Nicholl, 1994). In 1992/93 about 5 per cent of these were paid for by the NHS. As part of waiting-list initiatives in the late 1980s the first steps were taken to buy NHS services from the private sector. The introduction of the internal market in general, and general practitioner fundholding in particular, have accelerated this trend. In 1994 Virginia Bottomley stated that it did not matter who provided the services so long as they were paid for by the public purse.

Further developments increased private-sector interests in public hospitals. These were the introduction of capital charges in 1992 as part of the internal market, the extension of competitive tendering, which had been introduced in the 1980s for catering, cleaning and laundry, to other services such as computing and laboratory services, and the launch by the Treasury in 1992 of the Private Finance Initiative (Pollock and Gaffney, 1998).

When replying to a parliamentary question about the backlog of maintenance costs, the Department of Health admitted that a substantial reduction in the size of the NHS estate was anticipated over coming years along with a continuing ambitious programme of capital investment by the private sector as part of the private finance initiative (PFI) (Pollock *et al.*, 1997). The object of PFI is to bring private finance into large capital projects for the NHS, but the buildings will be built and largely owned by private companies and consortia, which will lease them and their services back to the NHS. The NHS, using its revenue funds, will pay for both the capital and the running costs but will not 'own' the facilities built under the initiative. A recent report has shown that the NHS and the Treasury are using public funds to subsidise the expansion of the private sector. The report also draws attention to the reduction in overall capacity and service provision as a result of the increased cost of the PFI hospital building programme compared with publicly funded alternatives (Gaffney and Pollock, 1997). It is likely that the rundown of NHS acute and community service provision will mirror the recent trends in long-term care.

Privatisation and the employment of NHS staff

Because so many NHS staff work part time, statistics about them are expressed as whole-time equivalents. Each employee is counted according to the proportion of whole-time hours she or he works. The numbers of whole-time equivalent ancillary staff in the hospital and community health services in England decreased from 124 270 to 66 760 between 1986 and 1995, after which a change in the way jobs are classified made trends difficult to monitor (Department of Health, 1997f) Many of these were replaced by staff employed by private contractors. There are no data about the number of staff these firms employ on NHS work nor about the extent to which nursing and other staff may have to take on some of the tasks previously carried out by ancillary staff.

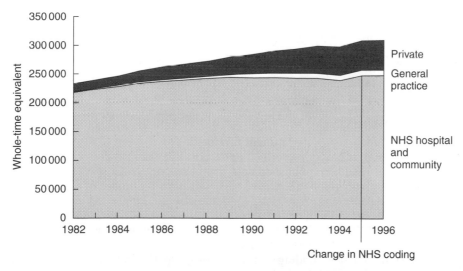

Figure 29.5 Whole-time-equivalent qualified nurses by sector, England, 1982–96
Source: Department of Health non-medical workforce census, bi-annual census of general medical practitioners and annual return KO36

The numbers of whole-time-equivalent qualified nurses working in NHS hospital and community services declined from a peak of 244 220 in 1989 to 238 780 in 1994. After that, the numbers appeared to increase slightly as a result of a change in the way nurses are classified. Senior nurses, who had been counted as managers, are now counted as nurses. The numbers of whole-time-equivalent nurses working in general practice rose from 1449 in 1982 to 9820 in 1996 (Department of Health, 1997f). Nevertheless, the most marked feature of Figure 29.5 is the increase in the number of qualified nurses working in private homes and hospitals from 12 208 whole-time-equivalents in 1982 to 50 810 in 1996.

Private sources of funding

As well as being funded by general taxation, health and social services and private care can be funded by charges to users. These may be met wholly or partly from private health care insurance. The trend since 1979 has been to increase charges and to introduce more charging policies for health and social care. The proportion of NHS income coming from charges for prescriptions, dental fees, private health care and other sources has always risen under Conservative governments and fallen under previous Labour governments (Webster, 1997). It rose from 2.0 per cent of total NHS expenditure in the UK in 1978/79 to 2.9 per cent in 1984/85 (Radical Statistics Health Group, 1987). After reaching a peak of 4.3 per cent in 1990/91, the proportion started to fall, reaching 2.0 per cent in 1996/97 (Office for National Statistics, 1998b). The main reason for this fall is that NHS trusts, as autonomous public corporations, no longer report income from charges.

There are other reasons why income from charges appears to have fallen. First, some drugs which could formerly be obtained only on prescription became available without prescription. This transfers the cost of the drug to the person paying and removes its cost from the NHS. Second, where drugs cost less than the NHS prescription charge, many general practitioners are prescribing them on private prescriptions. Third, eye tests and the provision of spectacles have been completely privatised and dentistry has been partially privatised. Finally, some other private charges, such as local charges for screening and management of subfertility, are not declared nationally and so do not appear in statistics.

The creeping privatisation of prescribing is particularly important. From 1979 onwards, prescription charges rose every year, usually well ahead of inflation. It has been estimated, using Department of Health Prescription Cost Analysis data for 1995/6, that over half of prescribed items cost less than the prescription charge (Department of Health, unpublished data from Prescription Cost Analysis). If the use of private prescriptions increases, this will have a number of effects. First, prescribing analysis and cost (PACT) data will no longer accurately reflect general practice prescribing patterns, reducing the value of indicators of prescribing derived from PACT data. Second, the net costs of drugs will rise because of 'cream skimming' of cheaper drugs. Third, private prescribing will act to increase inequalities in practice funding. Costs of drugs prescribed privately will not appear in practice prescribing budgets. This will favour practices in affluent areas where the demand for private prescriptions will be greater, as fewer people are exempt from charges. Such practices will therefore have lower NHS prescribing costs than practices in more deprived areas. Fundholders have been able to use savings on their prescribing budgets for other purposes (Heath, 1994).

Social services departments' income from charges was poorly documented throughout the 1980s. Charges accounted for 9.6 per cent of total social services expenditure in 1991/92 rising to 12.4 per cent in 1995/96 (Department of Health, 1996). These data may not reflect the extent to which charges were actually recovered, which is thought to be much lower. The potential to raise income from charges for residential care is taken into account in the central government allocation to local authorities. Social services charges are particularly regressive because they require payment from users of services. These are often the people with the highest need for health and social care. For many services, such as home care and domiciliary care, people increasingly have to pay from their Income Support for the domiciliary care provided.

Private health care insurance

Although in the original NHS Act, the then Minister of Health, Aneurin Bevan, allowed doctors to offer private health care in NHS pay beds, the private sector remained small, with only 120 000 people holding private insurance in 1950. The number had increased to over 2 million by 1972 and by 1992 approximately 6.4 million people – that is, 10.8 per cent of the population – were covered. These data are compiled in a variety of ways, from either surveys or returns made by third-party insurers, and their accuracy has not been evaluated (Laing, 1997). Insurance coverage can be measured as the numbers of subscriptions taken out each year, or in terms of the numbers enrolled at one point in time.

The uptake of private medical insurance is known to vary geographically, although data have not been published for some time. The most recent published data are from the 1987 General Household Survey. This estimated that 16 per cent of people in the Outer Metropolitan Area around Greater London and 15 per cent in the Outer South East were covered, compared with only 3 per cent in the North and 4 per cent in Scotland and Wales (Office of Population Censuses and Surveys, 1989). Uptake of private health insurance also varies strongly by socio-economic group. As might be expected, policy holding and coverage are highest among professional and managerial groups. In 1995, policy holding ranged from 21 per cent of men in professional occupations to 1 per cent of men in manual occupations (Office for National Statistics, 1997c).

What conclusions can we draw?

The patchy data available show that the privatisation of the NHS and social services has reduced the availability of care to most people in England. Similar forces were at work in the other three countries of the UK. Given the extent to which public funds have been used to subsidise investment in the private sector, better data should be available about how they have been spent. Better data are also required on the needs for health and social care and how they are met, together with the impact of changes in funding and charging policies on access to care. Compiling good data is no easy task. It is difficult to collect data in a way which reflects policy changes and anticipates their impact, while also allowing consistent time series to be compiled. We suspect that such data would seriously call into question the way in which these policies of privatisation are still being pursued under a Labour government.

30

Industrial Injury Statistics

Theo Nichols

Official statistics on industrial injury are not a valid measure of safety performance

In late 1997 when Frank Davies, Chairman of the UK's Health and Safety Commission (HSC) presented the 1996/97 Annual Report, he reflected that the increases in fatalities at work provisionally estimated for that year had to be taken seriously, even though at that point it was too early to judge whether the new statistics indicated a 'brief rise in this country's previously improving safety record or the start of an upward drift' (Health and Safety Commission, 1997, p. xi). The key purpose of this chapter is to critically assess the assumption that the British safety record had been improving from the mid-1980s. It is argued that it is doubtful it had improved, if by this is meant that a reduction in the Health and Safety Executive (HSE) statistics had come about as a consequence of improved attention to safety matters in British industry.

What is safety?

It is an important feature of the history of advanced capitalist societies that they have generally evidenced rises both in the level of productivity and in safety. As far as safety, rather than safety and health, is concerned – for safety and health may move in contradictory ways and it is safety alone which is the focus here – the contribution of several different factors has to be reckoned. One of these is the role of increased investment and, especially important in manufacturing where machinery is a more common immediate cause of injury, increased mechanisation and automation. As far as fatality rates are concerned, long-term improvements in medical facilities, both on site and by way of general public provision, can make for improvement. Also – and generally this is the more so the longer the time period that is considered – a part is played by increased expectations about how people should be treated at work and, just as important, increased expectations about the rights and confidence they have outside it. However, whatever the importance of each of these possible contributory factors, the emergence of Thatcherism as a short-termist, anti-labour political regime at the start of the 1980s should be sufficient warning that there is nothing inevitable about safety improving over time. In the first half of the 1980s the health and safety of workers in British manufacturing did not benefit but, rather, suffered from the much celebrated process whereby British industry allegedly became 'leaner and fitter' (Nichols, 1986). Moreover, any attempt to depict capitalism as a naturally beneficent force which decade by decade inexorably advances safety at work is, apart from anything else, likely to encounter severe problems once the consequences are admitted of the conditions under which it contemporaneously operates on a world scale – both in the Third World, and also in what used to be called the Second World, as this itself now takes a capitalist turn.

In reviewing the course of safety over time it is important to distinguish improvements in the level of industrial safety from what is ordinarily implied by reference to an improvement in safety performance. The problem is that the term 'safety performance' tends to carry the implication that something has been accomplished, achieved or consciously attained. Often there is the further implication that the performance has been orchestrated by management. Yet when injury statistics suggest improvements in the level of safety it is necessary to question whether this is a function of better safety performance by operational managers or of the operation of other forces that are largely outside their control which, in this instance, have had the effect of improving industrial injury rates. Such forces may lead to the decline of more dangerous branches of industry (for reasons other than a regard for workers' safety) and to the replacement of high-risk machinery with modern technology (again not necessarily as the consequence of a specific intention to improve safety). It is with such considerations in mind that we now turn to recent evidence about the level of safety in British industry, and to the question of whether this indicates that there has been an improvement in safety performance, not over the past hundred years, nor even the past quarter of a century, but over the most recent period for which statistics are available.

Following some comments on the adequacy of recent British administrative statistics, it will be considered what has happened to safety in Britain over the years 1986/87 to 1994/95 – a consideration that will necessarily concern the question of whether, especially in this period of rapid economic change, improvements in official injury statistics may be a misleading guide to improvements recorded in safety performance in the sense indicated above.

Safety statistics

Briefly, there are at least three problems with using the administrative statistics to assess safety. The first concerns the manufacturing fatality rate. The technical virtues of this rate are that it does unequivocally relate to injuries at work and that we can be reasonably sure that the injuries it is intended to record are actually recorded. The problem is that the numerator in this rate has now become exceptionally small. In 1960 for example, there were 300 fatalities a year in British manufacturing. In 1994/95 the number of fatalities had, thankfully, dropped yet lower to 46.

The second problem concerns the over-three-day injury rate. The nub of the problem is that this rate is dominated by minor injuries and that it has long been thought that decisions made by workers about whether to take time off for relatively slight injuries, and how much time they take off, are socially mediated (Chief Inspector of Factories Annual Report for 1945 (1946), p. 8; Chief Inspector of Factories Annual Report for 1967 (1968), p. x). This suggests that such a rate is inappropriate as an indicator of safety; or as the HSE has put it, the 'over three day injury statistics are intrinsically unreliable as a measure of safety performance' (Health and Safety Executive 1987, p. 58).

The third problem concerns the major injury rate. The technical virtue of the major rate has often been held to be that it both avoids the small-number problem of the fatality rate and provides a better indicator of safety than the three-day rate. However, in 1990 evidence from the Labour Force Survey (Health and Safety Commission, 1991, p. 56; Stevens, 1992; and *see also* Chapter 3) confirmed what several commentators had suspected: that the non-fatal rate, and with this also the major rate, is subject to significant under-reporting. Only about one-third of all reportable non-fatal injuries at work were actually reported under RIDDOR (the Reporting of Injuries, Diseases and Dangerous Occurrences Regulations) in 1990, although some variation was found between sectors. As many as 70 per cent of reportable non-fatal injuries were reported in energy, a mere 15 per cent in agriculture, with manufacturing and construction reporting around 40 per cent. Taken as it stands, then, the major rate now looks less valid as an indicator of the level of safety than it was intended to be (Nichols, 1994). Fortunately, though, the HSE not only included questions to elicit self-reports of injury from the 1990 Labour Force Survey, it went on to do so in 1993/94 and 1994/95. The information that was obtained in this way leaves no doubt that employers' reporting rates for all non-fatal injuries to employees in manufacturing remain very poor. In 1994/95, for example, only 45 per cent of such injuries were reported in the construction industry and only 37 per cent in services. In manufacturing, the situation, although better, was still far from satisfactory, the level of reporting having risen from 43 per cent in 1989/90 to 49 per cent in 1993/94 to 54 per cent in 1994/95 (Health and Safety Commission, 1995a, p. 1). However, reductions in the extent of under-reporting are to be welcomed, and if the information now available on under-reporting continues to be collected on a regular basis, in future years it may permit corrections to be applied to the officially published major rate and thus allow better estimates of changes over time.

Changes in the economy

Whatever the virtues and defects of the administrative statistics, some of the changes that have occurred in the structure of the UK economy in recent years do not inspire much confidence that any reduction they might record in the injury rate for the economy as a whole can be uncritically interpreted as evidence of improved safety performance. According to the HSC itself, whereas there was a decrease in the employee fatality rate for industry as a whole between 1985 and 1994/95, 'a large part of the decrease' was accounted for by changes in the pattern of employment. Such a conclusion is not surprising. Between 1986/87 and 1993/94 there was a 10 per cent rise in the number of employees in the service industries (which generally have a low rate of injury). By constrast, there was a fall in the number of employees in hazardous sectors: in energy there was a drop of 36 per cent, in agriculture of 20 per cent, in manufacturing of 16 per cent, in construction of 11 per cent, and in the coal extraction industry, which typically had a very high fatality rate, employment decreased in excess of 80 per cent (Health and Safety Commission, 1995b, p. 23; *see also* Chapter 38).

There is some sign of similar processes being at work within the manufacturing sector. For example, in both 1986/87 (the earliest date from which a recent consistent time series can be constructed) and 1994/95 the most dangerous sub-order of manufacturing industry was that concerned with the extraction and preparation of metalliferous ores (1980 SIC 21/23). This had a combined fatal and major injury rate of about three times that of manufacturing as a whole at both dates (for example, a rate of 498 per 100 000 employees in 1986/87 compared to one of 147 per 100 000 for manufacturing as a whole). But the number employed in this sub-order fell from around 195 000 in April 1986 at the start of the period to around 26 000 by March 1995 at its end, a fall of nearly 90 per cent, which was greatly in excess of the fall for manufacturing as a whole.

Industries do not necessarily undergo uniform changes in each of their product markets and production technologies. It is therefore possible that high- and low-risk subsectors of particular industries have been differentially affected by the broader processes of restructuring indicated above. Other changes may also occur which the very broad shifts in employment discussed so far will not bring to light. There can be changes in the distribution of higher- and lower-risk occupational groups (for example, in the proportion of operatives at risk as against that for administrative, technical and clerical workers). Changes can also occur in the total hours of work of those employees who are in higher risk categories. The British administrative statistics take neither possibility into account. The official injury rates are calculated by attributing the same weight to each and every 'employee'. No account is taken of whether the employees are managing directors or

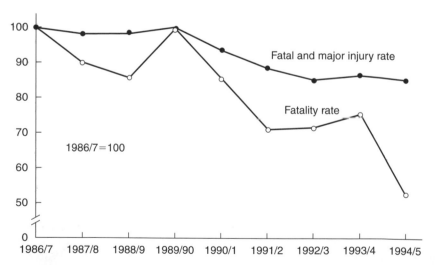

Figure 30.1 Health and Safety Executive (HSE) manufacturing fatal injury rate and HSE fatal and major injury rate, 1986/87–1994/95 (per 100 000 employees)
Source: *Health and Safety Statistics*, various years

operatives, or of whether they work part time, full time or work extra hours through overtime.

Figure 30.1 plots the movement of the official manufacturing fatality rate and of the combined fatal and major injury rate over the past decade. The combined employee fatal and major rate for manufacturing registers a decrease. So does the fatality rate, more erratically but also apparently more markedly. It might be quickly surmised that an improvement must have occurred in safety performance.

In the case of the fatality rate, one problem is that, despite the appearance of some sharp decreases in injury rate in Figure 30.1, what we are dealing with here is a change from something like 2.1 fatalities per 100 000 in 1986/87 to 1.2 fatalities per 100 000 in 1994/95 (Health and Safety Commission, 1995b, p. 44). Reductions in fatalities at work are very much to be welcomed. But, because changes in industrial composition can have such major effects on the distribution of danger, it would be unwise to attribute a difference such as this to an improvement in safety culture, increased safety performance or the like.

The decline in the fatality rate (and in the combined fatal and major injury rate) from 1989/90 onwards roughly coincides with the downturn in the economic cycle. An earlier enquiry into the relation between economic fluctuation in British manufacturing and the behaviour of injury rates in the period up to 1985 suggested that the relation was a pro-cyclical one (Nichols, 1989). This again therefore indicates the need for caution in the attribution of decreases in the official statistics to improved safety performance.

Further reason to search behind the story that the official statistics appear to tell is provided by Figure 30.2. This focuses more closely on the HSE rate for fatal and major injuries, which is of course based on all employees, and plots alongside

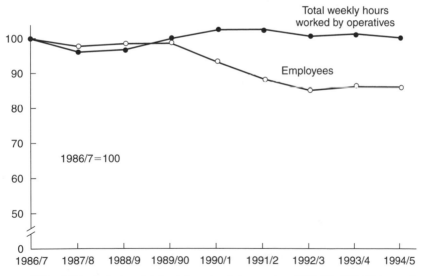

Figure 30.2 Manufacturing fatal and major injury rate, 1986/87–1994/95: employees versus total weekly operative hours
Source: Employee injury rate as for Figure 30.1. Total weekly hours worked by all operatives from Central Statistical Office. Here and in Table 30.1, annual series for operatives have been rebased on HSE financial years, i.e. April to March

Table 30.1   Average hours worked per operative and total mass of operative hours in British manufacturing 1986/87–1994/95

	1986/7	1987/8	1988/9	1989/90	1990/1	1991/2	1992/3	1993/4	1994/5
Average hours per operative	100.0	101.0	101.5	101.0	100.8	100.0	99.8	99.4	100.1
Total mass of operative hours	100.0	101.4	102.3	100.1	89.5	76.7	72.0	71.4	71.5

Source: CSO

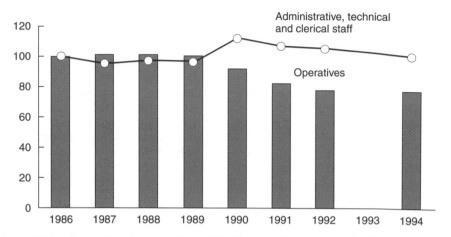

Figure 30.3   Occupational composition of British manufacturing, 1986–94. Note: no data are available for 1993 or after 1994. Employment as at September each year; 1986 = 100
Source: Central Statistical Office

it a recalculated rate for fatal and major injuries which uses a denominator derived from the Central Statistical Office time series for total weekly hours worked by all operatives in manufacturing. It thus seeks to take approximate account both of the changing number of those in an occupational group which is most at risk (operatives) and of the length of their exposure to risk (the total mass of hours they work per year). Compared to the improvement apparently indicated by the original administrative statistics, the recalculation spells a non-change result for this injury rate and thus for safety performance. This outcome is for the most part a function of a decrease in the number of operatives employed rather than in the hours that they worked. Average hours per operative in manufacturing changed little in the period under review and were only very slightly higher in 1994/95 than in 1986/87 (Table 30.1). But as can be seen from Figure 30.3, from the turn of the 1990s onwards the number of operatives underwent a decline compared to that of administrative, technical and clerical staff. In other words, the HSE rate would seem to have benefited from changes in the occupational composition of the labour force: fewer indians (in the guise of operatives), more chiefs (at least in the guise of administrative, technical and clerical staff). The category 'administrative, technical and clerical staff' is a somewhat polyglot one, but it is overwhelmingly non-manual, including 'directors, managers, superintendents, foremen who supervise other foremen, professional, scientific, technical and design staff, draughtsmen, salespeople and office staff'.

The impact of legislation

Over the past decade there has been considerable publicity about health and safety associated with the build-up to, and the implementation of, European Union (EU) directives, particularly the 'six-pack'. The 'six-pack' was a bundle of regulations that came into force from 1993. In a move towards compliance with an EU Framework Directive (Directive 89/391/EEC), the Management of Health and Safety at Work Regulations set out minimum requirements for managerial controls for safeguarding health and safety, including consultation with employees; and five daughter directives related to health, safety and welfare in the workplace: the Workplace (Health, Safety and Welfare) Regulations; the Manual Handling Regulations; the Health and Safety Display Screen Regulations; the Personal Protective Equipment at Work Regulations; and the Provision and Use of Work Equipment Regulations. The heightened publicity about health and safety that these measures stimulated may have contributed to the reduction in under-reporting that has been identified by the HSC and will mean that some of the major injury rates reported in Figures 30.1 and 30.2 do not fully acknowledge the extent to which injury rates have actually improved. Against this, although it is known that the reporting of all non-fatal injuries in manufacturing improved in 1993/94, the extent of under-reporting is not known for years prior to 1989/90 nor for some years since. Nor is specific evidence on the extent of the under-reporting of major injuries, as opposed to all non-fatal injuries, available at industry level (for example, with specific reference to manufacturing as opposed to, say, construction). In any case, it is still difficult, in the light of the more general considerations about injury rate determination that were introduced earlier – especially those that relate to industrial and occupational distribution – to be sanguine that the past decade in British manufacturing has witnessed a general improvement in safety performance.

The fact that any heightened safety awareness in the 1990s has been predominantly generated by the need to comply with an EU directive is but the tip of a larger iceberg. During the 1990s, positive intervention in UK health and safety legislation has come about largely through increased influence from the EU. Just as surely, attempts to weaken the safeguards for British workers have come from the UK government. Indeed, in the light of the political priorities and the economic stance that Conservative governments adopted between 1979 and 1997, it would have been remarkable had this led to an improvement in safety performance in the way that an uncritical interpretation of the HSE's administrative statistics on injury rates might initially suggest.

Labour government

The 1997 Labour government still has to prove itself. There is talk of higher fines being introduced for health and safety offences and the introduction of a charge of corporate manslaughter. As against this, Labour has already committed itself to the cuts in HSE spending in 1997 and 1998 that were scheduled by the previous government. In the words of the new Labour Minister responsible for health and safety at work, 'We have been lumbered with . . . significant cuts this year, and even bigger cuts in HSE resources next year' (Eagle, 1997, p. 6). In due course the evaluation of safety performance under Labour will require the same sort of analysis as is attempted in this chapter. A particular difficulty for any such evaluation is that changes in RIDDOR took effect in 1996/97 so that, among other things, some hitherto over-three-day injuries were reclassified to major injury status, including certain dislocations, fractures and amputations. This change will have lessened three day and over injury rates and increased the major injury rates for 1996/97. Even so, the 1996/97 figures give no cause for complacency. The number and rates of major injuries are up across the board. Fatalities to employees in manufacturing are estimated at 57, up from the comparable figure of 42 for 1995/96 (Health and Safety Executive, 1997).

Social scientists, and others, might like to note that the official time series used in this chapter for average weekly hours worked per operative and for total weekly hours worked by all operatives have now been discontinued. So too has the time series for the manufacturing engagement rate, which was used in the exploration of the relation between business cycles and injury rates in Nichols (1989), and similarly the time series for the number of operatives and of administrative, technical and clerical staff, which is used in Figure 30.3.

Acknowledgements

This chapter was extracted from Theo Nichols (1997): *The Sociology of Industrial Injury*. London: Mansell, 199–205.

PART VI
Assessing Education

THE NEXT GOVERNMENT INITIATIVE?

31

What's Worth Comparing in Education?

Ian Plewis

Debates about standards are often fruitless because they focus on impossible tasks

Pupils in state schools in England and Wales are now assessed, or tested, more intensively than previous generations have been. From the ages of 5 to 16 – the compulsory years of schooling – pupils are required to be tested at the ages of 7, 11 and 14, they are increasingly likely to receive some kind of 'baseline' assessment when they start school, and the great majority go on to sit General Certificate of Secondary Education (GCSE) examinations at 16. The majority of 16-year-olds now remain in full-time education until they are 18, at which point many sit A-level examinations.

It is far from clear whether, and how, this assessment juggernaut improves pupils' learning, especially because the assessments at 7, 11 and 14 which were introduced after the 1988 Education Reform Act have become more and more 'summative' and less and less 'formative' over the years. In other words, a system which was intended, at least in part, to provide pupils, teachers and parents with information which would help pupils' learning in school has been replaced by one which governments and local education authorities use to monitor and to evaluate schools, and the education system as a whole, sometimes in a rather punitive way.

Nevertheless, a vast amount of data is generated by this system of national assessment, and various summaries of these data are published. The collection of so much educational data inevitably leads to the generation of educational statistics, leading in turn to various kinds of comparisons, and hence to inferences about the way in which teachers, schools and the educational system as a whole are functioning. The inevitability of the process does not, unfortunately, imply that all the inferences are necessarily sound ones. One reason for this is that measuring pupils' educational attainments and achievements is not as simple as measuring their heights and weights. Another is that schools are not like football teams: the schools with the best results are not necessarily the best schools, a point elaborated by Harvey Goldstein in Chapter 32.

In this chapter, I consider four kinds of comparisons, each of which links to public debates in education. These are:

- comparisons between different educational measures at a single point in time;

- comparisons of educational attainments over time, i.e. between different cohorts of pupils;
- comparisons of pupils' educational attainments at different ages; and
- international comparisons of educational attainments.

My main conclusion is that, rather than trying to answer the unanswerable, statisticians should try to persuade policy-makers and others to reformulate their questions for each of these four situations.

Comparing different measures at a single point in time

Here we are concerned with questions such as:

1. Comparing pupils' performance across different A-level subjects. In 1995, three times as many candidates got an A in A-level Physics compared with Business Studies. However, using a measure of 'relative ratings' described by FitzGibbon and Vincent (1994), Physics appears to be the most 'difficult' A-level. These issues of comparability surfaced in the review of qualifications for 16 to 19-year-olds (Dearing, 1996).
2. Comparing levels attained by pupils across the National Curriculum subjects on the ordered 10-point scale introduced by the Education Reform Act in 1988. In the 1996 Key Stage 1 national assessments for 7-year-olds, 6 per cent scored above level 2 – the expected level – in Writing, whereas 15 per cent did so in Number.
3. Comparing levels across Attainment Targets within National Curriculum subjects such as Mathematics. In the 1995 Key Stage 1 national assessments, 20 per cent scored below level 2 and 15 per cent above level 2 in Number, whereas for Handling Data the corresponding percentages were 25 and 10.

If we were to use a biological analogy, then comparisons 1 and 2 are akin to comparing upper and lower limb strength, which, because we use our arms and legs for different activities, is not terribly useful. However, comparison 3 is more like comparing left- and right-eye vision, which can be valuable, especially when you want a prescription for glasses.

It might be argued, on the basis of statistics like those referred to in 1, that some A-levels are soft options – what we might call the 'media studies' syndrome. Turning to 2, some might argue that the data imply that not enough time is spent teaching writing, or too much time is spent on number, in the early years. However, before we leap to conclusions of this kind, there are a number of statistical issues which we must address when trying to interpret these kinds of findings.

When comparing A-levels, we need to think about the effects of sampling and self-selection, because different students choose different subjects and different mixes of subjects. Consider two students of apparently equal ability. One student gets a B in Physics while the other gets an A in English. This does not necessarily imply that the student who took Physics would necessarily have got an A in English, had he or she taken English rather than Physics. We also need to consider the effects of subject mixes across the different examination boards, because, for historical reasons, A-level curricula, and hence the examinations themselves, are not uniform across the country.

When comparing results for different parts of the mathematics curriculum, we need to consider developmental issues: is Handling Data, as construed by the architects of the National Curriculum, inherently more difficult for young children than Number? And for all three of the above situations, we must think about the quality of teaching: how it might vary across subjects, and what effects variability in quality might have on student outcomes. Indeed, one more interesting approach to understanding and interpreting these educational statistics would be to develop measures of teaching quality across the curriculum, which might help us to account for some of the observed differences.

As well as thinking about the technical issues linked to these questions, we also need to consider why these questions are seen to be important in the first place. There appears to be a desire to make all 16–19 qualifications comparable in order to make entry to higher education fairer. On the other hand, applicants' offers are determined by a mixture of criteria. For example, many universities do not recognize an A-level in General Studies. Trying to make A-level grades more equivalent *across subjects* than perhaps they are at present would not appear to be especially important. What is more important is to make grades equivalent *within subjects across examination boards*, especially in the light of the scandal surrounding high grades given to candidates, many from public schools, taking English with the Oxford and Cambridge Schools Examination Board (*Times Educational Supplement*, 1997). Another concern is linked to league tables of schools' performance: if A-level grades are not comparable, how can we rank institutions on their overall A-level performance? On the other hand, supposing that adjusting A level grades will create fairer league tables is rather like trying to handicap international athletes by asking them to run barefoot at a school sports day.

Comparing performance across time: between-cohort comparisons

Here we are interested in questions such as how and why the distribution of pupils' performance in public examinations changes across time. We know that 12 per cent of those children born in 1958 got at least two A-levels, and that this rose to 20 per cent of the 1978 birth cohort. Also, 11 per cent of candidates got a grade A in A-level in 1989, 16 per cent in 1995 (OFSTED, 1996d). A biological analogy here is looking at between-cohort differences in height.

Competing explanations have been put forward to account for these trends in performance. Some would argue that the trend reflects a genuine improvement in *educational* standards, whereas others argue for a decline in *marking* standards over time, sometimes called grade inflation, such that A-levels have become easier. (It is possible for there to be some truth in both of these explanations.) We should be aware that when the word 'standards' is used on its own in this context, it is often done in a rather confusing way, so that the phrase 'standards have fallen' sometimes means that performance levels have fallen and sometimes that the assessments have become easier.

The statistical issues to be addressed here include changes in the gender mix (an increasing proportion of girls are taking A-levels); gender dispersion (boys' results are more variable than girls'); changes in the labour market such as a decline in apprenticeships; changes in examination board mix; changes in subject mix; changes in institutional mix (variations in the proportions taking A-levels in state and independent schools, sixth-form colleges and further education colleges); changes in the form of examinations; and changes in the information available to teachers about syllabuses and what is expected of candidates.

Rather than being prepared to recognise that just as young people are getting taller, perhaps they are also getting cleverer, there is a widespread view that A-levels used to be really difficult and now 'nobody knows what a gerund is any longer'. A more useful approach is to ask whether the match between A-level curricula on the one hand, and higher education courses and employers' requirements on the other, could be better than it is (and, perhaps, was twenty years ago). Some progress was made in this direction as part of the Dearing review of qualifications for 16 to 19-year-olds. Perhaps a more fundamental issue is to what extent the courses leading to qualifications provide students with the tools to be able to participate fully in a democratic society, and how courses (and qualifications) should change both to reflect and to contribute to change in the future.

Comparing educational attainments at different ages

The questions here are to do with how well pupils of different ages are doing in national assessments, and whether we can draw any inferences from age differences. A biological analogy is the calculation of height norms at different ages.

We find, from various reports produced by the Schools Curriculum and Assessment Authority (SCAA, now QCA, the Qualifications and Curriculum Authority), that, for English in 1996, 78 per cent of pupils reached at least level 2 at the end of Key Stage 1, 57 per cent reached at least level 4 at the end of Key Stage 2, and 57 per cent reached at least level 5 at the end of Key Stage 3. These levels are the ones which 'average pupils' are expected to reach. Table 31.1 gives more detail (*see also* Chapter 34).

We might wish to infer from the data in Table 31.1 either that pupils' performance in English declines relative to some notional standard as they get older, or that pupils at the end of Key Stage 1 in 1996 were performing at a much higher level than the corresponding pupils were 4 and 7 years earlier. The first of these inferences seems unlikely, the second is implausible.

The important statistical issue here concerns the validity of the 10-point scale for national assessment. The data presented in Table 31.1 suggest that the scale

Table 31.1 Percentage of cohort at each level: English, 1996

Key stage	Level									Total
	0	1	2	3	4	5	6	7	8	
1[a]	3	18	48	30	0	n.a	n.a	n.a	n.a	99
2	0	1	6	30	45	12	0	n.a	n.a	94
3	0	0	2	10	23	31	18	7	1	92

Sources: Schools Curriculum and Assessment Authority: (a) *Standards at Key Stage 1, English and Mathematics*; (b) *Standards at Key Stage 2, English, Mathematics and Science*; (c) *Standards at Key Stage 3, English* (all published 1997)

[a] Refers to reading. (*See* Chapter 33 for a chapter concentrating on reading)

does not have the properties claimed for it. Indeed, the lack of research into the measurement properties of the scale is scandalous, and makes an extraordinary contrast with the emphasis placed on inadequately measuring school performance with league tables. It is also important to be aware of the weaknesses in the way in which data in SCAA reports have been presented. From the reports referred to above, we find that when we add up the percentages in Table 31.1 across the different levels, the totals are 94 for Key Stage 2 and 92 for Key Stage 3 rather than 100 as we would expect, and which we find (apart from what could be rounding error) for Key Stage 1. There are no explanations for why so many data are apparently missing, and what effects these missing data might have on the results. The issue is, however, an important one because so much is now being made about reaching specified targets, especially at the end of Key Stage 2 (Department for Education and Employment, 1997a).

A potentially more useful approach to studying changes with age is to focus on individual change, just as it can be more useful, when evaluating their health, to look at how children are growing, rather than at how tall they are at a particular age. In other words, we may find out more by taking a longitudinal or developmental perspective. This is especially so if we focus on relative change rather than on absolute change, as illustrated by Figure 31.1.

Here we plot attainment at age 11, say, against attainment at age 7 for individual boys and girls. We see that, for any level of attainment at age 7, girls are

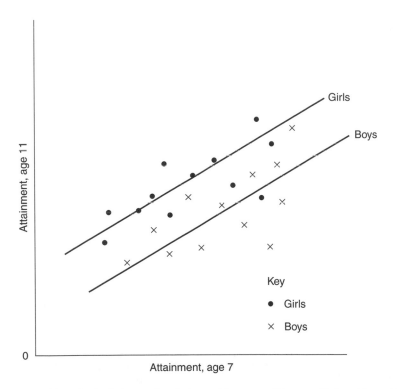

Figure 31.1 Hypothetical change in attainment from age 7 to age 11 for girls and boys

performing at a higher level, on average, at age 11. We could then use information like this on the way inequalities develop with age to devise intervention programmes with the aim of reducing or eliminating these kinds of inequalities, which can also be found for other variables such as social class and ethnic group. Unfortunately, education inequalities do not receive a high priority in the development of education policies. Moreover, the official statistics on education tell us very little about inequalities of this kind.

International comparisons

Here we are concerned with between-country differences in mathematics and science performance as revealed by, for example, the recent Third International Mathematics and Science Study research (Keys *et al.*, 1996). For example, the median score at age 13 for mathematics for Japan is 572 points whereas for Sweden it is 497. A biological analogy is that Swedes are, on average, taller than the Japanese.

The statistical issues here include the way in which the target population is defined in each country, and whether comparisons are vitiated by differences in response rates. It is also important to take account of variability in scores both within and between schools in any one country, because this could have a substantial effect on inferences about mean differences between countries. If variability between schools is ignored, as it usually is in studies of this kind, then mean differences between countries will appear to be estimated more precisely than is really the case. In other words, published sampling errors will be too small. Another important question is how well the test is able to take account of curriculum differences between countries, an issue sometimes known as the 'opportunity to learn'.

Turning to some of the inferences which have been drawn in England from this international study, we find a good deal of official hand-wringing about the fact that pupils in Singapore are apparently much better at mental arithmetic than English students are. Policy-makers and others then fall into the 'ecological fallacy' trap of supposing that teaching methods in Singapore must therefore be better than teaching methods in England. This is no more helpful than it is to suppose that because Swedes are taller, therefore the Swedish diet is better than the Japanese diet. We see in Figure 31.2 how this ecological fallacy can arise. We can plot mean attainment in, say, mental arithmetic, for each classroom in the study against the hours of whole-class teaching in mental arithmetic in that classroom. We see, in this hypothetical example, that attainment is higher in Singapore than it is in England, and also that there is more whole-class teaching in Singapore than in England. However, *within* both England and Singapore, there is no relation between the amount of whole-class teaching and attainment. Thus any differences in mean attainment between the two countries cannot be directly attributed to this particular aspect of teaching methods.

Government-funded international studies of this kind can be used in a negative way against teachers, or to buttress an educational system in need of reform, and hence to support a political agenda with roots elsewhere. A more positive approach would be to recognise that investigations such as the Third International Mathematics and Science Study are not an end in themselves but a stimulus to

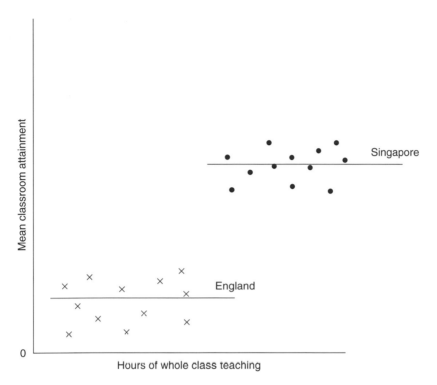

Figure 31.2 Illustrating the ecological fallacy

further research. For example, can we establish whether there are any causal links between differences in teaching methods, classroom organisation and so on, and differences in attainment and progress, and, if so, how can we make use of this information to improve learning?

Conclusions

Some of the data set out earlier lead to questions and puzzles, and statisticians have a lot to contribute by way of unravelling these puzzles. However, rather than engaging in debates about education, the terms of which often seem to be characterised by a longing for the past combined with a wish to denigrate teachers at every opportunity, perhaps the most important contribution statisticians can make is to suggest that the way in which questions are often posed in public debates is not necessarily the best way, and that there are other, potentially more useful, questions which should be asked. Moreover, answers to these questions might turn out to be easier to obtain, and could be used to improve learning for all pupils.

Other publications develop some of the issues raised in this chapter. For comparisons at a particular point in time, especially comparing A-levels, *see* critiques by Goldstein and Cresswell (1996) and Newton (1997). Inter-cohort comparisons of reading standards have been a controversial area for many years, leading to 'educational panics' based on rather flimsy evidence. Goldstein (1983) reviews some of the methodological issues. The problems associated with setting targets for attainment at different ages are raised by Plewis and Goldstein (1997). The logic of making the kinds of comparisons illustrated by Figure 31.1 are set out by Plewis (1985). The importance of using a multilevel approach for the analysis of all kinds of educational data, and the link between multilevel modelling and the ecological fallacy (Figure 31.2), are set out by Plewis (1997). The 'opportunity to learn' idea is described by Plewis and Veltman (1996).

32

Performance Indicators in Education

Harvey Goldstein

Knowing about uncertainty in performance indicators prevents their misuse

The major impetus behind the extensive collection and publication of performance indicators for schools in England and Wales came with the Education Reform Act of 1988. Prior to that it had been possible to collect public examination data on schools, but although these were sometimes published locally, such data were not the subject of systematic rankings or 'league tables'. Since 1988 the collection and publication of average results for schools has become routine and extends all the way from test scores at the age of 7 to A-level examination results at the end of schooling.

The principal official support for the 1988 legislation was a report produced by a task force (Task Group on Assessment and Testing, 1987). This report sketched out a national framework for educational assessment and testing, and made particular recommendations on the publication of test scores and examination results. It recommended that aggregated (mean) scores for each school should be published, but only 'as part of a broader report by that school of its work as a whole' (para. 132). That broader report was to include aggregate data about the nature of the school catchment area, and other 'contextual' data the school considered relevant to interpreting its results. The TGAT report briefly discussed the issue of whether school results should be adjusted for factors outside its control and argued that it 'would be liable to lead to complacency if results were adjusted, and to misinterpretation if they were not' (para. 133). It went on to conclude that 'results should not be adjusted for socio-economic background but the [school] report should include a general statement . . . of the nature and possible size of socio-economic and other influences which could affect schools in the area'.

Two important issues arise immediately from these highly ambiguous, and somewhat contradictory, statements. First, they betray a considerable level of ignorance about the purpose and nature of 'adjustments', and I will return to this in the next section. Second, they neatly illustrate how a statistical issue, namely how to construct and interpret a statistical model explaining school performance, can be dealt with by the use of plausible, but fairly meaningless, language. In my final section I shall return to this issue of how politicians and others are able to appropriate aspects of mathematical and statistical terminology in order to avoid confronting the complexities of real data and the need to be clear about the caveats and limitations of such data. I should of course make it plain that such appropriations and consequent distortions are not necessarily deliberate and may often be done for what are perceived of as the best of motives; my concern is more with the fact that we inhabit a culture which tolerates such incoherence.

In 1992 the first national league tables of public examination results were compiled by the then Conservative government and published in the main national newspapers, with annual reports thereafter. This policy has now been extended to test scores at 7 and 11 years, and is a policy endorsed and extended by the Labour government elected in 1997. These tables, with a minimum of contextual information, have had a profound effect on schools (Ball and Gewirtz, 1996) which see themselves as competing for 'customers' in an educational marketplace. The Citizen's Charter (UK Government, 1991) formalized this by extending the TGAT recommendations to require schools and colleges to 'publish their annual public exam results in a common format' so as to provide 'easier comparison of results between schools'. The official justification was 'to help parents choose a school'. Interestingly, as a result of widespread concern, in 1995 the government officially endorsed the principle of contextualisation when it recommended the use of 'value added' comparisons among schools (Department for Education, 1995), and subsequently a report was produced by a government quango, the Schools Curriculum and Assessment Authority (1997), which sought to operationalise this. Nevertheless, this implicit admission of inadequacy has not affected the practice of publishing 'raw' or unadjusted comparisons. For reasons I will go into below, even if implemented, the use of value-added data would not resolve most of the underlying problems.

Performance indicators and public accountability

The basic political justification for the publication of league tables, whether in education, health or social services, is that the 'public' has a right to know something about the performance of publicly funded institutions. Such knowledge can then be used, for example by government itself, in deciding whether to take action against institutions perceived to be 'failing', or by parents, say, in choosing a school for their child. For information of this kind to be useful it clearly has to meet certain quality standards. It should be reliable enough to make useful distinctions among institutions and it should be valid in the sense that it really does reflect the qualities claimed for it, whether these are standards of educational delivery, health care or social service provision. In the remainder of this chapter I shall show that current performance indicators fail on both these counts: they are unreliable and they distort the underlying reality.

It is of some interest, given the shaky intellectual foundations of current performance indicators, to ask why they have been promoted with such vigour by successive governments. This is not the place for a detailed analysis, but a few observations may be useful. First, there has been a great centralising tendency of governments since the late 1970s. To some extent obscured by the free-market rhetoric of Thatcherism, there has been, nevertheless, especially in education, a transfer of power from local to central government and at the same time a transfer from professionals, i.e. teachers, to government or quango employees and also to other bodies such as governors. Performance indicators have served to control both schools and the teachers within them as an external yardstick which has forced adherence to a nationally imposed curriculum and testing regime.

Second, and closely related to centralisation, governments have often been suspicious of educational professionals as potentially 'subversive', users of unfamiliar language and with a perceived great influence over future citizens. The reaction has taken the form of attempting to simplify educational debates. Thus, the 1997 Labour administration has made much of 'standards', by which it means achievements on readily understandable tests and examinations. It appears to be relatively uninterested in the real complexities surrounding teaching and learning. The rhetoric of 'standards' has even extended to the setting of 'targets' for schools to achieve certain test scores several years in advance with little concern for whether this is really feasible (*see* Plewis and Goldstein, 1997, for a critique).

Whatever the reasons for the politicians' interest in performance indexes, it does seem fairly clear that these satisfy a deeply felt need and this therefore makes a proper debate about their status very important.

How should we compare schools?

In this section I want to take some examples to show how problematic this issue of school, or any other institutional, comparisons really are. I shall draw heavily on a technical review (Goldstein and Spiegelhalter, 1996) which sets out the issues in some detail, as well as other research which has been extensively replicated.

The principal argument against examination league tables is that the performance of a school is determined largely by the pre-existing achievements of the students when they enter it. Since schools differ markedly in this respect – for example, some schools are highly selective – it is impossible to judge the quality of the education *within a school* solely in terms of final 'outputs'. There are also, however, problems which apply to 'value-added' tables, and I will show how initial expectations that these could provide a more sensitive indicator of school performance have failed to materialise. Furthermore, attempts to adjust 'raw' results using average socio-economic background or even average intake scores of students are inadequate. Thus Woodhouse and Goldstein (1989) showed that attempts to do this resulted in highly unstable rankings, and that small and essentially arbitrary decisions about how to formulate the statistical models led to very different conclusions. Thus, the suggestion in the appendix to the 1997 education White Paper (Department for Education and Employment, 1997a) that an adequate adjustment can be made using the percentage of pupils in a school having free school meals, is invalid. Proper adjustments can only be made, at the very least, if *individual student* intake scores are taken into account.

Nevertheless, even if an adjustment can be made using individual student data, several difficulties remain. The first problem is that typically only a single figure is reported, such as the overall percentage of high GCSE grades. Yet schools may be 'differentially effective'. Thus, for example, two schools may perform equally well on average but one may have poor performance in mathematics and good performance in English and the other vice versa. Likewise, where value-added tables are concerned, some schools may exhibit relatively good performance for initially (on intake) poorly achieving students and produce relatively weak performance for initially highly achieving students, and vice versa for another school (Goldstein *et al.*, 1993). A second problem, with both 'raw' and value-added tables, is that the percentages or scores produced for each school typically have a large margin of error or 'uncertainty' associated with them. This problem is even more acute when individual subjects or departments within schools are the focus of interest, since the sometimes small numbers of students involved mean that very little can be said about any individual department's performance with reasonable accuracy. In the extreme case, for some A-level subjects there may be only two or three students involved, and any generalisation, even over a number of years, from such small numbers is extremely hazardous.

Figure 32.1 illustrates this general problem. It is taken from a survey of some 400 schools and colleges with A-level results where value-added scores are calculated by adjusting for the GCSE performance of the candidates (Goldstein and Thomas, 1996). Each bar corresponds to a value-added score for a school. The fewer the number of students in a school the longer is the bar and the more uncertainty is attached to the value-added score. The bars are ordered from left to right on the school's value-added score (the mid-point of the bar). The bar lengths are chosen in such a way that two schools can be judged as being reliably separated only if their bars do not overlap. Thus, for example, the bars from the extreme left-hand school and the extreme right-hand school do not overlap and we may conclude that there is a real underlying difference between their value-added scores. In this figure, however, for some three-quarters of all possible comparisons of pairs of institutions, it is not possible to make such a separation. In other words, finely graded value-added comparisons are of limited value since in most cases we will find no difference.

Another problem with all of these tables is that they refer inevitably to a cohort of students who began their education at those institutions many years earlier. Thus, for example, GCSE results published for a particular year refer to a cohort starting at their secondary schools some 5 years previously: given that schools can change markedly over time, there will be some uncertainty over the use of those results to predict the performance of future cohorts.

A further problem arises from recent research (Goldstein and Sammons, 1997a) which shows that the primary school attended by a child exerts an important influence on GCSE performance and that this should therefore be taken into account when producing value-added tables. Also, there are other factors, such as sex, ethnic origin and social class background, all of which are known to be associated with performance and progress throughout secondary schooling and which therefore will affect the interpretation of any rankings. Finally, there are several practical problems associated with producing any kind of performance tables based upon test or examination results, perhaps the most important being that during the course of a period of schooling, say from 11 to 16 years, many students will change schools. To ignore such students is likely to induce considerable biases into any comparisons, yet to include them properly would require enormous efforts at tracing them and recording their results.

Taking all these caveats together we can see that attempts to rank educational institutions are fraught with difficulty. Even with extensive and good-quality information, there are some inherent limitations which preclude the use of rankings other than as initial *screening instruments* to isolate possibly high- or low-achieving institutions or departments which can then be further investigated;

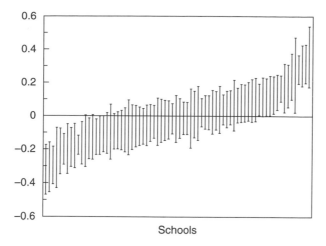

Figure 32.1   A-level scores: pairwise (95 per cent) uncertainty levels for a random sample of schools and colleges for students in the middle (50 per cent) GCSE score band. The data refer to the 50 per cent group of students who have an average GCSE overall subject score in the middle range. Schools are to be judged as statistically significantly different at the 5 per cent level only if they have intervals which do not overlap

bearing in mind that the information is historical. These caveats apply particularly to the public presentation of comparative tables. For internal 'school improvement' purposes, however, schools can often extract useful information from knowing where they are ranked among schools with similar characteristics, especially where detailed information about differential effectiveness is available. For such information to be useful it is essential that it be presented with all the necessary caveats and that it remain confidential to the school so that it can be properly evaluated in context. Several local education authorities are now beginning to develop such schemes. In Hampshire, for example, over 100 primary schools are taking part in a system where value-added results for different curriculum subjects and for different kinds of pupils are provided for each school in terms of progress between school entry and age 7 and between age 7 and age 11. The schools use these to help themselves to understand their strengths and weaknesses in comparison with other schools in the local authority, bearing in mind the inherent limitations of such analyses. (Chapter 35 examines other developments in Hampshire's education.)

A detailed discussion of the ethics of performance indicators and some suggested guidelines is given by Goldstein and Myers (1996), who also discuss the issue of how *particular* measures, such as examination results, have come to assume a dominant role in evaluating the performance of schools. They point out in particular that if comparisons among schools are to be attempted, it is very important to provide users with careful descriptions of all the limitations.

Social manipulation of statistical information

I have already referred to the fact that statistical information can become ambiguous and incoherent when its terminology is used for political or similar purposes.

In fact, this is a much more serious problem than that of mere abuse for particular purposes. It symptomises a cultural attitude towards quantitative information which informs discussion of social issues at all levels.

Statistical analysis has two key components: first, the modelling or summarisation of a set of data in terms of a small number of 'parameters', for example an average examination score; and second, a statement about the precision of the summary measures. Thus, for example, it is a common practice when reporting some opinion poll results to quote a percentage in favour of a course of action, plus or minus a 'margin of error' due to sampling fluctuations. In the previous section this error was expressed in terms of confidence intervals. We saw how the use of such intervals prevented any precise comparisons among institutions. Apart from their use as crude screening devices or as additional pieces of information for use by schools for improvement purposes, league tables, of whatever kind, have severe limitations.

Absorbing uncertainty when making judgements appears to create severe problems and is a major difficulty in conveying statistical results. This uncertainty may arise from sampling variation, as I have already discussed, or it may be due to difficulties in finding reliable measurements, or to lack of response from certain pupils in schools, etc. All of this is familiar territory to experienced statisticians, and forms a part of most research reports. Yet all too often such caveats are ignored. It is as if there is an assumption that numerical results *must* accurately reflect reality. The common view of mathematics and mathematically based science is that it deals only in those things which have accurate numerical representations. It is particularly unfortunate that this misunderstanding often accords with the common demand from politicians for justification of a position on the basis of 'hard facts', which can tolerate no uncertainty – and this makes any change much more difficult to envisage. What seems to be required is a cultural shift in attitude along with a positive attempt to incorporate a fuller understanding of statistical information and uncertainty into education at all levels.

It would not be too difficult to set out guidelines for the reporting of such things as performance indicators (Goldstein and Myers, 1996), and perhaps the single most important innovation would be the *mandatory* inclusion of uncertainty estimates. There are very few instances where this would not be possible. It can be readily justified in terms of freedom of information, on the grounds that such uncertainty estimates are a key component of any publication and that it is misleading to withhold them. In a democratic society there would seem to be little excuse for refusing to provide citizens with the caveats which are implicitly attached to public indicators of performance. A requirement to make such information prominently available could be incorporated into any new freedom of information legislation, and this would constitute a very important step towards mitigating some of the more harmful effects of performance indicator publication that we have seen.

33

Can Trends in Reading Standards Be Measured?

Pauline Davis

The use of the National Curriculum's Standard Assessment Tasks (SATs) for assessing trends in reading standards

School Standards Minister Estelle Morris today welcomed improved 1997 National Curriculum test results for 7, 11 and 14 year olds in English . . . . These results show our continuing highlighting of the importance of literacy . . . – and primary school homework – is clearly having a helpful effect in the classroom.

(Department for Education and Employment, 1997d)

How can we know if such a claim is true? Are children's literacy standards improving or declining? Are children from all sections of British society exhibiting a common trend, or are some groups of children showing an improvement in standards while others are showing a decline? Why are the National Curriculum tests assessed as they are? Why do the test results take the form they take and why did they come into existence when they did? This chapter considers reading standards in conjunction with the Standard Assessment Tasks (SATs) of the National Curriculum of England and Wales in order to explore some possible answers to these questions, and in particular to consider the detection of variations in national reading standards over time (an issue raised for study in Chapter 30).

National Curriculum

The reasons for the introduction of the National Curriculum and SATs can be usefully described in a historical context, but a detailed review is beyond the scope of this chapter (*see* Chitty, 1989). However, it was in the 1980s that the supporters of policies typified by those of the Thatcher government commonly became known as the 'New Right'. The New Right rejected the social democratic-type policies, which had previously been accepted by Conservative and Labour governments alike since the introduction of the modern welfare state. 'New Right' policies were summarised by Gamble (1988), as 'free economy/strong state'. This bedding down of neo-liberalism, advocating freedom of choice, the individual, minimal government intervention and *laissez-faire* economics, seems to lie somewhat uneasily with neo-conservatism that supports social control, the nation and the strong state (*see* Chitty, 1989): both can be seen clearly in the 1988 Education Reform Act. This important Act, perhaps most notably, introduced local management of schools (LMS), whereby schools became responsible for their own budgets, setting up what many have described as a quasi-market in education (*see* Ball, 1993; Ranson, 1993; Stokes, 1995); the National Curriculum; and open enrolment, whereby parents have a right to place their child in the school of their choice (unless, of course, the school is full). Chitty (1989, p. 218) writes:

> Despite the marked reservations of the neo-liberals, there is a very real sense in which the National Curriculum is not necessarily incompatible with devolution of power or with market principles . . . the Curriculum does, after all, act as justification for a massive programme of national testing at 7, 11, 14 and 16 which will in turn, provide evidence to parents for the desirability or otherwise of individual schools.

Following in the footsteps of Margaret Thatcher, the governments of both John Major and Tony Blair have endorsed the National Curriculum as an integral part of their respective educational policies. One of its purposes is to provide a vehicle through which to monitor standards in education.

Assessment of reading standards is a highly sensitive political issue. The mere hint of a decline is taken as an invitation to criticise teachers, education policies and governments. Particularly in the age of 'education, education, education', when the education system is becoming increasingly intertwined with perceived economic needs, there is a need for governments and policy-makers to demonstrate that their policies are working by showing a rise in educational standards. Before the implementation of the National Curriculum, reading standards in schools tended to be assessed by using a variety of tests, some of the most commonly used being Schonell, Young's Group Reading Test, and the Salford Sentence Reading Test (Shorrocks, 1993, p. 60). Usually, these tests were conducted by the class teacher as a means of monitoring a child's progress, or for surveys to compare children's reading abilities. These tests are still widely used by teachers as a supplement to the National Curriculum.

The introduction of SATs was extremely turbulent, as witnessed by the boycott of the Key Stage 3 tests in 1993, and this shows the importance of having willing and committed data collectors for gathering social statistics. The imposition of the National Curriculum was not smooth, and the early SAT results were too unreliable and incomplete to use.

The National Curriculum divides children's education into four Key Stages (KSs) depending on their school year, KS1 for years 1 and 2, KS2 for years 3 to 6, KS3 for years 7 to 9, and KS4 for years 10 and 11. Figure 33.1 illustrates the structure of the National Curriculum for Key Stage 1. Within each Key Stage, with the exception of some pupils with special educational needs (*see* Chapter 35), pupils are expected to work within a range of recordable levels of achievement covering specified subject areas, such as English or Mathematics, which are broken down into Attainment Targets (ATs). For instance, reading is covered by AT2 for English and at KS1 it is recorded at four levels: working towards level 1 (WT 1), level 1, level 2 and level 3, whereas at KS4 the levels run from 3 to 10 inclusive. The levels are specified by Statements of Attainment. For instance, to obtain level 1 in reading,

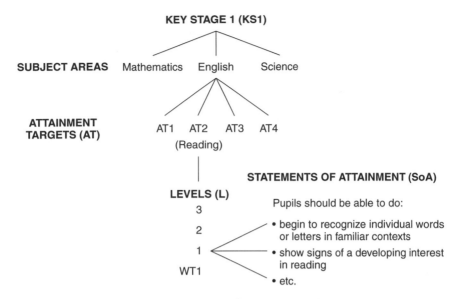

Figure 33.1 Simplified structure of National Curriculum Key Stage 1

pupils should be able to:

 a) recognise that print is used to carry meaning, in books and in other forms in the everyday world.
 b) begin to recognise individual words or letters in familiar contexts.
 c) show signs of developing an interest in reading.
 d) talk in simple terms about the content of stories, or information in non-fiction books.

(Department of Education and Science, 1989b)

As well as being continuously monitored throughout a Key Stage, each child's progress is both teacher-assessed and assessed using SATs at the end of the Key Stage, in order to determine the level of attainment. This type of test is referred to as criterion-referenced, as to achieve any given AT level all the specified criteria must be met. Criterion-referenced tests are qualitatively different from normative tests where the assessments are made with regard to the other children in the cohort (e.g. Schonell's test, or O and A levels). This is a distinction that will be elaborated on, in order to reflect whether or not SATs provide a suitable tool with which to measure trends in reading standards at a national level.

Tests

All test results can be used to make either a norm-referenced interpretation or a criterion-referenced interpretation. The term 'norm' relates to 'normal' or, in other words, typical or average. Norm-referenced interpretation involves comparing a pupil's score with the average score of some relevant group of people. A criterion-referenced interpretation is made when we compare a pupil's score with scores, each of which represent a distinct level of performance in a specific content area. Meaning is obtained by describing what the pupil can do in various graduations in an absolute sense. A test that lends itself to measuring content tends to be called criterion-referenced, whereas a test that lends itself to comparison between pupils tends to be called norm-referenced. It is important to realise that all tests can be used for either or both types of interpretation (*see* Ebel and Frisbe, 1991, p. 34).

A 'good' test needs to be reliable (we need to know that were the test taken again, we would draw the same conclusions), valid (the test measures what it is supposed to measure) and have minimal measurement error. A test is based on a model of the phenomena concerned and so there will always be some degree of error. Whether or not a test meets these conditions depends very much on its purpose. Much literature examines the reliability and validity of SATs in measuring individual pupil achievement, but this will not be dwelt on in this chapter, as the intention is to consider the detection of trends in national reading standards, rather than whether or not the structure of the National Curriculum provides the best framework in which to learn and teach reading.

Obtaining a valid analysis of test results over time is very difficult. One problem is that the same test cannot be given twice, for obvious reasons. Issuing a different test immediately begs the question of whether or not the tests are in fact comparable. Would the same child meet exactly the same criteria in both tests? This becomes more problematic as the number of tests to be compared increases. Related to this is the notion of 'teaching to the test'. As the years pass, teachers become finely tuned to the specific requirements of the test, and pupils become used to the test format, and so there must be some doubt as to whether improvement in test performance reflects a 'real' improvement of the construct being measured or just an improvement in ability 'to do the test'. This can be seen with IQ tests, where a person's IQ increases as he or she takes more tests (Ebel and Frisbe, 1991). As with other tests, this is likely to be true of SATs. In addition, SATs link the notion of an average child to official 'average' attainment targets. This may well produce a tendency for teachers to identify their 'average' pupils with the official opinion. This is supported by Thomas and Davis (1997), who, when investigating reading standards in primary schools, found that teachers perceive reading relative to the average standard of the other children in the class. In some cases, children perceived by their teachers as having reading difficulties in one school (in an area of high socio-economic status) were reading better than the best readers in certain other schools (in areas of low socio-economic status). Thus, because children, especially children in primary school, tend to attend local schools along with other children from their neighbourhood, comparisons between children based on indicators of socio-economic status may underestimate the differences between socio-economic groups.

SATs assessed

The main strength of SATs is, perhaps, their formative purpose: teaching is linked by official programmes of study, via ATs with the teacher assessments and SATs, and provides guidance as to which objectives still need to be covered. This is a common feature of criterion-referenced tests and goes a long way towards explaining their popularity with planners in recent years. However, when they are used in order to monitor reading standards, the use of SATs becomes more problematic. Detection of changes over time requires a 'norm-referenced' interpretation. To be effective in detecting small changes the test needs to be measured by a long scale (that is, it must have a wide range of possible scores).

SATs have a very crude measurement scale, and particularly so at KS1, where there are only four possible levels to choose from: working towards level 1, level 1, level 2 and level 3. Such a scale lends itself to criterion-referencing, but it is inadequate for detecting differences over time as it is not sensitive enough to detect subtleties. For instance, in reading, the jump between a level 2 reader and a level 3 reader is particularly large, and so the range of standards of a level 2 reader is correspondingly large. Pumfrey and Elliott (1991) have shown that when SATs are used to measure the spelling standards of 7- and 8-year-olds, their performances were virtually the same. Pumfrey and Elliott go on to say, 'it is difficult to believe that this approach to assessment provides educationally valuable information since it is known from many other sources that children's spelling ability improves markedly between these ages' (p. 79). This lack of sensitivity in-built within SATs is precisely why the 'norm-referenced' tests remain so valuable when assessing reading standards. These tests often give a measure of children's reading ages. This is especially useful when working with beginner readers, as a difference in reading ability with young children of just a few months can be highly significant.

Another problem with using SAT measurement scales to detect changes over time is that the data are ordinal. That is, levels 1 to 10 reflect a sliding scale of achievement, but, for example, the difference in achievement between a level 4 and a level 5 reader is not the same as for a level 1 and a level 2 reader. Norm-referenced tests use an 'interval' rather than an 'ordinal' measurement scale, a form of construction that allows far more powerful analyses to be performed. For instance, for ordinal data the 'median' average is the preferred measure of location, whereas for interval data the 'mean' average is preferred. Statistically this is an important consideration as the level of data determines the types of analyses that can legitimately be performed, and hence the information that can be obtained. This viewpoint is supported by Pumfrey and Elliott (1991), who state, in the context of measuring changes in reading standards over time, 'To date, SATs are technically incapable of providing this quality of information.'

Norm-referenced tests carry with them their own problems. Norm-referenced reading tests tend to measure limited aspects of reading, for example being able to say particular words rather than being able to understand their meanings. In addition, they may have an in-built tendency to underestimate improvement as the language tested becomes out of date. However, these problems can be reduced by careful test construction and, in spite of the problems associated with them, norm-referenced tests have the potential to provide reliable and useful data.

The National Curriculum statistics have limited use for those intent on gaining an understanding of trends in reading standards, although this is one of their purposes. At Key Stage 1, the English results are broken down into the attainment targets, so that for Reading, percentages of children attaining level 2 or above are provided. This is not the case at any other Key Stage and so there are no official national statistics for reading except at Key Stage 1. Throughout the Key Stages, percentages are given separately for girls and boys, but statistics are not provided for other categories of interest, such as social class or ethnicity. Such statistics are

Table 33.1   Overview of 1996/97 SAT results in English, for girls and boys: percentage obtaining government targets

	Key Stage 1		Key Stage 2		Key Stage 3	
	1997	*1996*	*1997*	*1996*	*1997*	*1996*
English	—	—	63	58	56	57
Speaking and listening	—	—				
Reading	80	78				
Writing	80	79				
Spelling	62	—				

Source: *DFEE News 292/97*

not available despite, for example, the body of evidence linking social class with educational achievement. For instance, Sammons (1995, p. 123) shows that 'social class remains a very important prediction of later academic achievement and ... the gap between non-manual and other social class groups increases steadily throughout the school career'. The failure to include social difference groupings in official statistics enables 'measuring reading standards' to be reduced to a technical, rather than a social, problem; raising standards becomes a case of obtaining higher figures in official statistics, and in doing so effectively distances the policy-makers from the everyday experiences of teachers, parents and children.

Trends in standards

Table 33.1 refers to the figures associated with the minister's statement quoted at the beginning of this chapter, in which improved 1997 National Curriculum test results are welcomed. An overall figure is not available for the English SATs because 'Speaking and Listening' is only teacher-assessed. The 1997 figure for 'Reading' at Key Stage 1 is 80 per cent, compared with 78 per cent in 1996. For Key Stage 3 English, 56 per cent of pupils achieved government targets in 1997 compared with 57 per cent in 1996. The minister's claim is based on these figures, and the Key Stage 3 statistic, for 14-year-olds, is lower in 1997 than in 1996. Even if a comparison between 1996 and 1997 figures were valid, using a change of 1 or 2 per cent to claim that policies are working is highly suspect, as all tests, however good, are subject to a certain degree of measurement error (*see* Chapter 32).

Assessing reading standards is controversial for reasons outlined above. Several studies have shown that children's reading abilities have changed little since 1945. On the other hand, the Secondary Heads Association argues that there has been a noticeable deterioration since 1992 (Budge, 1997). This view is supported by Davies and Brember (1997), who found that although the reading scores remained fairly static between the years 1989 and 1995, there was a marked deterioration at age 11:

> The number of poor readers increased substantially while the proportion of very good readers dropped. The percentage of 11-year-olds who scored less than 85 in the Primary Reading Test rose from 10% in 1989 to a high of 18%

in 1994. At the other end of the ability spectrum, the performance of children scoring more than 115 fell from 22% in 1989 to only 7% in 1994.

This study used the Primary Reading Test (a norm-referenced test). The limitations of educational research mean that direct comparison of figures from separate studies is rarely appropriate. Studies tend to use area-specific samples, and differences between local education authorities tend to be large. Indeed, the number of studies concerned with assessing reading standards is small. A national survey which collects interval level data is necessary if we are to have confidence in the figures we are provided.

The main problem with using SATs to monitor change is that SATs utilise ordinal-level data of limited range. Now that the National Curriculum has taken root, and national annual tests are routine, given the will it would be relatively inexpensive and easy to collect the information necessary to monitor reading standards effectively. If existing norm-referenced tests were not deemed acceptable, a new test could be designed and standardised which would yield the quality of information needed. Then trends in reading standards could be measured.

34

Inspecting the Inspection System

Nicola Brimblecombe

OFSTED inspection statistics give a misleading picture of schools in this country

Introduction: the philosophy and politics of inspection

Inspection in its present form, under the auspices of the Office for Standards in Education (OFSTED), was implemented by the 1992 Education (Schools) Act and added to in subsequent Acts. Prior to that, inspection was carried out on a more or less random basis by Her Majesty's Inspectors (HMIs) and by local education authority (LEA) adviser/inspectors. The 1992 Act set up a system whereby all schools would be subject to 'independent inspection' (OFSTED, 1993) once every 4 years against pre-determined criteria. The results of those inspections would be made public and the school would be required to draw up an action plan, within 40 days of the inspection report, to address the 'key issues for action' raised in the report. This represented a huge change in inspection policy and, consequently, in inspection practice for schools.

The present inspection system contributed to strengthening the market form within the education system, a tenet of Conservative – and, in particular Thatcherite – policy on education. This represented an ideological shift from the idea of education providing equity to that of it providing excellence – a change from postwar thinking on, and concerns for, education. This is evident in the language of OFSTED inspection. OFSTED's aim is 'improvement through inspection' (OFSTED, 1993, p. 5). Improvement is defined as the way in which schools 'raise standards; enhance quality; increase efficiency' (OFSTED, 1994, p. 6).

The role of parents, in particular parental choice, is one of the ways in which OFSTED intends to achieve its aims and another of the ways in which the market has been brought into education. As Ball (1994) puts it, in terms of market relations, the school has become the producer, the parents the consumer and the children and their performances the commodities. Parents as consumers make choices about schools and act accordingly. One requirement of the market and of consumer choice is information. OFSTED inspection reports, sent to parents and available to the public, are one such source of information; league tables are another. The theory is that, on the basis of the information available, parents will make informed choices between 'good' and 'bad' schools. They will then withdraw their child from, or not send their children to, the bad or ineffective schools, which will then be forced to improve or close down, whereas the good or effective schools will thrive. Thus schools will improve and resources will be allocated to schools not by bureaucratic decision-making, but by the market-led choices of parents. Inspection's role in this is to supply not only more information but also standardised information so that better comparisons can be made. Improvement can also be brought about, theoretically, via two other mechanisms: by identifying the strengths and weaknesses of individual schools and requiring them to draw up and implement an action plan based on those strengths and weaknesses, and by building up a picture of schools overall in England and Wales and basing advice, exemplars of good practice and policy on that picture.

That, then, is the stated theory underlying OFSTED inspection. What, however, is the reality, and, in particular, what is the reality of the ways in which the statistics underlying this system are generated, interpreted and used or abused?

Constructing the statistics

The claims made for the results of OFSTED inspection and their potential effects are far-reaching. However, there are a number of questionable assumptions about the inspection data on which these claims are based.

Context

Inspection makes little consideration of the context, in particular the socio-economic context, of schools when making judgements and comparisons. Although some details, such as the proportion of students receiving free school meals and the number of children for whom English is not their first language, are included in the introduction to the inspection report, those factors are not taken into account when assessing the workings of the school and the achievements of its teachers and students. However, using those factors as a proxy for ability or prior level of learning, whether mentioning them obliquely in the introduction or using them when considering the teaching and learning, itself has problems. Many commentators (e.g. FitzGibbon and Stephenson, 1996) feel that, if we are to continue with this form of inspection, it should at least take into account prior achievement of the students, the so-called value-added approach.

The implications of this lack of consideration of context are many, and particularly impact on those schools that are already disadvantaged in terms of resources and that have students from deprived backgrounds. It is no coincidence that they are the ones that often come out poorly, as research has repeatedly shown that material disadvantage is associated with educational disadvantage (e.g. Smith and Noble, 1995). If inspection's contribution to the realisation of market forces in education works as is intended, then students, and therefore resources, will be taken away from those schools; the situation will thus worsen for the school, thus disadvantaging the students in it. Although schools requiring 'special measures' (that is, those that are considered failing or borderline failing) are given extra funding, schools in the long term do not tend to recover from a bad report and will lose students and therefore resources, as schools are funded by the number of students. In addition, many schools receive 'bad' reports but do not require special measures. This will be compounded by the fact that the better-off parents in that school have more means to remove their children (for example, they are more able to take their children to a school other than the local one because of greater access to a car or because they can afford alternatives to state education). This process is not limited to the potential effects of inspection, but applies to all aspects of the marketisation of education and the role of parents as consumers within that system. For example, Adler and Raab (1988, in Ball, 1994) found that in Scotland most of the schools that lost numbers as a result of parental choice were in the least prosperous housing areas; those that gained were in affluent areas. Ball (1994), among others, argues that choice and the market advantages the middle classes and disadvantages the working classes.

Validity

A number of researchers have queried the accuracy of the picture of the school that inspectors are seeing (Wilcox and Gray, 1996; Wragg and Brighouse, 1995; OFSTIN, 1996; FitzGibbon and Stephenson, 1996; Brimblecombe *et al.*, 1996). In particular, inspection is a very stressful experience and has far-reaching implications for the school and the teachers within it. Dramatic newspaper headlines about failing teachers and schools do nothing to reduce this pressure. Neither does the format of the inspection, which allows inspectors to come into a classroom without warning at any point in the lesson. Hence teachers' and students' behaviour and, in many cases, the lesson itself is, unsurprisingly, not as usual. This obviously alters the representativeness of what the inspectors observe in the classroom. Ian Plewis and Harvey Goldstein also examine the validity of comparing schools in Chapters 31 and 32 respectively.

Judgements made about teaching and learning in a school are based on inspectors' subjective, qualitative evaluations which are then given a number, turned into percentages and averages and presented as 'objective' statistics. This spurious quantifying of qualitative data is what underlies many of the findings which are outlined below, such as '80% of pupils had reading ages below their chronological age' (OFSTED, 1996a). Like many people filling in questionnaires, OFSTED inspectors have a marked aversion to ticking boxes at the far ends of the scale, with the result that the majority of grades are clustered around the middle. This too skews the findings.

There are also difficulties with agreement between inspectors' judgements. Inspectors come from a variety of backgrounds and have different foci, expectations of what constitutes 'quality' education, different politics and different philosophies. While this is to some extent standardised by the framework (for the inspection of schools), what the observer observes is very much affected by his or her perceptual framework. This causes difficulties with reaching 'corporate agreement' about, say, teaching at Key Stage 3, which is exacerbated by the fact that many of the inspection teams have not met before the first day of the inspection. In addition, if the verdict on a lesson is dependent on who observes it, this undermines the reliability of the judgements made.

Interpretation

There are a number of issues related to how far we can trust the inspection data themselves. More crucially, however, is how those data are interpreted and how they are used. OFSTED is an 'independent government department . . . independent from the Department for Education' (OFSTED, 1993). However, its Chief Inspector (Chief Her Majesty's Inspector; CHMI) Chris Woodhead has revealed himself on numerous occasions to have had a pro-Conservative, right-wing bias (for example, in his publication for Politeia, a right-wing think tank: Woodhead, 1995) which is reflected in his, and OFSTED's, interpretation of the statistics. In addition, it is difficult to see how an organisation set up by the Conservative government of 1979–97 as part of its plans for the marketisation of education and based on its philosophy for how children should be taught, can be independent. With the best will in the world (and even regardless of the politics of the head of the organisation), it is constrained by its own structure and framework. Research shows that in fact inspectors do subscribe to both the inspection framework and the Conservative government's 1980s and 1990s market ideology (Jeffreys and Woods, 1998). A number of examples are used below to indicate problems with the interpretation of the statistics and the implications of those interpretations.

Primary-school teaching

In 1996, the annual report of the CHMI Chris Woodhead stated that 'it is evident that overall standards of pupil achievement need to be raised in about half of primary [schools]' (OFSTED, 1996b). This 'finding' was arrived at by a particular interpretation of the inspectors' judgements. Inspectors rate various aspects of a school on a range from 1 to 7 with 4 used for the neutral rating: that neither strengths nor weaknesses were dominant.

In 1996, unlike in previous years, the CHMI denoted this middle grade as a negative. Had the CHMI counted it as neutral, the result would have been that around 10 per cent, and not 50 per cent, of primary schools were in need of improvement. On the basis of this annual report, the Secretary of State for Education announced the need for a return to basics in primary education and in particular in primary teacher training, a view on education favoured not only by the Conservative government in power at that time but also by the CHMI himself.

Shortly after this, an unpublished OFSTED report praising primary teacher training and saying that primary trainee teachers do not rely on outdated methods was leaked. The leaked draft said that the quality of the training to teach English was 'good or very good in over half of the 44 courses inspected' (*Times Educational Supplement*, 5 July 1996). However, the released report does not mention this finding, despite the fact that it is evident from the statistical table, nor several other findings mentioned in the leaked report. The released report was made available via a press release only, rather than as a press release and published report, and was tagged onto an announcement that primary initial teacher training was to be inspected again. The new inspections would focus on 'the student's ability to teach reading and number-work' (OFSTED, 1996c). This is despite the fact that the original leaked report does include statistics on reading and number work.

Thus OFSTED has the power to interpret and report statistics in a certain way, and it has a wide audience for those interpretations and reports. OFSTED is one of the few organisations producing national statistics about schools at a time when there is a high demand for stories about education from the media. When that body is, in theory, set apart from the government or any other political organisation, so giving the impression of being independent, its output has even greater impact. While the credibility and presumed independence of OFSTED may have been eroded to some extent in the eyes of the public, newspaper headlines that one in five 7-year-olds cannot read have a major impact on the readers and serve to justify political decisions in education.

Setting the criteria

Even more powerful is the ability to make the judgements in the first place. Those who decide what we mean by quality of teaching and learning, who set the framework under which schools are assessed, decide the future of education. The individual inspectors then have to make judgements within that framework. Hence when we are considering whether, for example, quality of teaching is related to level of resources in schools, it is OFSTED's definition of quality of teaching which is used to find the answer to that question. As Silver (1994) argues, the question of who makes the judgements on effective schools is an important one, as judgements are conditioned by the source of their definition and the power to make them.

Class size

OFSTED's report on class size and the quality of education published in 1995 entered this arena as a supposedly objective report on the subject: 'Analysis of inspection data differs from research into class size in that it is based on disinterested judgements by inspectors for whom the contentious issue of class size was not uppermost in their minds' (OFSTED, 1995, p. 16). That the issue of class size was not 'uppermost in their minds' is one of the bases on which the research was described by Day (1996) as 'unreliable' and 'methodologically suspect'.

   Whether the judgement of the inspectors was disinterested might be debatable; however, the main issue is that of the framework under which they make those judgements. As Day says:

> Its uncritical use of its own inspection results as research data means that it conducts the debate entirely within its own 'official' definition of what constitutes quality teaching and learning. It selects from the available information to support a narrow, predetermined 'official' definition of quality.
>
> (Day, 1996)

In addition, it presents that definition of quality as neutral and devoid of political and philosophical constructs. OFSTED found 'no simple link between the size of a class and the quality of teaching and learning within it' although 'small class sizes are of benefit in the early years of primary education . . . [and] pupils with SEN [special educational needs] generally make greater progress when they are taught in small classes'. Despite contradictory contemporaneous research from both the UK and the USA (Day, 1996; Bennett, 1996) it was the OFSTED research that received the highest-profile publicity. Its findings, in particular those about the early years of primary school, were used to support the interests of the Conservative government then in power, in this case not to meet the extra costs of reducing class sizes in state schools across the age range.

However, OFSTED's main findings are not supported by, or at least not empha-sized by, the data in the report itself. If we look at secondary schools, for which OFSTED found no clear links between class size and quality of teaching and learning, as can be seen from Table 34.1, class size in secondary schools (as well as in primary schools) is related to attainment level of students on entry, which is itself related to the socio-economic circumstances of the students. The smaller class sizes are due to local funding formula, falling rolls for lower-achieving and therefore less popular schools, streaming of classes according to ability and addi-tional funding, for example Section 11 funding (funding for students with English as a second language; often used to provide extra teachers). This clouds the issue, as the effects of a smaller class may well be balanced out by the ability of the stu-dents within it; and similarly for larger classes. A student of higher ability and greater ability to supplement school learning with home learning may not suffer so much in a large class. The report shows this to be the case. Once the statistics are subdivided according to the ability of the students, no relationship is found between quality of teaching and learning and class size for higher-ability students; however, as Figure 34.1 shows, there is a relationship for those of lower ability. The main findings of the report, rather than taking this initial ability into account, blame teaching methods. Once again the finger is being pointed at individual teachers, schools and colleges rather than the socio-economic circumstances of the students.

Table 34.1   Class size[a] in comprehensive secondary schools, by ability grouping[b]

Year group	Number of classes	Class size				
		Median	Lower quartile	Upper quartile	Percentage 25 and under in the class	Percentage over 30 in the class
Key Stage 3						
Upper Ability	4 600	29	27	31	16%	27%
Middle Ability	3 526	27	24	29	37%	11%
Lower Ability	5 135	18	14	23	86%	1%
Mixed Ability	11 319	26	24	28	45%	5%
Key Stage 4						
Upper Ability	4 757	26	23	29	45%	11%
Middle Ability	3 339	24	20	27	67%	3%
Lower Ability	4 782	17	13	21	94%	0%
Mixed Ability	4 508	22	19	25	76%	3%

Source: OFSTED (1995)

[a] Lessons are restricted to English, mathematics, science, geography and history. These lessons invariably take place with the whole class present in the classroom

[b] Ability grouping based on inspector judgements. Ability of group defined relative to the number of pupils on the school roll

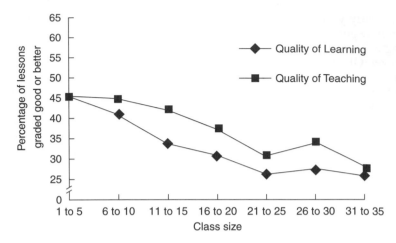

Figure 34.1 Percentage of lessons graded good or better by class size: secondary schools, lower ability
Source: OFSTED (1995, p. 36)

Implications

As we have seen, OFSTED was set up with the purpose of and responsibility for improving education in England and Wales and was legitimated by being given a veneer of independence. Invested with a great deal of power, it has the ability to determine what is perceived as quality education and what education should achieve. It makes the judgements based on its own criteria and reports or suppresses those findings in a way that is selective and that is unquestioning of the validity of the data and the basis on which the judgements were made.

OFSTED is one of the few sources of public information about schools. However, it is only the reports which are public. The data in the reports are aggregate data – aggregated from individual lessons in school reports and aggregated from schools in issue-based reports – in order to prevent identification of individuals. Thus the raw data on which the judgements are based are not available for public scrutiny or academic verification. OFSTED has the budget to produce such information, and, because of the limited information available and the desire for such information, OFSTED has had an easy route to the media and the public and has gained a great deal of authority, all of which have added to its power as information provider, information creator and information disseminator. This information has far-reaching consequences. Individual schools are required to draw up an action plan based on the findings and show how they will implement it. The only basis of challenge for the school is on factual grounds, and not on either interpretation grounds or on the criteria by which it was judged. If parents are put off by a report, the school may well lose funding and have to close. In some cases (so-called 'failing schools'), staff are replaced, and in Hackney, for example, an overhaul of the LEA took place following a report by OFSTED for the Labour government in 1997 which claimed that the education services in the LEA were 'in turmoil' (OFSTED, 1997). More widely, policy decisions and publications setting out what makes good practice in schools are based on OFSTED reports. Thus

OFSTED has the power to determine the course of education in England and Wales. This power was in the past used to support the prevailing Conservative government's ideology and policy on education and the continuing drive towards the marketisation of the education product. It appears that the Labour government that came into power in 1997 is unquestioning of this organisation and the judgements it makes, and appears to be supporting the continuation of inspection in its present form. However, we must question what OFSTED inspection is providing, for what purposes and for whose benefit, particularly given that, in an education system short of resources, OFSTED had in 1998 an annual budget of £150 million.

35

Special Statistics for Special Education

Cecilio Mar Molinero

Special education statistics measure willingness to provide rather than need

Class sizes, curriculum design and teaching methods are designed in such a way that children of average ability, in an environment conducive to learning, can acquire knowledge, maturity and skills at a pace which is compatible with their stage of development. Unfortunately, there are some children for whom normal provision is not sufficient. There are many reasons why this might be the case. Examples might include children whose first language is not English; physically disabled children; children whose intellectual ability lags behind what is expected for their age; and emotionally disturbed children who exhibit disruptive behaviour. In all these cases extra resources, such as special needs assistants (SNAs) and/or special equipment, are required in order to ensure that the child is given reasonable opportunity to develop and learn.

All the above examples are cases of special educational needs (SEN). There are many more ways in which SEN can arise. They all have in common a requirement for extra resources. Often the extra resources can be provided in the mainstream school. Occasionally, the needs of the child will be so great that provision needs to be made for the child to attend a 'special school' on a day or residential basis.

There has long been a debate between the desirability of educating a child with SEN with his or her peer group in the mainstream, perhaps with extra support, and educating the same child in a specialist school. This is the debate between integration and segregation. There are advantages and disadvantages to each alternative. For example, a child with moderate learning difficulties (MLD) may be ignored by the teacher or bullied by peers in the mainstream despite extra support. In this case more might be achieved in a special school.

The difference in cost betwen special schools and mainstream education is quite substantial. For example, in 1990 Hampshire Local Education Authority (LEA) in southern England estimated that a child in mainstream secondary education cost an average of £1265 per annum (Table 35.1). Providing extra support in mainstream added about £100 for a child with no statement of need (*see* next section) and about £1500 if the child had a statement. If a special unit in the mainstream was involved, the average extra cost was £3000. Providing a place in a day special school added £4500. The extra cost of a place in a residential

Table 35.1   Cost per year of secondary education

Type of provision	Cost per year
Mainstream provision	£1 265
Mainstream with SEN but no statement	£1 370
Mainstream with statement	£2 891
Unit in mainstream	£4 565
Special school	£5 900
Non-LEA provision	£13 000

Source: Hampshire County Council (1990)

school could be as high as £12 000. The message is clear: special education is very expensive.

The Warnock Report

Before the 1981 Education Act, 'handicapped children' used to be allocated to one of 11 categories and separate special provision was made for each type of handicap. It was felt that categorising children in this way was inappropriate, that it was wrong to remove them from their communities and from the social interaction with other children, that SEN should be tackled at mainstream school in the first instance. To investigate all these issues, a committee of inquiry was created under the chairmanship of Lady Warnock. The committee reported in 1978. Many of its recommendations found their way into the 1981 Education Act.

The Warnock Committee took the view that a child is in need of special education when he or she makes demands on the education system which are above what is expected from a 'normal' child. They argued that a special need should be addressed when it arises, speculated that about 20 per cent of all children have special needs at some time during their school life, and that resources should be provided in order to address such a need. They also accepted that some children will always make such demands on the mainstream school that it would be better to make separate provision for them. They observed that about 2 per cent of all school children attend special schools and presumed that there will always be a need for such schools.

In their search for a procedure to ensure the availability of resources, the committee proposed that some SEN children, those with the greatest needs, be given a 'statement of need'. This statement was to be agreed between all those concerned: parents, teachers, education officers, health services, school pyschologists, social services officers, and any other relevant authority. The statement was to specify the provision to be made, and there would be a legal obligation to provide the resources called for in it. These proposals were accepted, found their way to the 1981 Act and were later regulated in the Code of Practice for special education needs (Department for Education, 1994). In this way a statutory procedure was put in place which required a great deal of administrative effort and committed LEAs to open-ended expenditure.

LEAs were not given any extra resources to deal with the cost of their new statutory obligations. Funds had to be found from within existing budgets. It did not take long for LEAs to realise that Warnock's call for the needs to be addressed as soon as they were identified, and within the usual environment to which the child belongs, could be interpreted as a call for integration. This interpretation suggested that efforts should be made to reduce the number of children sent to special schools.

Social trends

It has long been known that the demand for special education is related to social background. Examples are Mar Molinero and Gard (1987), Appendix 2 in Audit Commission and Her Majesty's Inspectorate (1992), and Chapter 20 of this volume. It was initially thought that the SEN problem was in the process of resolving itself. The 1980s were times of declining school rolls. This was a reflection of the fall in fertility rates. Since the fall in the birth rate had been most pronounced in the working classes, the ones that contributed most to special education, it was expected that demand for special education would decline faster than rolls in the mainstream. It was, therefore, a surprise to discover that the number of children who required statements of special needs increased continuously, as can be seen from Figure 35.1. LEAs made an effort to integrate as many children with statements as possible in mainstream schools, and the percentage of all children who attend special schools has been decreasing, as seen in Figure 35.2. The total number of children has been increasing, and rolls in special schools refused to decline, as can be seen in Figure 35.3. Thus LEAs had to cope with increased SEN expenditure in special schools, and increased SEN expenditure in the mainstream. This was bad news since local authorities faced a tight budgetary squeeze.

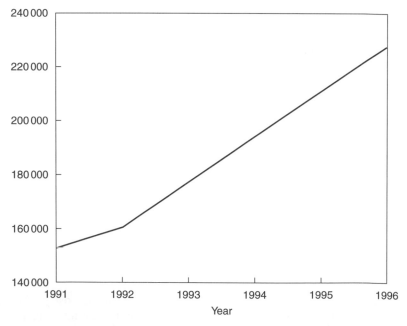

Figure 35.1   Number of children with statements in England
Source: Department for Education and Employment (1997c)

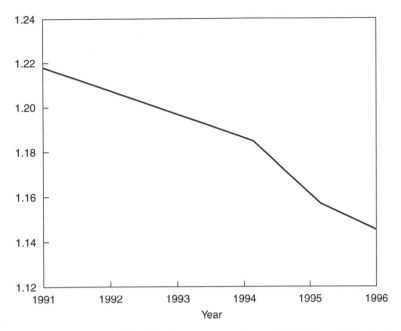

Figure 35.2 Children in special schools as a percentage of all children in England
Source: Department for Education and Employment (1997c)

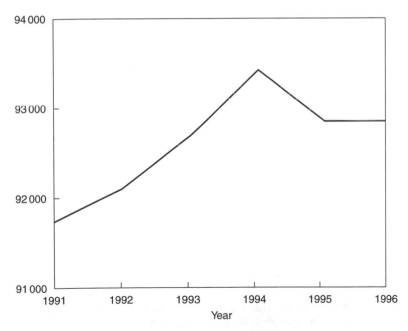

Figure 35.3 Rolls in special schools in England
Source: Department for Education and Employment (1997c)

A series of reasons have been suggested for the increase in the demand for SEN. It is possible that parents and schools had been using 'statementing' as a way of acquiring extra resources. But there are other forces at work. Other legislation consecrated parental choice. To help parents in their choice, schools were to publish statistics of academic results. It is possible that schools that had difficult children were pushing them out of the mainstream in order to improve their results. The pressure to do this can only have become stronger after the introduction of formula funding and the publication of league tables. Finally, the closure of mainstream schools may have resulted in increased demand for special education. The reasoning would be as follows. The largest drop in rolls took place in schools in deprived areas – because the combination of differential birth rates and parental choice worked against such schools (*see* Mar Molinero, 1988). But if deprived areas contribute most to special education needs, then by closing schools in deprived areas the problem of special needs in the mainstream is being transferred to schools in more affluent districts. One can imagine heads making every possible effort, under the pressure of parents, either not to admit children with special needs, or to ensure that these are given statements of need if acceptance cannot be avoided.

Rather than accept the new trend of increased incidence of SEN, LEAs made efforts to design policies that could reverse them. An example is provided by Hampshire's policy, which will be discussed next.

Review of special needs in Hampshire

In July 1989 Hampshire LEA started a review of special education needs. A document, *All Our Children*, was issued in 1990 summarising the findings of the review (Hampshire County Council, 1990). The document stated on its second page that finance was an important motivation for the review. It quoted a report by the District Auditor in which it was suggested that, compared with similar LEAs, Hampshire's SEN expenditure was 'in excess of that required'. The report of the District Auditor was never made public, so it is not possible to form a judgement on the quality of the comparison; for example, it is crucial to know whether the degree of deprivation in comparator LEAs was similar to the degree of deprivation in Hampshire. No reasons were even given for this auditor's report, which informed later policy, not to be made public.

All Our Children also mentioned that, despite implementation of the 1981 Act, aimed at integration, 'the number of children in separate, specialist provision had increased'. It echoed Warnock's conjecture that about 20 per cent of all children have an SEN at some stage in their development. This 20 per cent can be split in two groups: 18 per cent merely need extra support in the classroom, but

> approximately 2% of the total school population benefit from multi-disciplinary assessment which identifies their specific needs and indicates the most appropriate way in which those needs can be met. The result of the multi-disciplinary assessment is the production of a Statement of Special Educational Needs.
>
> (Hampshire County Council, 1990, p. 6)

That paragraph is important because it associates two different issues: the proportion 2 per cent with the number of statements. Two per cent is, of course, the proportion of children with SEN for whom separate provision was being made, as observed in the Warnock Report. There is no reason why this percentage should coincide with the percentage of children who have statements. The percentage of children with statements in Hampshire's schools was reported to be 2.9 per cent, while the percentage of children in special schools, including units in the mainstream, was stated to be 2.2 per cent. If SEN units in the mainstream were excluded, the percentage of children for whom separate provision was being made was 1.95 per cent.

All Our Children further argued that, since the percentage of children with SEN statements was 2.9 per cent, this was too high, that the number of statements of need should be reduced, and that an effort should be made to reduce the number of children in special schools. This would imply an increase of SEN children in the mainstream.

What was a statistic had become a target to be achieved. There was no further analysis to establish the reasons why the percentage was different from the national average, or whether the number of statements was appropriate to meet existing needs. The *All Our Children* team did not stop with 2 per cent as a target, but it suggested that 2 per cent should be the target for statements of need, and that of these, half the children should be educated in the mainstream. This could be achieved only through a wide programme of special school closures (p. 16). The implicit reasoning was that as long as there are places in special schools, attempts will be made to send children there. Thus, a reverse Say's law was formulated: supply was to be removed in order to force demand to go away.

In 1994, Hampshire County Council issued a Development Plan for the Education Committee which stated the explicit aim, 'to provide thorough assessments for approximately 2% of children and to complete these assessments within the specified time limits' (p. 29). Senior officers made public pronouncements where they indicated that they were following government guidelines in their attempt to reduce the number of statements to 2 per cent. Such advice, however, was never formally written in any document. When challenged to give reference to such government guidelines or advice, they answered that it was internal advice not released to the general public! There is indeed advice on planning special needs provision in, for example, Circular 22/89 issued by the Department of Education and Science (1989). But in this circular, and others, 2 per cent is quoted as an estimate, not as a target. At the time of the controversy, a draft code of practice was issued by the Department for Education (1993) for public consultation; it said that around 2 per cent of all children would require a statement, the 2 per cent being a national average which would be expected to vary between regions. Hampshire's proportion did not appear to be exceptional.

In view of the LEA's reluctance to disclose sources, a different approach was taken. Southampton elects two Members of Parliament. They were both requested to ask parliamentary questions on the matter of special needs provision. One of the MPs argued that there was legislation on statements of special needs, and that whether the implementation of the law was correct or not could only be established in the courts. The second MP was much more positive. He asked the Secretary of State for Education, 'What is the policy governing the percentage of children with statements of special educational needs and in what form is this policy conveyed to local authorities?' In his reply, the Secretary of State for Education quoted standard legislation on SEN, and repeated the general aims of SEN provision as discussed in Warnock. If targets had been set for LEAs on the subject of SEN, the minister was not prepared to disclose their existence. A more reasonable interpretation is that such targets were never set, and that officers were avoiding having to take the blame for controversial decisions by passing the responsibility further up the hierarchy. A further three parliamentary questions were put to the minister, and they all received similar replies.

But there is the issue of whether 2 per cent is an appropriate level for statements. What determines the percentage of statements that are made? *All Our Children* was particularly candid in this respect. In appendix A it said:

> The threshold at which the decision to issue a statement is made will depend upon the availability of resources to help children in mainstream schools, as well as policies about what types of provision are to be allocated through statements.

Thus, resources are attached to the child with problems on the basis of statements, but the decision to start the statement process depends on the resources available. There is no doubt that children with the most severe needs will always be protected by a statement; the problem is the limit at which a child does not become eligible. This is clearly a policy decision, and a policy decision that will depend on the particular LEA. In other words, the percentage of statements issued does not totally reflect the demand for SEN. This was also observed by the Audit Commission and Her Majesty's Inspectorate (1992), which commented on the differences that exist in the proportion of children placed in special schools and reflected on the lack of correlation between this proportion and an indicator of deprivation.

In order to establish whether, at least in Southampton, the process of issuing statements was considered to be satisfactory, and whether there was a backlog of statements, an MSc project was set up. The person conducting the study was particularly well qualified for the job. He was a senior civil servant in Tanzania with responsibility for education. He contacted all primary schools in Southampton and asked about children with statements, children for whom a statement was in the process of being issued, and children who would benefit, in the opinion of the headteacher, from being issued with a statement of special needs. The response was excellent: of about 50 primary schools in the city, only one refused to disclose the information. The survey demonstrated that there were both much hidden demand and a great deal of dissatisfaction both with the number of statements being issued, and with the time it took to complete the process to issue a statement. Input from educational psychologists, key professionals in this process, was considered to be insufficient. Schools wished there were more educational psychologists in order to be able to assess the needs of more children, and not only the needs of the most difficult ones (*see* Mamuya, 1992).

This MSc dissertation was sent to the LEA, which made no comments. Several months later, in the context of special school closures, it was also sent to the local press. One of the local newspapers, *The Advertiser*, reported the findings of the dissertation. The response from the public was overwhelming. This prompted the paper to run a two-page feature with the letters of irate parents who volunteered detailed horror stories. The LEA then felt obliged to reply and dismissed the survey as 'unrepresentative'.

What do the statistics measure?

There is very little information available on the subject of special education needs. Statistics on rolls in special schools have long been published. In recent times, information has also been published on the total number of statements, on the number of children with statements in mainstream schools, and on the number of children with special needs but without statements in mainstream education. An example would be a recent Department for Education and Employment Statistical Bulletin (1997c). Bulletins also identify trends. For example, the number of children with statements of need has been continuously increasing: in 1991 the average was 2 per cent, and in 1996 it was 2.8 per cent. During this time the proportion of children with statements placed in special schools has declined from 56 per cent to 41 per cent, and so has the percentage of children placed in special schools, which has declined from 1.3 per cent in 1991 to 1.2 per cent in 1996. However, when one looks at rolls in special schools these have increased. These trends can be seen in Figures 35.1–35.3. Given the available data, little else can be said. It is possible to make emotional pronouncements as to these numbers being high or low. But these numbers by themselves do not mean much.

Setting targets for integration, and for statements, makes good financial control sense, but little educational sense. If budgets drive the availability of technical expertise, technical expertise limits provision, and policy influences the amount of resources available, how can we know that a need is being met? How can we know the kind of need we face and how it is evolving?

But what should we measure? It may be useful to take a medical analogy. Imagine that it has been decided that a given percentage of the population will need hospital treatment at any stage. The number of hospital beds would be limited in this strange world in order to discourage GPs from sending their patients for treatment. Imagine further that if the demand for hospital treatment exceeds the target, steps are taken to limit diagnosis. There would be a public outcry. Any civilised society would like to have information by medical speciality, since it is not the same thing to treat a heart condition as asthma. We would also like to know what determines current trends. We would be asking questions about the effectiveness of various treatments for given conditions. We would like good diagnosis made by a sufficient number of independent professionals in order to establish clearly what treatment is needed and the urgency of the treatment. We might not have enough resources instantly to provide treatment for all those who require it and waiting lists might develop but we would not try to limit diagnoses in order to avoid queues! We would also be interested in budgets, but this is only one of the many aspects of health for which information would be required. It would certainly not be the only one. It is odd that in special education we are concerned with the number of diagnoses and with the total number of treatments – to use the medical metaphor – and that we do not collect information which would be considered crucial in health.

In an ideal world there should be a classification of SEN for statistical and analytical purposes. It would not be difficult to achieve. We already have schools specialising in the different types of need. Why are there not published statistics by type of SEN? It would also be desirable to see how the different types of SEN evolve over time.

There is also the issue of statements. The process of issuing a statement of need is bureaucratic, time-consuming, frustrating and expensive. Does it have to be this way? Why can the educational psychologist alone not recommend the appropriate SEN provision, having seen the child and having collected all the relevant information? There is a case for improving and standardising the production of statements, not for constraining their number. There is also a case for divorcing the production of statements from the consideration of the resources available.

Work is also needed on the measurement of educational progress. We know little about the effectiveness of special education both in the mainstream and in the special school. Is the extra cost of a special school justified by better results?

The above are issues of data definition, data quality and data analysis. There are still questions relating to value judgements. One could argue, for example, that some of the children on whom so many resources are invested will never acquire more than a bare minimum of knowledge and that scarce cash might be better spent in a different way. This would be equivalent to asking why we spend so much on patients who suffer from terminal diseases. Statistics cannot answer this question.

The 1997 Green Paper on special educational needs

In October 1997 the Department for Education and Employment issued the Green Paper *Excellence for All Children* (Department for Education and Employment, 1997b), which identifies many of the above issues as worthy of investigation. It comments, in particular, on possible inconsistencies between LEAs when defining SEN. The Green Paper takes an integrationist attitude. It suggests, with a great deal of sense, that special educational needs should be identified at an early stage, so that appropriate action can be taken before the child falls behind in his or her academic progress. It suggests that if appropriate and early action takes place, there will be less need for statements of need, and that the money thus saved can be used to better provide for SEN children. While claiming that no targets are to be set for the percentage of children with statements, the Green Paper suggests that there should be an expectation that it should decrease to 2 per cent. It is not inconceivable that this percentage will be interpreted as a target by LEAs, particularly since no action is proposed to end league tables, change the emphasis on academic results, or depart from formula funding. In other words, the dynamics that led to an increase in demand will remain unchanged while efforts will be made to reduce specialist provision. *Plus ça change, plus c'est la même chose.*

PART VII

Measuring Employment

PART VII
Measurement

36

Problems of Measuring Participation in the Labour Market

Anne E. Green

To understand how labour markets operate, a range of measures of participation are needed

The European map of unemployment (Figure 36.1) shows that according to the conventional unemployment rate measure – which is one element used in the allocation of European Union funds – the incidence of unemployment is relatively low in the UK. In order to understand how unemployment is measured, this chapter outlines conventional approaches to measuring participation and evaluates them in the context of key features of labour market change. It advocates the need for a wider range of statistics to capture the new reality of labour market (non-) participation and presents selected analyses using alternative measures of unemployment and non-employment.

Conventional approaches to measuring participation

In labour market terms the adult population is conventionally divided into three main categories: those in *employment*, the *unemployed* and the economically *inactive*. According to the categorisation defined by the International Labour Office (ILO) used in the presentation of Labour Force Survey (LFS) data (*see* Chapter 3) in 'official' publications, those in *employment* either have a paid job (of at least 1 hour's duration a week) in an employee or self-employed capacity, or are on government-supported training and employment programmes, or are unpaid family workers. Again on the ILO definition, the *unemployed* do not have a job but are actively seeking work and are available to take up a job. Remaining members of the population are classified as economically *inactive*.

While headline statistics on increases and decreases in absolute levels of employment and unemployment are commonly quoted as indicators of changing economic fortunes, for comparative purposes (to contrast the circumstances of particular subgroups of the population, areas, etc.) the usual practice is to translate absolute figures into rates. The *economic activity rate* – the economically active population (i.e. those in employment plus the unemployed) expressed as a percentage of the total population of interest – is the crucial measure of participation in the labour market. Conversely, the *inactivity rate* – the economically

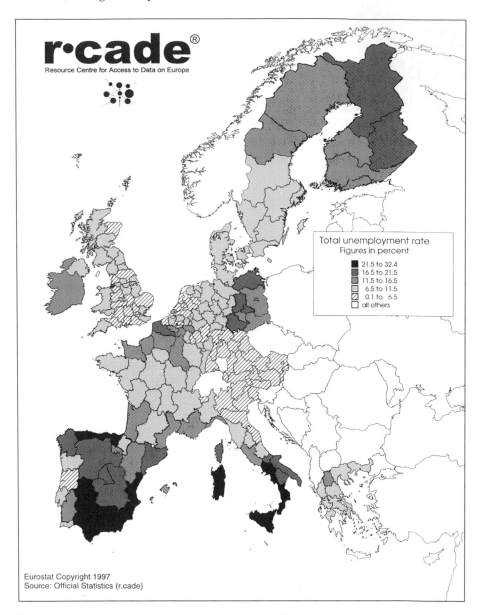

Figure 36.1   Harmonised unemployment within the European Union member states. Geographical resolution: UTS2. Time period: 1996 (Greece 1995)
Source: Official Statistics (r.cade). Eurostat copyright © 1997

inactive population expressed as a percentage of the total population of interest –
measures non-participation in the labour market. Perhaps the best known of the
commonly used measures is the *unemployment rate* – conventionally defined as
the unemployed population expressed as a percentage of the economically active

population. The unemployment rate is used variously as an economic indicator of labour market imbalance and as a social indicator of labour market distress, and has crucial policy significance as a key indicator in ranking of areas for allocation of funds and assistance. To date, somewhat less attention has been focused on the *employment rate* – those in employment as a percentage of the total population. However, there is increasing interest in both this indicator and its converse, the *non-employment rate* – the unemployed plus the inactive as a percentage of the population – which may be interpreted as a measure of 'joblessness'. Ray Thomas, in Chapter 37, provides some of the history behind this rising interest, and its implications.

It is possible to disaggregate each of the three main labour market categories – employment, unemployment and inactivity – into subcategories. Two such levels of disaggregation are illustrated in Figure 36.2. At the first level of disaggregation, those in *employment* are divided into those working full time, those working part time, those on government-supported training and employment programmes, and unpaid family workers. At the second level of disaggregation, those in part-time employment are further subdivided into those who could not find a full-time job and those who did not want or were not available for a full-time job. Within the *unemployed* category a distinction is made between those looking for full-time work (or who have no preference) and those looking for part-time work only. Turning to the economically *inactive*, at the first level of disaggregation a subdivision is made between those who want a job and those who do not want a job. At the second level of disaggregation those who want a job are subdivided between those not seeking work but available for work, those seeking work but not available and those neither seeking nor available for work.

By grouping together these first- and second-level subcategories it would be possible to generate a wider range of alternative labour market indicators (as illustrated in the third section of this chapter) than the conventional economic (in)activity and (un)employment rates generated by combining the first-level categories of those in employment, the unemployed and the inactive in different ways. Moreover, those disaggregations outlined in Figure 36.2 are not the only disaggregations that are possible: for instance, among those in *employment*, a distinction may be made between permanent and non-permanent working; the *unemployed* may be further subdivided according to duration of unemployment; and among the economically *inactive*, reasons why individuals are not seeking work may be identified.

Evaluation of conventional approaches in the face of labour market change

Over recent years there have been a number of important changes in the UK labour market, which in turn have prompted some users of labour market statistics to question the continuing applicability of conventional approaches to measuring participation, and the usefulness of the standard statistics outlined above. These aspects of labour market restructuring include a rise in the proportion of women in employment, as industrial and occupational employment changes – notably the rise in employment levels in services and the loss of jobs in manufacturing, and the reduction in the demand for manual labour – have favoured women

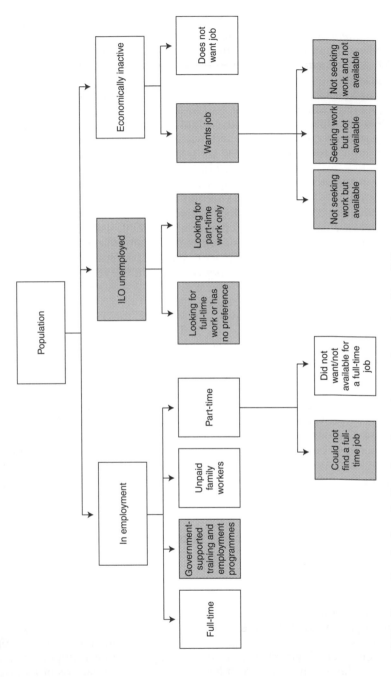

Figure 36.2 Conventional categorisation of employment, unemployment and inactivity

at the expense of men (Institute for Employment Research, 1997). While economic activity rates for women have risen – as more married women seek employment, and breaks for childbirth and child-rearing have become shorter – economic activity rates for men have declined (McRae, 1997). Higher 'staying on' rates in full-time education have led to a decrease in participation among younger age groups, while at the other end of the age spectrum a trend towards earlier retirement for some groups has contributed to a fall in participation rates among the older working-age groups. Hence, since the recession of the early 1980s, high levels of unemployment and non-participation have become entrenched among some subgroups of the working age population. At the same time, those in employment have witnessed a decline in the number of 'traditional' full-time permanent jobs and a rise in 'flexible' working – notably part-time and temporary employment (Meadows, 1996). In the face of the demise of 'jobs for life', some commentators have heralded the advent of 'portfolio careers' (which, as Humphrey Southall shows in Chapter 40, are not a new phenomena).

The result of these trends is a greater variety of patterns of work and non-work – with more individuals in irregular jobs and with discontinuous employment patterns. The changes in the labour market landscape have culminated in a 'blurring of boundaries' between employment, unemployment and inactivity (Green, 1997a). This new reality has been variously characterised as 'fuzzy' (Bryson and McKay, 1994), 'complex' and 'fluid': there are all kinds of 'grey areas' on the fringes of 'employment', 'unemployment' and 'inactivity' as conventionally defined (Nicaise *et al.*, 1995). Moreover, within the three main labour market categories this same ambiguity is evident: 'There are umpteen categories into which people can be put, varying degrees of part-time work, seeking work and not seeking work' (Professor David Bartholomew quoted in House of Commons Employment Committee, 1996, p. xxxi).

Particular concerns have been levelled about the validity of *unemployment* statistics and the applicability of conventional approaches to measuring unemployment in the face of the changes in the labour market landscape outlined above. In the mid-1990s these concerns prompted a Royal Statistical Society (1995b) review of the measurement of unemployment, and a House of Commons Employment Committee (1996) inquiry into unemployment and employment statistics. These reviews of the measurement of unemployment and employment statistics encompassed three main debates, focusing on:

- *data sources* – the relative merits of survey versus registered figures. The key debate here concerns the use of survey sources (such as the LFS) and administrative sources (such as the claimant count of unemployed). Accusations of 'fiddles' of the claimant count (for political and administrative purposes) were made throughout the 1990s (*see* Employment Policy Institute, 1993, for a review) and rejected (Royal Statistical Society, 1995b; *see also* Chapter 6).
- issues of *definition* – particularly the classification of borderline cases. The scope of the definition of unemployment adopted varies between sources (for example, those recorded as unemployed in the LFS and in the claimant count are distinct but overlapping groups of the population, such that estimates of the numbers and composition of the unemployed vary according to the scope of the definition applied). The debate also extends beyond who *is* counted as unemployed to who *should be* counted as unemployed.

- issues of *interpretation* of unemployment – as an indicator of the overall labour market situation.

Moreover, as regards interpretation, it has become increasingly apparent to a larger number of users of, and commentators on, labour market statistics that there is no straightforward one-to-one relationship between a change in unemployment and job losses or gains. Rather, interactions between labour supply and demand are such that an increase in unemployment in a particular local area is only one possible response to localised job loss; other responses include increases in inactivity (as Ray Davies shows in Chapter 38), or out-commuting and out-migration (Green and Owen, 1991). For example, application of a labour market accounting framework at the local level in the UK coalfields between 1981 and 1991 shows that for every 100 'surplus' male workers resulting from job losses in the coalfields coupled with the natural increase in the labour supply, none of the labour market adjustment was accounted for by an increase in unemployment. Instead, 38 became economically inactive, 27 migrated out of coalfield areas, 20 took up employment outside the coal industry, 11 went on government training schemes and 2 commuted out of the area (Beatty and Fothergill, 1996). This may be an extreme case, but it illustrates the complexity and variety of processes of labour market adjustment, and the conventional unemployment rate may measure only the tip of the iceberg of joblessness.

Towards a wider range of statistics to measure participation

The complexity of the new labour market landscape painted in the previous section is such that a wider range of statistics than those conventionally used is needed to capture the new 'reality'. To date, most debate has focused on measuring unemployment and joblessness. One group of researchers (Beatty *et al.*, 1997) has focused its efforts on generating a count of *real unemployment* to set alongside the claimant count measure at the local level. The approach taken was to add four groups of *hidden unemployed*:

- those who are unemployed and not claiming benefit – measured as the excess of those unemployed in the census of population (or the Labour Force Survey) over the unemployed claimant count;
- those on government schemes – who are classified as *in employment* in the conventional categorisation;
- excess numbers of early retirees – measured in relation to the number of early retirees in the English South East (one of the regions with lowest unemployment) in 1991 (when unemployment was at a trough); and
- excess numbers of permanently sick – using the number of permanently sick recorded by the 1991 census of population in the South East region as a benchmark.

According to this methodology, in January 1997 the 'real' level of unemployment was 3.95 million (14.2 per cent), compared with the claimant count figure of 1.84 million (7.1 per cent). Of course, it is possible to dispute the precise definition and measurement of the groups of 'hidden unemployment'; the key issue is that the level of 'real' unemployment is estimated to be substantially higher than the unemployment level as measured by the claimant count.

Since from both a conceptual and a practical perspective any single 'true' measure of unemployment is likely to be flawed, rather than adopt one alternative measure of unemployment, a more flexible (and perhaps a more realistic) approach is to manipulate available data sources to generate a battery of cumulative and/or overlapping measures to serve different objectives of labour market analysts. Such a suite of alternative measures may encompass both 'narrower' measures (e.g. the long-term unemployed) and 'broader' measures (such as some of those conventionally defined as inactive as well as those conventionally defined as unemployed). In the USA, the US Bureau of Labor Statistics publishes six alternative measures of unemployment, ranging from those unemployed for 15 weeks or longer as a percentage of the civilian labour force (U1 – the 'narrowest' measure) to the total unemployed plus all 'marginally attached' workers (i.e. those who want and are available for work and who have recently looked for a job regardless of their main reason for currently not looking) plus all people employed part-time for economic reasons as a percentage of the civilian labour force plus all 'marginally attached' workers (U6 – the 'broadest' measure). Hence, as well as the *unemployed*, the broadest measure includes some individuals who in a conventional categorisation would be included as *inactive* and some who would be included as *employed*.

In the UK context, the Employment Policy Institute (1996) has published alternative indicators of unemployment at the individual and household levels, while Green and Hasluck (1998) have also used the LFS to operationalise alternative indicators of labour reserve at the regional scale (*see* Table 36.1). The list of indicators U1–U7 in Table 36.1 is not intended to be definitive or exhaustive; rather it is illustrative of the range of possible indicators that can be operationalised. Indeed, the ILO unemployment rate (U1) is the 'narrowest' measure included in the list, while the 'broadest' measure (U7) incorporates the inactive who want a job and those individuals working part-time who could not find a full-time job in addition to the ILO unemployed. In spring 1995 the 'labour reserve' on U7 exceeded that in U1 by 3.4 million. This difference would have been even greater

Table 36.1 Alternative indicators of labour reserve in the UK

Indicator	Definition	Total[a], spring 1995
U1	ILO unemployed	2 436 000
U2	U1 + those employed on government-supported education and training programmes	2 727 000
U3	U2 + those inactive who want a job but are not seeking work because they believe no jobs are available[b]	2 820 000
U4	U3 + those inactive who want a job and are not seeking work[c] but available	3 615 000
U5	U4 + those inactive who want a job and seeking work but not available	3 897 000
U6	U2 + all those inactive who want a job	5 027 000
U7	U6 + those part-timers who could not find a full-time job	5 821 000

Source: Data derived from Labour Force Survey
[a] Males aged 16–64 years and females 16–59 years
[b] Discouraged workers (these are a subset of those who are inactive and want a job)
[c] For whatever reason – not just those who are discouraged workers

if the range were extended (as would be possible) to incorporate a 'narrower' measure of severe (long-duration) unemployment, and a 'broader' measure incorporating all the non-employed plus those in employment who are underemployed.

The debate concerning alternative measures of labour market participation and non-participation is now beginning to extend beyond unemployment and joblessness to *labour market attachment* more generally (Laux, 1996). There is growing recognition of the fact that in an increasingly diverse and dynamic labour market, conventional measures of employment, ILO unemployment and economic inactivity are insufficient. Rather, it may be assumed that different groups of the employed, the unemployed and the inactive have different degrees of attachment to the labour market.

As noted in the first section, even those working only 1 hour per week are categorised as 'employed' according to the ILO definition, yet those individuals working a small number of hours per week may be a very different group from those working 60 or more hours per week. Hence there is a case for further disaggregating the employed according to *hours worked*, at a more detailed level than the standard full time–part time distinction. Another important dimension of variation is that between those in *permanent* and *non-permanent* employment, and within the latter category there may be important variations in reasons for working on a temporary basis: some may be working on a temporary basis through choice, while others may not. Those working on a temporary basis because they could not find permanent employment and those working on a part-time basis because they could not find full-time employment (and those working longer hours than they want) may be considered to have 'unmatched aspirations' (i.e. their labour market arrangements do not match what they want).

The concept of '*(un)matched aspirations*' is also applicable to those individuals conventionally defined as economically inactive and unemployed. While those economically inactive individuals who do not want a job have 'matched aspirations' (and have least attachment to the labour market), those among the inactive who want a job (but who do not qualify as ILO unemployed because they do not meet the active job search and availability criteria to be so categorised) have 'unmatched aspirations'. By definition, all the unemployed also have 'unmatched aspirations'.

Figure 36.3 illustrates how LFS data may be used to examine changes in the labour market situation of men and women over the period of economic recovery from winter 1992/93 to winter 1995/96, and so provide insights into both changes in labour attachment and interactions between employment, unemployment and inactivity. Total employment (measured using the conventional definition) rose by 729 000 over the period, with the bulk of the increase being accounted for by employees. Over the same period unemployment fell by 640 000, with men accounting for two-thirds of this decrease. However, crucially, at the same time as the reduction in unemployment there was a 374 000 increase (of whom 279 000 were men) in the number of economically inactive classified as wanting a job. If this change in this group is combined with the change in ILO unemployment over the same period, the decrease in 'joblessness' is reduced to 275 000. The reduction in the number of inactive not wanting a job (accounted for solely by women) also points to an increase in the degree of attachment to the labour force.

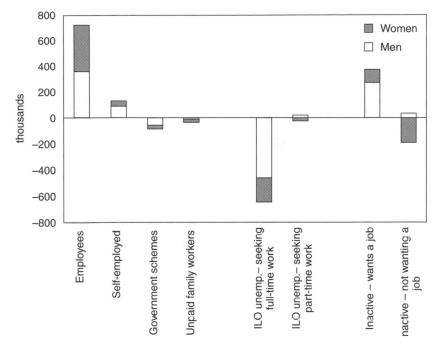

Figure 36.3 Change in labour market attachment, winter 1992/93 to winter 1995/96
Source: Labour Force Survey (reported in Laux, 1996)

Inter-area comparisons

Just as there is an increasing recognition of the growing complexity and diversity of the labour market, so there is a wider understanding of the fact that labour markets function in locally, regionally and nationally specific ways (Peck, 1996; *see also* Chapter 39). In turn, this underlines the importance of looking beyond single measures of 'employment', 'unemployment' and 'inactivity' to a suite of alternative measures to gain more meaningful insights into the complexity of the labour market situation.

To illustrate this point, Figure 36.4 shows alternative indicators of labour reserve (defined in Table 36.1) in three contrasting UK regions: Merseyside (a regional economy in long-term decline characterised by persistent high unemployment above the UK average), Wales (traditionally one of the more depressed regions of the UK but one which in recent years has witnessed a convergence in the unemployment rate towards the national average) and the South East excluding London (traditionally considered the most prosperous region). Together with the UK as a whole, these regions are ranked in Figure 36.4 in descending order on U1 (the ILO unemployment rate). What is immediately apparent is that while a slightly smaller percentage of the working-age population are categorised as ILO unemployed (U1) in Wales than in the UK, on all broader indicators (U2–U7) the labour reserve is proportionately larger in Wales than across the UK as a whole – with the differentials being notably large on U4 and U6, reflecting the particularly

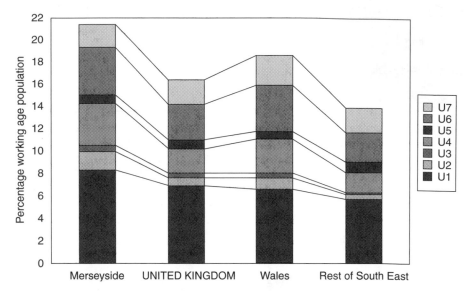

Figure 36.4   Alternative indicators of labour reserve for selected regions
Source: Labour Force Survey, spring 1995 (adapted from Green and Hasluck, 1998)

high rates of inactivity in Wales (Green and Owen, 1996; Beatty *et al.*, 1997; and *see also* Chapter 38). On all indicators, the labour reserve in Merseyside is larger than across the UK as a whole, while in the Rest of the South East the labour reserve is smaller than nationally on all indicators.

At the regional level in the UK, use of the conventional ILO unemployment rate tends to understate the size of the labour reserve to the greatest degree in the traditional 'high-unemployment' regions. Moreover, complementary analyses of a wider range of LFS data on labour market participation indicate that this broad regional geography of unemployment and non-employment is reinforced by a similar geography of underemployment and insecure employment (Green and Hasluck, 1998).

There is also scope for undertaking similar comparisons across European Union countries using data from the European Labour Force Survey (ELFS). However, international comparisons are not without their difficulties, given differences between countries in employment structures (including atypical, part-time, casual and informal working) and participation rates (Fagan and Rubery, 1996), which are in turn intimately linked with unemployment and social benefit regimes and the character and extent of passive and active labour market policies. Furthermore, although the ELFS sets out to ask internationally standardised questions, there is increasing recognition that such differences mean that it does not receive internationally standardised answers. Nevertheless, analyses operationalising alternative indicators of unemployment and non-employment at the national and regional level within the European Union reveal that the Netherlands, Italy and the UK are among the countries where ILO unemployment understates 'broader' unemployment (defined in this instance as those economically inactive who would like work) to the greatest degree, while in Spain – which has a high incidence of ILO unemployment – the understatement is much smaller (Green, 1997b). Now look again at Figure 36.1, the 'official' map of unemployment in Europe, and ask yourself what it shows!

Conclusion

As labour market developments and socio-economic changes more generally have led to a more complex situation in which there is greater ambiguity about what constitutes *employment*, *unemployment* and *inactivity*, the boundaries between these three main categories have become increasingly fuzzy. Hence conventional statistics provide an incomplete picture: for example, the unemployment rate provides only a partial picture of labour market disadvantage and joblessness.

This chapter has illustrated that in order to gain a more complete picture of joblessness, labour market slack, labour market attachment, etc., there is a case for adopting a wider range of indicators than those conventionally used. This might involve developing suites of alternative 'narrower' and 'broader' indicators to provide insights into the severity and extensiveness of particular labour market situations across the full (non-)participation spectrum. While on the one hand the existence of alternative indicators relating to a particular concept may confuse some users, on the other hand a greater range of (non-)participation statistics may help clarify the understanding and interpretation of labour market developments among labour market analysts.

Acknowledgements

This chapter reports some of the results of a research project 'Alternative Measures of Employment and Non-Employment' funded by the ESRC (grant reference number R000236608).

37

The Politics and Reform of Unemployment and Employment Statistics

Ray Thomas

Problems and solutions concerning two of the UK's best-known data series

Introduction

The UK has two statistical series for unemployment. The Labour Force Survey unemployment series, following International Labour Office (ILO) criteria, counts the numbers seeking work. The monthly Count of Claimants series, with antecedents going back to the nineteenth century, gives the numbers entitled to unemployment benefit.

These two series identify different unemployed populations, but neither series measures the extent of joblessness in the UK. This chapter points to the lack of links between the Count of Claimants and the LFS unemployment series, which creates serious problems for monitoring trends in the labour market, and proposes that compatibility should be achieved through the use of claimant records as a sampling frame for part of the Labour Force Survey. The chapter argues that the growth of joblessness in the UK has made it more difficult to measure unemployment. It is proposed that records of pay as you earn (PAYE) income tax deductions and National Insurance contributions should be used for the development of statistics of employment, of reliability comparable to that of the Count of Claimants, in order to improve monitoring of labour market conditions at national, regional and local levels.

The Count of Claimants

The Count of Claimants is one of the UK's most historical statistical series. Chapter 40 traces the series back to the middle of the nineteenth century. A degree of continuity is preserved over this period because the series consistently measures an insured population. Claimants were those who were entitled to receive unemployment benefit by virtue of having paid insurance contributions or trade union dues.

Figure 37.1 compares trends in the Count of Claimants series, and that for employment, for the period since 1971. The dominant feature of Figure 37.1 is the growth of structural unemployment, particularly in the 1980s. In this period the nature of the demand for labour in Britain became different – in terms of skills or geographical area or both – from the nature of the supply. But the statistics of Figure 37.1 do not do justice to the changes which have occurred. The figures for employment give an unrealistic impression of increases in the demand for labour, and the figures for unemployment grossly understate the growth of joblessness.

The notes on Figure 37.1 give some detail of ways in which the statistics distort the patterns of change. The figures for employment, for example, disguise the growth of part-time employment and the fact that the number of full-time jobs in 1997 was actually less than it was 26 years earlier. The figures for unemployment do not take into account the effects of limitation of unemployment benefit to the insured population; they do not take into account changes in the regulations determining entitlement to benefit, or changes in the ways in which these regulations have been applied.

Most of such exclusions from the Count of Claimants are acknowledged in the 'small print' of Office for National Statistics (ONS) statistical publications and other publications such as *How Exactly Is Unemployment Measured* (Office for National Statistics, 1998d). However, there has been a covert movement from the Count of Claimants in the 1980s and 1990s to other forms of social security benefit. This movement is documented in a study made at Sheffield Hallam University (Beatty *et al.*, 1997).

In the 1980s and early 1990s employment offices were encouraged to transfer claimants to incapacity benefit in order to reduce the Count of Claimants (Beatty *et al.*, 1997, p. 13). Over the period 1981–95 the numbers recorded as receiving long-term sickness benefit increased from 0.6 to 1.8 million (p.11). In some parts of the country, such as Merthyr Tydfil, Liverpool and Tyneside, more than 20 per cent of the male population of working age are classified as incapacitated (p. 14). By comparison with the proportion of the population of working age classified as permanently sick in more prosperous parts of the country, the Sheffield team estimated that in January 1997 the national 'excess' of permanently sick was 1.3 million (p. 23).

The term 'covert movement' seems justified because even the Department for Education and Employment (DfEE) does not seem to have been aware of what was happening. The DfEE gave full cooperation to the Royal Statistical Society (RSS) for the report it made in 1995 on unemployment statistics (Working Party, 1995, p. 405). But the RSS report shows no evidence of awareness of this movement. Neither does the report of the Parliamentary Select Committee made in the following year (UK Government, 1996).

In spite of these limitations, the monthly publication of the Count of Claimants dominates the picture of labour market conditions given to the public and to government alike. The Count is credible because short-term changes appear to sensitively indicate changes in labour market conditions. A regression of year-to-year changes in the Count of Claimants on employment for the period 1971–97 gives a coefficient of -0.73 ($r^2 = 0.77$). A change of 10 per cent in employment, in other words, has generally been associated with a change in the Count of 7.3 per cent in the opposite direction.

The major advantage of the Count of Claimant statistics is that they are available in detail for local areas (*see* Chapter 6). Since computerisation and the ending of the registrant system in 1983, claimants' statistics have been available for

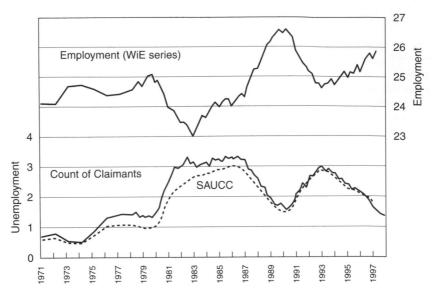

Figure 37.1 The Count of Claimants and employment, 1971–97 (millions)

Note: The employment statistics are the Workforce in Employment (WiE) series, produced mainly on the basis of postal surveys of employers registered for PAYE tax, which relate to a particular day in the month. From 1971 to 1977 the statistics relate to surveys conducted in June. From June 1978 the statistics relate to quarterly surveys conducted in June, September, December and March. The figures used are not adjusted for seasonal variation.

Employees classified as part-time amounted to 3.1 million in 1971 and increased gradually to 6.4 million by 1997. The number of employees classified as full-time was 18.3 million in 1971 and declined to 15.8 million in June 1997. Those classified as self-employed amounted to 2.1 million in 1971, peaked at 3.5 million in June 1990, and amounted to 3.1 million in June 1997. From June 1983 the WiE series include those on work-related government training programmes. The numbers on these programmes peaked at 456 000 in September 1989. The WiE series does not distinguish National Insurance contributors and others in employment. But Figure 37.2 suggests that the proportion of those covered by the WiE series paying National Insurance contributions fell substantially in the period to 1994.

The Count of Claimants (CoC) statistics used relate to June for the period 1971–77 and to March, June, September, December for the period from June 1978 to June 1977, without seasonal adjustment. The CoC statistics cover only those entitled to unemployment benefit, usually through having paid National Insurance contributions. The unit of unemployment benefit is the household, not the individual. If one member is working, or is a claimant, then the other member can claim only contribution-based benefit or National Insurance contribution credits. The largest single group who are unentitled in this way are women married to, or cohabiting with, men who are in employment or who are claimants. Recipients of most other social security benefits, such as incapacity benefit, cannot be claimants. People on government training programmes are excluded. Benefit is also refused to those who have broken rules – such as leaving a job without good reason, being dismissed for misconduct, refusing a job without good cause, or failing to sign on fortnightly.

Between 1979 and 1989 there were 30 changes in the rules affecting the compilation of the count. Until October 1982 the count included those who were registered at employment offices as seeking work whether or not they were claimants. But after October 1982 registrants who were not claimants were excluded – which reduced the count by 135 000. In 1981 and 1983 changes were made which excluded men over 60 who, according to the Department of Employment, 'mostly considered themselves retired'. These changes reduced the count at that time by about 200 000. In 1988 entitlement to unemployment-related benefits for 16- and 17-year-olds was ended, which reduced the count by 130 000 (Fenwick and Denman, 1995).

Such changes led to lack of continuity in the count and to charges that the statistics were being fiddled. The Department of Employment claimed that most of the other changes were small or inconsequential (Fenwick and Denman, 1995). The Seasonally Adjusted Count Consistent with Current Coverage (SAUCC) was produced in response to such criticisms. The SAUCC series, shown by a broken line in Figure 37.1, is claimed to be retrospectively consistent. The SAUCC series aims to measure what would have been the count had the regulations on entitlement to benefit been the same in the past as at the time the latest SAUCC series is compiled. But the SAUCC series does not cover changes in the way the regulations affecting entitlement to benefit are administered. The 1989 Social Security Act, for example, required claimants to prove that they are actively seeking work and introduced the idea of a permitted period after which level of remuneration would not be treated as good cause for claimants to refuse an employment opportunity. The Department of Employment states that this change did not have a 'significant' effect on the count (Fenwick and Denman, 1995, pp. 399–400), and this change is not taken into account in the compilation of the SAUCC series. Neither have the SAUCC statistics been changed in response to the introduction of the Job Seeker's Allowance in October 1996.

The WiE figures for the period 1978–91 are from the Abstract of Employment Statistics, 1994. The later WiE figures are from table 1.1 of *Labour Market Trends*. The Count of Claimants statistics come from *Labour Market Trends* and its predecessor, the *Employment Gazette*. The SAUCC series is not published but is available on request from the ONS.

postcode areas. Such statistics are used extensively by local authorities. But the use of these statistics for the identification of concentrations of unemployment has been handicapped by the lack of proper denominator statistics; that is, statistics for the number of residents of an area in employment. As a result, the identification of concentrations of unemployment has been concealed by the use of Travel to Work Areas (*see* Thomas, 1998; Turok, 1997; Webster and Turok, 1997).

Political and statistical background

These problems with unemployment and employment statistics are in part attributable to the unusual conditions, politically and statistically, of the period of Conservative administrations from 1979 to 1997. There was little governmental interest in unemployment statistics. The decline in the level of employment of the early 1980s was attributable to Margaret Thatcher's not-for-turning public-expenditure-cutting policies. But Thatcher herself believed that 'new technology would create jobs', that the level of unemployment 'was related to the extent of trade union power' and 'reflected past overmanning and inefficiency' (Thatcher, 1995, pp. 257, 272 and 292). The government's interest was limited to the Count of Claimants as an indicator of the cost of unemployment to the government.

The Government Statistical Service (GSS) in the 1980s was not well prepared to deal with the challenges presented by labour market conditions and such governmental policies. Statisticians had enjoyed a relatively privileged position within the Civil Service before Thatcher came to power. But one of Thatcher's first acts as prime minister was to appoint Sir Derek Rayner to 'tackle waste and ineffectiveness of government' (Thatcher, 1995, p. 30). The GSS was one of the first targets.

The pseudonymous account 'How official statistics are produced: views from the inside', written in the late 1970s, portrays the function of the GSS as primarily to serve the machinery of government itself – with little sense that government statisticians in any other way act in the public interest and try to serve society as a whole (Government Statisticians' Collective, 1979). The Rayner Review of the GSS reinforced this focus, revealing government statisticians as technicians, shy of political thinking, and isolated from groups outside the government such as social scientists who might have been able to give them support. The Director of the GSS at that time claimed that the needs of society still received 'some weight', but reported a 20 per cent cut in his budget, and confessed that the GSS was obliged to concentrate more heavily than before on serving the government of the day (Boreham, 1984, p. 64.1).

The main effect of Rayner's investigation of the GSS appears to have been to demoralise government statisticians. The Rayner demand that they justify the work they did could be answered only in terms of their doing only what they were told to do – which meant focusing narrowly on their immediate responsibilities without attempting to take a broader view. The GSS could justify most of its services in terms of uses within the Civil Service and to 'customers' outside. This atmosphere had a profound influence on the main developments of labour market statistics in the 1980s in the form of the Labour Force Survey.

The Labour Force Survey

The Labour Force Survey (LFS) was first conducted in 1979, and has continued annually since 1984 and continuously since 1992 (*see* Chapters 3 and 36 for more detail). Since 1992 the LFS has become the leading source of employment statistics preferred over the Workforce in Employment (WiE) series (shown in Figure 37.1), because the LFS series counts people rather than jobs. A remarkable feature of the LFS is that it has been designed for international purposes quite independently of the operation of the National Insurance and Count of Claimants systems.

One specific omission is that respondents in employment are not asked whether they pay National Insurance contributions. There is consequently no clear dividing line between what might be considered 'real' jobs and other jobs in the employment statistics produced from the LFS. The LFS follows the ILO definition in counting all respondents who worked more than 1 hour in the previous week as employed. It is known that the LFS employment statistics include more than a million who worked less than 10 hours (Naylor, 1994, p. 479), but such statistics are not published regularly, and there is no easy way to measure the growth of part-time employment over a particular period. And there is no way at all of measuring growth in the number of jobs paying less than the National Insurance contributions exemption limits.

Figure 37.2 indicates that over the period 1979 to 1994 the number of National Insurance contributors lagged the increase in employment as recorded by the LFS by about 2.5 million. Does this mean that the number of jobs falling below the National Insurance exemption limits (£57 a week in 1995) has increased by 2.5 million in this period? Does it mean that employers have been cheating their employees and not passing on payments to the Ministry of Social Security? This discrepancy is unresolved and may be unresolvable.

The LFS does ask respondents if they are claimants. But this question appears only very late in the questionnaire and does not produce answers which are consistent with the Count of Claimants itself. In recent years the number of claimants estimated on the basis of the LFS has been some 20 per cent below the Count itself – although it is believed that a significant proportion of respondents mistakenly state that they are receiving unemployment benefit.

The Alternative Measures of Unemployment series

The criteria used by the ILO/LFS unemployment series differ from those used by the Count of Claimants in three respects. First, as mentioned above, the ILO/LFS criteria classify all respondents working more than 1 hour a week as being in employment. But claimants are allowed to work a small number of hours at low pay without losing entitlement. Those claimants who are counted as in employment by the ILO/LFS criteria are called *claimants in employment* in what are called the Alternative Measures of Unemployment series – or the AMU statistics.

Second, to be counted as ILO/LFS unemployed, respondents have to have taken active steps to seek work in the previous 4 weeks. This means that the LFS series does not include, for example, those labelled as 'discouraged workers'. 'Discouraged workers' are LFS respondents who say that they would like to be in employment but who have not taken steps to seek employment because they do not believe that jobs are available. Such discouraged workers are examples of what are called *inactive claimants* in the AMU statistics.

Third, the ILO/LFS series covers the whole population of working age and so distinguishes between the jobless who are claimants, which the AMU statistics call *LFS claimants*, and those who are not claimants, called *LFS non claimants*.

Eight articles were published in the *Employment Gazette* between 1983 and 1993 which aimed to reconcile the Count of Claimants and the LFS series, and the results were published as the AMU series, which extended back to the 1984 LFS. But as a result of a recent investigation involving record linkage between the LFS and the Count, the historic AMU series has been declared as invalid by the ONS for the period 1984 to 1992 (Pease, 1997a, b). The outcome can be seen by comparison of the table published in the December 1997 issue of *Labour Market Trends* with that in any earlier issue.

The Labour Force Survey non-claimants group

Figure 37.2 uses unadjusted LFS responses rather than the AMU series in order to give a picture of the whole period. Figure 37.3 indicates that the size of each of the three subgroups which together constitute the Count of Claimants series – LFS *claimants, economically inactive claimants* and *claimants in employment* – mirror changes in the level of employment. In other words, the *economically inactive claimants* and *claimants in employment* groups, which are excluded from the LFS unemployment series, behave in accordance with their claimant label – as if they are measures of unemployment.

The number of LFS *non-claimants*, by contrast, is not responsive to the level of employment. Since 1984 the size of the LFS *non-claimants* group has been remarkably stable. The inconsistency between the LFS series and the Count implies that these statistics are subject to error; *see* note on Figure 37.3. But it is difficult to imagine that any pattern of error would change this contrast in the

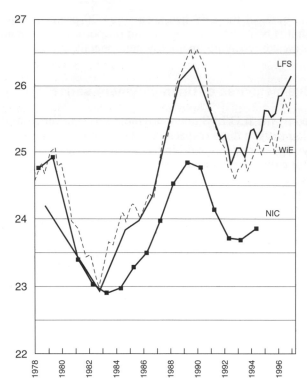

Figure 37.2 Employment in Britain, 1978–97, according to the Labour Force Survey, the Workforce in Employment series, and the number of National Insurance contributors (millions)

Note: The LFS employment series is based on surveys of households carried out over a 13-week period. Thus the LFS series measures the average number employed over a period of a quarter of a year. There was no LFS in 1980 or 1982, and the LFS was carried out only in the spring quarter (March, April and May) between 1984 and 1992, but after 1992 it has been conducted every quarter.

The NIC (National Insurance Contributions) series covers employees and the self-employed contributing at any time during the year (for the financial year, ending in April). It is to be expected, therefore, that the number recorded by the NIC series would exceed the number of contributors at any one point of time. The NIC series is published only about 18 months after the end of the financial year. For ease of comparison with the other series the NIC figures in Figure 37.2 have been centred on the previous September. The ONS states that 'a discontinuity' was introduced into the LFS employment series in 1983 (although this is not evident from comparison of the WiE and LFS series over 1981–84, and states that the discrepancy between the LFS and NIC series is 'much smaller' than is suggested by Figure 37.1 (Pease, 1998, pp. 62–3).

The LFS series covers all people working more than an hour a week, but more than 0.9 million people in 1997 had two or more jobs. The WiE series counts jobs rather than people, but in 1977 missed something like a million jobs which are not covered by the PAYE sampling frame. The differences between the LFS and WiE series are not fully resolved (*see* Thomas, 1997), but are subject to ongoing investigation by the ONS (*see* Perry, 1996; Pease, 1997c, 1998; Office for National Statistics, 1998c).

The LFS figures are from *Labour Market Trends*. The NIC series comes from the annual Social Security Statistics. The NIC series is subject to retrospective revision. This table gives the most recently published figure for each year.

behaviour between the LFS *non-claimants* group and that of all three of the claimant groups.

It is not reasonable to assume that members of the LFS *non-claimants* group are less likely to enter employment than claimants. Surveys conducted in the mid-1980s found that what were called non-registered seekers are actually more likely to enter employment than registered job-seekers (Gallie *et al.*, 1994). Nor is it reasonable to assume that people do not enter the group when they lose their jobs. The stability in size of the LFS *non-claimants* group is an indicator not of stability of composition, but rather that the group covers a transitory population.

A fall in the level of employment, for example, is likely to lead to an increase in the number of LFS *non-claimants*. But a fall in employment is also likely to lead to LFS *non-claimants* dropping out of the labour force, as has been acknowledged by the Department of Employment (*see*, for example, Woolford and Denham, 1993, p. 460). In an upturn in the economy, an increase in the level of employment is associated with some LFS *non-claimants* getting jobs. But the increase is also associated with an increase in the number of people looking for work. The LFS criteria capture entry to unemployment as well as volition to employment. There are encouraged workers as well as discouraged workers. The LFS *non-claimants* group is, in effect, the top slice of a reserve army of labour.

When people feel optimistic about getting work they will take active steps to look for it – and so become unemployed according to the ILO criteria. When people feel pessimistic about getting work they are (as widely measured by the ILO criteria) discouraged and not counted as unemployed. The ILO criterion essentially measures an attitude, and questions have to be asked about the reliability of the resulting statistics. To what extent do the attitudes measures really reflect labour market conditions? And to what extent do they reflect something analogous to what the pollsters call the 'feel-good factor', and what they might call, at other times the 'feel-bad factor'?

These questions are important because the behaviour of the LFS *non-claimants* group makes the LFS series itself less sensitive to changes in the level of employment than the Count of Claimants. The uncertain relationship to labour market conditions and relative insensitivity casts doubt on the value of the LFS series as an indicator of trends in unemployment.

It may be that the number of LFS respondents who say that they would like to be in regular employment is a more valid and reliable measure of the unused labour supply than the number who meet the ILO availability and seeking work criteria for unemployment. The Unemployment Unit, a group based in London which promotes better labour market policies, has long supported the use of this measure, and Chapter 36 relies extensively on this statistic. But the number of LFS respondents who say that they would like to be in regular employment is not a recognised or regularly published statistical series.

Reform of the Labour Force Survey

It cannot make sense to have two unemployment series, each costing millions of pounds to produce, which do not relate to each other. But record linkage between the LFS and claimant records, as used in the Pease study, is not a satisfactory method of reconciliation. This procedure does not formally break the pledge of

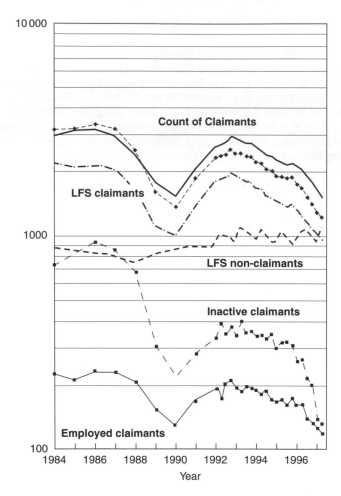

Figure 37.3 Alternative measures of unemployment, 1984–97 (thousands (log scale))
Note: The solid upper line in the figure shows the Count of Claimants. The broken line, visible above the solid line over the period 1984–87, and below the solid line after 1988, represents the grossed up number of Labour Force Survey respondents who say that they are claimants. The shortfall was 10 per cent in 1992 and had increased to 20 per cent by 1997.

For the period 1989–97 the Department for Education and Employment assumed that the shortfall was limited to the *claimants in employment* and *economically inactive claimants* groups. But new record linkage study published in 1997 (Pease, 1997a, b) found that this assumption was oversimplified. The linkage study found that the shortfall was understated in the sense that a significant number of respondents mistakenly stated that they were claimants. The study also found that the shortfall could be attributed to all three claimant groups, i.e. including LFS *claimants*. The statistics shown in the figure for individual unemployed groups relate to the number of responses given in the LFS and have not been adjusted to allow for the shortfall in the number stating that they are claimants.

The numbers for all three claimant groups may be understated for the years since 1989. Where the LFS *claimants* figures are understated, then the LFS *non-claimants* figures, which are obtained by subtraction from the total for LFS unemployment, are overstated. For the summer of 1996 the figure for LFS *claimants* could be understated by up to 200 000 and that for LFS *non-claimants* could be overstated by up to 200 000. But there is no clear basis for making any adjustments for years before or after 1996.

The sources of the shortfall in the LFS figures for the number of claimants are attributed to a variety of factors. There is some non-response to the LFS which is known to be associated with claimants, which is allowed for in the LFS grossing up procedures – but this may be underestimated. A shortfall in the LFS *claimants* group can be attributed to claimants getting only National Insurance credits being unaware that they are claimants. A shortfall in the *inactive claimants* and *employed claimants* group can be attributed to respondents not being willing to admit that they are claimants because they are not looking for work, or not being willing to admit that they are claimants because they are working.

The figures for the period 1984–92 are for the spring quarter (March to May). For the period spring 1992 to summer 1997, the figures cover all four quarters, and have not been seasonally adjusted. The source of all these data is Pease (1997a, b).

confidentiality given to LFS respondents, but it could well be said that the appearance of confidentiality has been violated.

A complementary way of obtaining information on claimants would be to use claimants' records as a sampling frame for part of the LFS. Interviews based on such a sample frame would be conducted in the light of acceptance by both respondent and interviewer of the respondent's claimant status. Such an up-front approach would be at the other extreme from that of the current LFS questionnaire.

Such a procedure makes sense in terms of the function of the LFS in measuring the total labour supply. Those who are claimant unemployed are of special interest because they are critical to the measurement of the total size of the labour force. A separate sampling frame would permit the use of a higher sampling fraction for claimants than for non-claimants. A higher sampling fraction could well be justified in terms of the high public cost of claimant unemployment, and the value of information which is not obtained through the administrative processes involved in the claimant system. Data from such a sampling frame would permit direct comparison between claimants and other forms of unemployment.

The LFS uses a postcode address file as a sampling frame. There may be difficulties in bringing together statistics from a sample of households with those from a sample of claimants for the production of representative statistics at a regional or national level. But the value of studies made on matters such as characteristics of the unemployed (Woolford *et al.*, 1994), flows between different states of economic activity (Laux and Tonks, 1996), and degrees of attachment to the labour market (Laux, 1996) would all be substantially enhanced if they also covered claimant status.

The LFS could also be modified in order to deal with the omissions and ambiguities identified in this chapter. For those in employment there needs to be a question on whether they pay National Insurance contributions – in order both to reconcile differences with the series produced by the Department of Social Security, and to monitor trends in employment, both jobs liable to National Insurance contributions, and jobs not liable. For those in part-time employment there need to be questions on volition to full-time employment in order to properly estimate the size of the underemployed population.

For those not in employment there needs to be a battery of questions to establish volition to employment and the nature of the obstacles to finding and taking employment. These questions could well be linked to a revision of the highly ambiguous question on availability (*see* Thomas, 1997).

Measuring changes in employment

Reconciliation of the Count of Claimants and the LFS series would represent a substantial advance in the investigation of unemployment problems, but an inescapable problem is that unemployment as unused labour supply is difficult to measure in a scientific way, because it inescapably involves matters of personal volition to employment. For much of the unused labour supply now available in the UK the volition to employment is interdependent with the condition of the labour market itself. If there are plenty of good jobs available people will be drawn into the labour market. If the jobs are unpleasant, poorly paid and relatively scarce, many people will be economically inactive. Even a restructured LFS will not surmount these problems.

It is not sensible, therefore, to depend on statistics of unemployment as indicators of trends in labour markets without also taking into account more valid and reliable statistics which could be produced for employment. Unlike unemployment, employment is a contract and is measurable with precision. The proper management of the national, regional and local economies requires the development of monthly statistical series for employment which are as reliable and detailed as those which can be obtained from the Count of Claimants. The best source for such statistics, as pointed out in the Rayner review in 1981, are the records of the PAYE and National Insurance contributions made every month to the Inland Revenue (Brimmer, 1981). PAYE records include the addresses of both place of employment and, albeit often out of date, place of residence. Both addresses are, or could be, postcoded. Claimant statistics were postcoded in the aftermath of the Rayner review – and there was no consultation about possible infringements of privacy in making the statistics available. The population paying PAYE tax covered in each postcode area would generally be much larger than the number of claimants, and it would not be reasonable to reject postcoding on grounds of invasion of privacy. The PAYE tax records could be used for the development of statistics of employment at national, regional and local levels.

38

Unemployment and Permanent Sickness in Mid Glamorgan

Roy Davies

Local statistics are difficult to compile but can produce important results

This chapter describes how someone living in a community dominated by a particular social issue can gain access to the statistical, historical and analytical resources needed to determine, for themselves, where the truth may lie. The example of relatively low statistics of unemployment being reported in the Welsh coalfields is taken here as an issue that a lone researcher can investigate and illuminate.

On learning of the problem of counting unemployment and permanent sickness I decided to investigate the position in my own county of Mid Glamorgan. These problems are explained in detail in the chapters in this book by Theo Nichols, Anne Green, Ray Thomas and Humphrey Southall (Chapters 30, 36, 37 and 40). I made use of local libraries in Cardiff, Swansea, Newport, Pontypridd and Llanelli. I also used the university libraries at Cardiff and Swansea, including the miners' library at Swansea.

I found the Mid Glamorgan research publication *Update* produced by researchers from that county, very helpful, but when the county was reorganised the publication was abandoned. Consequently I had to refer to the telephone directory to contact key persons in such agencies as Rhondda, Cynon, Taff, a new unitary authority, where I received great help from Michele Leyton Johnson, a research worker; Bro Taf Health Authority; the Library of Health Promotion Wales; and the Welsh Office. Current government publications were not always available, and I had to make do with excerpts from newspapers. Access to more specialist libraries such as the ones at the London School of Economics and the Newspaper Library at Colindale, London, was difficult. These difficulties made me aware of the importance of libraries and archives as learning institutions, and the value of librarians and archivists who know how and where information can be obtained. They demonstrate why we must ensure that we are not deprived of these resources.

Mid Glamorgan was a thriving industrial area at the turn of the twentieth century and a prosperous part of the UK with its wealth founded on the products of coal mining and its ancillary industries of iron, steel and tinplate. Most of these goods were exported and enriched the economy of the UK. But the human cost was awesome, with industrial injuries and diseases causing death, or permanent disability. (Theo Nichols in Chapter 30 shows how industrial injuries are still concentrated in such manual labour.) Those who did not work in the mines and heavy industries (which included wives and children) suffered from malnutrition, tuberculosis and dysentery, attributed to poor housing, overcrowding, unsanitary conditions and pollution. The conditions were clearly described in the reports of medical health officers of the time, which indicated the need for improved social conditions.

> Long before 1914, south Wales lagged behind most parts of Britain in terms of substandard working class housing, in urban overcrowding, in its health and hospital services, in the indices of industrial disease among workers, poverty and ill health among the old, malnutrition among children. (Morgan, 1981, p. 71)

Miners and others responded to the social strains that accompanied industrial developments by creating self-help organisations such as sick clubs and medical aid societies, which helped to lay the foundations of the welfare state shaped by Lloyd George, Beverage and Bevan. While they made a major contribution to the wealth of the UK and had some share of that wealth, they also experienced unemployment and its social consequences.

Today, with the exception of Tower colliery and the new industries attracted to the mining valleys, Mid Glamorgan still has a lower standard of living than most areas of the United Kingdom, with an average household income in 1992 of £13 509 compared with the UK average of £15 422 (Welsh Office, 1996). This lower standard of living is linked to the high rate of unemployment in these areas but, as the Sheffield studies of Beatty *et al.* (1997) have shown, the claimant figures on which unemployment is based underestimate the problem, as they do not take into account the number of people who do not work because they are in the permanently sick category (*see also* Chapters 6, 36 and 37).

I examined the figures for the groups in my own unitary authority of Rhondda, Cynon, Taff, and found that in each of the age groups over the age of 45, there were more workers registered as permanently sick than were registered as unemployed. When I looked at comparable statistics for the Vale of Glamorgan, a neighbouring non-mining community, this pattern was not evident: the proportion of permanently sick workers in the older age groups was only half that in the corresponding age groups in the Rhondda.

Beatty *et al.* (1997) have drawn attention to the way in which unemployment in the old coalfields communities is underestimated because of the exclusion of the workers classified as permanently sick. They argue that most of these workers are not genuinely sick but are so classified as a result of connivance between worker and doctor, to avoid the means test entailed in the Job Seeker's Allowance. I have some reservations about these findings: they do not deal with the legacy of sickness caused by accidents and illness such as emphysema and bronchitis acquired by many miners while working in the coal mines; neither does the study deal with the historical and political considerations of mining communities who had to cope with the hardships of unemployment in the past. Chapter 27 demonstrates how sick men in manual work quickly lose their jobs.

Health of the valley people

A doctor from Merthyr Tydfil, a mining valley community, made a significant comment in a television documentary dealing with health in Wales when he said that in Merthyr four out of ten men were taking tablets for heart conditions (Vincent Kane, *Health in Wales*, BBC2, 20 November 1996):

> Julian Hart found that between 1968 and 1972 there was a higher number of deaths in the mining valleys attributed to coronary heart disease and strokes than in other regions of Wales and in 1967 he referred to demands on valley GPs being exceptionally high with consultation rates at least twice the average for the UK (Hart, 1975, p. 185).

The 1995 study of health in Wales found that the mining communities of Mid Glamorgan and Gwent had the two worst death rates for respiratory and cerebrovascular diseases, as documented in the Bro Taf Health Authority Health Profile for 1996. Another revealing study was the Welsh Health Survey (Welsh Office, 1995), which analysed the responses of the 500 000 people who answered a questionnaire about their physical and mental health.

Illness and despair

An analysis of the scores of 280 000 respondents in the Welsh Health Survey provided a measure of their physical and mental health components, which indicated that people from the industrial areas of the new unitary authorities had higher scores (considered themselves as less healthy) than those living in urban and rural areas; and people from the mining valleys of Merthyr Tydfil, Rhondda, Cynon, Taff, Caerphilly, Torfaen and Blaenau Gwent had the highest scores (Welsh Office, 1995). It is reasonable to assume that this condition is rooted in unemployment. The poverty of many workers and the hopeless search for jobs can induce despair. Although some employers such as B & Q will employ older workers, those workers over 40 had difficulty finding jobs either because they lacked the skills required or because they despaired of finding a job. Frequent rejection can sap confidence, undermine morale and affect mental health. Evidence of a high rate of mental illness in Mid Glamorgan was discovered in the 1980s. *Social Trends* in 1996 found that in the whole of the UK 'around a quarter of unemployed people had a neurotic disorder a week before they were interviewed' (*Social Trends*, p. 135, 1996, HMSO). This finding suggests that unemployed workers in the mining communities are also affected by this condition, especially when one considers the traumatic experience of the miners' dispute of the 1980s and the mass redundancies that followed. The younger age group, those aged 20–34, experience 23.4 per cent unemployment. Many young people pursued one or two government training courses and failed to obtain work. Many in their early thirties referred to themselves when interviewed as being 'past it'.

Historical and political considerations

Unemployment and sickness have always been contentious issues in the mining areas of south Wales. When the National Health Insurance (NHI) Act of 1911 was introduced, the administrators focused on workers claiming sick pay when they were fit enough to work and doctors were scolded for not being firm with patients. The newspapers in Wales raised the issue of malingering (*Western Mail*, 17 July 1913). The government responded by introducing medical referees to supervise the judgement of the doctors.

The NHI medical officers reprimanded many doctors in the medical aid societies, which were prominent in south Wales, with laymen in control of many of their affairs. In the 1930s, unemployment meant that in most mining areas one out of every three miners was unemployed and in the Rhondda one out of two. Some of the doctors' replies to NHI officials are worth noting: 'How difficult it is to send a man back to work in conditions which were responsible for his original illness' and 'My district has been hit by unemployment with its repercussions on National Health Insurance. There has been a depression of spirit and I am afraid two suicides.' While these comments are echoes from the past, they reflect the conditions of the mining industry (PRO/MH/49/21,1, Tredegar Medical Aid Society).

It should also be remembered by people who advocate insurance as the answer to the financial problems of the National Health Service (NHS) that when the NHI was in deficit, many unemployed miners were removed from the insurance scheme and lost their medical benefits. They were put in the hands of the Public Assistance Committee doctors, with the associated low standard of Poor Law care (*see* Chapter 40 concerning Poor Law statistics and Chapter 29 concerning privatisation of the NHS).

Controversy around malingering has also existed under the NHS. The *South Wales Echo* came out against malingering on 31 October 1961 under the headline 'Surgery shams: 60% are just dodging their work'. The article read:

> Tonight in surgeries all over South Wales a steady shuffle of patients will keep the doctors busy . . . and in each there will be 60% shamming, 60% who are not seeking succour but a certificate, a paper chit that will excuse them from work for another week. THEY ARE THE WORKSHY. A harsh picture? Perhaps. It was drawn for me by a Cardiff doctor. Other doctors confirmed it. Others were more charitable – they cut shammers down to 30%.

Julian Hart, then a doctor in Glyncorwyg, a mining village in south Wales, rebutted the article, and pointed out that illness in the mining communities was exceptionally high (Hart, 1975, p. 14). The view of the *Echo* in 1961 was echoed in the *Sun* as this chapter was being written when an editorial on the welfare state attacked 'spongers' (*Sun*, 21 December 1997).

Coping with unemployment and sickness

Although connivance does exist, not enough attention has been given to the legacy of sickness caused by working underground, particularly industrial diseases such as emphysema and bronchitis, and the poverty caused by unemployment, which is recognised as a contributory factor to ill health and depression. Neither does it bring out the historical and political circumstances that influence workers' attitudes and responses to unemployment and sickness benefits. Workers who have been brought up in a culture where unemployment has meant a low income and low living standards have been forced to survive as best as they can. When unemployment benefit was greater than sick benefit in the 1930s, workers would prefer 18 shillings (90p) unemployment benefit to 15 shillings (75p) sick benefit. A Cwmavon doctor in 1932 had this to say in a report to a NHI medical officer: 'I see patients going to the Labour Exchange rather than claim sickness benefit, as unemployment benefit pays more than sickness benefit' (PRO, MH 49/21 Cwmavon Medical Aid Society).

The two benefits have varied from time to time, and while workmen and doctors might have connived for the best deal for the patient, the problem is rooted in the welfare benefit structure. Indeed, there is the view that the principal conniver for the past 18 years has been the government, which many times changed the way unemployment has been measured and has accepted invalidity and long-term sickness benefit to make the unemployment figures more acceptable (*see* Chapters 6 and 37). In any case, poverty and honesty are not always compatible. Some politicians might share this view, as Frank Field implied in his alleged comment in the debate on the future of the welfare state: 'means testing is discouraging saving, taxing and honesty and penalising effort' (*Guardian*, 21 October 1997).

Moreover, small research teams in the unitary authorities lack the resources they previously had and confine their work to narrow sectional interests; a section dealing with economic questions cannot provide answers on social care. While there is goodwill among officials, who are doing an excellent job in their own sphere, their job description inhibits them from crossing departmental boundaries. If these difficulties exist for officials, how much more difficult it is for someone who wants information on community care, until that person is personally involved.

The reorganisation of health authorities has merged deprived areas with affluent ones. It is important that the differences between the areas are clearly delineated, as they present different problems. In fairness to Bro Taf Health Authority, it does take account of the differences within its domain.

Rhondda, Cynon, Taff County Council has paid close attention to the findings of Beatty *et al*. (1997) and is determined to attract new industries to the area and, in preparation, train the workers with the technical skills via a system of technical education (Rhondda, Cynon, Taff Economic Development Committee, July 1997).

This chapter shows how important it is to train those in the younger age groups as they still have the will and potential to learn new skills. The tendency of workers over 45 to ease themselves out of employment has been noted by others; Samuel Brittan (1996) drew attention to this problem throughout the UK and focused upon the proportion of men in the age group 55–64 still in employment, which was 86 per cent in 1977 and fell to 65 per cent in 1992. When there are fewer jobs, older men experience strain in the job market and many opt out of it by choosing invalidity benefit. Beatty *et al*. (1997) have shown that this pattern is a particular problem in the coalfield areas, but it may well be much more widely spread.

39

Voodoo Economics: 'Art' and 'Science' in Employment Forecasting

Jamie Peck

Forecasting employment is pointless

Welcome to forecaster world. Here, life is predictable; change occurs only gradually, and usually by small increments visible only to the trained economist; and tomorrow is certain to be pretty much like today. Future economic conditions can be predicted with near certainty because key variables such as unemployment, interest rates and manufacturing investment all move along smooth trend curves, while also being related to one another in a stable fashion. The economy, in other words, works like a very reliable machine. Trained forecasters, like skilled mechanics, say they know this machine so well that they can determine in advance how certain labour-market 'outcomes' (such as employment levels by occupation, unemployment rates, wage levels) will follow from specific economic conditions (including exchange rates, inflation, the level of public expenditure). There is no need for guesswork or gambling in forecaster world, for here forecasting is a predictive 'science'.

Meanwhile, in the real world, the economy repeatedly proves to be much more unpredictable, economic trends and processes much more uncertain. At the time of writing, Asian stock markets are collapsing, sending economic shock waves around global markets. For all the media punditry about the end of the Asian 'miracle', it is clear nobody knows what will happen next, and neither can they predict the long-term effects on the 'real' economics of Asia or elsewhere. While economic 'experts' may guess at this, they cannot be certain how this will play out, even over the next 48 hours, let alone over the next five or ten years. In fact, the economic 'machine' is anything but reliable, breaking down during some periods, overheating in others. It does not trundle steadily along the tracks at a predictable speed. The reality is more like a ride on a badly maintained roller-coaster. Certainly, this has been the recent experience in the British labour market: the recession of the early 1980s saw unemployment rise to over 3 million, as a manufacturing shake-out hit the north and west of the country especially hard; by the late 1980s the economy was booming again, at least in the south and east, and unemployment fell to a new 'low' of just over 1.5 million; then recession hit again, this time hitting the service sector and the South East first, and unemployment rose quickly again to over 3 million; by the mid-1990s, growth was back and unemployment was soon tumbling to 1.5 million (although *see* Chapters 36 and 37). In the process, the 'rules' of the labour market had also changed, following the deregulation drive of the 1980s, the introduction of restrictive trade union legislation, social security reforms, and so forth, such that the 'flexible' labour market of the late 1990s was clearly very unlike its 'rigid' 1970s precursor.

Thus the labour market in reality could hardly be more different from the forecasters' view of it as a highly predictable domain in which the policy environment remains pretty stable, while underlying trends curve gently and slowly. Even ignoring the fact that forecasters have a very poor record in predicting recessions (often the time when labour market practices, processes and patterns change most radically), their inability to anticipate policy shifts renders their views of the future highly questionable. Changes in policy can exert a profound influence on the subsequent development of labour markets. If we look back (as does Chapter 40), the history of labour markets is often better told in terms of momentous policy shifts and political interventions. Key moments in the UK experience, for example, include the 'shock treatment' of monetarism in the early 1980s, the miners' strike of 1984–85, the Lawson boom of the late 1980s (and for that matter the 'Lawson bust' of the early 1990s), and the exit from the European Exchange Rate Mechanism on 'Black Wednesday' in 1992. For all their importance, these phenomena are assumed into insignificance in forecaster world, where they are dismissed as 'exogenous variables'. Because forecaster world is a place of economic purity, there is no role for 'political', 'social' or 'institutional' influences. As employment forecasters explain, these are (somewhat conveniently) regarded as 'outside' the model:

> Most models are set up on the basis that behaviour is fixed. This itself may be a questionable assumption, especially in the social sciences. Behaviour may alter in response to new events or changes in exogenous variables (that is variables whose own determination lies outside the system being modelled). . . . In producing any particular forecast a view must be taken about any exogenous factors that may be important . . . [G]overnment policy is frequently regarded as exogenous. (Wilson and Briscoe, 1991, p. 14)

An employment forecast for, say, 2005 or 2010, tells us what the labour market will look like *assuming there are no changes in policy and no 'unforeseen' events*. Until 1997, then, employment forecasts for the UK would have been drawn on the assumption of a perpetual Conservative government (*see* Chapter 45). Following Labour's victory in 1997, they must assume a succession of Labour administrations, operating steady-as-she-goes economic policies. Employment forecasts, in other words, are 'tales of the expected'. They tell us what the labour market will look like if the trends stay on course and if the rules remain unchanged. One thing we *can* be sure of, however, is that the trends will bend and the policies will change. These are not 'freak' events. They represent business as usual in a complex, unpredictable and recession-prone labour market. Clever forecasters can take into account some of these changes, such as interest rate changes: 'punch that 1 per cent cut into the computer and this is what will (or *may*) happen to manufacturing employment'. But like you, me and the Bank of England, they really do not know in advance – let alone *years* in advance – when and in what direction interest rates will change. They also cannot be sure how a change in interest rates will affect, say, the level of unemployment or job growth in retailing, because these are immensely complex and essentially unpredictable relationships. And interest rates are not, of course, the only things that affect employment levels which are difficult to estimate: the balance of payments, strikes, the skill level of the workforce, the rate of investment in new machinery, relative wage levels and welfare policy all present problems for the forecaster because their complex interactions are intrinsically *un*predictable.

Indeed, some of the biggest changes in future labour market conditions can be anticipated to follow 'exogenous' shocks such as a stock market crash or the UK's adoption of the new European currency, the euro, the precise timings of which are difficult to predict even for experienced speculators like George Soros. In forecaster world, of course, events like this do not happen. There are no earthquakes, no crop failures, no elections, no Black Wednesdays. Instead, every day is a variant of Dull Tuesday. There are no surprises around the corner.

Exploring forecaster world, or 'Trust me, I'm an economist'

Forecasting is a fundamentally orthodox economic practice, central to the profession's self-image as the only 'true science' of the social sciences. Economic orthodoxy has it that the labour market operates in the same way as a conventional commodity market: the supply and demand of labour are balanced out by the price (wage) mechanism; market participants have perfect knowledge and function as utility maximisers; competitive principles dominate; state policy, collective action, institutional influences and so forth are all dismissed as 'exogenous', external and not especially important factors (*see* Marsden, 1986). Stripped in this way of its social and institutional underpinnings, the labour market is rendered an arena of 'pure' and unsullied economic relations.

The problem, needless to say, is that real-world labour markets do not work in this simplistic fashion (Peck, 1996). They are in fact highly institutionalised, being shaped in complex ways by social conventions, legislative frameworks, policy interventions, power relations, and so on. Correspondingly, they are simply not as mechanistically predictable as orthodox economists would have us believe. And this is not just an issue for employment forecasters – as labour markets are admittedly among the most complex, locally variable and socially structured of all economic institutions – but applies to the entire sphere of economic prediction. For all the economics discipline's aspirations to the status of a predictive science, economic forecasts remain deeply fallible, and for good reason.

> Whilst it is sometimes possible to hazard an informed guess concerning the future, it should be pointed out that, in social science, predictive accuracy is partly a matter of luck or chance and can never be founded solely on a deterministic model. [S]ocial systems depend on knowledge and we cannot predict what future knowledge will be. Furthermore, predictions of the future are often confounded by the familiar phenomenon of the self-fulfilling prophecy, and other problems related to the fact that the observer is part of the system being observed. (Hodgson, 1988, p. 43)

So, while economic forecasting inevitably involves guesswork, these are guesses which can have real economic (and policy) consequences. Currency speculation is just that, speculative, but even *speculation* on the part of a key player in the system, like the Chancellor of the Exchequer, might be expected to cause a reaction in the markets.

Policy decisions 'informed' by forecasts can have profound consequences. The political fixation with 'balancing the budget' in the USA, for example, has for all its dubious foundations helped to justify deep cuts in social welfare expenditures. It is not just that the economic principles upon which the notion of a balanced budget are founded are highly questionable (*see* Block, 1996), the economic forecasts that underpin its arithmetic are also wildly unreliable. When in 1996 President Clinton authorised savage cuts in welfare provision, experts in the Congressional Budget Office – who have no shortage of good data or fancy predictive models – were forecasting a government deficit of about $170 billion for fiscal year 1997. Because these 'experts' underestimated the strength of economic growth in the USA, however, the deficit for fiscal 1997 turned out to be just $40 billion, not $170 billion (Cassidy, 1997). On this trend, the US government's budget deficit would quietly disappear of its own accord around the year 2000, a fact which exposes the folly of Clinton's balanced budget agreements with the Republicans, and the regime of public austerity which they have justified.

The point is not that the folks at the Congressional Budget Office should have done better, but rather that they cannot be *expected* to know the economic future, no matter how elaborate their predictive models. If this manifest inability to forecast even economic fundamentals of the largest economy in the world just 12 months in advance tells us anything, it is that forecasters are not only fallible, they are fallible big-time. And the task is no easier for smaller economies like the UK. In general, the smaller the economic unit, the greater is its vulnerability to economic 'shocks'. The uncertainty is multiplied when one is trying not only to forecast the economic fundamentals, such as GDP growth, but also to estimate the effects of these macro-economic shifts on labour markets. In an employment forecast, one is both predicting the future in terms of the economic fundamentals *and* predicting the relationship between these fundamentals and the labour market. If the first task is difficult, coupled with the second it is close to impossible.

Let's make it easy to begin with. Perhaps the most basic of labour market variables is the headline rate of unemployment. At the national level, projecting forward the underlying level of unemployment ought to be one of the more straightforward of the forecaster's tasks ... or is it? Figure 39.1 compares the unemployment forecasts produced by one of the most respected organisations in the field, the National Institute for Economic and Social Research (NIESR), with the actual outcomes since the beginning of the 1980s. Drawn 2 to 3 years in advance, the NIESR forecasts are actually quite cautious in comparison to many employment forecasts (which may have a 5- or 10-year time horizon), but still they struggle to stay with the trend. In fact, the NIESR forecasts typically *follow* the trend rather than *predicting* it. When it comes to the most difficult thing in forecasting, predicting when the trend will bend, the NIESR forecasts usually lag behind reality. To the extent that they tell us anything about the timing of a downturn or a recovery, they do so usually *after* this has started to happen in reality.

These are dilemmas which all forecasting houses share. As Table 39.1 shows, the leading UK forecasters were in 1996–97 anticipating quite different levels of unemployment for the year 2001 and beyond, and the further into the future one

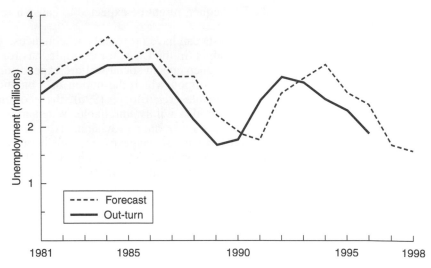

Figure 39.1   Forecast and actual unemployment, National Institute of Economic and Social Research, 1981–98. Based on data from the *National Institute Economic Review* (January, various years)

Table 39.1   'Pick a forecast, any forecast': selected UK unemployment forecasts 1996–2010 (in millions)

Forecaster	Date of forecast	1996	2001	2006	2010
BSL	October 1996	2.121	2.058	1.990	
IER	December 1996	2.1	1.8	1.5	1.5
Cambridge Econometrics	June 1997	2.1	1.5	1.5	1.5
NIESR	July 1997	2.1	1.3		

Source: Meager and Evans (1997) United Kingdom European Employment Observatory, *Trends* **29**, p. 59.

seeks to forecast, of course, the more unpredictable the labour market becomes. An unemployment forecast for 2010, for example, must at least implicitly take into account the results of two or more general elections, the decision to adopt (or not) the euro, the effectiveness of welfare-to-work policies, and a bewildering array of other, equally unfathomable factors (including the changing definition of unemployment described in Chapter 37). Part of the problem is that government policy decisions are themselves bound up in complex ways with the changing state of the economy. Interest rates, for example, must be set in accordance with a relatively short-term picture of the economic 'fundamentals' because even the Bank of England does not know with any precision where inflation and GDP growth are tending (*see* Figures 39.2 and 39.3). What is more, the Bank's *own* decisions will shape the macro-economic environment: applying the interest-rate 'brake' too sharply could itself trigger a recession. And the effects of a recession, of course, would ripple through other parts of the government policy agenda, such as welfare-to-work. Rising unemployment tends to discredit welfare-to-work policies, whether they are 'working' or not (*see* Nightingale and Haveman, 1994), with the consequence that recessions are often associated with a shift in policy.

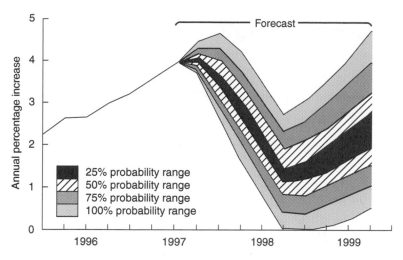

Figure 39.2 Forecast change in gross domestic product, 1996–99. Based on data from the *Economist* (15 November 1997)

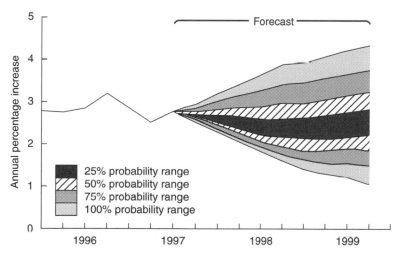

Figure 39.3 Forecast change in inflation, 1996–99. Based on data from the *Economist* (15 November 1997)

The nature of this shift cannot be predicted – the welfare-to-work policy effort, for example, could be redoubled or it could be abandoned altogether – while the net effects on the level of unemployment are near to unknowable. So, while in 1997 the NIESR published estimates of the likely impact of the New Deal welfare-to-work programme for young people, suggesting that it will generate 250 000 net new jobs per year along with annual savings (after three years) of £2 billion (quoted in the *Observer*, 29 June 1997, p. B1), the *actual* impact of the scheme will be contingent on the underlying state of the economy and the effectiveness of local programmes up and down the country, none of which was even on the drawing-board when the NIESR's forecast was produced.

Figure 39.4 'And so, extrapolating from the best figures available, we see that current trends, unless dramatically reversed, will inevitably lead to a situation in which the sky will fall.'

If it is difficult to predict national shifts in unemployment, think how much more difficult it is to predict changes at the local level. The kinds of unexpected events which may fortuitously cancel one another out at the national level, as a factory closes in Burnley while another opens in Bracknell, clearly do not have this 'equilibrating' effect at the local level. Forecasting unemployment in Burnley or Bracknell is inevitably more difficult because at this local level 'small' events can have big effects. These difficulties are multiplied still further where the objective is not only to forecast the aggregate levels of employment and unemployment, but to predict the sectoral *composition* of employment by industry or occupation. Changes in the composition of employment by industry or occupation are the complex outcome of a multitude of decisions concerning, for example, investment in new technology, corporate strategies, the organisation of domestic work, training and recruitment policies, exchange rate levels, social security systems, staying-on rates in full-time education, and so forth. Whether they are explicit about it or not (most are not), employment forecasts must involve judgements about all these factors. Some of course may be more predictable than others: the size of the cohort of 16-year-olds in a particular area is relatively predictable quite a few years in advance, though the proportion of young people entering the labour market each year will tend to vary along with local labour market conditions, the availability of education and training opportunities, and so on. Usually the forecaster plays safe by assuming that nothing will change very much, though this does not make their forecasts any more reliable. After all, 'assuming' that an area's rate of participation in full-time post-16 education will remain frozen at 41 per cent is just as much a 'forecast' (or guess) as predicting that it will rise to 51 per cent.

Doubtless part of the problem here stems from the cultural status of economics, coupled with the spurious degree of accuracy which tends to be associated with numerical data. Quantification itself seems to be a source of assurance for some policy-makers, even if the bases of this quantification remain dubious, as must always be the case with economic forecasts (*see* Chapters 11–13). Forecasts may *look* reliable, but behind the facade of scientific respectability is a labyrinth of guesswork and questionable assumptions. Whereas a sociologist might *imagine* what the future might be like, an economist can *forecast* it. Neither, of course, has access to a crystal ball, but one has the legitimacy of 'science' and quantitative data. Recall Kelvin's dictum: 'When you cannot express it in numbers, your knowledge is of a meagre and unsatisfactory kind' (quoted in McCloskey, 1986, p. 7).

Perhaps society places confidence in economists in part because there is a *desire* for the economy to be predictable and orderly. Unlike other spheres of life, such as football, which are known to be subject to all manner of vagaries and uncertainties, we seem to want to believe that the economic future is rather more reliable. But why should we place more confidence in an economist's estimate of the number of construction jobs in Nottingham in the year 2003 than we would in a football pundit's estimate of Nottingham Forest's final position in the 2002–03 league? Even if the football pundit created an elaborate computer program which could estimate the result of every match between now and then, few would have any confidence in the 'predictions' it produced. This is because we all understand football to be a very unpredictable game and because football pundits are often shown to be wrong. The economy, though, is just as unpredictable; economic forecasters just as fallible. The basic difference between economic forecasting and football punditry is that economists are able to hide their guesswork behind the legitimacy of science, and if really cornered retreat to the high ground of econometric jargon. A common ruse is to climb inside the 'black box' of the forecasting model, to bewilder doubters with talk of log-linear specifications and omitted variable bias. Anything, in fact, but admit that their guesses are no better than ours. As a concession to the cynics, forecasters will also often admit that theirs is as much an 'art' as a 'science' (*see* Bolling, 1996), though in practice few, if any, are prepared to drop the mantle of scientific respectability. To do so would be to erode their professional status in the eyes of clients in the policy-making world.

> Fortune tellers have used the stars, palms and tea leaves (amongst other things!) to attempt to peer into the future. In more recent times, the pre-eminence of the 'scientific' method and the development of computers has resulted in the dramatic growth of formal forecasting of socio-economic trends. . . . Everyone plans, even someone who simply lives from day to day with no apparent thought for the future is implicitly assuming that tomorrow will be much the same as today. Of course, there is much uncertainty about what precisely may happen. Forecasting provides a means of trying to minimise this uncertainty by building a picture of what tomorrow might be like, given information about today and the recent past. (Wilson and Briscoe, 1991, p. 12).

But adopting 'formal' methods and using computers does not make the future any more predictable. Forecasters' models are only as reliable as the assumptions on which they are constructed and the guesses that are fed into them. The problem of the unpredictability of economic life is endemic in all forecasting work, no matter how precise or convincing the results look. Guesses in, guesses out.

Forecasting in the real world, or 'Sorry, wrong number'

If problems of employment forecasts run so deep, why do policy-makers find them so irresistible? Much of the reason for this may originate in the anti-interventionist culture established in the 1980s: the belief was that the market is right, therefore the market had to decide. Forecasting played a part in this ideology because it was one of the means by which the policy-making process was subordinated to supposedly inevitable market forces. The purpose of policy was to run with the grain of the market (or even mimic the market itself), not to buck the trend. 'Our forecast tells us that manufacturing activity will have all but disappeared from our area by 2005, so let's concentrate on one of the growth areas – tourism. What we need is an industrial dereliction theme park.' Ironically, this kind of backwards planning (in which policy tools are used to reinforce anticipated market developments) may help to confirm the predictions of the forecasters *despite the fact that:* (a) their forecasts may be wrong; and (b) all government policy is treated as exogenous to the forecasting model! It would, of course, have been dreadfully *outré* in the 1980s to take the same forecast as the rationale for a manufacturing reinvestment strategy, though in retrospect perhaps more sensible.

Forecasting has also proved antithetical to planning in the sense that it has focused attention on the supposed final outcomes of employment restructuring without enquiring into the causal processes involved. Without understanding the mechanisms involved, it is impossible to intervene in a sensible way. By presenting a plausible economic future as an inevitable economic future, forecasters and their clients in the policy-making world have played a part in stifling debate over alternatives. The future ceases to be contestable; policies can only oil the wheels of the economic machine as it rolls along its fixed course. Forecasters will often present themselves as servants of the planning process (*see* Heijken, 1994), but more often than not their activities undermine purposive planning. Planning should be about making the future, not succumbing to it.

It would be different, of course, if forecasters were right more often, or they could accurately factor in the effects of different policy options. Instead, forecasters usually give us bad guesses about the future based on the assumption that behaviour and policy are somehow set in stone. When in 1989–90 the first round of Training and Enterprise Councils (TECs) were established, for example, the initiative was predicated on the expectation that unemployment would continue to fall, accentuated by the 'demographic timebomb' (the sharp drop in the school-leaver population due in the early 1990s). Employment forecasts commissioned by many of the new TECs, with projections through to 1994 and beyond, confirmed this optimistic scenario. Accordingly, the new TECs proposed policies designed for a tight labour market: mobilising women returners for the workforce, tackling skills bottlenecks, and so on. Conceived in good times, the TECs were, however, born in bad (Gapper and Wood, 1991). Recession returned. In this case, the bend in the trend clearly wrong-footed the forecasters. Policies designed on the basis of their advice turned out to be quite inappropriate for the depressed labour market conditions of the early 1990s, reinforcing the short-termism in the TEC regime itself (Peck, 1993).

The NIESR had originally predicted a further fall in unemployment (of 100 000) over the period 1990–91. In fact, unemployment rose by 700 000. Local-level forecasts were generally much wider of the mark. Many of them expensively obtained through consultants, these forecasts were self-evidently worse than useless as aids to planning. In their defence, forecasters will argue that while they cannot be expected to foretell the future precisely, a carefully drawn forecast is 'better than nothing'. How? Planning in accordance with a conservative and unrealistic view of the future is clearly not going to be good planning. In this sense, the forecast is the 'do nothing' scenario; it represents what will happen to the labour market if the policy-makers fall asleep at the wheel. But why not scrap the plans for the industrial dereliction theme park and set up a long-term venture capital fund? Why not, in other words, set out to defy the forecast by planning proactively rather then defensively? Appropriately modest short-term projections of the likely impacts of policy might have a role in this process, but should be used only in situations where all the assumptions and parameters are on the table for all to see.

Forecasting, in spite of the protestations of its practitioners to the contrary, has played a role in the process whereby active planning has been subordinated to market passivity. Economic policy-makers should be working towards a vision of how they and their political leaders *want* the economy to be, not simply facilitating market trends. One may want to argue about these visions and about how to achieve them, but at least then the discussion will be about political priorities and policy measures, not just submitting to some inevitable future. Surely the worst thing to do in this situation is to hitch policies to the 'do nothing' scenario of the employment forecast. Forecasters, in this sense, are economic reactionaries; forecasting itself is an intrinsically political process, although its association with formal economic methods tends to suggest otherwise. As with everything in politics, it has to be struggled over. Leaving it to economic technocrats with their black-box models is not enough. Their readings of the economic tea-leaves may *look* reliable, but it is time – given the scale of the guesswork behind them – that they were exposed for what they are. Our guesses really *are* as good as theirs.

40

Working with Historical Statistics on Poverty and Economic Distress

Humphrey Southall

Statistics change with society but, with care, comparisons can be made

For most of the twentieth century, the accepted way of measuring whether economic conditions are getting better or worse, and which regions have the greatest economic problems, has been the unemployment rate. Methods for calculating it are highly contested; under the Conservative governments of 1979–97 repeated 'technical' revisions reduced the reported rate and gave rise to critical views of its use as an economic indicator (*see* Chapter 37).

This chapter is concerned not with the past twenty years but the past 150. How best can we study the timing and the geography of hardship over so long a period? This is partly a matter of sources but also raises broader questions about the relationship between statistics and the changing society they document.

Records of unemployment and pauperism

From 1912 onwards, an unemployment rate is available, gathered through employment exchanges and the National Insurance system (Garside, 1980). As discussed below, National Insurance covered only a few industries until 1920, while middle-class occupations were added only in the 1940s. The series plotted in Figure 40.1 also excludes pre-1939 'juveniles under 16', farm workers and domestic servants. The county-level data used in Figure 40.2 were computed from the *Local Unemployment Index*, published monthly between 1927 and 1939 by the Ministry of Labour. The percentages reported are problematic because the count of those registering as unemployed included under-16s and over-65s, and farm workers, while the divisor did not, but the rates used here were adjusted to compensate.

If the study of twentieth-century economic hardship has concentrated on unemployment, nineteenth-century research has focused on the Poor Law. From the creation of the New Poor Law in the 1830s onwards, central government agencies published voluminous statistics on numbers of 'paupers'. These recipients of Poor Relief were stigmatised, for example losing the right to vote. The system developed its own language for categorising paupers, but Figures 40.1 and 40.2 are

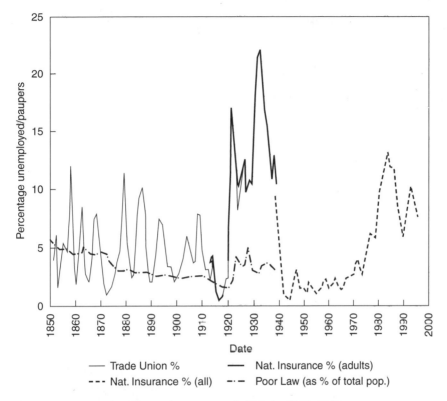

Figure 40.1   Unemployment and pauperage in Britain, 1850–1996
Source: The trade union series for 1851–80 comes from Mitchell and Deane (1962) and
thereafter from *British Labour Statistics Historical Abstract* (Department of Employment
and Productivity, 1971). The Poor Law series is taken from Williams (1981, table 4.5). The
1913–39 National Insurance series for adults comes from *British Labour Statistics* (table
160). The National Insurance series for all workers was taken from: (1939–47) *British
Labour Statistics* (table 161); (1948–68) *British Labour Statistics* (table 165); and
thereafter from either *Social Trends* or the *Employment Gazette*. This last series is an
annual average of the monthly claimant count for both men and women in the UK

concerned simply with the total number of paupers as a percentage of population;
Figure 40.1 plots the average of counts for 1 January and 1 July while Figure 40.2
uses January data only.

Twentieth-century Poor Law statistics are easily available up to 1939, if little
studied. A national pre-1911 unemployment time series is widely known, created
by government statisticians but based on trade union data (*see*, for example,
Mitchell and Deane, 1962). However, the underlying trade union records have
been almost completely neglected even though they provide more detail. Of
course, without a 'laying-on of hands' by government statisticians they are 'unofficial', but their history is fascinating.

The earliest known trade union scheme was operated by the London Tin-Plate
Workers (1798), but early unions were tiny and mainly helped jobless workers
through the so-called tramping system, granting them a meal, a pint of beer and a
night's lodging in each town as they searched for work. Reflecting a nationwide

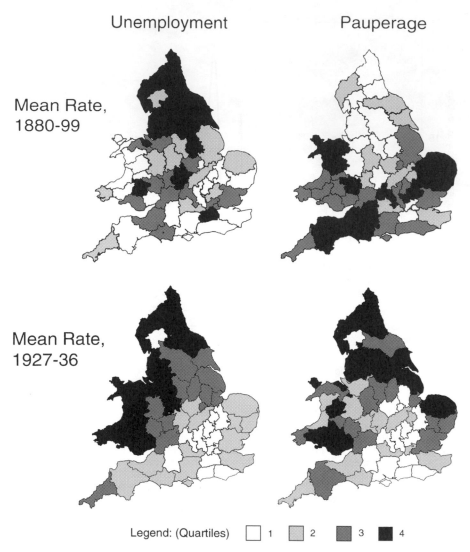

Figure 40.2   County-level measures of economic distress

recession in the late 1840s, and a more integrated labour market promoted by the new railways, the major artisan unions introduced weekly payments to their unemployed, and *Monthly Returns* based on numbers signing the 'Vacant Book' in each branch: the Amalgamated Engineers from 1851, the Ironfounders from 1854, the Carpenters and Joiners from 1863 and so on. Even in the 1890s almost all unions paying unemployment benefit were in engineering and shipbuilding, construction and printing. Figure 40.1 shows the official series for 'all unions reporting', while Figure 40.2 was computed from the returns of the two largest, the Engineers and the Carpenters, using January and July only.

Figures 40.1 and 40.2 combine trade union and National Insurance unemployment rates with Poor Law statistics to cover a very long period indeed – but what conclusions can we draw? Let us consider first the pre-1914 period and then the recession years of the 1920s and 1930s.

Distress ecologies

Pre-1914, the trade union series show an almost regular oscillation, peaking every 8 to 10 years – the trade cycle – while pauperage shows a slow downward trend, accelerating in the 1870s. We can just detect a set of cyclical peaks, typically lagging those in the unemployment series by about a year; the Poor Law peak in the early 1860s is much the largest. Similarly, the maps of unemployment and pauperage in Figure 40.2 suggest two completely different geographies: trade union unemployment was highest in the north, while paupers were concentrated into a belt from Dorset to Norfolk.

What we have here are the records of two very different systems of relief. The Poor Law was primarily intended to relieve farm labourers' families and was well suited to areas where most workers were only fully employed at seed time and harvest; the band from Dorset to Norfolk was where grain production was most dominant. The system was funded via a local property tax, hence it can be seen as a mechanism whereby landowners collectively supported their workforce through winters. Conversely, in northern towns it had been imposed only following great resistance and served only as a last resort for the truly desperate (Driver, 1993). A government statistician commented in the 1890s that 'distress ... will be more acute towards the end of a cycle of depressed trade ... owing to the gradual exhaustion ... of savings and resources' – which explains the lagged relationship between the peaks in unemployment and pauperage (Southall, 1991).

Other features of the Poor Law series must similarly be related to policy. The New Poor Law was designed to achieve greater central control, especially over the perceived abuse of 'outdoor relief' – cash handouts rather than relief in the workhouse. The national Poor Law Board supervised local guardians, and even though the highest levels of outdoor relief were in agricultural districts, they applied greatest pressure on urban guardians: agricultural labourers unemployed in winter were somehow legitimate, urban workers jobless owing to recessions were not. The one exception was the cotton famine of 1862–63, when the American Civil War interrupted Lancashire's supply of cotton, causing vast hardship: the Board suspended the Outdoor Relief Regulation Orders, leading to the blip in Figure 40.1 (Rose, 1977). Conversely, in 1871 the Poor Law Board was replaced by the Local Government Board, which launched a crusade against out-relief, hence the sharper decline (Williams, 1981).

The trade union series is less problematic as its behaviour was not driven by administrative changes but we must still bear in mind its origins. The union schemes were created by unionists to meet their own needs. Entry to the artisan trades was limited by apprenticeship, and the unions were often even more selective, admitting only men under 40 and excluding the unhealthy. Consequently, structural unemployment was almost completely absent. Further, unemployment benefit rules were subtly linked to the unions' industrial role; what the rules typically said was:

Should any free member be thrown out of employment under circumstances satisfactory to the branch to which he belongs . . . or non-free member be withdrawn from his situation by [the union], and continue out for three consecutive days, he shall be entitled to the sum of ten shillings per week. [Amalgamated Engineers' rules, 1868]

This benefit, which was as much as an unskilled man's weekly wage, covered men losing their jobs because their employer simply had no further work for them. However, 'circumstances satisfactory to the branch' also covered men refusing pay cuts, while 'withdrawing men from their situation' meant strike action. In fact, the unions issued annual wage books which laid down 'standard rates' for each town, which was the rate below which members were entitled to refuse work. These rates were pushed up during booms but cut in recessions, to prevent the unions' funds being exhausted by benefit payments. The pre-1914 data used by Phillips in his famous study of the relationship between wage inflation and unemployment are largely based on union wage and unemployment data – so we should perhaps interpret the Phillips Curve as a micro-economic phenomenon (Phillips, 1958).

The Poor Law and union benefits were both adaptations to particular economic environments: agriculture, particularly grain farming, which inevitably created seasonal unemployment; and artisans who sought to protect their high wages during the recessions which were endemic to the capital goods industries of the 'workshop of the world'. Once we understand this, the resulting statistics provide real insights into the lives of these groups. However, at least two other sets of environments and adaptations are less easily studied.

First, artisans' skills fitted them to work for many different employers and made them mobile. However, other workers had acquired skills 'on the job' that were relevant only to a particular workplace. In such sectors, most obviously in textiles and mining, industrial communities were often isolated and dominated by a single employer, typically an owner–manager living locally – Mr 'Ardcastle at t'mill. Such workers had little mobility, but equally, employers needed to sustain 'their' community to retain a workforce for better times. Here the normal response to recession was not unemployment for the few but short-time working for the entire firm. This can sometimes be studied locally, but there were no national statistics.

Second, in the great cities many workers, generally with poorly defined skills, could rely on neither trade union nor employer. In utmost desperation they could turn to the Poor Law, but mostly they moved between different occupations depending on circumstance. Sometimes that occupation meant 'employment' – labouring in a factory, or casual dock work – but often they scavenged a living: collecting and selling firewood, running messages, casual prostitution, petty crime (Stedman-Jones, 1971). This was the world of Dickens's *Oliver Twist* (1837), or Mayhew's *London Characters* (1870), and modern notions of 'unemployment' are inappropriate.

Given this diversity of economic ecologies, how can we best measure overall patterns of distress? There are statistics which tell us something of those outside the union schemes and the Poor Law, such as records of small debt cases and even the marriage rate – which resembled the trade union data (Southall and Gilbert, 1996). However, any real understanding of economic distress in pre-1914 Britain requires use of a range of statistical sources and qualitative data.

What changed in the inter-war Depression?

The figures show clearly that both unemployment and pauperage moved to a higher level. However, how much of the change in unemployment was due to the shift from trade union to National Insurance sources? Similarly, was the reversal of the long-term downward trend in numbers on poor relief yet another consequence of administrative changes?

Many writers have claimed that the First World War marked the end of a golden age, particularly for the north of England (for examples, *see* Southall, 1988). Two pieces of evidence support this erroneous view. First, most contemporary poverty researchers were London-based and concentrated on the East End – 'Darkest London'. The most extreme disparities in pre-1914 Britain were within London; for example, in 1911 the two urban districts with the highest and lowest numbers of domestic servants, a measure of affluence, were Hampstead and Bethnal Green. However, this tells us nothing about the *regional* disparity between north and south. Second, if we see the Poor Law system as the precursor of National Insurance, as do most histories of the welfare state, and then wrongly compare pre-war pauperage with inter-war unemployment – i.e. the top right and bottom left maps in Figure 40.2 – we do find a striking reversal in the geography of distress.

Once we understand that there were two parallel systems, contributory insurance and means-tested poor relief, we must examine change in each. The change in insurance is most obvious: a voluntary union scheme was replaced by a compulsory state system. However, the continuities were substantial: the union schemes grew during the 1900s to include non-artisan sectors, while the state scheme of 1911 borrowed much of its detail from the unions and was largely limited to building, shipbuilding, engineering and vehicles – sectors already covered by union schemes. In 1920 it was extended to all employees over 16 excluding, principally, agriculture, domestic service and non-manual workers earning over £250 per annum; by 1931, the system covered approximately two-thirds of the working population.

The government continued to gather trade union statistics until 1926, and Figure 40.1 shows that between 1913 and 1926 the union and National Insurance statistics behaved very similarly. It therefore seems fairly clear that unemployment *was* much higher in the inter-war period, and more sustained. However, two provisos should be noted. First, while the union schemes pre-1914 were not only largely limited to artisans but also selective, the unions doubled in size between the 1900s and 1920s. Pre-1911, the ability of unions to deny blacklegging members their sick pay and superannuation, and to grant unemployment benefit to members rejecting wage cuts, strengthened them in industrial conflict. Post-1911, industrial strength depended far more on sheer size, while sickly or incompetent members were less of a problem provided they were admitted to new classes of membership outside the benefit system; increasingly, unions withdrew from welfare provision and became more purely industrial organisations. Therefore, in comparing pre-1914 union rates with the early 1920s, we are not entirely comparing like with like.

Second, once National Insurance became almost universal, other mechanisms for dealing with reduced labour demand became uneconomic. Industries which relied on short-timing working were still required to pay into the scheme, but men had to be unemployed for at least three *consecutive* days to claim – so short-timers got nothing out. Unsurprisingly, 'unemployment' became increasingly prevalent in sectors which had previously relied on short-time, although even in the 1930s short-time was common in the Midlands, Lancashire and Yorkshire, while casualism and underemployment were still common on Merseyside, for example (Whiteside, 1991).

If we accept that the pre-1914 trade union unemployment data are *broadly* comparable with the inter-war National Insurance data, Figure 40.2 shows considerable continuity in the geography of unemployment. This is particularly true when we examine sub-county rates: in April 1909, rates among engineers were 30.9 per cent in Howden on Tyneside and 24.2 per cent in Greenock, while the *Labour Gazette* gave a rate of 38.8 per cent for shipbuilding on Wearside; meanwhile, new centres of light engineering in the Midlands had much lower rates: 4.1 per cent in Derby, 3.8 per cent in Coventry. Broadly speaking, the geography of unemployment in the inter-war period is best described as an intensification of pre-1914 patterns, not a reversal. Other statistical evidence, for example on infant mortality and overcrowded housing, shows clearly that the late nineteenth century was scarcely a 'golden age' for the north.

The Poor Law is associated with the nineteenth century, but the inter-war Depression gave it new life. Unlike National Insurance, it was locally administered by elected guardians, which meant that by the 1920s it was often Labour-controlled. While the initial National Insurance system was modelled on the self-financing union systems, with clear entitlements, during the 1920s the depth of the recession and its extension to other groups meant that it ran at a large loss, and consequently politicians intervened to limit expenditure. This blurred the divide between insurance and welfare, and meant that in some areas a generous Poor Law was preferable to an insurance system which, for example, denied millions of claims on the grounds that claimants were not 'genuinely seeking work'.

The Poor Law therefore became the last resort for workers who had exhausted their six months' entitlement to unemployment benefit, and for women and many others who almost automatically failed the 'seeking work' requirement, while remaining the first resort for those outside the insurance system, notably agricultural workers, and for those outside the labour market such as the chronically sick and the elderly. The headings used in classifying Poor Law statistics reveal a system slowly adapting to an industrial society. Pre-1914, the dominant divisions were, first, into indoor and outdoor relief, and then into able-bodied, not able-bodied and lunatics – implicitly defining the causes of their being on relief in terms of personal characteristics (*see* Chapter 23); pre-1858, they tabulated the marital status of adults and the legitimacy of children. In 1884, they began to divide able-bodied male paupers into those 'relieved on account of their own sickness, accident or infirmity' and those 'relieved for other causes', while in 1913 the able-bodied/not able-bodied distinction was dropped altogether. From 1922, those on 'domiciliary relief' – i.e. outside workhouse institutions – were divided into those relieved 'on account of unemployment' and others, mainly the sick.

The inter-war Poor Law map covers all paupers and shows a system in transition. High rates in East Anglia and parts of the south-west are a relic of earlier patterns, but now the highest rates appear in the industrial north, and in south Wales. The system had clearly changed, accepting industrial unemployment as legitimate grounds for relief. Mapping those relieved 'on account of unemployment' still more closely resembles the map of inter-war National Insurance data, but even here there are differences, notably much higher rates in east London. Note that while the pre-1900 maps cover quite separate groups of people, there is a complex overlap between the unemployed count, based on numbers signing on at labour exchanges, and the Poor Law statistics, based on benefit recipients.

The Poor Law has never really gone away, but it has been nationalised. In 1934, responsibility for everyone on poor relief through unemployment was passed to an Unemployment Assistance Board, justified by the 'irresponsibility' of local authorities. Then the post-war Beveridge reforms eliminated local democracy completely: Poor Relief and the 'dole' became National Assistance, Supplementary Assistance and now Income Support – as many name changes as Sellafield, but essentially the same system. Accountability seems to have been thrown out along with democracy: very detailed Poor Law statistics were published as part of a system of surveillance of supposedly spendthrift local administrations, but statistics on the detailed geographical distribution of recipients of Income Support are far less easily available.

Unemployment and the labour market

Statistics of unemployment and poverty were and are moulded by the rules of relief systems; the government does not always act as an impartial gatherer of facts. However, there is a still deeper problem inherent in any use of statistics spanning long periods of time: statistics are a social product, and just as society changes, so does the meaning of social statistics. We may be able to graph unemployment for Britain from 1851 to the present, but the profound changes in *employment* over that period must affect the measurement of *unemployment*. The superficial criticism of pre-1911 statistics is that they cover only unionists in certain sectors, but the real question is whether unemployment was a meaningful concept in other sectors. Rural landowners, through the Poor Law – and mill- and mine-owners through short-time – knew they had to sustain their localised workforces; while the urban poor had to sustain themselves somehow, rarely having the continuing status we understand as 'employment', and in the deepest recession had somehow to scavenge an income; they might well starve to death but were never 'unemployed'.

In the mid-twentieth century, labour law and welfare rules made 'employment' a near-compulsory part of earning a living while macro-economic 'full employment' policies made it almost universally attainable. Unsurprisingly, unemployment became the dominant and unproblematic measure of society's economic success. However, since the 1960s economic restructuring has transformed labour markets and made unemployment once again a highly contested measure. More and more workers are self-employed, not employees; work for a succession of small firms rather than a single lifetime employer; work part-time or flexible hours. Both the numerator and the denominator of the unemployment rate become more and more problematic, and, arguably, tinkering with unemployment measures serves to distract attention from this deeper transformation. However, these are not new challenges. All these problems and more are to be found in the analysis of historical statistics. Back to the future!

Acknowledgements

This chapter draws on research carried out in collaboration with David Gilbert of Royal Holloway, University of London. The database of pre-1914 statistics was funded by the Leverhulme Trust, and extended into the inter-war period with support from the Nuffield Foundation.

PART VIII
Economics and Politics

'And so, extrapolating from the best figures available, we see that current trends, unless dramatically reversed, will inevitably lead to a situation in which the sky will fall.'

41

Measuring the UK Economy

Alan Freeman

When reworked, UK national accounts show a lack of investment by profit-makers

The emperor's tailor

Schumpeter (1994, p. 7) offers an innocuous case for treating economics as a science:

> A science is any field of knowledge in which there are people, so-called research workers or scientists or scholars, who engage in the task of improving upon the existing stock of facts and methods and who, in the process of doing so, acquire a command of both that differentiates them from the 'layman' and eventually also from the mere 'practitioner'.
>
> Since economics uses techniques that are not in use among the general public, and since there are economists to cultivate them, economics is obviously a science within our meaning of the term.

This remains the image which the economics profession presents to the public. However, it is not without problems. It is not self-evident that the use of 'techniques ... not in use among the general public' by 'so-called research workers' qualifies something as a science. Any specialism has its own techniques, often arcane – for example, astrology. What distinguishes science from the merely esoteric is a means of judging its results.

Of course, economists check their own results, often rigorously. However, so did the Spanish Inquisition. The common factor is that both qualify only their own specialists to administer the checks. In a nutshell, economics judges its own results, and the idea that the uninitiated might judge for themselves is simply not entertained, as testified by the breathtaking presumption of an *Economist* leader entitled, of all things, 'The failure of economics' (23 August 1997, p. 11):

> Crucial ideas about the role of prices and markets, the basic principles of microeconomics, are uncontroversial among economists. These are the first ideas that politicians and the public need to grasp if they are to think intelligently about public policy.

Perhaps the first idea the economists need to grasp, if they are to think intelligently about anything at all, is that what they take to be uncontroversial might also be wrong.

The emperor's tailor cannot judge the emperor's suits. The medical profession is highly specialised, but offers an independent test of its methods, namely whether they cure patients. Economics has never been known to cure anything. In practice, when challenged, it seems to offer only three means of demonstration: proof by expertise, proof by authority, and, under duress, proof by obscurity. This chapter seeks to restore the scientific principles of *independent verification* and *transparency*. Since both computers and statistics are now easy to obtain, an average person can, with some hard work, reconstruct the 'plain facts' of the economy, and so study – and judge – policies and their outcomes. The problem seems daunting only because every economic fact seems to lurk in a maze of theory. This is what I hope to redress: I hope to explain how the facts are constructed, and so clear the way to an independent judgement on what they really are. My purpose in this chapter is not to construct a new definitive authority, but to restore to the public the authority stolen from it, by providing it with the means and the right to look at the data differently.

On the basis of the reconstruction suggested below – one among a number of different possible presentations – a radically different insight into the structure of the UK economy emerges, as can be seen from Table 41.4 showing the distribution of national output and the allocation of property incomes in 1980 and 1994.

What are national income statistics about?

I deal exclusively with macro-economic statistics from the National Income Accounts. These principal measures of the economic state of the nation are published in the 'Blue Book' (Central Statistical Office, 1995); its companion, *Sources and Methods*, explains how these are calculated from raw data. The accounts adhere to an international standard, the UN's *System of National Accounts* (United Nations, 1993).

Measurement of output

Economic statistics, like all others, are derived from raw data. The results are presented as absolute facts, following in some self-evident way from the nature of the raw data. However, the derivation expresses a choice, as we can see by considering the most basic economic concept as it appears in the accounts: *output*, or what the economy produces.

Example 1. Domestic work is not treated as part of output. The accountants say it adds no value because it is not paid a wage.

Example 2. However, payment does not alone qualify something as a contribution to output. Pensioners are not considered productive because they do not work. Their money is treated as a *transfer* from someone else.

Example 3. But work is not an essential qualification. Landlords, who arguably work no more than pensioners, are considered to contribute to output by

supplying a factor of production (a piece of land), and their rent is said to measure the size of their contribution.

Example 4. Nor is payment essential: two-thirds of all rent is not paid to anyone
but is 'imputed'. If you own your house, you are deemed to rent it from yourself:
the bigger your house, the better off is the nation. In 1994 imputed rent of owner-
occupied buildings (series CDDF in CSO (1995), table 4.1) was £35 115 million
and total rent income (series DIDF, table 1.4) was £56 793 million.

These four choices, taken together, cannot be squared with any notion that the
data themselves inform us what to do with them. Paid and unpaid labour are both
included, as are both active and idle people. Nor are practical problems the issue:
for example, since what is good enough for houses is good enough for housewives
it would be simplicity itself to impute the value of domestic labour – for example,
as the charge made by a typical firm selling the same services. This is not done,
because the accounts arc based on a distinct, theoretical classification external to
the data, and this is what actually determines the accountants' choices.

Productive and unproductive activities and their relation to the accounts

Is there any 'correct' choice of categories for the accounts? Personally, I think the
choice of statistics varies with purpose, so that while a full audit of our resources
should include both nature and unpaid labour, I would personally start with a
study of our capitalists – who command these resources with money – and the
way they use this money to reproduce these resources. Since they can do so only
by hiring labour, the labour itself can reasonably be treated as their only universal
productive resource.

However, nothing is sacred about any version of the accounts. Economics bases
a form of intellectual terrorism, erecting a claim to unchallengeable 'hard science'
status, on facts which it constructs out of the very theory it sets out to test. Science tests theories against facts, but this calls for a level playing-field: it is illegit-
imate to appoint a player as referee. The public needs to confront the facts
presented by each theory with *external* evidence which the profession has not had
the chance to tamper with. For this it requires different versions of the facts, each
corresponding to the theoretical perspective under scrutiny.

This is not pure relativism. Each theory results in a transformed version of
the accounts but these alternative transformations are not arbitrary. They are subordinate to overarching accounting principles, dictated by the logic of a money
economy.

The most fundamental choice, as the examples suggest, is to determine what
counts as output. This results from a conceptual classification, either explicit or
implicit, of all economic activities into those which *add value* and those which
do not; into *productive* and *unproductive* activities (or, in the language of neo-
classical theory, those which are factors of production, and those which are not).
This is the decisive accounting distinction which everyone must make, regardless
of their theory.

A further subdivision is evident. Some activities, like domestic labour, are altogether absent from the accounts. However, there are unproductive factors such as
pensions, in which money changes hands and must be accounted for but which do

not contribute to output. These are treated as *transfers*; they consume income produced elsewhere. These choices are politically and ideologically sensitive, since a transfer appears parasitical.

Measurement and disposal of income

If output measures what people produce, what measures their consumption or *income*? Orthodox theory preaches that income and value-added are identical. Thus rent is the just reward of the landlord, profits of the property-owner, and wages of the worker, each in proportion to the value they add. The accounts, informed by this view, even designate their primary output table the 'factor income' accounts (table 1.4 in Central Statistical Office, 1995b). A second 'expenditure' table (Central Statistical Office, 1995b, tables 1.2, 1.3 and chapter 4) records what these rewards are spent on – but not *who spends it*. The classes of persons defined in the income accounts are amalgamated and treated alike as 'consumers'. We can see what wage-earners or property-owners add to output, but we can see neither their *actual* income nor what they do with it.

This conflicts with the treatment of tax, recorded as a deduction from income or *transfer*, introducing a clear if sheepish distinction between the value which a factor adds, and its eventual or 'disposable' income. The *net tax* calculation proposed by Shaikh and Tonak (1994) categorises all taxes to show from which income they are deducted, and all benefits to the class that receives them, so exhibiting the transfers which the state effects between different types of income (*see also* Fazeli (1996), Shaikh and Tonak (1994) and Freeman (1991)).

But the state is not the only agent that transfers incomes. In principle, as we have remarked, the very concept of transfer depends on what is considered productive. If we deem that any given revenue is not a contribution to output, then we must treat it as a transfer: something which does not add to output but moves it from one class of person to another.

Transfer or product? The strange case of interest payments

Sources and Methods (Central Statistical Office, 1985, p. 88) explains a dilemma 'which has always caused some difficulty in national accounting statistics': measuring what the banks do. Normally, a firm's contribution to GDP is

> measured by its 'net output' – the excess of its receipts from the sale of goods and services over its operating expenditure on purchasing goods and services from other enterprises. .... However the application of this definition to financial companies and institutions produces a paradoxical result.

Profits are normally defined as the excess of sales over costs. However, while bank costs are considerable (marble, security, banquets, etc.), their only sales are 'bank charges and commissions received from depositors'. Hence their output is small and their profits negative, because

> banks derive much of their income by lending money at a higher rate of interest than they pay on money deposited with them; and in the national accounts interest receipts and payments are regarded as transfers and not as receipts and payments for a financial service.

In 1994 the banks recorded gross trading profits of −£10 839 m but the non-bank (industrial and commercial) sector recorded £102 028 m. Total value-added by both sectors is £91 189 m, being the £102 028 m less £10 839 m transferred to the banks (Central Statistical Office, 1995b, series: AIAD (table 5.4), AIFB (table 5.7), CIAC (tables 1.5, 5.1)).

Unfortunately, this partial outbreak of common sense is applied inconsistently, or the accounts would show *all* interest transfers, between consumers, each other, and the banks. We can get some indications from the household accounts, which in 1994 show that consumers received £34 bn in rent, dividends, and interest and paid £30 480 m (Central Statistical Office, 1995b, DJAO (table 1.4), GITP, GIUG (table 4.9)). However, consumers do not pay interest to themselves but evidently fall into two classes: those who pay interest and those who receive it. The monetary system thus *transfers* around £30 bn (8 per cent of wages) from wage-earners to property-owners. This is the beginning of a *true income* account, shown in Table 41.1.

But the story only starts here. If interest is a transfer, then why not profits? The accountants want on the one hand to classify profits as the legitimate reward of 'capital'. But the interest on capital is treated as a transfer. You can't have it both ways: if interest is not a source of value-added, capital cannot be a factor of production, the source of much discomfort.

There are two coherent solutions. The first, adopted in many countries and some UK tables, is to treat *all* property income as a part of output, imputing a sale to the banks equal to their interest receipts and called an 'adjustment for financial services'. However, an equally coherent alternative is to say that *no* property income adds value. In that case, all value is added by wage-earners and all profit income is a transfer out of this value. The accounts then read as in Table 41.2.

We may then further break down the income of the property-owners to show how they divide this parasitical income between investment, military activity, financial costs and the indulgences of the leisure classes, and the actual priorities accorded to these social needs by their private requirements and conduct.

Table 41.1 Income and value added for the personal sector with interest transfers

(All figures in £ million)	*Value added*	*Income*
Value added by wage-earners	£362 758	
Value added by self-employed	£63 655	
Interest transferred to property-owners		(£30 480)
Income of wage-earners		£395 933
Value added by property owners	£157 048	
Interest transferred by wage-earners		£30 480
Income of property-owners		£187 528
Totals	£583 461	£583 461

Notes: Wage-earners include self-employed from now on unless otherwise stated.
Source: Central Statistical Office (1995b). Employment and self-employment income: DJAU, DJAO (table 1.4), income from property calculated as sum of all other categories. Interest payments: GIUG (table 4.9). Transfers to and from the state omitted: net-tax research indicates that this is small

Table 41.2   Income and value added for the personal sector with transfers from all property income

(All figures in £ million)	Value added	Income
Value added by wage-earners	£583 461	
Income transferred as interest		(£30 480)
Income transferred to property		(£157 048)
Income of wage-earners		£395 933
Value added by property-owners	£000 000	
Interest transferred from wage-earners		£30 480
Income received from workers		£157 048
Combined income of the property-owners		£187 528
Totals	£583 461	£583 461

Gross, net and constant capital: what is output, really?

Can we just eliminate property income altogether and define output to be £362 758, the income of the wage-earners? No: and this highlights the existence of a real, scientifically significant constraint. The intrinsic problem of national accounting is that there are two conflicting definitions of output which must be reconciled. The reconciliation procedure is what forces the contradictions in the theory to the surface.

The output of an enterprise consists of the things it sells. If we add up all sales of new useful things, we get an alternative measure of the nation's output. Yet this commonsense definition does not appear in the accounts, which throw away the cost of the materials and machinery – called 'intermediate inputs' in the input–output accounts – used up in making the useful things. This remaining 'net output' is the total of all personal incomes, since these arise from the difference between sales and intermediate input. If we omitted profits altogether, we could not say where all the receipts from sales had gone. National accounts often use the word 'net' to mean output net of depreciation. I mean net output in the census of production sense, calculated by subtracting the value of materials purchased from the value of sales.

The most consistent way to define output is to add up the receipts from the sale of new useful things, in which case value-added is that part of these payments which makes up the income of persons. Because the accounts record only this personal income, they classify a substantial part of the income transferred between property-owners as an intermediate input, and promptly omit it. This leads to the paradoxical result that when we reclassify something as a transfer, our estimate of net output will *increase*.

Example 5. Company cars are often recorded as an expense of production. If we designate them as a perk – a disguised part of wages – then this increases reported national output. Suppose a company previously wrote its accounts (in condensed form) as in Table 41.3a. The bottom part is value-added and the top part is cost. Our minor change in theoretical perspective leads to Table 41.3b. Although gross output is the same, net output has risen by £20 000. The sums of money which figure in the accounts pay for a whole thing, and not just the value-added in it. If a

Table 41.3a Company cars as cost of production

Car trips	£20 000	
Other materials	£100 000	
Total inputs		£120 000
Wages	£30 000	
Profits	£10 000	
Value-added		£40 000
Output		£160 000

Table 41.3b Company cars treated as a perk

Materials	£100 000	
Total inputs		£100 000
Wages		
In money	£30 000	
In kind (cars)	£20 000	
Profits	£10 000	
Value-added		£60 000
Output		£160 000

company car is suddenly treated as unproductive, then a previously suppressed expense suddenly appears like the bottom half of an iceberg.

Banks are like the company car. As the accountants observe, they produce no traded commodity but merely circulate what is created elsewhere. But they certainly get paid. The accounts make a partial concession by recording their interest receipts as a transfer, but their real cost is the whole of the money they receive. If we determine that they add nothing to gross output, then we cannot treat their costs as an intermediate input. Profits properly stated thus include the full cost of the financial sector since this is what it actually costs the rest of society to support the activities of the banks.

The results

The results are shown in Table 41.4. For reasons of space some details of the calculation cannot be reproduced here, but can be supplied by the author on request.

Table 41.4 Distribution of the product of the UK, 1981 and 1994

(£ million)	1980	Percentage of output	1994	Percentage of output
Value added by wage-earners and the self-employed	247 339		751 510	
Of which income of the value producers:	155 924		426 413	
Income from employment	137 783	53.0	362 758	44.7
Income from self-employment	18 141	7.0	63 655	7.9
Of which income of the property-owners:	91 475		324 737	
Gross investment (of which capital consumption)	41 561 (27 952)	16.0 (10.8)	100 075 (68 150)	12.3 (8.4)
Personal *rentier* income (interest receipts)	8 795	3.4	30 480	3.8
Employment in banking sector	12 637	4.9	59 649	7.4
Other costs of banking sector	23 176	8.9	106 729	13.2
Remaining property income	17 943	6.9	87 453	10.8

Table 41.5 Number of workers involved in various branches of production

Employment in the UK	1980		1994	
Agriculture	380		265	
All mining and minerals	354		88	
Manufacturing	7 253		4 330	
Electricity, gas and water	368		223	
Construction	1 239		886	
Trade	4 257		4 671	
Transport and storage	1 056		862	
Post and telecommunications	423		366	
Total productive (generous to retail)		15 330		11 691
Finance	1 647		2 788	
Other services	1 584		2 178	
Total unproductive		3 231		4 966
State services	4 602		4 800	
Grand total		23 163		23 451

An independent check of our calculations which affords further insight is given by the structure of employment. The figures shown in Table 41.5 (from Central Statistical Office, 1995b, table 17.1) show the numbers of people involved in various branches of production over this period. In a dramatic structural shift, 4 361 000 people moved out of production in the normal sense, and into financial or commercial services. The category 'other services' here refers almost exclusively to financial services, although some caution is needed in general with the term 'services' since it tends to include a pot-pourri of productive services such as communications and transport, mixed in with financial and commercial activities.

Over these core years of the Conservative government, employment income from this point of view fell by 15 per cent from 53 per cent of output to 44.7 per cent. But this gain in income to the property-owning classes was not spent on increasing the productive capacity of the nation; over the same period the proportion of investment in output fell from 16.0 per cent to 12.3 per cent; that is, by almost a quarter – and it should be remembered that 1980 was a recession year. Yet at the same time *rentier* income rose from 3.4 to 3.8 per cent and the cost of the banking system from 8.9 to 13.2 per cent. Without these drains on the economy, investment would actually have *increased* by 1 per cent.

These figures are every bit as factual as those retailed by the pundits – but the story they tell is altogether different. When it comes to economic facts, never forget you have a choice.

Acknowledgement

I would like to acknowledge the helpful comments of contributors to the OPE-L electronic mailing list. Any errors are of course my own.

42

Household Projections: A Sheep in Wolf's Clothing

Richard Bate

Statistics should inform debate on planning policy, not substitute for it

Forecasts play a significant part in the town and country planning system in the UK, as local planning authorities need some idea of how many houses, offices, shops, quarries and other necessary developments to plan for. The weight that should be attached to forecasts for planning policy purposes is a matter of political judgement. During the 1990s central government gradually adopted more pragmatic approaches than before. For example, the government has abandoned its long-cherished reliance on the traffic growth forecasts as a basis for planning new road construction to increase road capacity.

The 'predict and provide' approach to using forecasts nevertheless survives elsewhere in land-use planning, and this chapter concentrates on the household projections as a key example. A roof over one's head is obviously fundamental to any society, so at first sight the supremacy of the household projections over other aspects of policy could have its attractions. However, it is the argument of this chapter that over-reliance on the projections has significant adverse effects and that alternative routes to planning for dwellings would be better. There is no implication that these projections were devised with intent to deceive. Far from it: the aim was enlightenment and to inform policy. The wider interest of the case therefore lies in how the statistics were allowed to reach a significance above their station, and in possible alternative methods to plan for housing.

Key features of the town and country planning system

The planning process decides how land will be used. Land is identified in plans as suitable for one or another kind of development, or for a variety of uses. Planning applications for individual schemes are decided against a background of these plans and detailed policies set out nationally by the government. Plans and planning authorities cannot force developers to invest or landowners to improve the environment, but they can present a set of sticks and carrots to increase the likelihood of desirable land uses coming about. The power to refuse proposals which do not conform sufficiently with policy is a big incentive to developers to cooperate.

The projections of household numbers are the most influential forecasts in planning. They inform policy on housing requirements and in turn on the need to identify additional land suitable for house-building. This is enormously important from a land-use planning perspective, because of the significance of housing in the planning process. The quantity and location of housing built has a major influence on the development of shops, offices, industry and leisure facilities, and the transport facilities and other infrastructure which bind them all together. Housing is therefore a significant motor in urbanisation, the combined effects of which consume considerable land. Official research argues that an additional 169 400 ha of rural land will be developed between 1991 and 2016 if the household projections are treated as a target and there are no changes to planning or other policies (Department of the Environment, 1996).

Reliability of the projections of household numbers

New household 'projections' are published every three years or so. The household projections (Department of the Environment, 1995b), based on data from 1992 and 1993, remain controversial despite considerable effort by the Office of National Surveys, the Government Actuary, the Department of the Environment, the Building Research Establishment and others to improve upon the previous version, based on 1989 data. The housing requirements of the additional 4.4 million households projected over the 25-year period have aroused considerable debate, sparking a government discussion paper reviewing the implications (Secretary of State for the Environment, 1996). Reviews of the household projections for planning purposes have been published elsewhere (e.g. SERPLAN, 1996). The robustness of the projections can be questioned for many reasons, including the following:

- The 1989-based projections assumed nil net international migration whereas the 1992-based projections assume that nearly 1 million people net will enter England by 2011, though the basis for the migration assumption is itself controversial (*see* Chapter 19).
- About 1 million people were not counted in the 1991 census, with poll tax avoidance in all probability playing a significant part in the problem: the actual geographical allocation of the missing million may not correspond with the projected distribution (*see* Chapter 2).
- Assumptions about internal migration within England for the whole projection period rely on census information which records movements in the year prior to census night. Although no single year is necessarily representative, the year 1990–91 in the depths of the recession may be particularly inappropriate for projection purposes to 2016.

The strategic problems with the projections are compounded by the usual statistical difficulties of figures being less reliable at the disaggregated local level and at greater distances into the future. This is significant for land-use planning purposes in view of government encouragement for plans to look at least 15 years ahead (10 years in the shire districts) (Department of the Environment, 1992b).

Role of the household projections in planning for housing

Conversion of household projections into estimates of requirements for dwellings is essential for planning purposes but requires a further string of assumptions (Green Balance, 1994). For example:

- What proportion of households will share a property, either voluntarily or because they cannot obtain their own home?
- What allowance should be made for second homes, tourist accommodation and dwellings owned by companies?
- What is the scope for discouraging 'under-occupancy', by persuading small households to move to smaller property, thereby freeing up larger dwellings for families?
- What proportion of the dwelling stock will be empty at any time, and how can more efficient use be made of this resource?

The existing stock of dwellings provides for most of the projected housing requirements, while the shortfall is the amount of additional provision required. The requirements are particularly sensitive to two assumptions:

- *Vacancy rates in the existing stock.* Small changes applied across the entire dwelling stock have a big impact on the new stock required. In the case of Devon, for example, the additional dwelling requirement between 1995 and 2011 is significantly different depending on the choice of vacancy rates, whether 4.9 per cent (used by the county council in its draft Structure Plan), 3.9 per cent (the estimated rate in Devon in April 1995) or 3.0 per cent (the government's target to be achieved by 2005); *see* Figure 42.1.
- *Internal migration.* Net migration is the (often) small difference between gross figures of in- and out-migration, so small changes in actual movement patterns can have a big impact on dwelling requirements, and tend to produce unpredictable and volatile results from one year to another.

The government attaches considerable weight to the household projections in shaping planning policy. Generally speaking, the larger the geographical area, the more relevant are the household projections as a guide to housing requirements (after making assumptions on the issues noted above). The level of housing pro- vision at the regional level, issued in 'Regional Planning Guidance' by the gov- ernment for each of the eight standard regions of England, is expected to be reasonably consistent with the projected household increase for that region. At county level, however, there are often good reasons, in terms of local trends and markets, environmental constraints and policy considerations, why housing requirement figures need not correspond exactly with the household projections. The government is committed to its Regional Planning Guidance figures being tested at the county level during the preparation of county Structure Plans (Green Balance, 1996). In practice, the government has been reluctant to allow county councils to adopt policies which would supply much less than their share of the additional housing land specified in Regional Planning Guidance.

The government is committed to ensuring that the planning system finds the additional housing land specified in Structure Plans, so that the requirements can be met if the household projections do in practice materialise. The task of identi- fying sites falls to district and unitary authorities. The government has imposed

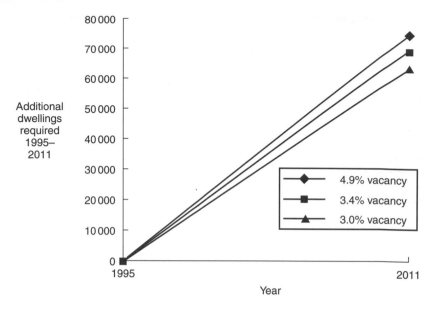

Figure 42.1   Devon County Structure Plan 1995–2011: dwelling requirements at different vacancy rates
Sources: Devon County Council Deposit Draft Structure Plan 'Devon 2011'; Green Balance

strict rules to ensure that there is a ready availability of sufficient land at all times (Department of the Environment, 1992a).

Effects of relying on the household projections

A demographic basis to planning for housing has one big practical attraction. The annual rate of births is known, so planners have about 20 years' notice of the main source of new housing requirements (subject to hefty assumptions on longevity, household size, the institutional population and migration). Unfortunately, there are also limitations in this approach, which eat away at the apparent benefit.

The first limitation is that household projections can be used to mask policy decisions on the location and scale of housing development. The general effect has been to reinforce past patterns of dwelling provision and inhibit change – a point also made in Chapter 39 on employment forecasts. Household projections reflect past trends in housing provision. The more households in an area, the larger will be the demographic roll-forward to generate households in future. These new housing 'requirements' are assumed to be inevitable. The provision of sufficient housing to satisfy them in each area is then assumed to be positively desirable. Failure to satisfy them could imply a local housing shortage which forced some households to migrate away from the area in which they emerged, possibly against their will. 'Zero net migration' is often assumed by central government, explicitly or otherwise, as the minimum level of housing supply at the county level for which planning authorities must find land.

This approach carries the implication that if housing development was desirable in a particular area in the past then it must be desirable to follow through in future the housing effects of that past development. The possibility is played down of deciding that additional dwelling provision should be at a lesser level than implied by the household projections, perhaps because development has reached a clear boundary such as a motorway or the edge of a special environmental area. The 'problem' is perceived as the existence of the limitation on further expansion, not the statistical mechanism of the household projection. The weight attached to the household projections is also not applied even-handedly: if planning authorities allow numerous dwellings to be constructed in an area attractive to migrants, the government has not been known to object on the basis that there was no expectation of demographic need for them in that area in the household projections. The household projections are therefore a tool which the government can use to force local authorities to allow housing development in their areas which for one reason or another the authorities might have preferred to turn away. More of the same may be the easy option, but it diminishes the role of planning, which is about making choices for public benefit in the way we use land.

The second limitation of the household projections is that they are not so wholly independent from planning policy as their origins in demography imply. There is evidence that additional land supply and housing development can fuel extra inward migration. Housing supply can influence the demand for it, as well as vice versa (Bramley and Watkins, 1995). The household projections and planning processes therefore show a degree of circularity. Land allocations based on household projections will to some extent be a self-fulfilling prophecy. Migration is a key variable in the preparation of the household projections, so this finding has important implications both for the projections and for the use of them by planning authorities. There is no control over who buys dwellings on the open market, whether local people or migrants. If, as the research suggests, migrants can acquire a substantial proportion of dwellings built in certain areas, then pre-existing housing needs will remain despite the house-building.

The third limitation of emphasising the role of household projections in shaping planning policy is that the planning system is insufficiently sophisticated to be able to respond properly to the implications of the projections. The projections neglect access to housing. In the private sector this is controlled by ability and willingness to pay. In the social sector it is controlled by the availability of re-lets and the number of new dwellings built. The scale of new social provision is fundamentally governed by the amount of money the government chooses to spend, not housing 'need'. This need for new social housing provision in England has been estimated at around 100 000 pa, plus more to deal with the backlog of unmet need, estimated at over 450 000 households in 1991 (e.g. households sharing against their will; Holmans, 1995). Actual provision of social housing falls far short of this – typically 40 000–70 000 pa – creating a mismatch with the demographic projections.

Planning controls cannot ensure that dwellings built will be occupied by the households whose existence was projected. Land identified by planning authorities as suitable for housing cannot be allocated on the basis of tenure, price or ownership of the dwellings to be built (because in land-use terms the effect of building dwellings will be the same regardless of these matters). The result is that planning authorities tend to make available more land for the social sector than can be used. Evidence to the House of Commons Select Committee on the Environment, Transport and Regional Affairs as amended has suggested that the surplus land arising from the government's Regional Planning Guidance could reach 800 000 plots (Wenban-Smith, 1998). In buoyant market areas this problem is compounded, as there is little to stop demand in the private sector taking up the land projected to be required in the social sector.

A fourth limitation on the usefulness of the household projections to planning is that short-term fluctuations in the housing market caused by economic stimuli swamp the longer-term demographic effects, particularly for the two-thirds of all households in or seeking owner-occupation. Economic issues affect not only households' access to housing, as noted above, but also whether and when some households form at all. There is some evidence that household formation responds to affluence, the housing market and the supply of housing, interest rates and unemployment (Bramley and Watkins, 1995; Petersen *et al.*, 1998). On this basis, the household projections should reflect economic assumptions. At present any connection is accidental, to the extent that a long-term increase in national wealth is reflected in changing patterns of household formation which are in turn accommodated in the household projections.

The planning system has no means of distinguishing the role of economic issues affecting household formation from other influences. Authorities are then uncertain whether short-term departures from trends are just blips or the start of a change in the trend. The price of getting these planning judgements wrong is high. A sustained undersupply of land could cause a rise in house prices. An oversupply, however, might allow developers to cherry-pick favourable locations, perhaps causing damage to the environment in places which could have been protected, and hampering efforts to attract developers to recycle degraded urban sites and other more awkward priority locations for housing.

In summary, using household projections to guide housing land supply and new housing development seems at first eminently sensible. However, when examined in more detail the numbers look less helpful:

● There are statistical difficulties with the projections themselves.
● Unreliable assumptions are required to translate household numbers into dwelling requirements.
● Future patterns of households reflect strongly decisions made in the past.
● No attention is paid to affordability or access to housing.
● Planning controls are unable to ensure that the households projected to materialise will be the beneficiaries of land made available for building.
● Household projections are not independent of planning policy, particularly where substantial supplies of new dwellings generate migration.
● Economic effects on housing provision are more important than demography or land supply.

Why are we here?

The household projections based on 1989 and 1992 data revised previous figures upwards as new information became available. While the reliability of any projections can be questioned, deriving better ones is difficult. This has tended to bolster the credibility of the official projections, and ministers have been reluctant to depart from them as a basis for planning.

This still does not explain why the household projections have been treated as so important. Clearly there is a commitment to meeting housing requirements, and the projections are one measure of that, despite imperfections. On a broader political canvas, the household projections published by government can be interpreted as government control over local choice. The projections provide a convenient yardstick for measuring local planning authorities' performance. By elevating this consideration, the government effectively regulates one of the most influential aspects of local planning. This has not produced the opposition that might be expected because there is some support for it within local government itself. Although many authorities in areas experiencing development pressure have little desire to meet the projections for their areas, they have still less enthusiasm to find extra land to make up for shortfalls spilling over from a neighbouring 'under-providing' authority. The government's attachment to the household projections serves as a means of protection from their neighbours.

Alternative approaches to planning for housing

The way in which the household projections have been used in planning has institutionalised a business-as-usual approach to the supply of land for new housing. On paper the government's view, presented by the Secretary of State for the Environment in a statement accompanying publication of the household projections on 6 March 1995, is that the household projections 'represent just one of the factors to be taken into account by local planning authorities in England when arriving at figures for housing provision to be included in regional guidance and development plans'. However, in practice, policies have been put in place which ensure that the projections are easily the most important strategic determinant of future rates of land supply. The government has relied on the statistical projections to justify political decisions about the 'need' for housing.

The difficult decisions on how much land to allocate for house-building around the country could be approached from the opposite angle. The debate could – and arguably should – focus on the pattern of housing supply that is politically desirable for environmental, economic and social reasons, instead of on the projection forward of previous patterns of habitation. Why not decide what is wanted and plan to achieve it, with suitable incentives, instead of assuming that projections of past trends must be desirable? The household projections could build-in these policies as assumptions, assess the consequences and then inform the debate rather than control it. Land supply for housing could be managed through phasing land releases as required, rather than rely on the inherent uncertainty of projection-based planning. This would make for a more interesting political process, in which the discussion centred on the real issues rather than on how to interpret the household projections.

There should be a greater recognition of the socio-economic and housing market influences on household formation and on effective demand for dwellings. Planning urgently needs to match more accurately the land and dwellings supplied to the people who need them. More attention needs to be paid to identifying the forces which prompt migration and to the means of channelling these forces in the direction of more sustainable patterns of development. All these issues have a strong locational element which is central to planning practice. Taking more decisions at the regional rather than national level would be consistent with the 1997 Labour government's commitment to devolve power to the regions.

43

The Statistics of Militarism

Paul Dunne

Military statistics serve the interests of the military industry

With the end of the Cold War there has been a clear sea change in international relations which has allowed a marked reduction in worldwide military spending. The world is still a dangerous place but the fact that there is no longer superpower involvement in regional conflicts has reduced the necessity for huge military burdens as well as reducing the availability of military aid. This has allowed the pressures of needs other than those of the military to have a greater impact on the allocation of government spending. There is, however, still a debate over the economic effects of reductions in military spending and their effects on military capability. The vested interests in many countries argue that further cuts can lead to economic problems as well as reducing capability to dangerous levels. There are strong arguments against these assertions, but the debate is hampered by lack of transparency and reliable comprehensive statistics. While some of the secrecy and misinformation of the Cold War has gone, not all of it has. The fog of the Cold War has lifted but a mist remains.

Given that the potential for further cuts in military spending exists, it is important that the nature and extent of the military, military industry and the arms trade are understood. For this, transparent and comprehensive statistics are needed. It is also important that the role that statistics, and the way they are presented, can play in reproducing the power structure which supports the military is understood. This chapter attempts to help bring about such understanding. The next section considers the production of military statistics and how they do not reflect 'facts' but are constructed from theory and reflect social relations. The chapter then indicates the problems which make military statistics so difficult to obtain and interpret. This is followed by a discussion of the available sources of data and a brief overview of what they suggest about the trend in the military economy in the UK.

Military statistics

Military statistics, like other statistics, are not objective facts; they are constructed using theory and reflect social relations. They are also estimated using available information (there are no true statistics hidden in the military establishments waiting to be uncovered), and this available information can often be what is made available by the government. There are examples of where the information is either limited or distorted in the so-called interests of state security, but it is also important to bear in mind that the data often result from processes of estimation with great uncertainties involved. They may be wrong but they may be the best estimate.

In trying to understand military statistics one needs to be clear what is being measured. There is a tendency to conflate a number of concepts and assume that they directly reflect each other. For example, changes in military expenditure will lead to changes in the security forces and this can have an impact on capability. But it is not a straightforward relationship. All the stages are contingent on other factors, yet it is common for commentators to forget this. Following Smith and Smith (1983) we can represent the process in a diagram:

Concepts

Expenditure $\rightarrow$		Forces $\rightarrow$	Capability $\rightarrow$	Security
Cost/burden $\rightarrow$ of resources	Stock of arms, $\rightarrow$ personnel		Performance $\rightarrow$ in combat	Peace/ Success

Links

	Efficiency	Morale	Objectives, threats
	Relative costs	Logistics	Interests

Military spending is an input, not an output. It may produce security, but it is not enough simply to spend on the military to produce or increase security; the money could be spent badly. As the diagram shows, military expenditure will produce a stock of arms and personnel, which will then create military capability and finally a level of security. But there are factors which link these concepts, and the effectiveness of the inputs in producing the outputs is contingent on these links. The effectiveness of the stock of arms and personnel will depend upon the efficiency and relative costs in procurement; the capability of the forces will be their performance in combat, and this is influenced by morale and logistics. Finally, security will depend on exactly what the threats, objectives and interests of the state are.

Things are even more complex in the post-Cold War world. The role of the military keeps changing (peacekeeping, disaster relief), and what it requires (skills and equipment) for these new roles is different to that needed for the Cold War confrontation. Yet the vested interests will tend to provide convincing arguments to maintain some form of status quo and to reject any more radical changes. Thus weapon systems, such as Eurofighter, which were designed for a Cold War role are still developed at great cost.

Such examples illustrate how important it is to recognise that the military–industrial complex represents an important force in society. It is part of the creation of the military statistics and their interpretation (Dunne, 1990). Any information that has been provided still has to be treated with care. Old Cold War distortions, secrecy and propaganda may have been reduced but they still remain.

Problems with the data become more acute when cross-country comparisons are made. Countries often have different definitions and classifications, which can cause problems when making comparison or adding up military expenditure to get an estimate of the aggregate amount. The differences include how para-military forces are treated – whether they are included in the military or not – and similarly how nuclear/space activities are treated. There are other expenditures on armed forces, such as health services to families and pensions, which may or may not be included in military spending. Increasingly, technologies are coming from civilian development 'spin-in', rather than the other way round ('spin-off'), which was the case in the earlier post-Second World War period, and dual-use technologies, with both civil and military applications, are being developed. These can be treated differently in countries when measuring the extent of the military sector, and other parts of military spending can be hidden in other categories of government spending and in foreign aid. Chapter 5 discusses this further in the context of funding for research and development. There are also usually indirect subsidies of the arms trade (Campaign Against Arms Trade, 1996). In making international comparisons it is common to convert to US$, but exchange rates may be distorted by government policy and they can fluctuate widely.

There are problems with the prices used to measure the value which are addressed in the context of national accounts in Chapter 41. For most large weapon systems there are no real 'market' prices, just 'accounting' prices, because of government procurement practices and the subsidies involved both in domestic production and exports. This is still true despite the changes which have occurred. In attempting to measure in constant prices there are problems in choosing the inflation estimate with which to deflate the current price series. Some countries have price indices for the military sector, others simply use the GDP deflator. The differences between the two can be very marked – in the past up to 10 per cent. There are also problems of choosing a sensible base year. Moreover, there are hidden costs, such as with conscript armies, which will appear cheaper but have a high opportunity cost (the cost arising from not using the resources for another purpose) and may be less effective in providing capability than professional soldiers.

To overcome some of these problems, comparisons of military spending are often based upon the military burden, which is usually measured as the share of GDP devoted to military spending. This ratio has the advantage that it does not require adjusting for differing inflation rates or population levels.

Apart from the continuing secrecy and misinformation in issues related to the military, there is also an increasing problem of lack of transparency with the push towards dual-use technology. Producing goods with both military and non-military uses has the potential to make weapons production less viable. In addition, civil technology is increasingly being used in weapons systems which again can make weapons production and trade less visible. So when one is presenting figures for weapons production and trade there is great scope for creative accounting. It is possible to get the desired figure – either high or low depending upon the adjustments made and what is and is not included. As well as the reductions in the visibility of arms with dual-use technology and 'spin-in', weapons technology is being continually upgraded. Over time there are 'more bangs per buck'.

One further concern is that military research and development expenditure has fallen less quickly than overall expenditure, suggesting that countries may be maintaining their capability for producing weapons even when they are reducing the actual number they produce. This could lead to a 'capability race' replacing the 'arms race', where the competition is with a perceived threat rather than a real one.

Finding and interpreting data

There are a number of sources of information on military spending in the world. The Stockholm International Peace Research Institute's *Yearbook* provides the most consistent and, given the institute's impartiality, in some ways the most reliable data. In the 1992 *Yearbook* the authors do warn that although there seems to be an increase in the quantity of information provided relative to the past, with an increasing number of sources, there has been a qualitative decline and the reliability of the data has gone down. The problems caused by the changes in Eastern and Central Europe and the uncertainties of the allocations of the costs of the Gulf War require a specific warning about the interpretation of recent figures.

SIPRI uses the NATO definition as a guideline:

> Where possible, the following items are included: all current and capital expenditure on the armed forces, in the running of defence departments and other government agencies engaged in defence projects as well as space projects; the cost of paramilitary forces and police when judged to be trained and equipped for military operations; military R&D, tests and evaluation costs; and costs of retirement pensions of service personnel, including pensions of civilian employees. Military aid is included in the expenditure of the donor countries. Excluded are items on civil defence, interest on war debt and veterans' payments. Calendar year figures are calculated from fiscal year data where necessary, on the assumption that expenditure takes place evenly throughout the year. (*SIPRI Yearbook*, 1992, p. 269)

Other sources of data include *The Military Balance* from the International Institute for Strategic Studies, Brasseys; Jane's Yearbooks (various), from Jane's Publishing Company, London; and World Military Expenditure and Arms Transfers, US Arms Control and Disarmament Agency, US Government Printing Office, Washington, DC. This last publication is comprehensive but care must be taken as some data suggest that they reflect government views rather than verifiable information (*see*, for example, figures for arms exports to Iraq). Relevant data are also available from the North Atlantic Treaty Organization, the Organization for Economic Co-operation and Development, the International Monetary Fund and the United Nations. Useful summaries are provided by Sivard (annual). Other organisations which provide information and analysis include Saferworld; Campaign Against the Arms Trade (CAAT); Oxford Research Group; BASIC; and DFax. Internet sources are given on the Economists Allied for Arms Reduction (ECAAR) site at http://www.mdx.ac.uk/www/economics.

It is also possible to go back to the source figures in government publications to overcome the various deficiencies in the secondary sources. For an example of this approach *see* Scheetz (1991) on Argentina, Chile and Peru.

Despite attempts by the international organisations to provide accurate and comprehensive data on a consistent basis, there are still considerable problems with the data that need to be borne in mind when using it. SIPRI, although an impartial and careful collector of information, cannot report other than what it can find out, and as we saw in the previous section there can be numerous deficiencies in the data made available.

It is clear from this discussion that there are real problems in undertaking research on the military economy in any country and that these are exacerbated when international comparisons are attempted. However, the problems should not be overplayed. All empirical research has similar, though maybe not such severe, problems, and with care it is possible to undertake research. At the same time, the importance of understanding what is happening to the military cannot be overemphasised, particularly in the post-Cold War world, where the potential exists to reduce military spending and to achieve a peace dividend, but the arguments have still to be won.

What is happening

Bearing many of these limitations in mind we can build up an idea of the developments in military spending and military industry over time. The first thing to note is that the so-called Cold War was not particularly cold. It saw some 30–40 million people die as a result of about 150 conflicts, many proxy conflicts between the superpowers. At its peak in 1989, world military spending was $1089 billion (current prices), the world armed forces totalled 28 580 000, the share of world GNP devoted to the military was 4.6 per cent (it had been above 5 per cent in the early 1980s) and there were 5 armed force personnel per 1000 people (ACDA, 1994; Sivard, annual). At the same time the United Nations Development Programme (1994) was suggesting that in developing countries the chances of dying from social neglect (malnutrition, preventable disease, etc.) are 33 times greater than from a war started by external aggression.

With the end of the Cold War there has been a marked decline in military spending in most parts of the world. As Figure 43.1 shows, military spending as a share of GNP has seen a marked decline for both developed and developing countries. The world is still a dangerous place but the changed strategic environment presents the possibility of further cuts in the future. There has been a reduced superpower involvement in regional disputes, which has reduced the scale of conflicts; there have been increased economic pressures, including pressure from the World Bank and the IMF to cut military spending. In addition, the cuts which have so far occurred in advanced economies are not high by historical standards. The failure to disarm at the end of the Second World War was unprecedented and led to unprecedented peacetime buildup, as shown for the UK in Figure 43.2 (Dunne and Smith, 1990, discuss sources).

As a result of these changes there has been a huge decline in demand for weapons, both domestic procurement and exports, and a marked increase in competition between arms producers (Wulf, 1993). This has led to a restructuring of the top arms manufacturers, with takeovers and mergers, most notably in the recent wave of activity in the USA. SIPRI maintains a database of the top 100 arms producers in the world and has charted this restructuring in its *Yearbook*,

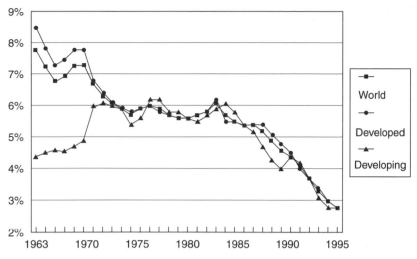

Figure 43.1 World military expenditures as a percentage share of gross national product, 1963–95

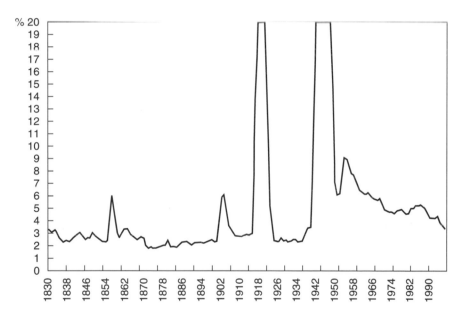

Figure 43.2 UK military burden as a percentage share of gross national product, 1830–1994

which has allowed the restructuring to be analysed. There are of course problems with the data. Many of the companies do more than produce arms and it is often difficult to get an accurate breakdown, either because they will not provide the information, or because the information that is provided is not correct. But it is the only attempt to develop a consistent set of information on the international

Table 43.1   Top UK arms producers, 1996

Company	Arms sales (£ million)	Sales (£ million)	Arms dependence	After-tax profits (£ million)	Total employees	1994 SIPRI rank among top 100 worldwide defence contractors[a]
British Aerospace	5 339	7 441	72%	311	42 000	3
GEC	2 854	11 147	26%	460	127 000	12
Rolls-Royce	976	4 291	23%	−44	42 900	30
GKN	961	3 337	29%	−42	30 000	59
Hunting plc	413	1 282	32%	−2.1	12 744	51
Racal Enterprises	351	1 185	30%	31.5	14 320	71
Vickers plc	345	1 198	29%	56.6	10 193	N/A
Babcock Rosyth Defence Ltd	298	646	46%	−24	9 029	N/A
Smiths Industries	230	1 008	23%	117.8	12 208	69
TI Group plc	222	1 757	13%	163.1	22 650	N/A
SEMA Group	185	927	20%	37 041	i0 641	N/A
Lucas Industries	185	2 989	6%	106.3	46 700	65
Cobham plc	169	270	63%	30.1	3 876	N/A
Devenport Management Ltd	168	201	84%	5.05	4 000	84
Vosper Thornycroft	147	241	61%	20.409	2 947	88
EDS-Scicon	100	400	25%	N/A	4 000	N/A
Alvis Industries Ltd	59	90	65%	7.798	503	N/A
Meggitt	48	256	19%	16.408	3 993	N/A
Fairey Group	27	247	11%	28.63	2 975	N/A
Logica International Ltd	24	338	7%	19.3	4 306	N/A

Source: Data drawn from annual reports for the 1996 fiscal year ending no later than 30/6/97 and/or company questionnaire responses
[a] Drawn from chapter 10 of the 1996 *SIPRI Yearbook*
N/A = not available

industry and represents a valuable resource. For the UK, Table 43.1 gives the data for UK companies and their world ranking in terms of size. After France, the UK is the largest arms producer in Europe, with British Aerospace the third largest arms-producing company in the world in 1994 (Dunne, 1993; Dunne and Smith, 1992, provide background). Within Europe, overcapacity is leading to pressures for consolidation which have so far been resisted or prevented by national governments.

Many of these recent developments are making it difficult to analyse the development in the international and UK arms market. After the Second World War most defence companies were domestic companies under the control of national governments, whether in public or in private ownership. The increase in privatisation of defence companies has gone alongside an increasing internationalisation in response to the decline in the market. International trade, international subcontracting and licensed production have been common. But recent developments, including cross-border mergers and acquisitions, joint ventures and inter-firm agreements such as co-production, management consortia, teaming arrangements and cross-equity ownership, represent a clear change in the structure of the industry. These developments are changing the nature of the military industry and could be removing accountability and control by national governments. The restructuring of European industry could make matters worse.

Conclusions

It is clear that the end of the Cold War has seen a marked change in military spending and has presented the opportunity for countries to achieve a peace dividend. The military–industrial complex is, however, a formidable opponent for peace activists and has fought a very successful action to prevent further cuts. If the opportunity for reducing the military burden and reorienting society to less militaristic postures and concerns is not to be lost, it is necessary for the peace movement to continue the pressure and win the ongoing debates. To do this requires information and statistics, and, while they are available, it is important that they are understood, that their limitations are recognised and that efforts are made to improve both collection and the method of analysis. This chapter is intended to provide a starting-point for anyone wishing to become involved in the process.

44

Counting Computers – or Why We Are Not Well Informed about the Information Society

James Cornford

The information industry largely determines the information that we have about it

Introduction

> As a form of wealth, focus of production and as a conception of value, information is a problematic category within our most basic ways of thinking about markets, property, politics and self definition. (Boyle, 1996, p. xi)

Around the world, politicians, corporate leaders and quite a few academics are hailing the emergence of an 'information society'. The emergence of a market for information and communication technologies – hardware, software and services for the domestic user – has supported, through its prodigious advertising spend, a growing specialist press. This takes the form of discrete magazines and television channels, and special supplements such as 'Online' with the *Guardian* newspaper, and special television programmes within mainstream channels. In all this media coverage, and in a plethora of government reports and corporate glossies, we are served up an endless diet of statistics relating to the 'information society' (*see*, for example, Motorola, 1996, and Department of Trade and Industry, 1996, for some of the better examples). These statistics are themselves mainly based on a number of 'market' or official research surveys.

We might, then, think that the information society is an area about which we are better and better informed. At the heart of the information society there is, however, a contradiction: we have very little useful information about the information society. In essence, the information society is not a well-*informed* society. This chapter explores some of the reasons for this apparent paradox. It argues that there are three main reasons for the lack of information *on* information:

- The first point concerns the commercial and political pressures (the interests of governments and corporations) to hype the information society and thus distort, if only by omission, the information, including statistical information, which is available.
- The second, hidden and more structural, problem concerns the tendency to substitute technologies for information, to count computers rather than the information that they are used to manipulate.

- The final point relates to the central dynamic of the information society itself, the transformation of information into a commodity which can be bought and sold – what is known as the commodification of information – which results in the removal of information from the public sphere and its conversion into private property.

Each of these points is illustrated through a brief case study of statistics relating to the Internet. A preliminary task, however, is to sketch out more fully exactly what we mean by the information society.

What is the information society? Codification and commodification

A satisfactory definition of the information society, indeed whether it is a useful term at all, is far from well established (*see* Webster, 1995). All human societies are, in a sense, information societies – they are all based on the use of information. What, then, is the point of using the term 'information society'? It is not so much that information is dramatically more important in the operation of the economy and society, but rather that its significance is becoming more and more *visible*. The advent of an information society, then, is based on two trends, both of which have a long history, but which recently have made the significance of information much more apparent.

First there is the codification of information. This trend goes back a very long way, but its new-found vigour is intimately related to the growing use of information and communications technology – computers and telecommunications – in an increasing number of spheres of life, from business and administration to education and leisure. Information technology, however much it may be described as 'smart' or 'intelligent', is not really very clever. In order for computers to manipulate and transmit information, that information has to be carefully structured, for example in a spreadsheet or database. In essence, information has to be extracted, or disembedded, from everyday life and *codified* in order for computers to be able to handle it. This process of structuring information, which is necessary to enable computers to work with it, forces us to think more carefully about information and makes us much more aware of it.

The second, and related, reason for the increasing visibility of information in modern societies – also a long-standing trend – concerns the increasing transfer of information from the personal or the public sphere to the market. This commodification of information takes a great many forms. The most spectacular examples concern the appropriation of traditional remedies by international drug companies or the patenting of an individual's genetic material (Boyle, 1996; Frow, 1996; Bettig, 1996).

These two trends work together, for example in the direct mail marketing industry. If you fill in one of those questionnaires sponsored by advertisers which asks hundreds of questions about you, your household and your purchasing preferences, those data are then codified into a database and the database is sold on to advertisers who will use it in targeting their direct mail shots. The loyalty cards that supermarkets have introduced in the UK in the mid-1990s are another example. Simple facts about your everyday choices in the supermarket, when combined in a structured way with information about your income and address, etc., become a very valuable property (Lyon, 1994; Gandy, 1993).

The information society is, then, the result of the intensification of these two trends – the codification of information associated with new technologies and the commodification of information associated with the rise of new information markets – and these features impart a particular set of biases and mystifications to many of the statistics that you will read about the information society.

Don't believe the hype

The first effect to distort the presentation of information relating to the information society is also the most obvious (and one that is attested to in a number of other chapters in this book). In short, the commercial interests of a lot of very large and very powerful technology and media firms are related to the need to sell both information technologies and information itself. It is directly in the interests of these companies to hype the development of the information society as they are positioned to be the prime beneficiaries of such a development.

This is a particularly serious issue as these companies – which include not only the manufacturers of hardware, software and telecommunications companies such as IBM, Microsoft or BT, but also large media companies such as News International with direct access to newspapers and television – are unlikely to publish stories or show programmes which question the value of, and therefore might lessen the demand for, the interactive services that they are developing and marketing. Even where companies do not have such direct access to the media, they have considerable influence through their advertising spending.

A more subtle distortion concerns what has been called 'the problem of tenses' (Friedman and Cornford, 1989) in the reporting of information technology-related issues. In reading the computer and information technology related press, there is an understandable focus on the 'latest', 'cutting-edge' technologies and practices. In part this arises from the way in which such magazines and newspapers are deluged with new hardware and software for review. Another cause is the sheer volume of advertising, most of which is, of course, for the latest technologies. But this focus on the 'leading edge' of new technologies and techniques extends deep into the editorial text. It is easy to get the impression that these 'leading-edge' technologies constitute the current norm in business and domestic settings. In reality, of course, most businesses and households are making do with technologies bought some years ago.

What counts? Information is not technology

A second problem concerns exactly what is counted. Information is a very slippery concept and difficult – perhaps impossible – to quantify. How much information is contained in an encyclopaedia, a novel, a CD or a CD-ROM? You can count the number of words or the megabytes of data, but does this tell you how much *information* those words or megabytes represent? What is more, information is not scalar. That is to say, it is very difficult to decide what constitutes more or less information. Do two encyclopaedias constitute more information than one encyclopaedia? Not really. The vast majority of the information in the second encyclopaedia will simply replicate that in the first.

These problems with the quantification of information mean that statistics concerned to measure progress towards the information society relate to almost anything but information itself. In the development of the information society idea, there have been a torrent of statistics relating to 'information sectors' (numbers and turnover of firms), information workers (employment statistics) or patents and copyright works (getting somewhat closer, but these only capture information that is codified and protected by intellectual property rights, a very select kind of information) (*see* Webster, 1995). Most of the statistics that one sees today, however, come down to counting computers.

Counting computers raises a number of questions. First, what counts as a computer? A Sony Playstation and a high-power Silicon Graphics Workstation are both computers in one sense – should they bouth count equally? The failure to answer this question leads to a number of significant statistical problems. As James Woudhuysen (1997, p. 357) points out, 'US government statisticians insist on treating PCs and laptops as equal, in power, to twenty-four hour number crunching mainframe computers', thus, for example, making any kind of meaningful judgement about the productivity implications of computers very hard to make.

More importantly, counting computers tells you very little about how much, for what and by whom they are used. In-depth studies of household computers only suggest the sheer variety of ways in which computers are used in the household (*see*, for example, Silverstone and Hirsch, 1992). Without that kind of detailed knowledge, it is very difficult to interpret what the presence of computers in the household actually means for the way it handles information.

The commodification of information

A further problem that arises in the reporting of the information society concerns the increasing 'commodification' of information, the way in which information is increasingly being appropriated as private property, protected by legal instruments (so-called intellectual property rights). This, of course, applies to statistics as much as any other type of information.

Intellectual property rights are intended to provide a reward for those who create information by enabling them to get an income from their work and to provide an incentive to create new information. The commodification information has to tread an uneasy path between rewarding the producers of information while at the same time not choking off the free flow of information (for example, in a public library or on licence fee funded television) that is necessary for the further creation of information. As critics point out, this theory is based on a nineteenth-century romantic vision of the author as the originator of information. This notion leads to an excessive level of control over the free flow of information, leading us to have '*too many* intellectual property rights; to confer them on the *wrong people*; and dramatically to undervalue the interests of both the *sources of*, and the *audiences for*, the information we commodify' (Boyle, 1996, pp. x–xi; emphasis in original).

The question of who is the 'author' of particular forms of information is often ambiguous. Take the example of a survey. In one sense, all those who participated in the survey and gave their time, usually free of charge, are the originators of the information. Yet the copyright to the survey rests with the survey organisation. What is more, the audience for a piece of information is often important in imparting value to that information (for example, the larger the number of people who use the same software as you, the cheaper it can potentially be and the better the quality of support you are likely to receive).

Intellectual property rights mean that you can be charged for information which you helped to originate or to which you have added value as a member of the audience. The information, in particular the raw data, belongs to the commercial organisations that have managed to assemble it. Without the opportunity for others to see that raw data, to analyse it in different ways, the claims of the owners can go more or less unchallenged.

Interrogating the Internet

The Internet – the worldwide network of computer networks based on the TCP/IP protocols – is perhaps the most widely diffused and certainly the most widely perceived manifestation of the 'information society'. A quick look in any newspaper or magazine will confirm that there is huge media interest in the Internet, and this is supported by a wide range of statistics. And yet, there is actually very little really reliable information about the Internet. A brief account of the way in which statistics about the Internet are generated and presented can well illustrate the three points made above.

First, it is clear that there is a growing industry making money out of the Internet and that a specialist Internet media has arisen. There were over 100 Internet service providers (ISPs) in the UK in 1996 (although there has been some shakeout in the market), including major corporations such as BT, News International and Microsoft. There are healthy profits to be made in selling Internet routers (the computers which direct traffic on the Internet), web server software (although the browsers such as Microsoft Internet Explorer and Netscape Navigator are given away free, the server-side software is often expensive), modems and other Internet-related software and services. This industry has begun to generate a substantial dedicated press, in addition to the general computer press which gives plenty of coverage to the Internet: the number of magazines in my newsagent with 'Internet' or 'Net' in their titles has risen to seven. National and local newspapers and television stations have all established a presence on the Net. All of these players have a vested economic interest in the continued growth of the Internet business.

The Internet also well illustrates the tendency to focus on easily countable hardware rather than information *per se*, in particular the much-quoted Internet host count statistics. In order to understand what these statistics mean one needs to understand what an Internet host actually is. It is, of course, a computer, but one of a particular type, one that can serve up information onto the Internet. This can be almost any type of computer from a large and powerful server (a computer that is dedicated to storing and serving up computer data), a router (a computer which switches Internet traffic from its source to its destination), to the laptop on which this chapter is being written (which can act as a server). A host can handle a lot of traffic of just a few bits of data per week.

According to the widely used early measurement, this figure had reached 19.5 million by July 1997 (*see* Figure 44.1). Confusion in estimating the number of

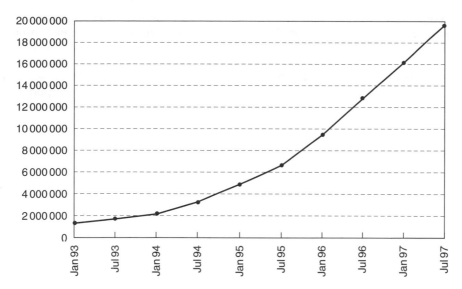

Figure 44.1   Growth in the number of Internet hosts (registered domain names).
Source: Network Wizards Internet Domain Survey, July 1997,
http://www.nw.com/zone/WWW/report.html

users from the number of hosts arises because a host can be a single computer
with only one or two individual users, or a university computer with thousands of
users. However, provisional 'guesstimates' have been established and an average
ratio of five users for every host is considered to be reasonable. On this basis, as
of July 1997, there are around 98 million Internet users worldwide. But again, in
reality, the situation is more complicated. We have to consider what we mean by
Internet 'users', as the figure is only an aggregate that does not differentiate
between varying levels of access to the Internet. So, while electronic mail is a
near-ubiquitous feature for the users included in this count, the services provided
with Internet access often do not extend beyond this.

There are a variety of factors creating this situation. Some of these relate to
price and technological concerns, but as both of these become less and less of an
obstacle to full connection, another factor is visibly slowing the uptake of full
Internet access. This is a reluctance on the part of owners and managers to allow
full access (particularly Web access) for their employees, for fear that they will
'surf' on company time. An indication, then, of the degree to which Internet
access is differentiated and restricted (for whatever reason) is provided by an earlier study by Texas Internet Consulting in 1994, which reported that of the estimated approximately 28 million total users worldwide, roughly only half this
figure, around 14 million, had full access to all the standard Internet services
(e-mail, WWW, ftp, etc. cited in *Screen Digest*, 1995, p. 83).

This focus on the numbers of Internet connections, rather than what they are
used for, can be extremely harmful when it reaches into the policy domain. For
example, the moves that have been made to 'wire up' British schools have, in spite
of some cautioning to the contrary, focused almost exclusively on funding the
hardware and the connections, without paying sufficient attention to the real costs

of computers and communications which are bound up in the maintenance of the equipment, the training of staff to make use of them and the organisational changes necessary to reap any benefit from them (*see*, for example, Tenner, 1996).

Having access is, of course, just one aspect of Internet *usage*. Where researchers have asked rather more probing questions about Internet usage they have come up with some rather interesting findings. For example, the report of a telephone survey carried out by the Department of Trade and Industry found that, while 92 per cent of those asked were aware of the Internet, just 25 per cent felt that it could be useful to them in their daily lives. As the consultants noted, 'it seems that the *benefits* of the Internet are a mystery to the majority' (Department of Trade and Industry, 1996, p. 3). Simple host-count statistics – counting computers – can, then, be highly misleading.

Finally, reliable information about Internet users is very expensive to buy. Although most Internet surveys ensure that snippets are released, so as to get a mention in the newspapers and trade magazines, the full reports, with the all-important interpretation of ambiguous data, are often very expensive. For example, *Retailing on the Internet: The Future for On-line Commerce*, published by the *Financial Times* in 1996 and described as a report which 'cuts through the hype providing an analysis of the latest issues and offers guidance on the real commercial issues to be reaped from the Net,' is priced at £320 or US$480.

Conclusion

For all the hype about the information society, we know remarkably little about information or the role that it plays in our society. The control over information that is exercised by governments and corporations, when added to intrinsic difficulties in dealing in a statistically valid way with such an intangible concept as information, generate a mass of obfuscation, mystification and half-truth.

45

The British Electoral System and the British Electorate

Ron Johnston, Charles Pattie and David Rossiter

Votes that count and votes that don't

> All voting is a sort of gaming, like checkers or backgammon, with a slight moral tinge to it, playing with right and wrong, with moral questions; and betting naturally accompanies it. The character of the voters is not staked. I cast my vote, perchance, as I think right but I am not vitally concerned that the right should prevail. I am willing to leave it to the majority. Its obligation never, therefore, exceeds that of expediency. Even voting for the right is doing nothing for it. It is only expressing to men feebly your desire that it should prevail.
>
> Henry David Thoreau, *Resistance to Civil Government*

The detailed results of British general elections are very unpredictable, however accurate the opinion polls at predicting each party's share of the vote. In 1992, for example, 41.9 per cent of voters supported the Conservative Party, which won 336 seats and a majority of 21 over all other parties. Five years earlier, it obtained 42.3 per cent of the votes, but gained 376 seats and a majority of 102; four years before that, 42.4 per cent of the votes brought 397 seats and a majority of 144. And then in 1997 Labour won 44.4 per cent of the vote, 419 seats and a majority of 179. And so we could go on: the UK electoral system appears to produce anomalous results. Why? To answer that we have to appreciate who the electorate are, where they live, and how they vote – if they do.

Where electors live and whom they vote for

These very different outcomes with virtually the same number of votes are a well-appreciated characteristic of the UK's electoral system – usually referred to as first-past-the-post. Geography is crucial to the system's operation, because it is based on territorially defined constituencies (659 in 1998). Chapter 46 examines how the geography of voting can be displayed. The Parliamentary Constituencies Act 1986 requires the four Boundary Commissions to review all constituencies every 8–12 years. When a review starts, an 'enumeration date' is chosen, and the electoral roll then in force provides the data which the Commissions use, however long they take over their task. (The English Commission's Third Review, for example, used 1976 data, and was reported in 1983; the constituencies introduced then were used for the 1983, 1987 and 1992 general elections.)

The rules the Commissions apply are ambiguous; a 1983 Court of Appeal interpretation was that they are not asked to do 'an exercise in accountancy' but simply to employ their judgement in weighing the various criteria, of which the most salient are:

1. There is a minimum number of constituencies for both Scotland (71) and Wales (35), and a minimum and a maximum for Northern Ireland (16 and 18), but no reference to England. Great Britain must have not substantially more or fewer than 613, which leaves 507 as the desideratum for England (it currently has 529). Because of this, the average constituency size in terms of population varies substantially between the three countries, and when the last review was completed in 1995 (using 1991 figures) the averages were:

| England | 68 626 | Scotland | 54 569 |
| Wales | 55 559 | Northern Ireland | 64 082 |

2. The requirement to match the constituency and local government maps varies by country.
3. When considering whether to recommend changes, the Commissions are required to take account of the disturbances that any changes may cause (to party organisation, for example) plus any local ties that might be broken, if changes are being proposed in order to create more equal constituencies. However, if the local government map has been changed, they must change the constituency map at their next periodic review.
4. 'Special geographical considerations' (size, shape and accessibility) can be employed to override the rules regarding size and local government boundaries, and are used for constituencies in sparsely populated areas (in Scotland and Wales only at present).

This *mélange* of rules results in variations between countries in average constituency size: Scotland and Wales are 'over-represented' relative to England and Northern Ireland. Constituencies also vary in size within countries. Within Greater London, for example, the three constituencies for the borough of Brent had average 1991 electorates of 57 105 and the two in nearby Islington averaged 55 047: neighbouring constituencies were much larger, however, with the two in Harrow averaging 73 992 electors. The need to change constituency boundaries is largely generated by population redistribution; during recent decades urban constituencies have lost electors over time whereas suburban, small town and rural constituencies have grown.

These differences can have important political consequences because of the geography of party support. A strong party in Scotland and Wales is likely to win relatively more seats than one strong in England, for example, as is a party which is strong in declining inner-city areas compared to one in the flourishing areas of the outer Home Counties.

The geography of winning and losing

This brings us back to the original question posed: 'Why can the same percentage of the votes win a party different numbers of seats?'. Some votes count much more than others, depending on where they are cast. Take a seat with three parties getting the following number of votes:

$$\text{A} \quad 30\ 000 \qquad \text{B} \quad 25\ 000 \qquad \text{C} \quad 10\ 000$$

A wins by 5000 votes, 4999 of which are surplus to requirements whereas 25 001 are 'efficient'; B's and C's votes are entirely 'wasted', since they bring no electoral reward. (If 4000 of C's supporters had voted for B instead, then A's 'surplus' would have been only 999.)

For any party, the fewer wasted and surplus votes the better (although no party will want to play it so tight that it wins every seat by one vote and so has no 'surpluses'). The relative percentages of the three types reflect: (a) the extent to which the definition of constituencies favours one party more than another; and (b) the degree to which a party successfully focuses its campaigning on those constituencies where 'votes matter'. Constituency definition is undertaken by independent commissions acting entirely non-politically, but the political parties can make representations (both written and, most importantly, orally at the public local inquiries that must be held to assess local opinion if their provisional recommendations are contested). The parties try to get the Commissions to recommend constituencies which will favour them electorally (using whatever non-political aspects of the rules they find best suit their cause). At the last review Labour performed much better than its opponents at this (Rossiter *et al.*, 1997a, b).

Local campaigns strongly influence electoral outcomes; the more active a party is in a constituency, relative to its opponents, the better its performance. The more canvassers and party workers in the offices and on the streets and doorsteps, and the more money spent on leaflets and posters, the more votes that are garnered (Denver and Hands, 1997). Labour's 1997 campaign focused on seats it needed to win; it spent little energy gaining more surplus votes in its safe seats, and little cultivating more wasted votes where it felt it could not win. The result was a very successful, focused campaign. This is illustrated by calculating not only the number of efficient, wasted and surplus votes for each party at the 1997 election, but also those that they would have won if the result had been reversed by a uniform swing across all constituencies; that is, instead of Labour getting 44.4 per cent and Conservatives 31.5 per cent, the figures had been 31.5 and 44.4 respectively (Table 45.1).

The Conservatives obtained one seat for every 58 017 votes in 1997, with nearly 60 per cent of their votes wasted and only 28.9 per cent efficient; Labour won a seat for every 32 338 votes, wasting only 20.8 per cent of its total and with nearly 10 per cent more 'efficient' votes than was the case with the Conservatives. If the situation were reversed, then the Conservatives would win 67 fewer seats than Labour (351 rather than 418) with the same vote total, because its electors were less likely to be in the right places. Labour would have wasted only 44.6 per cent of its votes if it came second, compared to 59.8 by the Conservatives when they did: when the Conservatives came second 28.9 per cent of their votes were efficient, but if Labour were in that position 36 per cent of its votes would have

Table 45.1 Wasted, surplus and efficient votes (percentage of each party's total) at the 1997 general election and if the result had been reversed between the Conservative and Labour parties

	Conservative		Labour		Liberal Democrat	
	Actual	Expected	Actual	Expected	Actual	Expected
Wasted	59.8	29.5	20.8	44.6	80.7	90.3
Surplus	11.4	28.2	41.0	19.3	4.6	1.4
Efficient	28.9	42.3	38.3	36.1	14.7	8.2
TOTAL	100	100	100	100	100	100
Seats won	165	351	418	253	46	24
Votes per seat	58 017	38 572	32 338	37 793	113 988	218 435
Surplus votes per seat won	6 591	10 870	13 244	7 285	1 429	3 099
Wasted votes per seat lost	12 022	13 374	12 157	10 766	7 016	7 581

Note: 'Expected result' refers to the situation if there had been a uniform swing from Labour to Conservative producing a reversal in their aggregate vote percentages

been. Thus Labour got a seat for every 37 793 votes when it 'had' 31.5 per cent of the votes cast, compared to the one per 58 017 for the Conservative Party when it had that percentage (i.e. in the 'real' election).

The situation for the Liberal Democrats, Britain's third party, is very different; without losing a single vote, the party would have lost 18 of its 46 seats if there had been a uniform switch of 6.5 per cent from Labour to Conservative. The same number of votes, all cast in the same places, would have produced a very different return depending on the distribution of the other votes between its two rivals. Even in the actual contest it won a seat for only every 113 988 votes, more than 3.5 times the ratio for Labour and nearly twice that for the Conservatives. Nevertheless, the Liberal Democrats, like Labour, very substantially increased the percentage of their votes cast for winning candidates, compared to recent elections (Figure 45.1).

Votes matter, but they matter more in some places than others. After the 1992 general election, the Conservatives had 65 more seats than Labour and an overall majority of 21; that majority would have been eliminated if just 10 278 voters in the 11 most marginal constituencies had preferred the second-placed candidate to the Conservative. Thus, in a country with some 44 000 000 registered electors, and perhaps 2 million more unregistered, the outcome of a general election can be determined by the decisions of less than one-tenth of 1 per cent of those people.

The UK electoral system is almost bound to produce biased results, which usually favour the two largest parties and disadvantage most others. And as the allocation of votes across the parties changes, so the shifts are magnified by the biased system: a switch of 9.2 percentage points from Conservative to Labour in the distribution of votes between 1992 and 1997 was reflected in a shift of 27.1 percentage points in the allocation of seats.

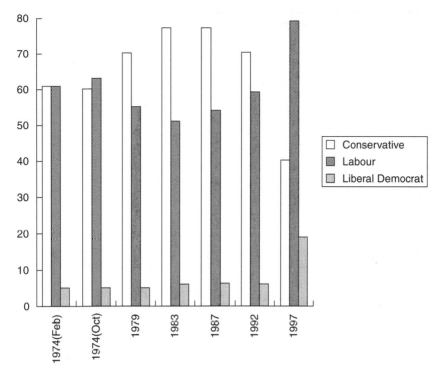

Figure 45.1 Votes for winners: the percentage of each party's votes which were cast for winning candidates

But what is the geography of change?

But are we right in estimating the amount of bias by assuming that change is uniform across all constituencies? The concept of swing has been used by political analysts and pundits for more than five decades: it plays a large part in the prognostications before an election and in the instant analyses on election night.

The assumption is that the change in the percentage allocation of votes between the two main parties is largely uniform across Britain's electoral map. Even in the 1950s and the 1960s, when those two parties between them won more than 90 per cent of the votes, this was an over-simplification. Butler and Stokes reported the mean and standard deviation for each inter-election period, and claimed that this validated the assumption of a uniform swing. (They define swing as 'the average of the Conservative gain and the Labour loss as a percentage of the total votes cast'.) But the volume of change actually varied quite considerably. The coefficient of variation (CV) – the standard deviation expressed as a percentage of the mean – is a simple index of the spread of values relative to the mean: a CV of 1.00 indicates that about two-thirds of the values are within 100 per cent of the mean value, on either side. Four of the values in Table 45.2 are less than 100, but the other three are 200 or higher. All cast considerable doubt on the assumption that change was geographically uniform.

Table 45.2   Inter-election swing in Great Britain, 1950–74

Period	Mean	Standard deviation	Coefficient of variation	N
1950–51	0.7	1.4	2.00	224
1951–55	1.8	1.4	0.78	339
1955–59	1.1	2.2	2.00	455
1959–64	3.5	2.4	0.69	370
1964–66	3.2	1.7	0.53	429
1966–70	4.4	2.1	0.48	468
1970–74 (February)	0.9	2.9	3.22	175

Source: mean and standard deviation, Butler and Stokes (1974, p. 121); coefficient of variation calculated by present authors. The swings were calculated only for those constituencies which were unchanged between the two elections in the pair and which had the same pattern of candidature from the three main parties at each of those contests

From 1970 on, the resurgence of the Liberals throughout Great Britain and of the nationalist parties in Scotland and Wales further eroded the validity and utility of the swing measure. In multi-party situations the net changes between parties (as illustrated for the main three by triangular graphs: see Chapter 46) hide a massive volume of gross inter-party flows: abstentions must be added because as many as 30 per cent of the British electorate do not vote (not to mention the 2 million or more people who could register as electors but do not). Flow-of-the-vote matrices indicate the gross shifts: Table 45.3 gives one for 1992–97 derived from a sample survey of some 13 000 electors, and that is incomplete since it cannot distinguish whether those who voted in 1992 but not 1997 left the electorate during the period (through death, migration, or non-enrolment) or decided to abstain.

The volume of change in voting preferences indicated by the first block of this table is considerable with, for example, fewer than 60 per cent of those who voted Conservative in 1992 doing so again in 1997. But are these percentages consistent across all constituencies? We can estimate the maximum-likelihood values of those flows for each constituency (Johnston *et al.*, 1988), using the 1992 and 1997 results plus the national flows as the constraints, via entropy-maximising procedures, producing the variations shown in Table 45.3B.

The substantial variations sustain our argument regarding differences across constituencies in campaign intensity and outcomes. The parties differentiated their effort according to their interpretations of its likely efficacy, and the results were differentiated as a consequence. Labour, for example, retained the support nationally of some 80 per cent of those who voted for it in 1992, but this figure fell as low as 51 per cent in one constituency, and whereas the Liberal Democrats on average retained the support of less than 60 per cent of their 1992 voters, more than 82 per cent remained loyal in one case.

Tactical voting helped the two opposition parties use their votes efficiently in 1997 (Figure 45.2). In Conservative-held seats where Labour was second, for example, the mean percentage of Labour loyalists (cell LL in Table 45.3B) was 83.2, whereas in those where the Liberal Democrats were second it was 75.8. Flows between parties were similarly differentiated: where Labour was second the LD flow averaged 2.5 per cent only, whereas where the Liberal Democrats were it was 5.6; the reverse flows (DL) averaged 2.07 and 12.2 respectively.

Table 45.3 The estimated 1992–97 flow-of-the-vote matrix (percentages of row totals)

A The national matrix

1992	1997					
	Con	Lab	LibDem	Nat	Other	A
Con	57.1	9.2	5.4	0.3	2.2	25.9
Lab	1.6	80.6	2.8	0.5	1.9	12.5
LibDem	3.2	14.8	58.7	0.6	2.9	19.8
Nat	3.0	13.2	1.6	67.9	2.2	11.3
Other	3.7	12.1	3.2	0.4	47.8	32.8
A	11.5	19.1	6.2	1.1	4.3	57.9

Key: Con – Conservative; Lab – Labour; LibDem – Liberal Democrat; Nat – nationalist parties (Scottish National Party and Plaid Cymru); Other – other parties; A – non-voters

B Summary statistics for variations in flows across 639 constituencies (excluding Tatton and West Bromwich West)

Flow	Minimum	Maximum	Mean	Standard deviation	Coefficient of variation
CC	27.0	67.1	53.8	8.3	0.15
CL	2.7	14.2	9.7	2.1	0.22
CD	2.3	12.2	5.3	1.7	0.32
CA	17.8	53.0	28.6	7.0	0.24
LC	0.6	4.8	1.8	0.6	0.33
LL	51.4	88.0	79.5	5.0	0.06
LD	1.3	17.2	3.4	2.0	0.59
LA	7.0	23.9	12.6	2.6	0.21
DC	1.2	5.6	3.2	0.8	0.25
DL	3.1	27.5	17.3	4.7	0.27
DD	29.2	82.5	53.1	9.5	0.18
DA	10.1	40.7	22.6	5.3	0.23
AC	3.9	19.6	12.1	3.4	0.28
AL	6.4	29.5	19.3	3.6	0.19
AD	2.3	18.8	6.5	2.5	0.38
AA	43.9	71.4	56.7	5.6	0.10

And who were the electors?

All reported percentages for British election results suggest a spurious accuracy, because of great problems with the denominator, the registered electorate. This figure can be quite misleading, such as the turnout figure of 71.2 per cent of registered electors in the 1997 general election.

Registration for electoral purposes is compulsory in the UK, but is not legally enforced. It is undertaken annually, by local governments (except in Northern Ireland). Forms are sent to every household in late summer, indicating those currently registered at that address: residents are asked to confirm if any are still there and eligible, to delete those who are not and to insert others who are (including those who will attain the age of 18 during the next 12 months). The returns are used to compile the new roll, which closes on 10 October (the 'qualification date':

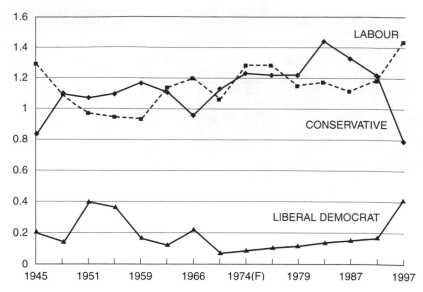

Figure 45.2   Ratio of seats to votes for each party since 1945.

again, Northern Ireland is an exception, where the date is 15 September). The roll becomes operative on 16 February: it provides the list of all qualified voters for the next year, and remains current for 16 months after the qualification date.

Much happens during that 16-month period: about 1.5 per cent of the adult population will die, and a further 13 per cent will move. The rolls are not updated, however, and in any case all contain substantial inaccuracies, because many households do not return their forms and checks are far from general: the degree to which non-respondents are purged from the roll varies substantially from area to area, for example. In addition, an unknown number of people, such as students, are (legally) registered on more than one roll, because they have several residences: they can choose where they will vote (it is illegal to vote more than once, but almost impossible to check this), and may do this in the constituency where they are least likely to 'waste' their vote. Supplementary rolls are maintained, and voters can register at any time. In the run-up to a general election, a local government may conduct a registration drive, for example, and the roll to be used for the election does not close until 2 weeks before polling day. The actual number registered as eligible to vote on election day is never recorded, however, let alone published; the official figures, used to calculate how much candidates can spend on their campaigns, are those published in February. (We collected data from Returning Officers in 426 constituencies and found that on 1 May 1997 in over 100 constituencies the eligible electorate was 500 or more larger than the official figure, and in two it was 3000 greater.)

We lack accurate figures on the electorate. The Office of Population Censuses and Surveys estimated that about 4.8 per cent (or 2.2 million people) enumerated in the 1991 census were not on the electoral roll (Heady *et al.*, 1996a). These were not evenly distributed across the population or country: people under 24 were twice as likely to be unregistered as those over 50, for example; Inner London residents three times as likely as those living in non-metropolitan areas; those who had moved in the previous year five times as likely as those who had not; New Commonwealth citizens six times as likely as UK/Republic of Ireland/Old Commonwealth citizens; and those living in privately rented, furnished accommodation 15 times more likely than those who owned their homes outright. We estimated where those missing voters were by Parliamentary constituency, at the time of the 1992 and 1997 general elections, suggesting that 31 per cent of them were concentrated in just 20 of the 641 new constituencies in Great Britain, all of them in urban areas and ten in Greater London (Dorling *et al.*, 1996b). Many of those seats were relatively marginal, and the missing voters, if registered, could have had a substantial impact on the general election result in 1992.

Some of those 2.2 million unregistered voters may have deliberately chosen not to be on the electoral roll for a variety of reasons: they may feel alienated from the democratic system and its political processes, for example, or fear that identifying themselves and their addresses on a public document may lead to negative consequences, such as liability to pay the council tax (the 'poll tax' before 1992). Others may just have been missed by the registration process, notably, but far from exclusively, the homeless.

Interestingly, those 2.2 million are not the largest number of people eligible to enrol but who have not. Some 3 million UK citizens living abroad, but who left the country less than 20 years ago, were enfranchised by the Conservative government in the late 1980s, undoubtedly in an effort to boost the party's support. Relatively few register (in the constituency where they were last resident within the UK) and then vote (by proxy); their votes can be crucial in marginal seats, however, and their registration is actively canvassed by the parties – especially in 'expatriate colonies' such as the Costa del Sol in Spain.

Summary

The British electoral system is based on an outdated method of identifying and enfranchising electors and a constituency system with very unclear rules as to where the boundaries should be drawn. As a consequence, the template within which elections are conducted contains many ambiguities and paradoxes which treat neither voters nor parties equally and fairly, and which can be exploited by the parties. The country needs an electoral system which is fair to all, and which is not dependent on artificially created constituencies produced using dubious, obsolete data. All votes should then be equally efficacious and elections would not be won or lost on the decisions of at most a few per cent of the voters who happen to live where how they decide to vote (or whether to vote) matters. Those who interpret the system must devise more sophisticated and meaningful measures of change than the crude and misleading ones currently employed, so that people can evaluate whether the system is working satisfactorily.

46

Illuminating Social Statistics

Graham Upton

There are better graphics than pie charts and bar charts!

This chapter is concerned with the use of diagrams to obtain a better idea of social science data. Considerable recent attention has been paid to the construction of good diagrams, with the three coffee-table books by Tufte (1983, 1990, 1997) setting the standard. Tufte's books are a wonderful guide to both brilliant and awful graphics. Tufte advocates maximising the ratio of information to ink – thus the exotic three-dimensional graphics so beloved by computer programs are anathema to Tufte.

Social science data are not well served by the traditional graphical methods (pie charts, histograms, bar charts, etc.), and so new types of diagrams have to be found. I hope that Tufte would not be too displeased by the diagrams used here, which represent my attempts at finding a remedy and will not be found in standard statistical packages.

A teaser!

Disraeli said, 'There are lies, damned lies and statistics', and this oft-quoted phrase has earned Statistics (and statisticians!) a bad press. The problem with statistics is *not* that they lie, but rather that they trip one up! Consider the following proposition:

> In England, in the 1997 General Election, the Conservative Party once again did extraordinarily well and gained the support of the majority of the country (54%). Labour came second with 34% with the Liberals bringing up the rear.

The Conservative Party was trounced in the 1997 general election, so surely this is nonsense? Well, yes it is . . . but then again there *is* a way in which it is true!

Read that opening paragraph again. Did it mention *people* anywhere – no! Instead a 'weasel phrase' has been used, namely 'the majority of the country'. We naturally interpret this as meaning the majority of the people – now read on!

Redrawing the map

Figure 46.1 represents the areas of English constituencies by circles centred on their centres of population. This serves to illustrate the fact that some constituencies are very much bigger than others! The small urban constituencies are dominated by supporters of the Labour Party, while it is in the huge rural constituencies that the Liberal Democrat and nationalist parties are most successful. The Conservatives won only 29 per cent of the English seats, but the average Conservative seat had an area 3.5 times as large as the average Labour seat, with the Liberal Democrat seats being bigger still. The misleading quotation in the previous section was quantifying area rather than population. For another type of political imbalance *see* Chapter 45.

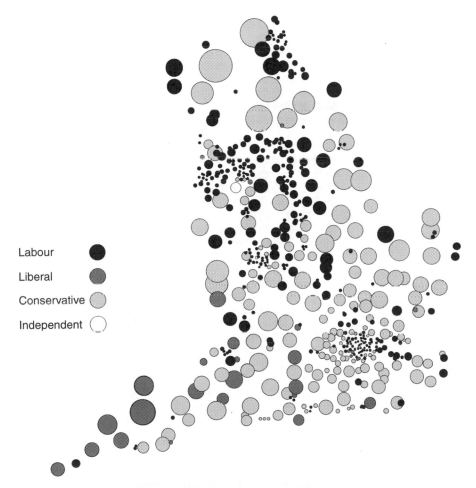

Figure 46.1   The outcome in England of the 1997 general election, with constituencies represented by circles scaled in proportion to the constituency areas. Constituencies are positioned according to their approximate centres of population and shaded to show the party of the elected Member of Parliament

The choice of colours was not arbitrary. Black was needed for Labour, so as to maximise the visibility of their small inner-city constituencies. White was a natural choice for the independent member for Tatton (the other independent was the Speaker of the House of Commons, Betty Boothroyd).

Figure 46.2 shows the rather startling effect of redrawing the map so as to give the same prominence to each constituency. The general outline is still recognisable,

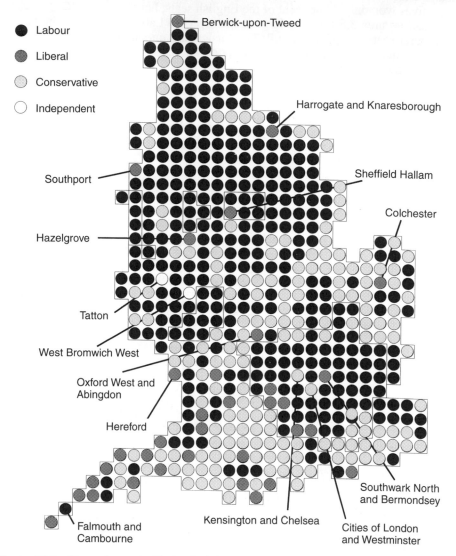

Figure 46.2 The outcome in England of the 1997 general election, with constituencies represented by circles of equal size on a cartogram. The constituencies are shaded to show the party of the elected Member of Parliament. The internal regions correspond to London, Birmingham, Liverpool, Manchester and Sheffield. Selected Liberal Democrat and unusual constituencies are labelled

I hope, but the importance of the 75 London seats is now apparent. The conurbations outlined are Birmingham, Liverpool, London, Manchester and Sheffield. Atypical seats (e.g. Liberal Democrat seats away from the south-west and Conservative seats in central London) are indicated.

Alternative methods for creating 'maps' of this type (known as *cartograms*) are described by Upton (1991) and Dorling (1991, 1996). Cartograms showing the results of the 1992 election and of the changes between the outcomes of the 1983, 1987 and 1992 elections are given by Upton (1994) and Dorling *et al.* (1996a).

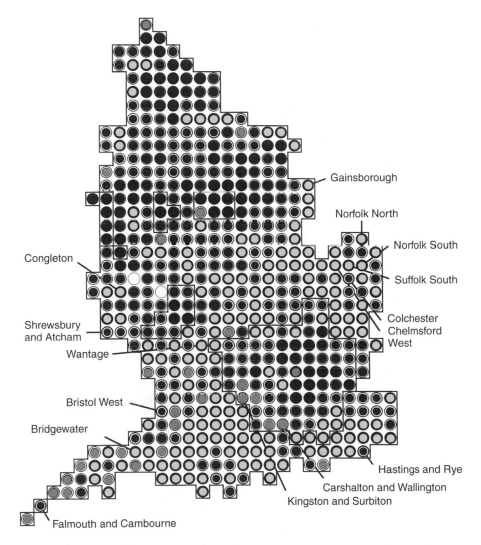

Figure 46.3 As Figure 46.2, but the constituencies are now represented by concentric circles, with the inner circle shaded to indicate the party that gained most votes. The successive outer rings show the second- and third-placed parties. All shaded areas are in proportion to the votes gained. Three-way marginals (*see text*) are labelled

Figures 46.1 and 46.2 indicated only the winning party in each constituency. Figure 46.3 provides more information: the winning party is indicated by the inner disc, with the other two parties arranged in order as outer rings. The areas of colour are in proportion to the votes obtained. We can see, for example, that the Labour Party was generally in third place in the south of England. The individually named seats are the 'three-party marginals' for which the difference in the votes gained by the first and third candidates was less than 10 per cent of the votes cast. The most marginal seat was Colchester (the author's own constituency).

Social background and political result

The previous figures have shown significant geographical aspects to the variations in the behaviour of the electorate. For example, Labour voters predominated in the inner cities and the north of England, while the Liberal Democrats were particularly strong in the south-west.

Can we relate these variations to differences in occupations, differences in housing or other background factors? Certainly we must be able to do so, since there are obvious resemblances between the inhabitants of the television series *EastEnders* and *Coronation Street*, and clear differences between these characters and those of *Emmerdale Farm*.

One way of seeing what is happening is to make further use of cartograms. Figure 46.4 shows the variations in the percentages of residents aged over 16 (a) who are non-white, and (b) who are of pensionable age. The highest proportions of non-white voters are in east and west London neighbouring constituencies, and in Birmingham. As the cartogram shows, in parts of Bradford and Leicester more than one voter in five is non-white.

The proportion of retired people is, as expected, highest in coastal constituencies, particularly those on the south coast, though the constituency with the highest proportion is Harwich (which includes the popular resorts of Clacton, Walton and Frinton).

Triangles

If we ignore the many candidates for the Referendum Party, the Raving Loony Party and similarly worthy organisations, the elections in England are three-party competitions. Triangles provide an illuminating way of displaying three-component data of this type.

Suppose the proportions voting for the Conservative, Labour and Liberal Democrat parties are denoted by x, y and z, where $x + y + z = 100\%$. An extreme case where one of x, y or z was equal to 0 would correspond to a point on one of the triangle's edges. If all three parties obtained the same number of votes, then this would be represented by a point at the centre (the centroid) of the triangle.

As z, the proportion voting Liberal Democrat, increases, so the distance of the point from the base of the triangle increases. If everyone voted Liberal Democrat then this would be represented by a point at the top of the triangle. Similar remarks hold true for the other parties. As an illustration, Figure 46.5 shows the location of the point corresponding to the outcome $x = 0.7$, $y = 0.1$ and $z = 0.2$.

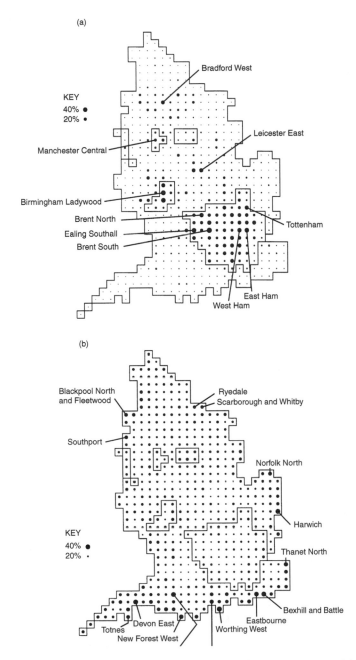

Figure 46.4 The variation from constituency to constituency in the percentages (a) classified as ethnically non-white, (b) classified as being of retirement age. Extreme cases are labelled

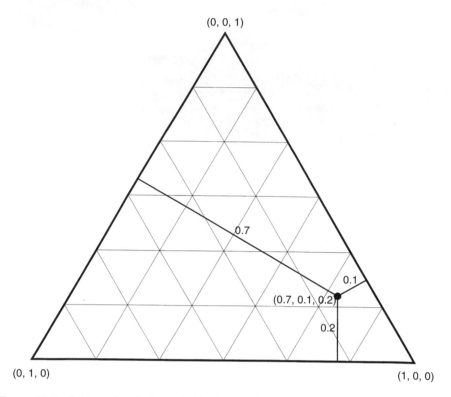

Figure 46.5   A triangular display suitable for showing the magnitudes of the components for any three-component characteristic. The point indicates the breakdown (0.7, 0.1, 0.2)

Graph paper for these triangular diagrams is commercially available. Ordinary graph paper can be used by plotting $(y - x)$ against z. Further details are given by Upton (1976) and Stray and Upton (1989).

Figure 46.6 displays the results of the 1997 election. Each constituency is represented by a dot shaded to indicate the constituency area. The relationship between voting outcome and constituency type is again apparent: the larger rural constituencies are mainly contests between the Conservative and Liberal Democrat parties, with the inner-city seats mostly claimed by Labour.

The crescent pattern, showing few seats that are contests between the Liberal and Labour parties, is a long-standing feature of British elections. Examples of voting triangles for previous elections are provided by Dorling *et al.* (1996a), Gudgin and Taylor (1979), Miller (1977), Stray and Upton (1989) and Upton (1976, 1991, 1994).

The three outlying Lab–Lib seats in 1997 (top left of the triangle) were Chesterfield, Rochdale, and Southwark North and Bermondsey. All three show the consequences of strong candidates: Chesterfield held by Tony Benn, Rochdale for many years the seat of Cyril Smith and the Southwark seat won in 1983 by the Liberal Simon Hughes at a by-election contested by the controversial Labour candidate Peter Tatchell.

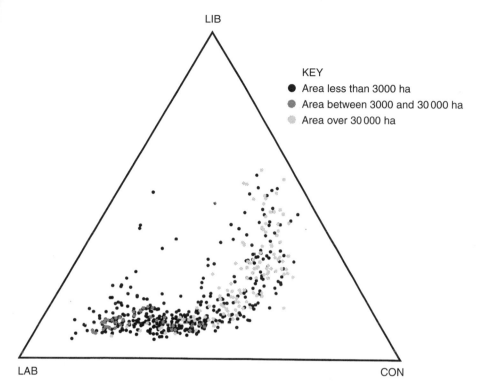

Figure 46.6   The outcome, in England, of the 1997 general election (ignoring votes cast for independents). Each point represents a constituency and is shaded so as to indicate the area of that constituency (*see* Figure 46.1). Three-party marginals are represented by points near the centre of the triangle

Although the main strength of Labour was in the city centres, Figures 46.1 and 46.6 have both demonstrated that Labour also holds a few more rural seats. With the exception of Norfolk North West, in 1997, these seats were in the north of England and they included Tony Blair's Sedgefield and two neighbouring Durham seats.

The same procedure can be applied to any three-category attribute. For example, we can categorise housing as being owner-occupied, rented from the local council or rented privately. Figure 46.7 shows the variation from constituency to constituency in the breakdown between these types. Each constituency is represented by a dot coloured as in Figure 46.1 to show the winning party. It is not very illuminating! This is because there is little variation in housing type between constituencies. Even so, four constituencies draw attention to themselves.

At the top of the clump, with very high proportions of privately rented properties, are two pale dots indicating unusual Conservative seats. Since for the most part Conservative supporters are owner-occupiers, these two seats must be in areas where house prices are exorbitant. This is the case, for these are the two Conservative seats in Inner London (*see* Figure 46.2).

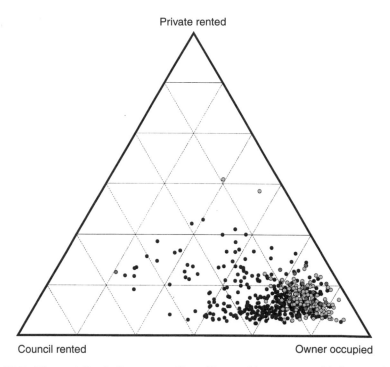

Figure 46.7 The variation in the composition of types of home ownership by constituency. The constituencies are shaded by the political affiliation of their Member of Parliament. The grid gives equal prominence to all 36 sections of the triangle

In the right-hand corner of Figure 46.7 there is a solitary black dot indicating a Labour seat. The constituency concerned is the Castle Point constituency consisting of Benfleet and Canvey Island in south-east Essex. In the 1992 general election the Conservatives had held the seat with a majority of 17 000: the swing to Labour in 1997 was one of the largest in the country. Figure 46.7 suggests that a subsequent swing away from Labour would see this seat return to its 'correct' allegiance.

The extreme left-hand dot in Figure 46.7 corresponds to the Southwark North and Bermondsey seat which was discussed earlier.

Figure 46.8 shows successive improvements on Figure 46.7. In Figure 46.8a we use some mathematics to shift the centre of the constituency galaxy to the centre of the triangle. It is still too concentrated a mass, so we use some more mathematics to exaggerate the differences. The final result is a diagram (Figure 46.8b) that shows clearly the anticipated links between housing type and political support.

A ring doughnut

The next type of diagram shows values displayed on the surface of a ring doughnut (properly termed a toroid). To get a two-dimensional display, the ring of the

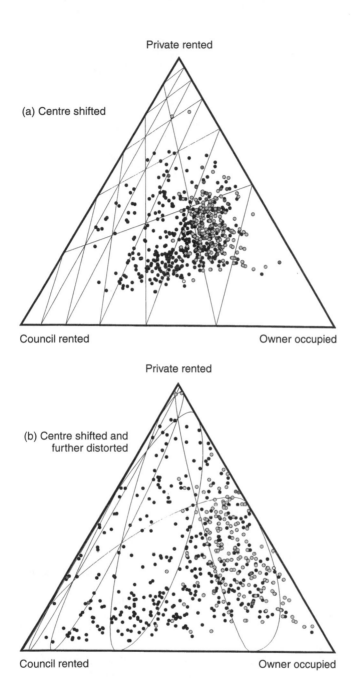

Figure 46.8 Two improved versions of Figure 46.7: (a) with the centre shifted, (b) with centre shift and subsequent distortion. The adjustments result in increased prominence for those sections of the original triangle that contained the bulk of the data. The relation between housing composition and political affiliation is clarified

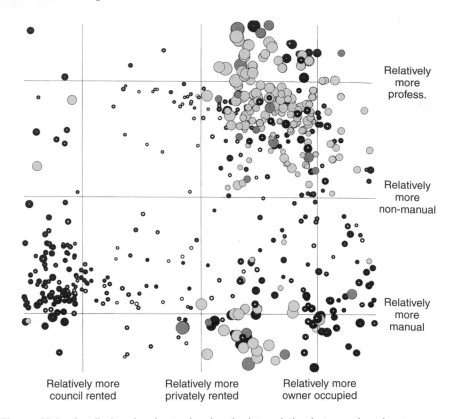

Figure 46.9   A split ring doughnut, showing the interrelation between housing type, occupation, constituency area (reflected by the circle area), political affiliation (the shading) and the proportion ethnically non-white (the area of the internal white circle)

doughnut is cut so as to give a tube, and the tube is then sliced so that the entire surface can be displayed in two dimensions as a square. What this all means is that the top of the square is actually joined (round the back) to the bottom and the left-hand side is similarly joined to the right-hand side. It's exciting stuff – if we walk off the top-right corner, then we find ourselves immediately walking on again at the bottom left!

Figure 46.9 is such a diagram. It combines the information about housing with similar information about job type (using the classifications professional, manual and other non-manual). The positions of the constituencies in this diagram are derived from their positions in Figure 46.8b and its housing counterpart. The areas of the circles indicate the physical areas of the constituencies. As usual, the circles are shaded to show political allegiance using the coding from Figure 46.1 (the independents are omitted). In addition, the percentages of those ethnically non-white are indicated by white central circles.

There are many features to be seen. In the bottom left is a compact group of constituencies that have relatively large proportions of manual workers and council houses. Most are outer-city constituencies, since the circles are quite small but are not the smallest. With the exception of Havant, all were won by Labour in the 1997 election.

The inner-city constituencies are characterised by relatively large numbers of rented properties. In Figure 46.9 these occur in a vertical band to the left of centre. Most of these constituencies have rather large proportions of voters classified as being ethnically non-white.

The large rural constituencies are clumped together at the top and bottom (remember that the top actually joins the bottom!) of the column to the right of the centre. These are constituencies with relatively low numbers of council houses and relatively few voters falling into the 'other non-manual' category. There is a notable collection of five Liberal Democrat seats (Brecon and Radnorshire, Somerton and Frome, Torridge and West Devon and Truro and St Austell at the top of the diagram and Devon North at the bottom). Six other Liberal Democrat seats in the collection at the bottom of the figure include, just to the left of centre, the other large seat of Berwick-upon-Tweed.

As we move around Figure 46.9, so there is a gradual transition in constituency size and affiliation. The two remaining groups are related to those already considered. There are a good number of suburban constituencies, with high proportions of professionals and low proportions of council houses, that are predominantly Conservative held. Finally, there is a small group of constituencies that have high proportions of owner-occupiers and high proportions of manual workers. It is the latter that is the more telling factor: these constituencies (such as Amber Valley, Burnley, Burton, Newark, Nuneaton and Pendle) are predominantly held by Labour.

Cobwebs

In the discussion so far, in addition to geographic aspects, I have referred to five sociological characteristics: the age structure, the ethnic blend, the housing breakdown, the job mixture and the constituency size in addition to the actual voting outcome. As a final *tour de force* I shall now interrelate these characteristics in a single 'cobweb' diagram.

For each characteristic, a constituency is classified as being in the top quarter of values for that characteristic (the upper quartile), the bottom quarter (the lower quartile), or the middle half.

By definition, one-quarter of the constituencies are classified as having a particularly high proportion of council houses. Similarly, by definition, one-quarter are classified as having a particularly high proportion of manual workers. If these two characteristics were unrelated then about one-sixteenth of the constituencies (i.e. about 33) would be classified as being high in both respects. In fact, no less than 71 are so classified, showing a very strong positive association between these links.

Figure 46.10 displays 18 nodes, three (labelled 'hi', 'mid' and 'low') for each of the six variables. Unexpectedly high frequencies are shown by black lines, with the widths of the lines indicating the importance of the links. I have noted the most important links in our previous discussion: these are the connections between occupation, housing and vote and the connection between constituency size and the percentage classified as non-white.

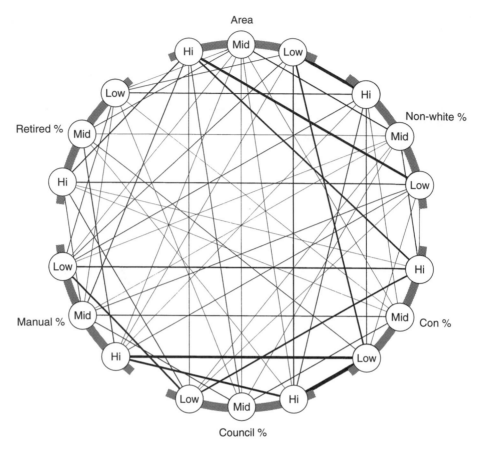

Figure 46.10 A cobweb diagram showing the links between the categories of six variables. Thicker lines correspond to stronger relationships

Summary

In this chapter I have used diagrams to reveal and confirm the existence of many links between sociological aspects of the population. We have seen that there are many relationships and that the variables are interrelated in subtle ways. What conclusions should we draw? The first is that complex data cannot be reduced to simple summary statistics without the risk of over-simplification. For example, if we looked only at housing and vote, then we might arrive at the simplistic view that Labour voting is more likely simply by virtue of living in a council house. If that were true, then the Conservatives might gain votes by eliminating rented accommodation; the recent policy of allowing renters to purchase their rented property from their council might be interpreted as meaning that the party had carried out exactly that analysis! The reality is different: people living in rented accommodation tend to differ in many other ways from those who own their accommodation.

Acknowledgements

This chapter has used the information on housing type and social class given by Waller and Criddle (1996), information on constituency areas kindly supplied by Danny Dorling and information on ethnicity and age structure made available by Danny Dorling and compiled by David Rossiter. The cobweb diagram is based upon an idea due to James Miller, who markets a related commercial product that he terms a *daisy*. For the FORTRAN program that produces the cobweb, contact Graham Upton (gupton@essex.ac.uk). For details of the daisy, view at http: \\www.daisy.co.uk.

47

Conclusion: Statistics and 'the Truth'

Stephen Simpson and Daniel Dorling

Social statistics are a social product

Statistics are a social product. In this chapter we discuss what this statement means for statistics and what it implies for people who want to change the world they live in. We draw on the chapters in this book to provide examples of how statistics are created and interpreted within society.

In a simple sense, statistics are a social product simply because they are produced by people. But they are also firmly located in the aims and tensions of the society that produces them – whether expressed by organisations of government, trade, or campaigns. In literature it is said that every text has a context. With statistics it is not just that what is discovered depends on the society from which those numbers are drawn. The methods and data that are used themselves are shaped by their context. Donald MacKenzie explores this context for the origins of modern statistics in Chapter 7.

Take homelessness. The official count of homeless people should, one would think, determine the amount of resources committed to tackle the problem. It is certainly taken by social commentators to indicate how the scale of the problem differs geographically. However, the official count of homeless people includes only those who apply successfully to local authorities for rehousing. This is reported quarterly to central government. As Rebekah Widdowfield shows in Chapter 22, it severely underestimates the scale of the problem while also overstating the numbers of people actually living on the street. The definition of these official statistics is being changed by government as we write, so we will have little idea of how the extent of homelessness is itself changing. It is very difficult to counter the official homelessness statistics, because other official statistics such as the national census also severely underestimate the homeless, as Ian Diamond shows in Chapter 2: those without a usual address are naturally difficult to count during any household enquiry. Homeless people can probably best recognise their conditions and count their number, but they have no incentive, nor the means, to do so. It remains a fact that homeless statistics do not estimate the number of homeless people.

The view that statistics are a product of their context is widely held. In *Demystifying Social Statistics* Dot Griffiths, John Irvine and Ian Miles (1979) distinguish four views of critics of statistical practice and more generally of scientific practice:

First, an 'anti-science' approach would dismiss every statistic because it 'inevitably turns people into objects to be manipulated and controlled and is contrary to basic human values'. This view is widespread as a social attitude. Some people would reject out of hand attempts to measure individuals' or society's progress as being fundamentally restrictive.

Second, an 'alternative technology' approach charges itself with 'convincing people of the social, environmental and health hazards of such high technologies as supersonic aircraft, nuclear reactors, factory farms and automated production lines, and by providing them with alternatives . . . small-scale and controlled by the individual or community'. Within statistics perhaps the most successful alternative approaches have been developed by the Community Operations Research Network described by Charles Ritchie, Ann Taket and Jim Bryant (1994). Jeff Evans and Ivan Rappaport pursue the statistical needs of communities and the concerned citizen in this book (Chapters 9 and 10). In these cases, alternative statistics are seen not as replacing traditional statistics, so much as tools necessary for research in settings where traditional methods are not appropriate.

Next, 'social responsibility' in science sees statistics 'as ethically neutral, asocial, bodies of knowledge and techniques: they could be used for good purposes or abused for bad ones'. This is a widespread attitude among statisticians themselves. Socially responsible statisticians argue for appropriate codes of practice for statistical work, aiming to limit the political misuse of statistics. They criticise and poke fun at abuses of statistics, with classic published examples in WJ Reichmann's *Use and Abuse of Statistics* (1964) and Darrel Huff's *How to Lie with Statistics* (1973), as well as a regular column in the Royal Statistical Society's newsletter.

The fourth view, labelled 'radical science' and to which we would subscribe, recognises the success of the first three views in achieving changes to statistical policy or practice, in particular circumstances, but goes on to point out that if statistics are a product of society then they are never neutral. They cannot be ignored nor can they be substantially changed unless society itself is changed. People who wish to make statistics more open (Simpson and Dorling 1994, Dorling 1995) more democratically linked to the priorities of local and other communities, less dominated by government and trade, are allies of anyone who wants to change society in these ways.

In order to understand why some statistics are believed and acted upon, while others are ignored, we need to know something about the process by which statistics in society are *paid* for. Four things are necessary to produce a well-known statistic:

- The statistic has to have a *purpose*, be seen to be lacking, be wanted, be commissioned, be called for, asked for, needed, demanded, by someone with the power and the money to express a need for it and ensure that it will be provided.
- The data have to be *assembled*, collected, brought together, gathered; this often requires resources, or control over an administrative system to extract the statistics from information already recorded.
- The statistic has to be *interpreted*, analysed, reasoned, understood, by researchers who use methods developed to answer previous questions, or who

develop new methods for the purposes of those who commission statistical work.

- The statistic has to be *described*, reported, presented, disseminated or communicated by the researchers or the commissioners to an audience which must be convinced of the results in order to support policy decisions.

These four processes are all required to produce a statistic. Social statistics have to be *paid* for. The processes are not always so clearly divided into four stages, but the reality is illustrated in detail in most of the chapters of this book; for example Paul Dunne in Chapter 43 describes the construction of statistics used to determine the development of military strength, while Alan Freeman in Chapter 41 reanalyses national economic accounts with a new purpose. Jeff Evans and Ivan Rappaport in Chapter 9 discuss the statistical needs of communities, while Ian Plewis in Chapter 31 seeks statistics that could make sense of educational achievement.

Statistics in the natural sciences are also a product of these processes. Statistics about society are more often used in the political processes of debating the formulation and implementation of policy. Inevitably they are themselves political. To talk of statistics as political because they are a social product does not invalidate them. In particular, we are not saying that statistics are invented to reflect someone's narrow interest: examples of inventions within research, such as Cyril Burt's twin studies to 'prove' hereditary intelligence, are famous but few (Norton and Mackintosh, 1995). In a democratic society different priorities call for different types of statistics to be collected, analysed and reported. Those who demand genuine and useful debate will also recognise that apparently conflicting statistics can be judged only if they are honestly collected and reported, and the processes of their construction are fully described.

The issues raised by this book are therefore of importance to research workers as much as to those who seek to understand the results of research. It seems useful to summarise those issues within the four stages of statistical research set out above; many of the issues may in fact apply to more than one stage.

Purpose: wanted, commissioned, called for, asked for, needed, demanded

Government statistics, by their nature, address existing government policies. They do not monitor new social issues, nor past and present issues that are not addressed by current policy. In this book Alison Macfarlane, Jenny Head and Allyson Pollock (Chapters 26 and 29) give examples from health and social care, Humphrey Southall, David Gordon and Pauline Heslop (Chapters 20 and 40) from poverty, and Christina Pantazis and David Gordon (Chapter 24) from crime. This is also the reason that only paid work is measured in the Labour Force Survey and the national accounts, and why the majority of government statistics are about business and trade.

People's priorities for statistics are often different from those of government. For instance, a common understanding of the local need for jobs would be based on the number of people in an area, while a company will focus on the skills that it needs. The difference between the jobs people want and the people that employers want creates unemployment, commuting and labour shortages, which are examined by government planners as problems without resolution as Jamie Peck makes clear (Chapter 39).

Government and other bodies can decide to commission a statistic as a means of doing nothing, or to give an image of doing something. While it is wise to plan research very carefully, feasibility studies and investigations are also a means to delay implementation of new approaches while the political support for them is gauged. How often is 'We shall investigate' the cry of those who should have known all along?

Assembled, collected, brought together, gathered

There is a huge degree of control by government over access to the raw materials needed to construct social statistics once they have been deemed to be lacking. Government administrative registers are often the only counts of the population; but they are also often particularly poor representations of social conditions. In this book Cecilio Mar Molinero (Chapter 35) discusses examples in special education. Ray Thomas and Anne Green (Chapters 36 and 37) demonstrate that unemployment statistics do not measure the number of people who need a job. David Gordon and Pauline Heslop (Chapter 20) refer to the poverty of poverty statistics, while Richard Bate and Rebekah Widdowfield (Chapters 42 and 22) show that housing statistics do not measure homelessness or housing need. Government undertakes a number of national social surveys. It is the only body to have the resources to do so regularly. Charlie Owen and Angela Dale (Chapters 3 and 4) discuss these government surveys in this book.

Social problems often become fragmented by specialist professional approaches. Statistical analysis is one of those professional approaches. Bound by rules developed explicitly in textbooks and journals, it provides some independence from political interference. This specialisation also removes statisticians from an understanding of the people and the issues they are investigating. Even more frequently, the researcher is much more familiar with a computer screen than with the human subjects of research. One of the most important conclusions of several chapters of this book is that a variety of research approaches are required to understand social development and needs. Dan Wright (Chapter 8), in discussing psychology, stresses that qualitative data about private events must supplement quantitative data that are more easily recorded, Diane Perrons (Chapter 14) does the same when talking of gender relations, and Mary Shaw (Chapter 18) regarding food surveys.

Interpreted, analysed, reasoned, understood

Results are not always categorised in the ways that would be most useful to policy consideration. Mel Bartley, David Blane and George Davey Smith (Chapter 27) illustrate how problems of categorisation influence studies of health. Nicola Brimblecombe (Chapter 34) shows how crude decisions about the categorisation of teachers determine official reports on schools. Ann Singleton (Chapter 19) shows how the categorisation of different people influences European migration statistics, Ian Miles (Chapter 5) shows how the categorisation of research statistics is used to disguise military funding. Peter Lee (Chapter 21) shows how different parts of Britain are labelled as deprived depending on the analysis used. Ray Thomas (Chapter 37) shows how choices of categories makes it appear that unemployment is either rising or falling. And Waqar Ahmad (Chapter 16) discusses the general problem of categorisation in relation to ethnicity.

Analysis of data always involves assumptions. Conservative assumptions are the norm because where there is doubt, it is often convenient to assume something acceptable to the status quo. Thus Ian Diamond (Chapter 2) refers to the fact that those missed from enquiries are often assumed to be the same as those who were not missed. Jamie Peck and Richard Bate (Chapters 39 and 42) refer to forecasts that assume that the future will be the same as the past, and Jay Ginn and Charlie Owen (Chapters 15 and 3) refer to the assumption that resources and consumption are spread evenly within families. Results are rarely tested for their robustness to alternative reasonable assumptions, but it is always reasonable to ask of research 'Would the conclusions have been different if the assumptions had been different?'

Statistics never reveal some precise truth simply waiting to be discovered, though this is how they are commonly perceived. The universal role of uncertainty is rarely recognised, yet it is a central part of any statistical study that relationships between variables are approximate and that true values are not known exactly. Harvey Goldstein (Chapter 32) illustrates the folly of assuming that pupils' school performances can be measured exactly. James Cornford (Chapter 44) shows how statistics about the Internet are riddled with uncertainty. Roy Davies (Chapter 38) shows how the labelling as sick of many unemployed men in the Welsh valleys adds uncertainty to economic indicators there. Ron Johnston, Charles Pattie and David Rossiter (Chapter 45) show how the margin of uncertainty in the British electorate and voter registration is sufficient to affect national election results. Theo Nichols (Chapter 30) illustrates how uncertainty over the meaning of industrial injury statistics allows for official inaction. Recognition of uncertainty would (certainly!) affect policy.

Disseminated, reported, presented, described, told, informed, communicated

Research reports often reflect the commissioners' view of the solution as well as of the problem and can for this reason be very restricted. Walid Abdul-Hamid (Chapter 23) shows how statistics have been used to make homeless people appear to have more of a mental health than a housing problem. Jane Elliott (Chapter 13) shows how statistical models are taken too literally. Ian Parker (Chapter 11) discusses how subjective studies are made objective, David Sibley (Chapter 12) argues that the use of statistics can damage understanding, while Pauline Davis (Chapter 33) shows how statistics about children's ability to read are seen all too readily as realities.

One cannot expect every report of research to include all the information that would allow each reader to judge the appropriateness of the design, the quality of the sample and collection, and whether different analyses might have given different conclusions. The demand from funders and media alike for concise and influential reports contradicts the tentative nature of most research results. It is reasonable to expect all research reports to highlight the limitations of their results, and to be treated with a large dose of scepticism if they do not.

The dissemination of social statistics does not reach very far. Headlines dominate. While technological advance makes vast sets of data available to professional researchers, very little is accessible to individuals and community associations. Access to data increasingly requires resources and computing skills. A prerequisite to wider access is clearer dissemination, explanation and interpretation of datasets. Government could stimulate such clarity by making data from its censuses and surveys more cheaply and easily available, and by offering alternative commentaries on social trends. The debate over the creation of a national statistical service independent of government should have access and criticism high on its agenda (*see* p. xxvi).

Finally, the authors of this book have raised some pressing issues for debates about social statistics in Britain today. Should we have open government and more access to data (Michael Blakemore, Chapter 6)? Do we need a question on income in the next census (James Nazroo, Chapter 25) and should we be asking about religion (Joanna Southworth, Chapter 17)? How vital are the General Household Survey and other government surveys (Charlie Owen, Chapter 3)? How should issues of inequality be explored and addressed through statistics (Anthony Staines, Chapter 28)? This book also illustrates how social statistics can be represented graphically. Graham Upton's chapter (Chapter 46) illustrates some innovative new methods of displaying multi-dimensional data. The presentation of clear and informative graphics is essential to good statistical dissemination (Dorling and Fairbairn, 1997).

Conclusion

This book has shown how a series of critical views can open up the black box of social statistics. Politicians quote statistics relentlessly and rely on them for their work. While some may abuse statistical practice with narrow and mystifying jargon, the origins and meaning of social statistics are usually capable of straightforward interpretation. With a little thought and a little knowledge you can produce statistics which are fair and just, or discredit other statistics which are neither fair nor just. Social statistics is not an area of knowledge best left to the social statistician, numerical sociologist or epidemiologist. The numbers which are produced under its heading concern, and are concerned with, all of us (*see* Chapter 1). The errors that are often made in their construction are obvious to many of us and have implications for most of us.

Statistics can also be developed to uncover injustice and prevent bad decisions being made. They can be used to change society as much as society changes them.

The critiques in this book attempt to clear away some of the mystery around social statistics. Clarity about the concepts and theories that are used to frame research questions is as important as clarity in the analysis and presentation of statistics. Faced with a statistical claim, it is fair to ask:

- Is that true? How do I know it is true? Where did it come from? Who wanted me to know that? What are the alternative explanations? Could I have done better?

Faced with a social problem that requires action, it is fair to ask:

- Who would want change and who would not? Who needs to be convinced of the nature of the problem? What extra information is needed to clarify the problem? What information is needed to clarify possible solutions? How would it best be collected and analysed so as to leave no doubt?

Appendix: a list of Radical Statistics publications

Books and pamphlets

Irvine J, Miles I and Evans J (eds) (1979): *Demystifying Social Statistics*. London: Pluto.

Jones D, Smith P, Kinnersley P and Radical Statistics Health Group (1982): *Two Statistical Methods for Detecting Hazards at Work*. London: Radical Statistics.

Radical Statistics (1978): *RAW(P) Deals: A Critique of 'Sharing Resources for Health in England'*. London: Radical Statistics.

Radical Statistics (1978): *Social Indicators: For Individual Well-Being or Social Control? (The Case of OECD)*. London: Radical Statistics.

Radical Statistics Education Group (1982): *Reading between the Numbers: A Critical Guide to Educational Research*. London: BSSRS Publications.

Radical Statistics Education Group (1987): *Figuring Out Education Spending: Trends 1978–85 and Their Meaning*. London: Radical Statistics. 20pp.

Radical Statistics Health Group (1976): *Whose Priorities? A Critique of Health and Personal Social Services for England*. London: Radical Statistics.

Radical Statistics Health Group (1977): *In Defence of the NHS: An Attack on Fee for Service Payments in Medical Care*. London: Radical Statistics.

Radical Statistics Health Group (1981) *The Unofficial Guide to Official Health Statistics*. London: Radical Statistics.

Radical Statistics Health Group (1985): *Unsafe in Their Hands: Health Service Statistics for England*. London: Radical Statistics.

Radical Statistics Health Group (1987): *Facing the Figures: What Really Is Happening to the National Health Service?* London: Radical Statistics. 191pp.

Radical Statistics Health Group and Local Radio Workshop (1980): *A Better Start in Life? Why Perinatal Rates Vary in Different Parts of the Country*. London: Radical Statistics. [audio cassette]

Radical Statistics Nicaragua Statistics Fund Subgroup (1987): *Statistics in Nicaragua: A Report*. London: Radical Statistics. 22pp.

Radical Statistics Nuclear Disarmament Group (1982): *The Nuclear Numbers*

Game: Understanding the Statistics behind the Bombs. London: Radical
Statistics. 95pp.
Runnymede Trust and Radical Statistics Race Group (1988): *Britain's Black
Population*. London: Heinemann Educational Books. 160pp.

Broadsheets

Radical Statistics (1997): *More Money for Running Costs?* London: Radical
Statistics.
Radical Statistics (1997): *More Money for Capital Projects?* London: Radical
Statistics.
Radical Statistics Groups on Performance Indicators and Education (1992):
Choosing Schools by Numbers? The Statistics of School Performance.
London: Radical Statistics.
Radical Statistics Health Group (1992): *A Growing Health Service?* London:
Radical Statistics.
Radical Statistics Health Group (1997): *More Patients Treated?* London: Radical
Statistics.
Radical Statistics Health Group (1997): *More Nurses and Doctors?* London:
Radical Statistics.

Journal articles: 'Watching the figures' pages in *Health Matters*

Radical Statistics Health Group (1991): Counting the cost of young lives. *Health
Matters,* **6**, 11.
Radical Statistics Health Group (1991): Residential care: a private takeover?
Health Matters, **7**, 17.
Radical Statistics Health Group (1991): Ignoring inequality. *Health Matters,* **8**, 7.
Radical Statistics Health Group (1992): Waiting lists: the long and the short of it.
Health Matters, **10**, 12–13.
Radical Statistics Health Group (1992): A growing health service? *Health
Matters,* **11**, 18–19.
Radical Statistics Health Group (1993): Fewer beds, but more patients? *Health
Matters,* **14**, 18.
Radical Statistics Health Group (1995): Figuring out league tables. *Health
Matters,* **20**, 5.
Radical Statistics Health Group (1996): Shorter waits for the clinic? *Health
Matters,* **25**, 17.
Radical Statistics Health Group (1996): Nurses up, managers down? *Health
Matters,* **27**, 7.
Radical Statistics Health Group (1997): Things can only get better? *Health
Matters,* **31**, 6.

Other journal articles

Radical Statistics Health Group (1991): Missing: a strategy for the health of the nation. *British Medical Journal*, **303**, 299–302.

Radical Statistics Health Group (1991): Let them eat soap. *Health Service Journal*, **14**, 25–7.

Radical Statistics Health Group (1992): NHS reforms: the first six months – proof of progress or a statistical smokescreen? *British Medical Journal*, **304**, 705–9.

Radical Statistics Health Group (1995): NHS 'indicators of sucess': what do they tell us? *British Medical Journal*, **310**, 1045–50.

Report

Public Health Alliance and Radical Statistics Health Group (1992): *The Health of the Nation: Challenges for a New Government*. Birmingham: Public Health Alliance.

Journal

Radical Statistics, published approx. 3 times a year. Current issues are available from The Radical Statistics Group, care of 10 Ruskin Avenue, Heaton, Bradford, DB9 6EB.

Date	No.	Date	No.	Date	No.
January 1975	1	June 1982	24	Winter 1990	47
June 1975	2	October 1982	25	Spring 1991	48
August 1975	3	February 1983	26	Winter 1991	49
October 1975	4	May 1983	27	Spring 1992	50
Spring 1976	5	October 1983	28	Summer 1992	51
Summer 1976	6	February 1984	29	Autumn 1992	52
Autumn 1976	7	July 1984	30	Chinese NY '93	53
Winter 1976	8	Spring 1985	31	Summer 1993	54
February 1977	9	Summer 1985	32	Autumn 1993	55
May 1977	10	Autumn 1985	33	Spring 1994	56
October 1977	11	Winter 1985	34	Summer 1994	57
Spring 1978	12	Spring 1986	35	Autumn 1994	58
Summer 1978	13	October 1986	36	Winter 1994	59
Autumn 1978	14	February 1987	37	S/Summer 1995	60
Winter 1978	15	June 1987	38	Winter 1995	61
September 1979	16	September 1987	39	S/Summer 1996	62
Winter 1979	17	January 1988	40	A/Winter 1996	63
May 1980	18	May 1988	41	Spring 1997	64
October 1980	19	October 1988	42	Summer 1997	65
February 1981	20	September 1989	43	Autumn 1997	66
May 1981	21	Winter 1990	44	Spring 1998	67
October 1981	22	Spring 1990	45	Summer 1998	68
February 1982	23	Autumn 1990	46	Autumn 1998	69

References

Abdul-Hamid W (1989): *The Mental Health Problems of Homeless People: London–New York Comparison*. Report to the Chadwick Trust Fellowship, London.

Abdul-Hamid W and Cooney C (1997): Setting up new medical services: the homeless. *Postgraduate Medical Journal*, **72**, 667–70.

Abdul-Hamid W, Wykes T and Stansfeld S (1995): The social disablement of homeless people. 2. A comparison with patients from long-stay wards. *British Journal of Psychiatry*, **166**, 809–12.

Abramson JH (1984): *Survey Methods in Community Medicine: An Introduction to Epidemiological and Evaluative Studies*. Edinburgh: Churchill Livingstone.

Ahmad W (1995): Review article: 'Race' and health. *Sociology of Health and Illness*, **17**, 418–29.

Ahmad WIU (1993): Making black people sick: 'race', ideology and health research. In WIU Ahmad (ed.) *'Race'and Health in Contemporary Britain*. Buckingham: Open University Press.

Ahmad WIU, Kernohan EEM and Baker MR (1989): Influence of ethnicity and unemployment on the perceived health of a sample of general practice attenders. *Community Medicine*, **11**, 148–56.

Ahmad WIU and Sheldon T (1993) 'Race' and statistics. In M Hammersley (ed.) *Social Research: Philosophy, Politics and Practice*. London: Sage.

Alcock P (1993): *Understanding Poverty*. London: Macmillan.

Aldridge J (1993): The textual disembodiment of knowledge in research account writing. *Sociology*, **27**, 53–66.

Allen J and Henry N (1995): Growth at the margins: contract labour in a core region. In C Hadjimichalis and D Sadler (eds) *Europe at the Margins: New Mosaics of Inequality*. Chichester: Wiley, 149–66.

Allison PD (1984): *Event History Analysis: Regression for Longitudinal Event Data*. Beverly Hills: Sage.

American Psychological Association (1996): *Working Group on Investigation of Memories of Childhood Abuse: Final Report*. Washington, DC: APA.

Anderson C and Loynes RM (1987): *The Teaching of Practical Statistics*. Chichester: Wiley.

Anderson M (1988): *The American Census: A Social History*. London: Yale University Press.

Andrews AY (1993): Sociological analysis of Jamaat-i-Islami in the United Kingdom. In R Barot (ed.) *Religion and Ethnicity: Minorities and Social Change in the Metropolis*. Kampen: Kok Pharos Publishing House.

Andrews B, Morton J, Bekerian DA, Brewin CR, Davies GM and Mollon P (1995): The recovery of memories in clinical practice: experiences and beliefs of British Psychological Society practitioners. *Psychologist*, **8**, 209–14.

Arber S and Ginn J (1991): *Gender and Later Life: A Sociological Analysis of Resources and Constraints*. London: Sage.

Arber S and Ginn J (1994): Women and aging. *Reviews in Clinical Gerontology*, **44**, 93–102.

Archard P (1979a): Vagrancy: a literature review. In T Cook (ed.) *Vagrancy: Some New Perspectives*. London: Academic Press, 11–28.

Archard P (1979b): *Vagrancy, Alcoholism and Social Control*. London: Macmillan.

Ardener S (ed.) (1975): *Perceiving Women*. London: Dent.

Arms Control and Disarmament Agency (1994): *World Military Expenditures and Arms Transfers 1991–2*. Washington, DC: ACDA.

Association of Metropolitan Authorties (1994): *A Survey of Social Services Charging Policies, 1992–94*. London: AMA.

Audit Commission (1986): *Making a Reality of Community Care*. London: HMSO.

Audit Commission and Her Majesty's Inspectorate (1992): *Getting in on the Act*. London: HMSO.

Axelson ML and Brinberg D (1989): *A Social-Psychological Perspective on Food-Related Behavior*. New York: Springer-Verlag.

Bachrach LL (1984): Interpreting research on the homeless mentally ill: some caveats. *Hospital and Community Psychiatry*, **35**, 914–17.

Bachrach LL (1992): The urban environment and mental health. *International Journal of Social Psychiatry*, **38**, 5–15.

Bahr H and Caplow T (1973): *Old Men Drunk and Sober*. New York: New York University Press.

Baker RM (1975): *Condorcet: From Natural Philosophy to Social Mathematics*. Chicago: University of Chicago Press.

Bakewell J (1996): Women and the media. Fawcett Annual Lecture, October. London: The Fawcett Society (45 Beech Street, London EC2Y 8AD).

Baldwin J (1979): Ecological and areal studies in Great Britian and the United States. *Crime and Justice*, **1**, 29–66.

Baldwin S (1977): *Disabled Children: Counting the Costs*. London: Disability Alliance.

Baldwin S (1985): *The Costs of Caring: Families with Disabled Children*. London: Routledge & Kegan Paul.

Ball S (1993): Education markets, choice and social class: the market as a class strategy in the UK and USA. *British Journal of Sociology of Education*, **14**, 3–19.

Ball S (1994): *Education Reform*. Buckingham: Open University Press.

Ball SJ and Gewirtz S (1996): School choice, social class and distinction: the realisation of social advantage in education. *Journal of Educational Policy*, **11**, 89–112.

Ballard R and Kalra V (1994): *The Ethnic Dimension of the 1991 Census: A Preliminary Report*. Manchester: University of Manchester.

Banister P, Burman E, Parker I, Taylor M and Tindall C (1994): *Qualitative Methods in Psychology: A Research Guide*. Milton Keynes: Open University Press.

Barclay GC, Tavares C and Prout C (1995): *Digest 3, Information on the Criminal Justice System in England and Wales*. Home Office Research and Statistics Department. London: HMSO.

Barnes B, Bloor D and Henry J (1996): *Scientific Knowledge: A Sociological Analysis*. London: Athlone; Chicago: University of Chicago Press.

Barot R (1993) Religion, ethnicity and social change: an introduction. In R Barot (ed.) *Religion and Ethnicity: Minorities and Social Change in the Metropolis*. Kampen: Kok Pharos Publishing House.

Barot R (ed.) (1996): *The Racism Problematic: Contemporary Sociological Debates on Race and Ethnicity*. Lewiston: The Edwin Mellon Press.

Bartley M and Owen C (1996): Relation between socioeconomic status, employment, and health during economic change, 1973–1993. *British Medical Journal*, **313**, 445–9.

Bassuk EL, Rubin L and Lauriat A (1984): Is homelessness a mental health problem? *American Journal of Psychiatry*, **141**, 1546–50.

Beardsworth A and Keil T (1997): *Sociology on the Menu: An Invitation to the Study of Food and Society*. London: Routledge.

Beatty C and Fothergill S (1996): Labour market adjustment in areas of chronic industrial decline: the case of the UK coalfields. *Regional Studies*, **30**, 627–40.

Beatty C, Fothergill S, Gore T and Herrington A (1997): *The Real Level of Unemployment*. Sheffield: Centre for Regional Economic and Social Research, Sheffield Hallam University.

Bebbington AC (1993): Regional and social variations in disability-free life expectancy in Great Britain. In JM Robine, CD Mathers, RM Bone and I Romieu (eds) *Calculation of Health Expectancies: Harmonisation, Consensus Achieved and Future Perspectives*. Colloque INSERM/John Libbey Eurotext, **226**, 175–91.

Bebbington PE (1990): Population surveys of psychiatric disorder and the need for treatment. *Social Psychiatry and Psychiatric Epidemiology*, **25**, 33–40.

Beishon S and Nazroo JY (1997): *Coronary Heart Disease: Contrasting the Health Beliefs and Behaviours of South Asian Communities in the UK*. London: Health Education Authority.

Bennett SN (1996): Class size in primary schools: perceptions of headteachers, chairs of governors, teachers and parents. *British Educational Research Journal*, **22**, 33–5.

Ben-Shlomo Y and Davey Smith G (1991): Deprivation in infancy or in adult life: which is more important for mortality risk? *Lancet*, **337**, 530–4.

Benzeval M, Judge K and Whitehead M (eds) (1995): *Tackling Inequalities in Health: An Agenda for Action*. London: King's Fund.

Beresford B (1995): *Expert Opinions: A National Survey of Parents Caring for a Severely Disabled Child*. Bristol: The Policy Press.

Beresford P (1974): Homelessness: a new approach. *Social Work Today*, **4**, 761–3.

Beresford P (1976): Poverty and disabled people: challenging dominant debates and policies. *Disability and Society*, **11**, 553–67.

Berrington A (1994): Marriage and family formation among the white and ethnic minority populations in Britain. *Ethnic and Racial Studies*, **17**, 517–46.

Berthoud R (1998): Defining ethnic groups: origin or identity? *Patterns of Prejudice* (in press).

Berthoud R, Lakey J and McKay S (1993): *The Economic Problems of Disabled People*. London: Policy Studies Institute.

Bettig RV (1996): *Copyrighting Culture: Political Economy of Intellectual Property*. Boulder, CO: Westview Press.

Bhaskar R (1989): *Reclaiming Reality: A Critical Introduction to Contemporary Philosophy*. London: Verso.

Bhrolchain NM (1990): The ethnic question for the 1991 Census: background and issues. *Ethnic and Racial Studies*, **13**, 543–67.

Billig M (1985): Prejudice, categorization and particularization: from a perceptual to a rhetorical approach. *European Journal of Social Psychology*, **15**, 79–103.

Blackburn S and Lincoln S (1992): Family Resources Survey, *Statistical News*, **99**, 14–18.

Blair PS, Fleming PJ and Bensley D (1996): Smoking and the sudden infant death syndrome: results from 1993–95 case-control study for confidential inquiry into stillbirths and deaths in infancy. *British Medical Journal*, **313**, 195–8.

Blalock HM (1979): *Social Statistics*, revised 2nd edn. London: McGraw-Hill.

Blane D (1985): An assessment of the Black Report's explanations of health inequalities. *Sociology of Health and Illness*, **7**, 231–64.

Blane D, Bartley M and Davey Smith G (1997): Disease aetiology and materialist explanations of socioeconomic mortality differentials. *European Journal of Public Health*, **7**, 385–91.

Blane D, Davey Smith G and Bartley M (1990): Social class differences in years of potential life lost: size, trends and principal causes. *British Medical Journal*, **301**, 429–32.

Blane D, Davey Smith G and Bartley M (1993): Social selection: what does it contribute to social class differences in health? *Sociology of Health and Illness*, **15**, 1–15.

Blane D, Power C and Bartley M (1996): Illness behaviour and the measurement of class differentials in morbidity. *Journal of the Royal Statistical Society A*, **159**, 77–92.

Blasi GL (1990): Social-policy and social-science research on homelessness. *Journal of Social Issues*, **46**, 207–19.

Blaxter M (1989): A comparison of measures in inequality in morbidity. In J Fox (ed.) *Health Inequalities in European Countries*. Aldershot: Gower, 63–78.

Block FL (1996): *The Vampire State*. New York: New Press.

Block G (1982): A review of validations of dietary assessment methods. *American Journal of Epidemiology*, **115**, 492–505.

Bolling GF (1996): *The Art of Forecasting*. Aldershot: Gower.

Bone M, Gregory J, Gill B and Lader D (1992): *Retirement and Retirement Plans.* London: HMSO.

Bone M and Meltzer H (1989): *The Prevalence of Disability among Children: Report 3.* London: HMSO.

Booth C (1899): *Life and Labour of the People in London.* London: Hutchinson.

Boreham J (1984): Official statistics in troubled times: the changing environment for producers and users. *Statistical News,* **64**, 1.

Bowling A (1991): *Measuring Health: A Review of Quality of Life Measurement Scales.* Buckingham: Open University Press.

Box S (1983): *Power, Crime, and Mystification.* London: Routledge.

Box S (1987): *Recession, Crime and Punishment.* London: Macmillan Education.

Boyle J (1996): *Shamans, Software and Spleens: Law and the Construction of the Information Society.* Cambridge, MA: Harvard University Press.

Bramley G and Watkins C (1995): *Circular Projections.* London: Council for the Protection of Rural England.

Brandon D (1974): Homelessness: the way ahead. *Community Care,* **8**, 16–19.

Brandon S, Boakes J, Glaser D and Green R (1998): Recovered memories of childhood sexual abuse: implications for clinical practice. *British Journal of Psychiatry,* **172**, 296–307.

Brannen J, Moss P, Owen C and Wale C (1997): *Mothers, Fathers and Employment: Parents and the Labour Market in Britain 1984–1994.* London: Department for Education and Employment.

Bravais A (1846): Analyse mathématique sur les probabilités des erreurs de situation d'un point. *Mémoire présenté par divers savants à l'Académie Royale des Sciences de l'Institut de France,* **9**, 255–332.

Brimblecombe N, Ormston M and Shaw M (1996): Gender differences in teacher response to school inspection. *Educational Studies,* **22**, 27–40.

Brimmer MJ (1981): *Review of Statistical Services in the Department of Employment and Manpower Services Commission.* London: Department of Employment.

British Psychological Society (1995): *Recovered Memories.* Leicester: BPS.

Brittan S (1996): *Capitalism with a Human Face.* London: Fontana.

Britton M (ed.) (1990): *Mortality and Geography: A review in the mid 1980s, England and Wales.* Series DS no. 9. London: HMSO.

Brodzki B and Schenk C (eds) (1988): *Life/lines: Theorising Women's Autobiography.* Ithaca, NY: Cornell University Press.

Bruegel I and Kean H (1995): Municipal feminism: relating gender and class to hierarchies, markets and networks. *Critical Social Policy,* **44**, 147–69.

Bruegel I and Perrons D (1998): Deregulation and women's employment: the diverse experiences of women in Britain. *Feminist Economics,* **4**, 103–25.

Bryman A and Cramer D (1996): *Quantitative Data Analysis with Minitab for Windows.* London: Routledge.

Bryman A and Cramer D (1997): *Quantitative Data Analysis with SPSS for Windows.* London: Routledge.

Bryson A and McKay S (1994): *Is It Worth Working?* London: Policy Studies Institute.

Budge D (1997): Time takes its toll on reading. *Times Educational Supplement,* 12 September, p. 10.

Bulmer M (1996): The ethnic group question in the 1991 Census of population. In D Coleman and J Salt (eds) *Ethnicity in the 1991 Census*, vol. 1: *Demographic Characteristics of the Ethnic Minority Populations*. London: HMSO.

Burghes L (1993): *One Parent Families: Policy Options for the 1990s*. York: Joseph Rowntree Foundation.

Burghes L and Brown M (1995): *Single Lone Mothers: Problems, Prospects and Policies*. London: Family Policy Studies Centre.

Burrows L and Walentowicz P (1992): *Homes Cost Less than Homelessness*. London: Shelter.

Butler DE and Stokes DE (1974): *Political Change in Britain*, 2nd edn. London: Macmillan.

CACI (1992): *The ACORN User Guide*. London: CACI.

Callan T, Nolan B and Whelan CT (1993): Resources, deprivation and the measurement of poverty. *Journal of Social Policy*, **22**, 141–72.

Calnan M and Williams S (1991): Styles of life and the salience of wealth: an explanatory study of health related practices in households from differing socio-economic circumstances. *Sociology of Health and Illness*, **13**, 506–29.

Campaign Against Arms Trade (1996): *Killing Jobs: The Arms Trade and Employment in the UK*. London: CAAT.

Carstairs V and Morris R (1989): Deprivation: explaining differences in mortality between Scotland and England and Wales. *British Medical Journal*, **299**, 886–9.

Carstairs V and Morris R (1991): *Deprivation and Health in Scotland*. Aberdeen: Aberdeen University Press.

Casey B, Metcalf H and Millward N (1997): *Employers' Use of Flexible Labour*. London: Policy Studies Institute.

Cassidy J (1997): The budget boondoggle. *New Yorker*, 11 August, pp. 4–5.

Castillo IY (1994): A comparative approach to social exclusion: lessons from Belgium and France. *International Labour Review*, **133**, no. 5–6.

Central Statistical Office (1985): United Kingdom National Accounts: Sources and Methods, Studies in Official Statistics no. 37. London: HMSO.

Central Statistical Office (1995a): *Social Focus on Women*. London: HMSO (in 1998 *Social Focus on Women and Men*).

Central Statistical Office (1995b): *United Kingdom National Accounts*. London: CSO.

Central Statistical Office (annual): *Social Trends*. London: HMSO.

Chadwick E (1965 [1842]) *Report on the Sanitary Conditions of the Labouring Population of Great Britain*. Edinburgh: Edinburgh University Press.

Chamberlain C and Mackenzie D (1992): Understanding contemporary homelessness: issues of definition and meaning. *Australian Journal of Social Issues*, **27**, 274–97.

Chambliss W and Seidman R (1971): *Law, Order and Power*. Reading: Addison-Wesley.

Chant S (1997): *Women-Headed Households: Diversity and Dynamics in the Developing World*. Basingstoke: Macmillan.

Chapman M and Mahon B (1988): *Plain Figures*. London: HMSO.

Charles N and Kerr M (1986): Food for feminist thought. *Sociological Review*, **34**, 537–72.

Charles N and Kerr M (1988): *Women, Food and Families*. Manchester: Manchester University Press.

Charlton J, Wallace M and White I (1994): Long-term illness: results from the 1991 census. *Population Trends*, **75**, 18–25.

Chartered Institute of Public Finance and Accountancy (1997): *Homelessness Statistics: 1995–96 Actuals*. London: CIPFA.

Chatfield C (1988): *Problem Solving: A Statistician's Guide*. London: Chapman & Hall.

Chesher A (1997): Diet revealed?: semiparametric estimation of nutrient intake–age relationships (with discussion). *Journal of the Royal Statistical Society A*, **160**, 389–428.

Chetwynd J (1985): Some costs of caring at home for an intellectually handicapped child. *Australian and New Zealand Journal of Developmental Disability*, **2**, 35–40.

Chief Inspector of Factories Annual Report for 1945 (1946). London: HMSO.

Chief Inspector of Factories Annual Report for 1967 (1968). London: HMSO.

Chief Medical Officer (1995): *Variations in Health: What Can the Department of Health and the NHS Do?* A report produced by the Variations Sub-Group of the Chief Medical Officer's Health of the Nation Working Group. London: Department of Health.

Chitty C (1989): *Towards a New Education System: The Victory of the New Right?* London: Falmer.

Choldin H (1994): *Looking for the Last Percent: The Controversy over Census Undercounts*. New Brunswick, NJ: Rutgers University Press.

Clearing House on Child Abuse and Neglect Information (1994): *Characteristics and Sources of Allegations of Ritual Child Abuse*. Fairfax, VA: CHCANI (Suite 350, 3998 Fair Ridge Dr., Fairfax, VA 22033).

Clemente F and Kleinman M (1977): Fear of crime in the United States. *Social Forces*, **56**, 519–31.

Clinard MB and Meier RF (1985): *Sociology of Deviant Behaviour*. New York: Holt, Rinehart & Winston.

Clode D (1985): When home is like no place. *Social Work Today*, **85**, 13–15.

Cockburn C (1985): *Machinery of Dominance: Women, Men and Technical Know-how*. London: Pluto.

Cohen J (1990): Things I have learned (so far). *American Psychologist*, **45**, 1304–12.

Cohen J (1992): A power primer. *Psychological Bulletin*, **112**, 155–9.

Cohen J (1994): The Earth is round ($p < .05$). *American Psychologist*, **49**, 997–1003.

Coleman D and Salt J (eds) *Ethnicity in the 1991 Census*, vol. 1: *Demographic Characteristics of the Ethnic Minority Populations*. London: HMSO.

Collins J (1993): Occupational pensions for the less well off: who benefits? *Watson's Quarterly*, **28**, 4–7.

Commission for Racial Equality (1997): *The Irish in Britain*. London: Belmont Press.

Commission on Social Justice (1994): *Social Justice: Strategies for National Renewal*. London: Vintage.

Confederation of British Industry and NatWest (annual): *CBI/NatWest Innovation Trends Survey*. London: CBI.

Connelly J and Crown J (eds) (1994): *Homelessness and Ill Health*. London: Royal College of Physicians.

Cook D (1997): *Poverty, Crime and Punishment*. London: Child Poverty Action Group.

Cook T and Braithwaite G (1979): A problem to whom? In T Cook (ed.) *Vagrancy: Some New Perspectives*. London: Academic Press, 1–10.

Cooper L (1986): Is there a case for community-based research? *Radical Statistics Newsletter*, **35**, 23–5.

Cooper P (1979): Problem of provision for the 'undeserving' homeless. MA dissertation, University of Manchester.

Cowan RS (1972a): Francis Galton's statistical ideas: the influence of eugenics. *Isis*, **63**, 509–28.

Cowan RS (1972b): Francis Galton's contribution to genetics. *Journal of the History of Biology*, **5**, 389–412.

Cowan RS (1977): Nature and nurture: the interplay of biology and politics in the work of Francis Galton. *Studies in the History of Biology*, **1**, 133–208.

Cox BD, Huppert FA and Whichelow MJ (eds) (1993): *The Health and Lifestyle Survey: Seven Years On*. Aldershot: Dartmouth.

Cox DR (1972): Regression models and life tables. *Journal of the Royal Statistical Society B*, **34**, 187–202.

Cox LH, McDonald S and Nelson D (1986): Confidentiality issues at the United States Bureau of the Census. *Journal of Official Statistics*, **2**, 35–160.

Cunningham P (1998): *Science and Technology in the UK*. London: Cartermill.

Curwin J, Slater R and Hart M (1995): *Numeracy Skills for Business*. London: Chapman & Hall.

Dale A, Arber S and Procter M (1988): *Doing Secondary Analysis*. London: Unwin Hyman.

Dale A and Davies RB (eds) (1994): *Analyzing Social and Political Change*. London: Sage.

Dale A and Marsh C (eds) (1993): *The 1991 Census User's Guide*. London: HMSO.

Darnton-Hill I (1988): Food habits methodology: some constraints. In SA Truswell and ML Wahlqvist (eds) *Food Habits in Australia: Proceedings of the First Deakin/Sydney Universities Symposium on Australian Nutrition*. Victoria: Renee Gordon.

Davey Smith G, Bartley M and Blane D (1990): The Black Report on socioeconomic inequalities in health 10 years on. *British Medical Journal*, **301**, 373–7.

Davey Smith G, Blane D and Bartley M (1994): Explanations for socio-economic differentials in mortality: evidence from Britain and elsewhere. *European Journal of Public Health*, **4**, 131–44.

Davey Smith G, Neaton JD, Wentworth D, Stamler R and Stamler J (1996): Socioeconomic differentials in mortality risk among men screened for the Multiple Risk Factor Intervention Trial. *American Journal of Public Health*, **86**, 486–96.

Davey Smith G, Shipley MJ and Rose G (1990): Magnitude and causes of socioeconomic differentials in mortality: further evidence from the Whitehall study. *Journal of Epidemiology and Community Health*, **44**, 265.

Davidoff P (1965) Advocacy and pluralism in planning. *Journal of the American Institute of Planners*, **31**, 331–7.

Davidson MJ and Cooper CL (1984): *Working Women: An International Survey.* Chichester: Wiley.

Davies H and Ward S (1992): *Women and Personal Pensions.* Manchester: Equal Opportunities Commission.

Davies J and Brember I (1997): *Monitoring Reading Standards in Year 6: A Seven Year Cross-sectional Study.* Manchester: School of Education, University of Manchester.

Davies RB, Elias P and Penn R (1992): The relationship between a husband's unemployment and his wife's participation in the labour force. *Oxford Bulletin of Economics and Statistics*, **54**, 145–71.

Dawes RM (1994): *House of Cards: Psychology and Psychotherapy Built on Myth.* New York: The Free Press.

Day C (1996): *Class Size Research and the Quality of Education.* London: National Association of Head Teachers.

Dearing R (1996): *Review of Qualifications for 16–19 Year Olds.* London: Schools Curriculum and Assessment Authority.

Deming W (1986): *Out of the Crisis.* Cambridge, MA: MIT Press.

Denver DT and Hands G (1997): *Modern Constituency Electioneering.* London: Frank Cass.

Denzin NK (1989): *Interpretative Biography.* New York: Sage.

Department for Education (1993): *Draft SEN Code of Practice and Draft Regulations.* London: DfE.

Department for Education (1994): *Code of Practice on the Identification and Assessment of Special Educational Needs.* London: Central Office of Information.

Department for Education (1995): *GCSE to GCE A/AS Value Added: Briefing for Schools and Colleges.* London: DfE.

Department for Education and Employment (1997a): *Excellence in Schools.* London: DfEE.

Department for Education and Employment (1997b): *Excellence for All Children: Meeting Special Educational Needs.* London: The Stationery Office.

Department for Education and Employment (1997c): *Special Education Needs in England: January 1996.* Statistical Bulletin 12/97. London: The Stationery Office.

Department for Education and Employment (1997d): *News 292/97.* London: DfE.

Department of Education and Science (1982): *Mathematics Counts: Report of the Committee of Inquiry into the Teaching of Mathematics in Schools* (The Cockcroft Report). London: HMSO.

Department of Education and Science (1989a): *Assessments and Statements of Special Educational Needs: Procedures within the Education, Health and Social Services.* London: HMSO.

Department of Education and Science (1989b): *English in the National Curriculum.* London: HMSO.

Department of Employment (1987): *A Short Guide to the Retail Prices Index.* London: HMSO.

Department of Employment and Productivity (1971): *British Labour Statistics: Historical Abstract 1886–1968.* London: HMSO.

Department of the Environment (1992a): *Housing*. Planning Policy Guidance Note 3. London: DoE.

Department of the Environment (1992b): *Development Plans and Regional Planning Guidance*. Planning Policy Guidance Note 12, paragraphs 5.16–17. London: DoE.

Department of the Environment (1992c): Households Accepted as Homeless (quarterly).

Department of the Environment (1994): *Information Note on Index of Local Conditions*. London: DoE.

Department of the Environment (1995a): *1991 Deprivation Index: A Review of Approaches and a Matrix of Results*. London: HMSO.

Department of the Environment (1995b): *Projections of Households in England to 2016*. London: HMSO.

Department of the Environment (1996): *Urbanization in England: Projections 1991–2016*. London: HMSO.

Department of the Environment, Transport and the Regions (1997a): *Information Bulletin (Statistics of Local Authority Activities under the Homelessness Legislation: England)*, no. 413/ENV.

Department of the Environment, Transport and the Regions (1997b): Rough sleepers in West End of London to get further help. Press release ENV/360, 17 September.

Department of Health (1988): *Community Care: Agenda for Action*. A report to the Secretary of State for Social Services. London: HMSO.

Department of Health (1990a): *Framework for Information Systems: Overview*. Working Paper 11. London: HMSO.

Department of Health (1990b): *Framework for Information Systems: Information*. London: DoH.

Department of Health (1990c): *Working for Patients. Framework for Information Systems: The Next Steps*. London: HMSO.

Department of Health (1991): *The Health of the Nation: A Consultative Document for Health in England*. Cm 1523. London: HMSO.

Department of Health (1992a): *The Health of the Nation: A Strategy for Health in England*. Cm 1986. London: HMSO.

Department of Health (1992b): More money to help mentally ill people sleeping rough. Press release H92/31, January.

Department of Health (1993): *Health Survey for England 1991*. London: HMSO.

Department of Health (1996): *Personal Social Services: A Historic Profile of Reported Current and Capital Expenditure, 1983–4 to 1993–4, England*. London: HMSO.

Department of Health (1997a): *The New NHS*. Cm 3807. London: The Stationery Office.

Department of Health (1997b): High quality, not 'bean counting' to be central NHS priority – Baroness Jay. Press release 97/234.

Department of Health (1997c): *Community Care Statistics 1997: Residential Personal Social Services for Adults, England*. Bulletin 1997/26. London: DoH.

Department of Health (1997d): *NHS Hospital Activity Statistics: England 1986 to 1996–97*. Bulletin 1997/20. London: DoH.

Department of Health (1997e): *Community Care Statistics 1996, England*. Bulletin 1997/8. London: DoH.

Department of Health (1997f): *NHS Hospital and Community Health Non-medical Staff in England: 30 September 1996*. Bulletin 1997/10. London: DoH.

Department of Health and Social Security (1980): *Inequalities in Health: Report of a Research Working Group* (The Black Report). London: DHSS.

Department of Health and Social Security (1981): *Growing Older*. Cmnd 8173. London: HMSO.

Department of Health and Social Security (1985): *Reform of Social Security*. Cmnd 9517. London: HMSO.

Department of Social Security (1990): *Social Security Statistics*. London: HMSO.

Department of Social Security (1993a): *Equality in State Pension Ages*. London: HMSO.

Department of Social Security (1993b): *Households below Average Income*. London: HMSO.

Department of Trade and Industry (1996): *Read All about It: A Survey of Public Attitudes into, Public Awareness of, Attitudes towards, and Access to Information and Communication Technologies*. DTI, Information Society Initiative, IT for All. London: DTI.

Department of Trade and Industry (annual): *R&D Scoreboard 1997*. Edinburgh: Company Reporting.

Derrida J (1976): *Of Grammatology*. Baltimore: Johns Hopkins University Press.

Dex S (1984): *Women's Work Histories: An Analysis of the Women and Employment Survey*. Research paper no. 46. London: Department of Employment.

Dex S and McCulloch A (1995): *Flexible Employment in Britain: A Statistical Analysis*. Equal Opportunities Commission Research Discussion Series 15. Manchester: EOC.

Diamond A and Goddard E (1995): *Smoking Among Secondary Children in 1994*. London: HMSO.

Dilnot A, Disney R, Johnson P and Whitehouse E (1994): *Pensions Policy in the UK: An Economic Analysis*. London: Institute for Fiscal Studies.

Dineen T (1996): *Manufacturing Victims: What the Psychology Industry is Doing to People*. Toronto: Robert Davies Publishing.

Disability Now (1995): Me and my computer. February, p. 13.

Disablement Income Group (1988): *Not the OPCS Survey: Being Disabled Costs More than They Said*. London: DIG.

Disablement Income Group (1990): *Short-Changed by Disability*. London: DIG.

Dorling DFL (1991): The visualisaton of spatial social structure. PhD thesis, Department of Geography, University of Newcastle upon Tyne.

Dorling DFL (1995): *A New Social Atlas of Britain*, Chichester: Wiley.

Dorling DFL (1996): *Area Cartograms: Their Use and Creation*. Norwich: University of East Anglia.

Dorling DFL and Fairbairn D (1997): *Mapping: Ways of Representing the World*. London: Longman.

Dorling DFL, Johnston RJ and Pattie CJ (1996a): Using triangular graphs for representing, exploring and analysing electoral change. *Environment and Planning A*, **28**, 979–98.

Dorling DFL, Pattie CJ, Rossiter DJ and Johnston RJ (1996b): Missing voters in Britain 1992–96. In DM Farrell, D Broughton, D Denver and J Fisher (eds) *British Elections and Parties Yearbook 1996*. London: Frank Cass, 37–49.

Downes D (1983): *Law and Order: Theft of an Issue*. Fabian Society no. 490. London: Fabian Society.

Downs C (1997): Pensions for an older population. *Benefits*, **18**, 9–12.

Doyal L (1979): *The Political Economy of Health*. London: Pluto.

Drever F and Whitehead M (eds) (1997): *Health Inequalities*. Series DS no. 155. London: Office for National Statistics.

Driver F (1993): *Power and Pauperism: The Workhouse System 1834–1884*. Cambridge: Cambridge University Press.

Dunne G (1997): Why can't a man be more like a woman?: in search of balanced domestic and employment lives. LSE Gender Institute Discussion Paper Series 3. London: London School of Economics.

Dunne P (1990): The political economy of military expenditure: an introduction. *Cambridge Journal of Economics*, **14**, 395–404.

Dunne P (1993): The changing military industrial complex in the UK. *Defence Economics*, **4**, 91–112.

Dunne P and Smith R (1990): Military expenditure and unemployment in the OECD. *Defence Economics*, **1**, 57–73.

Dunne P and Smith R (1992): Thatcherism and the UK defence industry. In J Michie (ed.) *1979–92: The Economic Legacy*. London: Academic Press, 91–111.

Dunnell C and Cartwright A (1979): *Medicine Takers, Prescribers and Hoarders*. London: Routledge.

Eagle A (1997): Health and safety is a government priority. Interview, *Safety Management*, November, 6–7.

Ebel R and Frisbe D (1991): *Essentials of Educational Measurement*. Englewood Cliffs, NJ: Prentice Hall.

Economist Intelligence Unit (1994): Slimming foods. *Retail Business*, **432**, 48–58.

Edwards G, Hawker A, Williamson V and Hensman C (1966): London's Skid Row. *Lancet*, **312**, 249–52.

Edwards G, Williamson V, Hawker A, Hensman C and Postoyan S (1968): Census of reception centre. *British Journal of Psychiatry*, **114**, 1031–9.

Ehrenberg ASC (1975): *Data Reduction*. Chichester: Wiley.

Ehrenberg ASC (1982): *A Primer in Data Reduction*. Chichester: Wiley.

Elford J *et al*. (1991): Early life experience and adult cardiovascular disease: longitudinal and case control studies. *International Journal of Epidemiology*, **20**, 833–44.

Elliot M (1996): *Attacks on Confidentiality Using Samples of Anonymized Records: An Analysis*. Proceedings of the Third International Conference on Statistical Confidentiality, Bled, Slovenia, October 1996.

Elpers JR (1987): Are we legislating reinstitutionalization? *American Journal of Orthopsychiatry*, **57**, 441–6.

Employment Policy Institute (1993): *The Curious Case of Falling Unemployment*. Employment Policy Institute Bulletin 7. London: EPI.

Employment Policy Institute (1996): Introducing the Employment Audit. *Employment Audit*, **1**, 1–14.

Enloe C (1996): Religion and ethnicity. In J Hutchinson and AD Smith (eds) *Ethnicity*. Oxford: Oxford University Press, 197–202.

Equal Opportunities Commission (1997): *Facts about Women and Men in Great Britain 1997*. Manchester: EOC.

Equal Opportunities Commission (various): Briefings on women and men in Britain: work and parenting. Manchester: EOC.

Equal Opportunities Commission (various): Briefings on women and men in Britain: pay, income and personal finance. Manchester: EOC.

Equal Opportunities Commission (various): Briefings on women and men in Britain: life cycle of inequality. Manchester: EOC.

Erikson R and Goldthorpe JH (1993): *The Constant Flux*. Oxford: Clarendon Press.

European Commission (1997): *European Science and Technology Indicators 1994*. Luxembourg: European Commission DG XII, EUR 15897EN.

Eurostat (1994a): *Asylum-Seekers and Refugees: A Statistical Report*, vol. 1: *EC Member States*. Luxembourg: Office for Official Publications of the European Communities.

Eurostat (1994b): *Asylum-Seekers and Refugees: A Statistical Report*, vol. 2: EFTA Countries. Report compiled by the Netherlands Interdisciplinary Demographic Institute at the request of the EFTA Secretariat in Luxembourg.

Eurostat (1998): *Europe in Figures*. Luxembourg: Office for Official Publications of the European Communities.

Evans A and Duncan S (1988): *Responding to Homelessness: Local Authority Policy and Practice*. Department of the Environment. London: HMSO.

Evans J (1982): After *Fifteen Thousand Hours*, where do we go from here? *School Organisation*, **2**, 239–53.

Evans J (1989): Mathematics for adults. In C Keitel *et al.* (eds) *Mathematics, Education and Society*. Reports and papers presented in Fifth Day Special Programme at the International Conference on Mathematics Education (ICME-6), Budapest, 27 July–3 August 1988. Science and Technology Education Document Series no. 35. Paris: UNESCO, 65–7.

Evans J (1992): Mathematics for adults: community research and 'barefoot statisticians'. In M Nickson and S Lerman (eds) *The Social Context of Mathematics Education: Theory and Practice*. London: Southbank Press, ch. 14.

Evans J and Thorstad I (1995): Mathematics and numeracy in the practice of critical citizenship. In D Coben (ed.) *ALM-1: Proceedings of the Inaugural Conference of Adults Learning Maths – A Research Forum*, 22–24 July 1994, Fircroft College, Birmingham. London: Goldsmiths' College, 64–70.

Fagan C and Rubery J (1996): The salience of the part-time divide in the European Union. *European Sociological Review*, **12**, 227–50.

Falkingham J (1989): Dependency and ageing in Britain: a re-examination of the evidence. *Journal of Social Policy*, **18**, 55–68.

Falkingham J and Victor C (1991): *The Myth of the Woopie? Incomes, the Elderly and Targeting Welfare*. WSP 55. London: Suntory–Toyota International Centre for Economics and Related Disciplines.

Farr W (1839): Letter to the Registrar General. In *The First Annual Report of the Registrar General*. London: Longman, Orme, Brown, Green & Longman.

Farrall LA (1970): The origins and growth of the English eugenics movement, 1865–1925. PhD thesis, Indiana University, Bloomington.

Fazeli R (1996): *The Economic Impact of the Welfare State and the Social Wage: The British Experience*. Aldershot: Avebury.

Fellegi LP (1972): On the question of statistical confidentiality. *Journal of the American Statistical Association*, **67**, 7–18.

Fenton S, Hughes A and Hine C (1995): Self-assessed health, economic status and ethnic origin. *New Community*, **21**, 55–68.

Fenwick D and Denman J (1995): The monthly claimant unemployment count: change and consistency. *Labour Market Trends*, **103**, 397–400.

Ferri E and Smith K (1996): *Parenting in the 1990s*. London: Family Studies Centre; York: Joseph Rowntree Foundation.

Feuerstein MT (1997): *Poverty and Health: Reaping a Richer Harvest*. London: Macmillan.

Feyerabend P (1987): *Farewell to Reason*. London: Verso.

Fiddes N (1991): *Meat: A Natural Symbol*. London: Routledge.

Filakti H and Fox J (1995): Differences in mortality by housing tenure and car access from the OPCS Longitudinal Study. *Population Trends*, **81**, 27–30.

Finch, S, Lowe C, Doyle W *et al.* (1997): *National Diet and Nutrition Survey: People Aged 65 or Over*. London: The Stationery Office.

Fine B, Heasman M and Wright J (1996): *Consumption in the Age of Affluence: The World of Food*. London: Routledge.

Fink A (ed.) (1995): *The Survey Kit*. London: Sage.

Fischer PJ and Breakey WR (1986): Homelessness and mental health: an overview. *International Journal of Mental Health*, **14**, 6–41.

Fisher K and Collins J (1993): *Homelessness, Health Care and Welfare Provision*. London: Routledge.

Fisher RA (1918): The correlation between relatives on the supposition of Mendelian inheritance. *Transactions of the Royal Society of Edinburgh*, **52**, 399–433.

Fisher RA (1930): *The Genetical Theory of Natural Selection*. Oxford: Clarendon Press.

FitzGibbon CT and Stephenson NJ (1996): Inspecting Her Majesty's Inspectors: should social science and social policy cohere? Paper presented to the European Conference on Educational Research, Seville, September 1996.

FitzGibbon CT and Vincent L (1994): *Candidates' Performance in Public Examinations in Mathematics and Science*. London: Schools Curriculum and Assessment Authority.

Fitzhugh Directory (1997): Independent healthcare and long term care. Financial Information, 1996–7.

Fitzpatrick R, Fletcher A, Gore S, Jones D, Speigelhalter D and Cox D (1992): Quality of life measures in health care: applications and issues in assessment. *British Medical Journal*, **305**, 1074–7.

Fleming PJ, Blair PS, Bacon C *et al.* (1996): Environment of infants during sleep and risk of the sudden infant death syndrome: results of 1993–95 case-control study for confidential enquiry into stillbirths and deaths in infancy. *British Medical Journal*, **313**, 191–5.

Forrest R and Gordon D (1993): *People and Places: A 1991 Census Atlas*. School for Advanced Urban Studies and Bristol Statistical Monitoring Unit.

Foster J and Hope T (1993): *Housing, Community and Change: The Impact of the Priority Estates Project.* London: HMSO.

Fox J and Goldblatt PO (1982): *1971–1981 Longitudinal Study: Sociodemographic Mortality Differentials 1971–75.* London: HMSO.

Fox J, Goldblatt P and Jones D (1990): Social class mortality differentials: artifact, selection or life circumstances? In P Goldblatt (ed.) *Longitudinal Study: Mortality and Social Organisation.* London: HMSO.

Fox J, Jones D, Moser K and Goldblatt P (1985): Socio-demographic differentials in mortality 1971–1981. *Population Trends*, **40**, 10–16.

Freedman S, Pisani R, Purves R and Adhikari A (1991): *Statistics*, 2nd edn. New York: Norton.

Freeman A (1991): National accounts in value terms: the social wage and profit rate in Britain 1950–1986. In P Dunne (ed.) *Quantitative Marxism.* Cambridge: Polity.

Friedman A and Cornford D (1989): *Computer Systems Development.* Chichester: Wiley.

Frow J (1996): Information as gift and commodity. *New Left Review*, **219**, 89–108.

Gaffney D and Pollock AM (1997): *Can the NHS Afford the Private Finance Initiative?* London: British Medical Association. Health Policy and Economic Unit.

Gallie D, Marsh C and Vogler C (1994): *Social Change and the Experience of Unemployment.* Oxford: Oxford University Press.

Galton F (1865): Hereditary talent and character. *Macmillan's Magazine*, **12**, 157–66, 318–27.

Galton F (1869): *Hereditary Genius.* London: Macmillan.

Galton F (1877): Typical laws of heredity. *Proceedings of the Royal Institution*, **8**, 282–301.

Galton F (1886): Family likeness in stature. *Proceedings of the Royal Society of London*, **40**, 42–73.

Galton F (1888): Correlations and their measurement, chiefly from anthropometric data. *Proceedings of the Royal Society of London*, **45**, 135–45.

Gamble A (1988): *The Free Economy and the Strong State: The Politics of Thatcherism.* London: Macmillan.

Gandy OH (1993): *The Panoptic Sort: The Political Economy of Personal Information.* Boulder, CO: Westview Press.

Gani J and Lewis T (1990): Statisticians in the service of the public. Letter to *RSS News and Notes*, December, p. 4.

Gapper J and Wood L (1991): Conceived in good times, born in bad. *Financial Times*, 2 April.

Garside WR (1980): *The Measurement of Unemployment: Methods and Sources in Great Britain 1850–1979.* Oxford: Blackwell.

Gay JD (1971): *The Geography of Religion in England.* London: Duckworth.

Georghiou L, Halfpenny P, Evans J, Hinder S, Nedeva M and McKinlay C (1997): *Survey of Teaching Equipment in Higher Education Institutions in England and Wales.* Manchester: University of Manchester, PREST/CASR.

Georghiou L, Halfpenny P, Nedeva M, Evans J and Hinder S (1996): *Survey of Research Equipment in United Kingdom Universities.* Manchester: University of Manchester, PREST/CASR.

Gershuny J, Buck N, Coker O *et al.* (1996): British Household Panel Survey. *Social Trends*, **26**, 27–36.

Giannelli PC (1995): The admissibility of hypnotic evidence in US courts. *International Journal of Clinical and Experimental Hypnosis*, **43**, 212–33.

Gigerenzer G (1993): The Superego, the Ego and the Id in statistical reasoning. In G Keren and C Lewis (eds) *A Handbook for Data Analysis in the Behavioral Sciences: Methodological Issues.* Hove: Lawrence Erlbaum, 311–39.

Gilbert B (1966): *The Evolution of National Insurance in Great Britain.* London: Michael Joseph.

Gilbert N (ed.) (1993): *Researching Social Life.* London: Sage.

Gill R (1993): Justifying injustice: broadcasters' accounts of inequality in radio. In E Burman and I Parker (eds) *Discourse Analytic Research: Repertoires and Readings of Text in Action.* London: Routledge.

Ginn J (1993): Vanishing trick: how to make married women disappear from pension statistics. *Radical Statistics*, **55**, 37–43.

Ginn J and Arber S (1991): Gender, class and income inequalites in later life. *British Journal of Sociology*, **423**, 369–96.

Ginn J and Arber S (1993): Pension penalties: the gendered division of occupational welfare. *Work Employment and Society*, **71**, 47–70.

Ginn J and Arber S (1995) Moving the goalposts: the impact on British women of raising their state pension age to 65. In J Baldock and M May (eds) *Social Policy Review No. 7.* London: Social Policy Association, 186–212.

Ginn J and Arber S (1996): Patterns of employment, pensions and gender: the effect of work history on older women's non-state pension. *Work Employment and Society*, **10**, 469–90.

Ginn J and Arber S (1997): Changing patterns of pension inequality in later life. Paper presented to the annual conference of the British Society for Gerontology, Bristol, 19–21 September.

Glendinning C (1983): *Unshared Care: Parents and Their Disabled Children.* London: Routledge & Kegan Paul.

Glover J and Arber S (1995): Polarization in mothers' employment. *Gender, Work and Organisation*, **2**, 165–79.

Goddard E (1996): *Young Teenagers and Alcohol in 1996.* London: The Stationery Office.

Goldblatt PO (1988): Changes in social class between 1971 and and 1981: could these affect mortality differentials among men of working age? *Population Trends*, **51**, 9–17.

Goldblatt PO (1989): Mortality by social class 1971–1985. *Population Trends*, **56**, 6–15.

Goldblatt P (ed.) (1990): *1971–1981 Longitudinal Study: Mortality and Social Organisation.* London: HMSO.

Goldstein H (1983): Measuring changes in educational attainment over time: problems and possibilities. *Journal of Educational Measurement*, **20**, 369–77.

Goldstein H and Cresswell M (1996): The comparability of different subjects in public examinations: a theoretical and practical critique. *Oxford Review of Education*, **22**, 435–42.

Goldstein H and Myers K (1996): Freedom of information: towards a code of ethics for perfomance indicators. *Research Intelligence*, **57**, 12–16.

Goldstein H, Rasbash J, Yang M *et al.* (1993): A multilevel analysis of school examination results. *Oxford Review of Education*, **19**, 425–33.

Goldstein H and Sammons P (1997a): The influence of secondary and junior schools on sixteen year examination performance: a cross-classified multilevel analysis. *School Effectiveness and School Improvement*, **8**, 219–30.

Goldstein H and Sammons P (1997b): The influence of secondary and junior schools and college performance. *Journal of the Royal Statistical Society A*, **159**, 149–63.

Goldstein H and Spiegelhalter DJ (1996): League tables and their limitations: statistical issues in comparisons of institutional performance. *Journal of the Royal Statistical Society A*, **159**, 385–443.

Goldstein H and Thomas S (1996): Using examination results as indicators of school and college performance. *Journal of the Royal Statistical Society A*, **159**, 149–63.

Goodman A and Webb S (1995): *The Distribution of UK Household Expenditure, 1979–92*. London: Institute for Fiscal Studies.

Gordon D and Forrest R (1995): *People and Places II: Social and Economic Distinctions in England*. Bristol: School for Advanced Urban Studies, University of Bristol.

Gordon D and Pantazis C (eds) (1997): *Breadline Britain in the 1990s*. Aldershot: Ashgate.

Gordon D, Parker R and Loughran F (1996): *Children with Disabilities in Private Households: A Re-analysis of the OPCS' Investigation*. Report to the Department of Health, School for Policy Studies, University of Bristol.

Gough D and Wroblewska A (1993): *Services for Children with a Motor Impairment and Their Families in Scotland*. Glasgow: Public Health Research Unit, University of Glasgow.

Government Statistical Service (1996a): *Government Statistical Service Committee on Social Statistics Annual Report on Major Social Surveys 1995/6*. London: ONS.

Government Statistical Service (UK) (1996b): *Longitudinal Social Statistics: A Guide to Official Sources*. London: GSS (SPH) Secretariat.

Government Statisticians' Collective (1979): How official statistics are produced: views from the inside. In J Irvine, I Miles and J Evans (eds) *Demystifying Social Statistics*. London: Pluto, 130–51. Reprinted in Hammersley M (ed.) (1993): *Social Research*. London: Sage, 146–65.

Graham H (1984): *Women, Health and the Family*. Brighton: Wheatsheaf.

Graham S (1987): The extra costs borne by families who have a child with a disability. *SWRC Reports and Proceedings No. 68*. University of New South Wales.

Green A (1994): *The Geography of Poverty and Wealth*. Coventry: Institute for Employment Research, University of Warwick.

Green AE (1997a): Exclusion, unemployment and non-employment. *Regional Studies*, **31**, 505–20.

Green AE (1997b): Unemployment and non-employment in Europe: insights using alternative measures. Paper presented to the European Urban and Regional Research Network 'Regional Frontiers' Conference, Frankfurt an der Oder, September 1997.

Green AE and Hasluck C (1998): (Non-)Participation in the labour market: alternative indicators and estimates of labour reserve in UK regions. *Environment and Planning*, **30**, 543–58.

Green AE and Owen DW (1991): Local labour supply and demand interactions in Britain during the 1980s. *Regional Studies*, **25**, 295–314.

Green AE and Owen DW (1996): *A Labour Market Definition of Disadvantage: Towards an Enhanced Local Classification*. Department for Education and Employment Research Series 11. London: The Stationery Office.

Green Balance (1994): *The Housing Numbers Game: A Campaigners' Guide*. London: Council for the Protection of Rural England.

Green Balance (1996): *Testing, Testing!* London: Council for the Protection of Rural England.

Gregory JR, Collins DL, Davies PSW, Hughes JM and Clarke PC (1995): *National Diet and Nutrition Survey: Children Aged 1½ to 4½ Years*. London: HMSO.

Greve J and Currie E (1990): *Homelessness in Britain*. York: Joseph Rowntree Foundation.

Griffiths D, Irvine J and Miles I (1979): Social Statistics: towards a radical science. In Irvine J, Miles I and Evans J (eds) *Demystifying Social Statistics*. London: Pluto, 339–81.

Gross E (1990): The body of signification. In J Fletcher and A Benjamin (eds) *Abjection, Melancholia and Love: The Work of Julia Kristeva*. London: Routledge, 80–103.

Grove WM and Meehl PE (1996): Comparative efficiency of informal (subjective, impressionistic) and formal (mechanical, algorithmic) procedures: the clinical–statistical controversy. *Psychology, Public Policy and Law*, **2**, 293–323.

Groves D (1987) Occupational pension provision and women's poverty in old age. In C Glendinning and J Millar (eds) *Women and Poverty in Britain*. Brighton, Wheatsheaf, 199–217.

Grunberg J and Eagle PF (1990): Shelterization: how the homeless adapt to shelter living. *Hospital and Community Psychiatry*, **41**, 521–5.

Gudgin G and Taylor PJ (1979): *Seats, Votes and the Spatial Organisation of Elections*. London: Pion.

Gunter B (1995) *Television and Gender Representation*. London: John Libbey.

Gupta S, de Belder A and O'Hughes L (1995): Avoiding premature coronary deaths in Asians in Britain: spend now on prevention or pay later for treatment. *British Medical Journal*, **311**, 1035–6.

Hacking I (1975): *The Emergence of Probability*. Cambridge: Cambridge University Press.

Hakim C (1996): *Key Issues in Women's Work: Female Heterogeneity and the Polarisation of Women's Employment*. London: Athlone.

Hamnett C (1992): *Inheritance in Britain: The Disappearing Billions*. Stratford-upon-Avon: PPP Lifetime.

Hampshire County Council (1990): *All Our Children: Special Needs Review*. Winchester: Education Committee, Hampshire County Council.

Hampshire County Council (1994): *1994–1996 Development Plan for the Education Committee*. Winchester: Hampshire County Council.

Hancock R and Weir P (1994): *More Ways than Means: A Guide to Pensioners' Incomes in Great Britain during the 1980s*. London: Age Concern Institute of Gerontology.

Hand DJ (1987): A statistical knowledge enhancement system. *Journal of the Royal Statistical Society A*, **150**, 334–545.

Hannay DR (1979): *The Symptom Iceberg*. London: Routledge & Kegan Paul.

Harding S (1995): Social class differences in mortality of men: recent evidence from the OPCS Longitudinal Study. *Population Trends*, **80**, 33.

Harding S and Maxwell R (1997): Differences in mortality of migrants. In F Drever and M Whitehead (eds) *Health Inequalities: Decennial Supplement No. 15*. London: The Stationery Office.

Harré R (1981): The positivist–empiricist approach and its âlternative. In P Reason and J Rowan (eds) *Human Inquiry: A Sourcebook of New Paradigm Research*. Chichester: Wiley, 3–17.

Harré R and Secord PF (1972): *The Explanation of Social Behaviour*. Oxford: Blackwell.

Harrington C and Pollock AM (1998): Decentralisation and privatisation of long term care in the UK and the USA. *Lancet*, **351**, 1805–8.

Harris P (1986): *Designing and Reporting Experiments*. Milton Keynes: Open University Press.

Hart J (1975): Health in the valleys. In PH Ballard and E Jones (eds) *The Valleys Call*. Ferndale: E. Ron Jones Publications, 184–98.

Hartman C (1984): *Critical Perspectives in Housing*. Washingtion, DC: Temple University Press.

Hartmann H (1986): The unhappy marriage of marxism and feminism. In L Sargent (ed.) *The Unhappy Marriage of Marxism and Feminism*. London: Pluto, 1–41.

Harvey D (1973): *Social Justice and the City*. London: Arnold.

Haskey J (1996): The ethnic minority populations of Great Britain: their estimated sizes and age profiles. *Population Trends*, **84**, 33–9.

Hattersley L and Creeser R (1995): *Longitudinal Study 1971–1991: History, Organisation and Quality of Data*. London: HMSO.

Heady P, Bruce S, Freeth S and Smith S (1996a): The coverage of the electoral register. In I McLean and DE Butler (eds) *Fixing the Boundaries*. Aldershot: Dartmouth, 189–206.

Heady P, Smith S and Avery V (1996b): *1991 Census Validation Survey: Quality Report*. London: The Stationery Office.

Health and Safety Commission (1991): *Health and Safety Commission Annual Report 1990/91*. London: HSC.

Health and Safety Commission (1995a): *Annual Report 1994/95*. London: HSE Books.

Health and Safety Commission (1995b): *Health and Safety Statistics 1994–95*. London: HSC.

Health and Safety Commission (1997): *Annual Report 1996/97*. London: HSE Books.

Health and Safety Executive (1987): *Health and Safety Statistics 1984–85*. London: HSE.

Health and Safety Executive (1997): *Safety Statistics Bulletin 1996/97*. London: HSE.

Heath A and Ridge J (1983): Social mobility of ethnic minorities. *Journal of Biosocial Science*, supplement, **8**, 169–84.

Heath I (1994): The creeping privatisation of NHS prescribing. *British Medical Journal*, **309**, 623–4.

Heijken JAM (ed.) (1994): *Forecasting the Labour Market by Occupation and Education.* Norwell, MA: Kluwer Academic.

Her Majesty's Treasury (1996): *Financial Statement and Budget Report 1997–98.* HC 90. London: HMSO.

Hickman M (1996): Racism and identity: issues for the Irish in Britain. In T Ranger, Y Samad and O Stuart (eds) *Culture, Politics and Identity.* Aldershot: Avebury.

Hills J (1993): *The Future of Welfare: A Guide to the Debate.* York: Joseph Rowntree Foundation.

Hills J and Mullings B (1990): Housing: a decent home for all at price within means. In J Hills (ed.) *The State of Welfare.* Oxford: Clarendon Press.

Hilts V (1973): Statistics and social science. In RN Giere and RS Westfall (eds) *Foundations of the Scientific Method: The Nineteenth Century.* Bloomington: University of Indiana, 206–33.

Himmelweit S (1995): The discovery of unpaid work. *Feminist Economics*, **1**, 1–20.

Hindess B (1973): *The Use of Official Statistics in Sociology.* London: Macmillan.

Hirdes JP and Forbes WF (1992): The importance of social relationships, socioeconomc status and health practices with respect to mortality among healthy Ontario males. *Journal of Clinical Epidemiology*, **45**, 175–82.

Hoch C (1987): A brief history of the homeless problem in the US. In R Bingham, R Green and S White (eds) *The Homeless in Contemporary Society.* Beverly Hills, CA: Sage.

Hodgson GM (1988): *Economics and Institutions.* Cambridge: Polity.

Hoggart K (1995): Political parties and the implementation of homeless legislation by non-metropolitan districts in England and Wales. *Political Geography*, **14**, 59–79.

Hollowell J (1997): The General Practice Research Database: quality of morbidity data. *Population Trends*, **87**, 36–40.

Holmans A (1995): *Housing Demand and Need in England 1991–2011.* York: Joseph Rowntree Foundation.

Holmes T (1912): *London's Underworld.* London: Dent.

Hope T (1986): Council tenants and crime. *Home Office Research Bulletin*, **21**, 46–51.

Hope T and Shaw M (eds) (1988): *Communities and Crime Reduction.* London: HMSO.

Horn R (1981): Extra costs of disablement: background for an Australian study. *SWRC Reports and Proceedings No. 13.* University of New South Wales.

Horwath C (1990a): Socio-economic and behavioural effects of the dietary habits of elderly people. *International Journal of Biosocial and Medical Research*, **11**, 15–30.

Horwath C (1990b): Food frequency questionnaires: a review. *Australian Journal of Nutrition and Dietetics*, **47**, 71–6.

House of Commons Employment Committee (1996): *Unemployment and Employment Statistics.* London: HMSO.

House of Commons Health Committee (1996a): *Long-Term Care: Future Provision and Funding*. Third report, session 1995–96, vol. 1. HC 59–1. London: HMSO.

House of Commons Health Committee (1996b): *Public Expenditure on Health and Personal Social Services*. HC 698. London: HMSO.

Howitt D and Cramer D (1997): *A Guide to Computing Statistics with SPSS for Windows*. Hemel Hempstead: Prentice-Hall.

Huff D (1973): *How to Lie with Statistics*. London: Penguin.

Hutson S and Liddiard M (1994): *Youth Homelessness: The Construction of a Social Issue*. Basingstoke: Macmillan.

Hutton S, Kennedy S and Whiteford P (1995): *Equalisation of State Pension Ages: The Gender Impact*. Manchester: Equal Opportunities Commission.

Illsley R (1955): Social class and selection and class differences in relation to stillbirths and infant deaths. *British Medical Journal*, **ii**, 1520–4.

Illsley R (1986): Occupational class, selection and the production of inequalities in health. *Quarterly Journal of Social Affairs*, **2**, 151–65.

Independent Healthcare Association (1995): *Acute Hospitals in the Independent Sector*. London: IHA.

Institute for Employment Research (1997): *Labour Market Assessment: Review of the Economy and Employment 1996/97*. Coventry: Institute for Employment Research, University of Warwick.

Irvine J, Miles I and Evans J (eds) (1979): *Demystifying Social Statistics*. London: Pluto.

Jacobs E and Worcester R (1991): *Typically British*. London: Bloomsbury.

Jarman B (1983): Identification of underprivileged areas. *British Medical Journal*, **286**, 1705–9.

Jarman B (1984): Underprivileged areas: validation and distribution of scores. *British Medical Journal*, **289**, 1587–92.

Jarvis H (1997): Housing, labour markets and household structure: questioning the role of secondary data analysis in sustaining the polarization debate. *Regional Studies*, **31**, 521–33.

Jarvis L (1998): *Teenage Smoking Attitudes in 1996: A Survey of Knowledge and Attitudes of 11 to 15 Year Olds in England*. London: The Stationery Office.

Jeffreys B and Woods P (1998): *Teachers under Inspection*. London: Falmer.

Jenkinson C, Coulter A and Wright L (1993): Short form (S-F)36 health survey questionnaire: normative data for adults of working age. *British Medical Journal*, **306**, 1437–40.

Jessop B (1995): The regulation approach, governance and post-Fordism: alternative perspectives on economic and political change? *Economy and Society*, **24**, 307–33.

Johnson P, Conrad C and Thomson D (eds) (1989): *Workers versus Pensioners: Intergenerational Justice in an Ageing World*. Manchester: Manchester University Press.

Johnston RJ, Pattie CJ and Allsopp JG (1988): *A Nation Dividing?* London: Longman.

Jones T, Maclean B and Young J (1986): *The Islington Crime Survey*. Aldershot: Gower.

Joseph Rowntree Foundation Income and Wealth Group (1995): *Inquiry into Income and Wealth*. York: JRF.

Kane E (1987): *Doing Your Own Research: How to Do Basic Descriptive Research in the Social Sciences and Humanities*. London: Marion Boyars.

Kass G (1980): An exploratory technique for investigating large quantities of categorical data. *Applied Statistics*, **29**, 119–27.

Kaufman JS, Cooper RS and McGee DL (1997): Socioeconomic status and health in Blacks and Whites: the problem of residual confounding and the resilience of race. *Epidemiology*, **8**, 621–8.

Kauppinen L (1995): Statement on behalf of the World Federation of the Deaf, the World Blind Union, the International League of Societies for Persons with Mental Handicap. Rehabilitation International and Disabled People's International. *Disability Awareness in Action Newsletter*, **25**, 2.

Keighley J (1992): Sex discrimination and private insurance: should sex differences make a difference? *Policy and Politics*, **202**, 99–110.

Kendell RE (1975): *The Role of Diagnosis in Psychiatry*. Oxford: Blackwell Scientific.

Keys W, Harris S and Fernandes C (1996): *Third International Mathematics and Science Study – First National Report: Part 1*. Slough: National Foundation for Educational Research.

Kiernan KK and Estaugh V (1993): *Cohabitation: Extra-marital Childbearing and Social Policy*. London: Family Policy Studies Centre.

Kinsey R (1984): *First Report of the Merseyside Crime Survey*. Liverpool: Merseyside County Council.

Kinsey R, Lea J and Young J (1986) *Losing the Fight against Crime*. Oxford: Blackwell.

Klein R (1988): Acceptable inequalities. In D Green (ed.) *Acceptable Inequalities? Essays on the Pursuit of Equality in Health Care*. London: Institute of Economic Affairs.

Koegel P (1992): Through a different lens: an anthropological perspective on the homeless mentally ill. *Culture, Medicine and Psychiatry*, **16**, 1–22.

Kotlikoff L and Sachs J (1997): Privatising social security: it's high time to privatise. *Brookings Review*, **15**, 16–19.

Kranzler GD and Moursund JP (1995): *Statistics for the Terrified*. Englewood Cliffs, NJ: Prentice-Hall.

Kuh D, Wadsworth MEJ and Yusuf EJ (1994): Burden of disability in a post war birth cohort in the UK. *Journal of Epidemiology and Community Health*, **48**, 262–9.

Kunst AE and Mackenbach JP (1994): The size of mortality differentials associated with educational level in nine industrialized countries. *American Journal of Public Health*, **84**, 932–7.

Labov W and Waletzky J (1967): Narrative analysis: oral versions of personal experience. In J Helm (ed.) *Essays on the Verbal and Visual Arts*. Seattle: University of Washington Press, 12–44.

Laing W (ed.) (1997): *Laing's Review of Private Healthcare and Directory of Independent Hospitals, Nursing and Residential Homes and Related Services*. London: Laing & Buisson.

Lambert C, Jeffers S, Burton P and Bramley G (1992): *Homelessness in Rural Areas*. Salisbury: Rural Development Commission.

Lanning KV (1992) *Investigator's Guide to Allegations of 'Ritual' Child Abuse.* Quantico, VA: Behavioral Science Unit, Naitonal Center for the Analysis of Violent Crime, FBI Academy, Quantico, Virginia 22135.

Laux R (1996): Measuring labour market attachment using the Labour Force Survey. *Labour Market Trends,* **104**, 407–13.

Laux R and Tonks E (1996): Longitudinal data from the Labour Force Survey. *Labour Market Trends,* **104**, 175–88.

Lea J and Young J (1984): *What Is to Be Done about Law and Order?* London: Pluto.

Lee P, Murie A and Gordon D (1995): *Area Measures of Deprivation: A Study of Current Methods and Best Practices in the Identification of Poor Areas in Great Britain.* Birmingham: Centre for Urban and Regional Studies, University of Birmingham.

Leech K (1989): *A Question in Dispute: The Debate about an 'Ethnic' Question in the Census.* London: Runnymede Trust.

Lees S (1996): Unreasonable doubt: the outcomes of rape trials. In M Hester, L Kelly and J Radford (eds) *Women, Violence and Male Power,* Buckingham: Open University Press, 99–116.

Le Fanu J (1993): *A Phantom Carnage: The Myth that Low Income Kills.* London: Social Affairs Unit.

Leicester M and Pollock AM (1996): Is needs assessment working? *Public Health,* **110**, 109–13.

Leon DA (1988): *Longitudinal Study, 1971–75: Social Distribution of Cancer.* London: HMSO.

Levitas R (1996a): Fiddling while Britain burns: the 'measure of unemployment'. In R Levitas and W Guy (eds) *Interpreting Official Statistics.* London: Routledge.

Levitas R (1996b): The concept of social exclusion and the new Durkheimian hegemony. *Critical Social Policy,* **16**, 5–20.

Levitas R and Guy W (eds) (1996): *Interpreting Official Statistics.* London: Routledge.

Lewis P (1994): Being Muslim and being British. In R Ballard (ed.) *Desh Pardesh: The South Asian Presence in Britain.* London: Hurst.

Lidstone P (1994): Rationing housing to the homeless applicant. *Housing Studies,* **9**, 459–72.

Lilley P (1992): Speech to the International Conference on Social Security 50 Years after Beveridge, York, 27 September, published as a press release.

Lindsay DS and Read JD (1995): 'Memory work' and recovered memories of childhood sexual abuse: scientific evidence and public, professional, and personal issues. *Psychology, Public Policy and Law,* **1**, 846–908.

Linstead S (1994): Objectivity, reflexivity and fiction: humanity, inhumanity and the science of the social. *Human Relations,* **47**, 1321–39.

Lloyd-Sherlock P and Johnson P (1996): *Ageing and Social Policy: Global Comparisons.* London: Suntory–Toyota International Centre for Economics and Related Disciplines.

Loach I (1976): *The Price of Deafness: A Review of the Financial and Employment Problems of the Deaf and Hard of Hearing.* London: The Disability Alliance.

Lodge Patch IC (1970): Homeless men. *Proceedings of the Royal Society of Medicine*, **64**, 437–41.

Loftus EF (1993): The reality of repressed memories. *American Psychologist*, **48**, 518–37.

Loftus EF (1997): Creating false memories. *Scientific American*, **277**, 70–5.

Loftus GR (1993): A picture is worth 1000 *p*-values: on the irrelevance of hypothesis-testing in the microcomputer age. *Behaviour Research Methods, Instruments and Computers*, **25**, 250–6.

Loveland I (1990) Homelessness in the USA. *Urban Law and Policy*, **9**, 231–76.

Lundeberg O (1991): Causal explanations for class inequality in health: an empirical analysis. *Social Science and Medicine*, **32**, 385–93.

Lupton D (1996): *Food, the Body and the Self*. London: Sage.

Lynch JW, Kaplan GA, Cohen RD, Tuomilheto J and Salonen JT (1996): Do cardiovascular risk factors explain the relationship between socioeconomic status, risk of all-cause mortality, cardiovascular mortality, and acute myocardial infarction? *American Journal of Epidemiology*, **144**, 934–42.

Lynch JW, Kaplan GA and Shema SJ (1997): Cumulative impact of sustained economic hardship on physical, cognitive, psychological and social functioning. *New England Journal of Medicine*, **337**, 1889–95.

Lyon D (1994): *The Electronic Eye: The Rise of Surveillance Society*. Cambridge: Polity.

Mabbett D (1997): *Pension Funding: Economic Imperative or Political Strategy*. Discussion Paper no. 97/1. Uxbridge: Brunel University.

McCloskey DN (1986): *The Rhetoric of Economics*. Brighton: Wheatsheaf.

McCormick A, Fleming D and Charlton J (1995): *Morbidity Statistics from General Practice: Fourth National Study 1991–92*. Series MB5 no. 3. London: HMSO.

McCrossan L (1991): *A Handbook for Interviewers*. London: HMSO.

McDowell L (1997): Re-examining economic and social polarisation. *Area*, **29**, 172–3.

Macfarlane A (1990): Official statistics and women's health and illness. In H Roberts (ed.) *Women's Health Counts*. London: Routledge, ch. 1.

Macfarlane AJ (1992): Monitoring targets or target monitoring: the role of health statistics in setting a strategy. In *The Health of the Nation: Are We on Target?* Birmingham: Public Health Alliance and Radical Statistics Health Group.

Macfarlane AJ and Mugford M (1984): *Birth Counts: Statistics of Pregnancy and Childbirth*. London: HMSO.

Macintyre S (1997): The Black Report and beyond: what are the issues? *Social Science and Medicine*, **44**, 723–45.

Mack J and Lansley S (1985): *Poor Britain*. London: Allen & Unwin.

MacKenzie D (1976): Eugenics in Britain. *Social Studies of Science*, **6**, 499–532.

MacKenzie D (1981): *Statistics in Britain, 1865–1930: The Social Construction of Scientific Knowledge*. Edinburgh: Edinburgh University Press.

McKenzie J, Schaefer RL and Farber E (1995): *The Student Edition of Minitab for Windows*. Reading, MA: Addison-Wesley. [Book and software package]

Macourt MPA (1995): Using census data: religion as a key variable in studies of Northern Ireland. *Enviroment and Planning A*, **27**, 593–614.

McRae S (1997): Household and labour market change: implications for the growth of inequality in Britain. *British Journal of Sociology*, **48**, 384–405.

McWilliams P (1984): *Personal Computers and the Disabled*. London: Quantum Press.

Malpass P (1986): *The Housing Crisis*. London: Routledge.

Mamuya WPM (1992): Managing education for children with moderate learning difficulties: a case study of Southampton primary schools in the UK. Unpublished MSc dissertation, Department of Accounting and Management Science, University of Southampton.

Mandelson P (1997): *Labour's Next Steps: Tackling Social Exclusion*. London: Fabian Society.

Marchant P, Rai T and Humphreys A (1996): *Guide to Surveys*, 4th edn. Leeds: Learning Support Services, Leeds Metropolitan University.

Mar Molinero C (1988): Schools in Southampton: a quantitative approach to school location, closure and staffing. *Journal of the Operational Research Society*, **39**, 339–50.

Mar Molinero C and Gard JF (1987): The distribution of special education (moderate) needs in Southampton. *British Journal of Educational Research*, **13**, 147–57.

Marmot MG, Adelstein AM, Bulusu L and OPCS (1984): *Immigrant Mortality in England and Wales 1970–78: Causes of Death by Country of Birth*. London: HMSO.

Marmot MG, Davey Smith G, Stansfield S *et al.* (1991): Health inequalities among British civil servants: the Whitehall II study. *Lancet*, **337**, 1387–93.

Marsden D (1986): *The End of Economic Man?* Brighton: Wheatsheaf.

Marsh C (1988): *Exploring Data*. Cambridge: Polity Press.

Marsh C (1993): Privacy, confidentiality and anonymity in the 1991 census. In A Dale and C Marsh (eds) *The 1991 Census User's Guide*. London: HMSO, 111–28.

Marsh C, Skinner C, Arber S *et al.* (1991): The case for samples of anonymised records from the 1991 census. *Journal of the Royal Statistical Society A*, **154**, 305–40.

Marshall EJ and Reed JL (1992): Psychiatric morbidity in homeless women. *British Journal of Psychiatry*, **160**, 761–8.

Marshall M (1989): Collected and neglected: are Oxford hostels for the homeless filling up with disabled psychiatric patients? *British Medical Journal*, **299**, 706–9.

Martin B and Salter A (with others) (1996): *The Relationship between Publicly Funded Basic Research and Economic Performance*. Brighton: Science Policy Research Unit, University of Sussex.

Martin J and White A (1988a): *The Prevalence of Disability among Adults: Report 1*. London: HMSO.

Martin J and White A (1988b): *The Financial Circumstances of Disabled Adults Living in Private Households: Report 2*. London: HMSO.

Matthews R (1997) Shock-horror statistics. *Teaching Statistics, ASLU Supplement 1997*, 9–11.

Mayhew PP and Maung NA (1992): *Surveying Crime: Findings from the 1992 British Crime Survey*. Home Office Research and Statistics Department Research Findings no. 2. London: HMSO.

Mayhew P, Maung NA and Mirrlees-Black C (1993): *The 1992 British Crime Survey*. A Home Office Research Planning Unit Report. London: HMSO.

Meadows P (1996): *Work Out – or Work In?: Contributions to the Future of Work Debate*. York: Joseph Rowntree Foundation.

Meager N and Evans C (1997): United Kingdom European Employment Observatory. *Trends*, **29**, 58–63.

Meehl P (1967): Theory-testing in psychology and physics: a methodological paradox. *Philosophy of Science*, **34**, 103–15.

Meehl P (1978): Theoretical risks and tabular asterisks: Sir Karl, Sir Ronald, and the slow progress of soft psychology. *Journal of Consulting and Clinical Psychology*, **46**, 806–34.

Meehl P (1997): Credentialed persons, credentialed knowledge. *Clinical Psychology: Science and Practice*, **4**, 91–8.

Meltzer H, Gill B, Pettigrew M and Hinds K (1995): *The Prevalence of Psychiatric Morbidity among Adults Living in Private Households*. OPCS Survey of Psychiatric Morbidity, Report 1. London: HMSO.

Meltzer H, Smyth M and Robus N (1989): *Disabled Children: Services, Transport and Education: Report 6*. London: HMSO.

Mennell S, Murcott A and Van Otterloo A (1992): *The Sociology of Food: Eating, Diet and Culture*. London: Sage.

Metcalf H and Leighton P (1989): *The Under-utilisation of Women in the Labour Market*. Brighton: Institute of Manpower Studies.

Middleton D and Edwards D (eds) (1990): *Collective Remembering*. London: Sage.

Milburn N and Watts RJ (1986): *International Journal of Mental Health*, **14**, 42–60.

Miller RJ (1982): *The Demolition of Skid Row*. Lexington: D.C. Heath.

Miller WL (1977): *Electoral Dynamics*. London: Macmillan.

Ministry of Education (1959): *Report of the Central Advisory Council for Education, England* (The Crowther Report) [15 to 18]. London: HMSO.

Ministry of Health (1950): *Report of the Ministry of Health for the Year Ending 31 March 1949*. Cmnd 7910. London: HMSO.

Minkler M (1986): 'Generational equity' and the new victim blaming: an emerging public policy issue. *International Journal of Health Services*, **16**, 539–51.

Minkler M (1991): Generational equity and the new victim blaming. In M Minkler and C Estes (eds) *Critical Perspectives on Aging: The Political and Moral Economy of Growing Old*. New York: Baywood, ch. 5.

Mirrlees-Black C, Mayhew P and Percry A (1996): *The 1996 British Crime Survey*. Issue 19/96, Home Office Statistical Bulletin. London: Research and Statistics Directorate.

Mischler EG (1986): *Research Interviewing: Context and Narrative*. Cambridge, MA: Harvard University Press.

Mitchell BR and Deane P (1962): *Abstract of British Historical Statistics*. Cambridge: Cambridge University Press.

Modood T, Beishon S and Virdee S (1994): *Changing Ethnic Identities*. London: Policy Studies Institute.

Modood T, Berthoud R, Lakey J *et al.* (1997): *Ethnic Minorities in Britain: Diversity and Disadvantage*. London: Policy Studies Institute.

Montgomery S, Bartley M, Cook D and Wadsworth M (1996): Health and social precursors of unemployment in young men in Great Britain. *Journal of Epidemiology and Community Health*, **50**, 415–22.

Moore DS (1997): *Statistics, Concepts and Controversies*, 4th edn. New York: Freeman.

Moore J (1987): The future of the welfare state. Speech to a Conservative Political Centre Conference at the Chichester Rooms, London.

Moore PG (1990): The skills challenge of the nineties. *Journal of the Royal Statistical Society A*, **153**, 265–85.

Morgan K (1981): *Wales, 1880–1980: Rebirth of a Nation*. Oxford: Oxford University Press and Wales University Press.

Morrell D and Wale C (1976): Symptoms perceived and recorded by patients. *Journal of the Royal College of General Practitioners*, **26**, 389–403.

Morris JN, Blane D and White I (1996): Levels of mortality, education and social conditions in the 107 local education authority areas of England. *Journal of Epidemiology and Community Health*, **50**, 15–17.

Morris S, Szuscikiewicz J and Preston M (1996): *Statistics for the Terrified, version 3.0*. Oxford: Radcliffe Medical Press. [Software package]

Morris T (1957): *The Criminal Area*. London: Routledge & Kegan Paul.

Mossman D and Perlin ML (1992): Psychiatry and the homeless mentally ill: a reply to Dr. Lamb. *American Journal of Psychiatry*, **149**, 951–7.

Motorola (1996): *Prepared for the Future: The British and Technology – The Motorola Report*. London: Motorola.

Murgatroyd L and Neuberger H (1997): A household satellite account for the UK. *Economic Trends*. **527**, 63–71.

National Audit Office (1990): *The Elderly: Information Requirements for Supporting the Implications of Personal Pensions for the National Insurance Fund*. HC 55. London: HMSO.

National Center for Health Statistics (1994): *Health United States 1993*. Hyattsville: Public Health Service.

National Consumer Council (1995): *Charging Consumers for Social Services: Local Authority Policy and Practice*. A report on charging for adult social services in England. London: NCC.

National Food Survey 1995 (1996): *Annual Report on Food Expenditure, Consumption and Nutrient Intakes*. Ministry of Agriculture, Fisheries and Food. London: The Stationery Office.

Naylor K (1994): Part-time working in Great Britain: an historical analysis. *Employment Gazette*, **102**, 473–84.

Nazroo JY (1997a): *The Health of Britain's Ethnic Minorities: Findings from a National Survey*. London: Policy Studies Institute.

Nazroo JY (1997b): *Ethnicity and Mental Health: Findings from a National Survey*. London: Policy Studies Institute.

Neale J (1997): Homelessness and theory reconsidered. *Housing Studies*, **12**, 47–61.

Nelson M and Naismith DJ (1979): The nutritional status of poor children in London. *Journal of Human Nutrition*, **33**, 33–45.

Nesbitt S (1995): *British Pensions Policy Making in the 1990s: The Rise and Fall of a Policy Community*. Aldershot: Avebury.

Newton PE (1997): Measuring comparability of standards between subjects: why our statistical techniques do not make the grade. *British Educational Research Journal*, **23**, 433–50.

NHS Executive (1996): *Collection of Data from General Practice: Overview*. Leeds: NHS Executive.

NHS Management Executive (1992): *The NHS Reforms: The First Six Months*. London: Department of Health.

Nicaise I, Bollons J, Dawes L *et al.* (1995): *Pitfalls and Dilemmas in Labour Market Programmes for Disadvantaged Groups and How to Avoid Them*. Aldershot: Avebury.

Nichols T (1986): Industrial injuries in British manufacturing in the 1980s. *Sociological Review*, **34**, 290–306.

Nichols T (1989): The business cycle and industrial injuries in British manufacturing over a quarter of a century. *Sociological Review*, **37**, 538–50.

Nichols T (1994): Problems in monitoring the safety performance of British manufacturing at the end of the twentieth century. *Sociological Review*, **42**, 104–10.

Nichols T (1997): *The Sociology of Industrial Injury*. London: Mansell.

Nightingale DS and Haveman RH (eds) (1994): *The Work Alternative: Welfare Reform and the Realities of the Labor Market*. Washington, DC: Urban Institute Press.

Nord M and Luloff AE (1995): Homeless chidren and their families in New Hampshire. *Social Service Review*, **69**, 461–78.

Norton WW and Mackintosh NJ (eds) (1995): *Cyril Burt: Fraud or Framed?* Oxford: Oxford University Press.

Norusis MJ (1994): *SPSS Advanced Statistics 6.1*. Chicago: SPSS Inc.

Oakley Λ (1989): Smoking in pregnancy: smokescreen or risk factor? Towards a materialist analysis. *Sociology of Health and Illness*, **11**, 311–35.

O'Callaghan B and Dominian L (1996): *Study of Homeless Applicants*. Department of the Environment. London: HMSO.

Office for National Statistics (1996): Topics to be included in the testing programme. Questions Working Group Paper QWG 16, 27 March 1998. Titchfield: ONS.

Office for National Statistics (1997a): *How Exactly Is Unemployment Measured?* London: ONS.

Office for National Statistics (1997b): *Health in England, 1996: What People Know, What People Think, What People Do*. A survey of adults aged between 16 and 74 in England, carried out by the Social Survey Division of the ONS on behalf of the Health Education Authority. London: The Stationery Office.

Office for National Statistics (1997c): *Living in Britain: Results from the 1995 General Household Survey*. London: The Stationery Office.

Office for National Statistics (1998a): *Guide to Gender Statistics*. London: ONS.

Office for National Statistics (1998b): *Annual Abstract of Statistics 1998*. London: The Stationery Office.

Office for National Statistics (1998c): *How Exactly Is Employment Measured?* London: ONS.

Office for National Statistics (1998d): *How Exactly Is Unemployment Measured?* London: ONS.

Office for National Statistics (1998e): *Official Statistics: A Matter of Trust*. London: The Stationery Office.

Office for National Statistics (annual): *Guide to Official Statistics*. London: HMSO.

Office of Population Censuses and Surveys (1986a): Estimating the size of the ethnic minority populations in the 1980s. *Population Trends*, **44**, 23–7.

Office of Population Censuses and Surveys (1986b): *Occupational Mortality, Decennial Supplement 1979–1980, 1982–1983*. London: HMSO.

Office of Population Censuses and Services (1989): *General Household Survey 1987*. Series GHS no. 17. London: HMSO.

Office of Population Censuses and Surveys (1990): *Registration: Proposals for Change*. Cm 939. London: HMSO.

Office of Population Censuses and Surveys (1991): *Small Area and Local Based Statistics*. London: HMSO.

Office of Population Censuses and Surveys (1993): *General Household Survey 1991*. London: HMSO.

OFSTED (1993): *Corporate Plan 1993–94 to 1995–1996*. London: Office for Standards in Education.

OFSTED (1994): *Improving Schools*. London: HMSO.

OFSTED (1995): *Class Size and the Quality of Education*. London: Office for Standards in Education.

OFSTED (1996a): *The Teaching of Reading in 45 Inner London Primary Schools*. London: Office for Standards in Education.

OFSTED (1996b): *Annual Report of Her Majesty's Chief Inspector of Schools*. London: HMSO.

OFSTED (1996c): Sharper focus on primary teacher training planned by OFSTED. OFSTED News Release, 11 July 1997.

OFSTED (1996d): *Standards in Public Examinations 1975 to 1995*. London: Office for Standards in Education and Schools Curriculum Assessment Authority.

OFSTED (1997): Hackney's education services in turmoil, says Ofsted report. Ofsted News Release, 18 September 1997.

OFSTIN (1996): *A Better System of Inspection?* Northumberland: Office for Standards in Inspection.

Olney J (1980): *Autobiography: Essays, Theoretical and Critical*. Princeton, NJ: Princeton University Press.

Open University (1983): *MDST242: Statistics in Society*. Milton Keynes: Open University Press.

Organisation for Economic Co-operation and Development (1993): *Frascati Manual: The Measurement of Scientific and Technological Activities*. Paris: OECD.

Orwell G (1986 [1933]) *Down and Out in Paris and London*. London: Penguin.

Pahl J (1989): *Money and Marriage*. Basingstoke: Macmillan.

Parker I (1994): Reflexive research and the grounding of analysis: social psychology and the psy-complex. *Journal of Community and Applied Social Psychology*, **4**, 239–52.

Parker I (1997): Discursive psychology. In D Fox and I Prilletensky (eds) *Critical Psychology: An Introduction*. London: Sage, 284–98.

Paterson G (1988): The caring for the old that leaves me cold. *Sunday Telegraph*. 10 January.

Paul D (1995): *Controlling Human Heredity*. Atlantic Highland, NJ: Humanities Press.

Paulos JA (1988): *Innumeracy*. London: Viking Penguin.

Pearce D and Thomas F (1990): The 1989 census test. *Population Trends*, **61**, 24–30.

Pearson K (1881): Anarchy. *The Cambridge Review*, **2**, 268–70.

Pearson K (1888): *The Ethic of Freethought*. London: Unwin.

Pearson K (1892): *The Grammar of Science*. London: Scott.

Pearson K (1896): Mathematical contributions to the theory of evolution, III: Regression, heredity and panmixia. *Philosophical Transactions of the Royal Society of London A*, **187**, 253–318.

Pearson K (1900a): *The Grammar of Science*, 2nd edn. London: Black.

Pearson K (1900b): Mathematical contributions to the theory of evolution, VII: On the correlation of characters not quantitatively measurable. *Philosophical Transactions of the Royal Society of London A*, **195**, 1–47.

Pearson K (1901): *National Life from the Standpoint of Science*. London: Black.

Pearson K (1902): Prefatory essay. The function of science in the modern state. *Encyclopaedia Britannia*, 10th edn, vol. 32, pp. vii–xxxvii. Volume 8 of new volumes.

Pearson K (1903): On the inheritance of the mental and moral characters in man, and its comparison with the inheritance of the physical characters. The Huxley Lecture for 1903. *Journal of the Anthropological Institute of Great Britain and Ireland*, **33**, 179–237.

Pease P (1997a): LFS estimates of unemployment-related benefits: the results of an ONS record linkage study. *Labour Market Trends*, **105**, 455–60.

Pease P (1997b): Methodological paper on the record linkage study between the Labour Force Survey and claimant administrative records. Available from Labour Market Division, Office for National Statistics, London.

Pease P (1997c): Comparisons of sources of employment data. *Labour Market Trends*, **105**, 511–16.

Pease P (1998): The Labour Force Survey in the dock. *Radical Statistics*, **67**, 62–4.

Peck J (1993): The trouble with TECs . . . a critique of the Training and Enterprise Councils initiative. *Policy and Politics*, **21**, 289–305.

Peck J (1996): *Work-Place: The Social Regulation of Labor Markets*. New York: Guilford.

Peck J and Tickell A (1996): The return of Manchester men: men's words and men's deeds in the remaking of the local state. *Transactions of the Institute of British Geographers*, **21**, 595–616.

Pendergrast M (1996): *Victims of Memory: Incest, Accusations and Shattered Lives*, UK edition. London: HarperCollins.

Perrons D (1998): Maps of meaning: gender inequality in the regions of Western Europe. *European Urban and Regional Studies*, **5**, 13–25.

Perry K (1996): Measuring employment: comparison of official sources. *Labour Market Trends*, **104**, 19–27.

Petersen W, Pratten C and Tatch J (1998): *An Economic Model for the Demand and Need for Social Housing: Technical Report on a Feasibility Study*. London: Department of the Environment, Transport and the Regions.

Phillimore P, Beattie A and Townsend P (1994a): *Health and Inequality: The Northern Region 1981–1991*. Newcastle: University of Newcastle upon Tyne.

Phillimore P, Beattie A and Townsend P (1994b): Widening inequality of health in northern England, 1981–91. *British Medical Journal*, **308**, 1125–8.

Phillips AW (1958): The relation between unemployment and the rate of change of money wages in the U.K., 1861–1957. *Economica*, NS **25**, 283–99.

Phoenix A and Owen C (1996): From miscegenation to hybridity: mixed relationships and mixed-parentage in profile. In B Bernstein and J Brannen (eds) *Children, Research and Policy*. London: Taylor & Francis, 111–35.

Piachaud D, Bradshaw J and Weale J (1981): The income effect of a disabled child. *Journal of Epidemiology and Community Health*, **35**, 123–7.

Plewis I (1985): *Analysing Change*. Chichester: Wiley.

Plewis I (1997): *Statistics in Education*. London: Arnold.

Plewis I and Goldstein H (1997): Excellence in schools: a failure of standards. *British Journal of Curriculum and Assessment*, **8**, 17–20.

Plewis I and Veltman M (1996): Opportunity to learn maths at Key Stage One: changes in curriculum coverage 1984–1993. *Research Papers in Education*, **11**, 201–18.

Plummer K (1983): *Documents of Life*. London: Allen & Unwin.

Pocock SJ, Shaper AG, Cook DG, Phillips AN and Walker M (1987): Social class differences in ischaemic heart disease in British men. *Lancet*, **ii**, 197–201.

Polanyi L (1985): *Telling the American Story: A Structural and Cultural Analysis of Conversational Story Telling*. Norwood, NJ: Ablex.

Polk K (1957): Juvenile delinquency and social areas. *Social Problems*, **5**, 214–17.

Pollock AM (1995): Where should health services go?: local authorities versus the NHS. *British Medical Journal*, **310**, 1580–4.

Pollock AM, Gaffney D (1998): Capital charges: a tax on the NHS. *BMJ*, **317**, 157–8.

Pollock AM, Dunnigan M, Gaffney D, Macfarlane A and Majeed AM (1997): What happens when the private sector plans hospital services for the NHS: three case studies under the private finance initiative. *British Medical Journal*, **314**, 1266–71.

Poortvliet W and Lane T (1994): A global trend: privatisation and reform of social security plans. *Geneva Papers on Risk and Insurance Issues and Practice*, no. 72, July.

Popper K (1959): *The Logic of Scientific Discovery*. London: Hutchinson.

Popper K (1963): *Conjectures and Refutations: The Growth of Scientific Knowledge*. London: Routledge & Kegan Paul.

Porter T (1986): *The Rise of Statistical Thinking, 1820–1900*. Princeton, NJ: Princeton University Press.

Power C, Manor O, Fox AJ and Fogelman K (1990): Health in childhood and social inequalities in health in young adults. *Journal of the Royal Statistical Society*, **153**, 17–28.

Power C, Manor O and Fox J (1991): *Health and Class: The Early Years*. London: Chapman & Hall.

Power C and Matthews S (1997): Origins of health inequalities in a national population sample. *Lancet*, **350**, 1584–9.

Prescott-Clarke P and Primatesta P (eds) (1997): *Health Survey for England, 1995*. A survey carried out on behalf of the Department of Health. HS no. 5. London: The Stationery Office.

Preston S (1984): Children and the elderly in the US. *Scientific American*, **251**, 44–9.

Priest RG (1976): The homeless person and the psychiatric services: an Edinburgh survey. *British Journal of Psychiatry*, **128**, 128–36.

Pumfrey P and Elliott C (1991): A house of cards? *Times Educational Supplement*, 3 May, p. 12.

Purdie E (ed.) (1996): *Guide to Official Statistics*. London: HMSO.

Quadagno J (1990): Generational equity and the politics of the welfare state. *International Journal of Health Services*, **204**, 631–49.

Radford J and Stanko E (1996): Violence against women and children: the contradictions of crime control under patriarchy. In M Hester, L Kelly and J Radford (eds) *Women, Violence and Male Power*. Buckingham: Open University Press, 65–80.

Radical Statistics Education Group (1987): *Figuring Out Educational Spending*. London: Radical Statistics.

Radical Statistics Health Group (1987): *Facing the Figures: What Really Is Happening to the National Health Service*. London: Radical Statistics.

Radical Statistics Health Group (1991a): Missing: a strategy for the health of the nation. *British Medical Journal*, **303**, 299–302.

Radical Statistics Health Group (1991b): Let them eat soap. *Health Service Journal*, **101**, 5278, 25–7.

Radical Statistics Health Group (1992): The NHS reforms: the first six months – proof of progress or a statistical smokescreen? *British Medical Journal*, **304**, 705–79.

Radical Statistics Health Group (1995): NHS 'indicators of success': what do they tell us? *British Medical Journal*, **310**, 1045–50.

Radical Statistics Health Group (1999): *Official Health Statistics: An Unofficial Guide*. Edited by Susan Kerrison and Alison Macfarlane. London: Arnold.

Rahkonen O, Lahelma E and Huuhka M (1997): Past or present?: childhood living conditions and current socioeconomic status as determinants of adult health. *Social Science Medicine*, **44**, 327–36.

Ramsay M (1983): City-centre crime. *Home Office Research Bulletin*, **16**, 5–8.

Randall S, McBrearty T and Mordecai J (1986): *Single Homelessness in London 1986*. A report by the Single Homelessness in London Working Party, a joint GLC and London boroughs working party. London: Greater London Council.

Ranson S (1993): Markets or democracy for education. *British Journal of Educational Studies*, **41**, 333–52.

Realeat Survey Office (1995): *The Realeat Survey 1984–1995: Changing Attitudes to Meat Consumption*. London: Realeat.

Redfern P (1987): *A Study of the Future of the Census of Population: Alternative Approaches*. Luxembourg: Office for Official Publications of the European Communities.

Reed A, Ramsden S, Marshall S *et al.* (1992): Psychiatric morbidity and substance abuse among residents of a cold weather shelter. *British Medical Journal*, **304**, 1028–9.

Rees P and Dale A (1997): *The Case for Census Questions for Use by the Academic Community*. London: Economic and Social Research Council/Joint Information Systems Committee.

Reicher SD (1982): The determination of collective behaviour. In H Tajfel (ed.) *Social Identity and Intergroup Relations*. Cambridge: Cambridge University Press.

Reichmann WJ (1964): *Use and Abuse of Statistics*. London: Penguin.

Reid F (1975): *The Incomes of the Blind: A Review of the Occaptional and Financial Problems of Blind People of All Ages*. London: The Disability Alliance.

Rex J (1993): Religion and ethnicity in the metropolis. In R Barot (ed.) *Religion and Ethnicity: Minorities and Social Change in the Metropolis*. Kampen: Kok Pharos Publishing House.

Rhondda, Cynon, Taff Economic Development Committee (1997): *Report of Chief Executive*. Clydach Vale, Rhondda: Rhondda, Cynon, Taff Unitary Authority.

Ringen S (1987): *The Possibility of Politics*. Oxford: Oxford University Press.

Ringen S (1988): Direct and indirect measures of poverty. *Journal of Social Policy*, **17**, 351–66.

Ritchie C, Taket A and Bryant J (eds) (1994): *Community Works*. Sheffield: Pavic Publications.

Ritson C and Hutchins R (1991): *The Consumption Revolution*. In JM Slater (ed.) *Fifty Years of the National Food Survey 1940–1990*. London: HMSO.

Roberts HV (1990): Applications in business and economic statistics. *Statistical Science*, **5**, 372–402.

Robins K (1996): *Into the Image: Culture and Politics in the Field of Vision*. London: Routledge.

Robinson WS (1950): Ecological correlations and the behaviour of individuals. *American Sociological Review*, **15**, 351–7.

Römkens R (1997): Prevalence of wife abuse in the Netherlands: combining quantitative and qualitative methods in survey research. *Journal of Interpersonal Violence*, **12**, 99–125.

Room G (1995): Poverty in Europe: competing paradigms of analysis. *Policy and Politics*, **23**, 103–13.

Rose D and O'Reilly K (eds) (1997): *Constructing Classes*. Swindon: Economic and Social Research Council.

Rose ME (1977): Rochdale man and the Stalybridge riot: the relief and control of the unemployed during the Lancashire cotton famine. In AP Donajgrodzki (ed.) *Social Control in Nineteenth Century Britain*. London: Croom Helm, 185–206.

Rosenthal R (1966): *Experimenter Effects in Behavioral Research*. New York: Appleton-Century-Crofts.

Rosenthal R and Rubin DB (1994): The counternull value of an effect size: a new statistic. *Psychological Science*, **5**, 329–34.

Roseveare D, Leibfritz W, Fore D and Wurzel E (1996): *Ageing Populations, Pension Systems and Government Budgets: Simulations for 20 OECD Countries*. Paris: Organization for Economic Co-operation and Development.

Ross SM (1996): *Introductory Statistics*. New York: McGraw-Hill.

Rossiter DJ, Johnston RJ and Pattie CJ (1997a): Estimating the partisan impact of redistricting in Britain. *British Journal of Political Science*, **27**, 319–31.

Rossiter DJ, Johnston RJ and Pattie CJ (1997b): Redistricting and electoral bias in Great Britain. *British Journal of Political Science*, **27**, 466–72.

Rowlands O, Singleton N, Maher J and Higins V (1997): *Living in Britain.* London: The Stationery Office.

Royal College of Psychiatrists Working Group (1997): Reported recovered memories of child sexual abuse: recommendations for good practice and implications for training, continuing professional development and research. *The College Psychiatric Bulletin*, **21**, 663–5.

Royal Statistical Society (1995a): The measurement of unemployment in the UK. *Journal of the Royal Statistical Society A*, **158**, 363–417.

Royal Statistical Society (1995b): *Report of the Working Party on the Measurement of Unemployment in the UK.* London: Royal Statistical Society.

Royal Statistical Society and Institute of Statisticians (1979): Submission to the Committee of Enquiry into the Teaching of Mathematics in Schools under the Chairmanship of Dr W.H. Cockcroft. *Teaching Statistics*, supplement **1**(3) (September).

Rubery J (1992): *The Economics of Equal Value.* Research Discussion Series no. 3. Manchester: Equal Opportunities Commission.

Rubery J, Smith M, Fagan C and Grimshaw D (1996): *Women and the European Employment Rate: The Causes and Consequences of Variation in Female Activity and Employment Patterns in the European Union.* European Network on the Situation of Women in the Labour Market. Report for the Equal Opportunities Unit, DG-V. Brussels: European Commisssion V/995/96-EN.

Rudat K (1994): *Black and Minority Ethnic Groups in England: Health and Lifestyles.* London: Health Education Authority.

Rumeau-Rouquette C (1978): Case-control studies. In WW Holland and L Karhausen (eds) *Health Care and Epidemiology.* London: Henry Kimpton.

Ryan BF and Joiner BL (1994): *Minitab Handbook*, 3rd edn. Belmont, CA: Duxbury Press.

Salt J (1996): Immigration and ethnic group. In D Coleman and J Salt (eds) *Ethnicity in the 1991 Census*, vol. 1: *Demographic Characteristics of the Ethnic Minority Populations.* London: HMSO.

Salt J and Singleton A (1993): *Comparison and Evaluation of the Labour Force Survey and Regulation 311/76 Data as Sources on the Foreign Employed Population in the EC.* Report to the Migration Statistics Working Party, Eurostat, November.

Salt J and Singleton A (1995): *Analysis and Forecasting of International Migration by Major Groups.* Report to Eurostat, February.

Samet JM, Humble CG and Skipper BE (1984): Alternatives in the collection and analysis of food frequency interview data. *American Journal of Epidemiology*, **120**, 572–81.

Scambler G (ed.) (1991): *Sociology as Applied to Medicine*, 3rd edn. London: Baillière Tindall.

Scheetz T (1991): The macroeconomic impact of defence expenditures: some econometric evidence for Argentina, Chile Paraguay and Peru. *Defence Economics.* **3**, 65–81.

Schols CM (1886): Théorie des erreurs dans le plan et dans l'espace. *Annales de l'Ecole Polytechnique de Delft*, **2**, 123–78 (first published in Dutch in 1875).

Schools Curriculum and Assessment Authority (1997): *The National Value Added Project*. London: SCAA.

Schumpeter JA (1994): *History of Economic Analysis*. London: Routledge.

Scott-Heide W (1984): Now for the feminist menopause that refreshes. In G Lesnoff-Caravavaglia (ed.) *The World of the Older Woman: Conflicts and Resolutions*. New York: Human Sciences Press, 162–74.

Screen Digest (1995): The Internet. Special feature, April.

Secretary of State for the Environment (1996): *Household Growth: Where Shall We Live*? Cm 3471. London: HMSO.

Sen A (1989): Gender and co-operative conflicts. In I Tinker (ed.) *Persistent Inequalities: Women and World Development*. Oxford: Oxford University Press.

Senior PA and Bhopal R (1994): Ethnicity as a variable in epidemiological research. *British Medical Journal*, **309**, 327–30.

Serlin RC and Lapsley DK (1993): Rational appraisal of psychological research and the good-enough principle. In G Keren and C Lewis (eds) *A Handbook for Data Analysis in the Behavioral Sciences: Methodological Issues*. Hove: Lawrence Erlbaum.

SERPLAN (1996): *Technical Analysis of the 1993/92-based Population and Household Projections*. RPC 2980.

Shaileh AE and Tonak EA (1994): *Measuring the Wealth of Nations: The Political Economy of National Accounts*. Cambridge: Cambridge University Press.

Shanks NJ (1981): Consistency of data collected from inmates of a common lodging house. *Journal of Epidemiology and Community Health*, **35**, 133–5.

Sheaffer R and Becker J (1995): Recovered memories cross the oceans. *Skeptical Inquirer*, **19**(4) [from http: //www.csicop.org/si/9507/memory.html].

Sheldon TA and Parker H (1992): Race and ethnicity in health research. *Journal of Public Health Medicine*, **14**, 104–10.

Shorrocks D (1993): *Implementing National Curriculum Assessment in the Primary School*. London: Hodder & Stoughton.

Shotter J (1993): *Cultural Politics of Everyday Life*. Buckingham: Open University Press.

Sibley D (1981): *Outsiders in Urban Societies*. Oxford: Blackwell.

Sibley D (1984): A robust analysis of a minority census: the distribution of travelling people in England. *Environment and Planning A*, **16**, 1279–88.

Sibley D (1990): Urban change and the exclusion of minority groups in British cities. *Geoforum*, **21**, 483–8.

Silver H (1994): *Good Schools, Effective Schools: Judgements and Their Histories*. London: Cassell.

Silverstone R and Hirsch E (eds) (1992): *Consuming Technologies: Media and Information in Domestic Spaces*. London: Routledge.

Simpson S (1991): Community surveys. *Radical Statistics Newsletter*, **48**, 10–13.

Simpson S and Dorling DFL (1994): Those missing millions: implications for social statistics of the undercount in the 1991 census. *Journal of Social Policy*, **23**(4), 543–67.

Sinfield A (1978): Analyses in the social division of welfare. *Journal of Social Policy*, **72**, 129–56.

Sinfield A (1993): Reverse targeting and upside down benefits: how perverse policies perpetuate poverty. In A Sinfield (ed.) *Poverty, Inequality and Justice*. Edinburgh: Edinburgh University Press, 39–48.

Singleton A and Barbesino P (1998): The production and reproduction of knowledge on international migration in Europe: the social embeddedness of social knowledge. In F Amthias and G Lazarides (eds) *The Soft Belly of Fortress Europe*. Aldershot: Avebury.

Sivard RL (annual): *World Military and Social Expenditures*. Washington, DC: World Priorities.

Slater JM (ed.) (1991): *Fifty Years of the National Food Survey, 1940–1990*. London: HMSO.

Smaje C (1995): Ethnic residential concentration and health: evidence for a positive effect? *Policy and Politics*, **23**, 251–69.

Smith AM and Baghurst KI (1992): Public health implications of dietary differences between social status and occupational category groups. *Journal of Epidemiology and Community Health*, **46**, 409–16.

Smith D (1977): *Racial Disadvantage in Britain*. Harmondsworth: Penguin.

Smith D and Smith R (1983): *The Economics of Militarism*. London: Pluto.

Smith T and Noble M (eds) (1995): *Education Divides: Poverty and Schooling in the 1990s*. London: Child Poverty Action Group.

Smyth M and Robus N (1989): *The Financial Circumstances of Families with Disabled Children Living in Private Households: Report 5*. London: HMSO.

Snow DA, Baker SG, Anderon L and Martin M (1986): The myth of pervasive mental illness among the homeless. *Social Problems*, **33**, 407–23.

Social Security Advisory Committee (1992): *Options for Equality in the State Pension Age: A Case for Equalising at 65*. London: HMSO.

Soni Raleigh V and Balarajan R (1992): Suicide and self-burning among Indians and West Indians in England and Wales. *British Journal of Psychiatry*, **161**, 365–8.

Sorlie PD, Backlund E and Keller JB (1995): US mortality by economic, demographic and social characteristics: the National Longitudinal Mortality Study. *American Journal of Public Health*, **85**, 949–56.

Southall HR (1988): The origins of the depressed areas: unemployment, growth, and regional economic structure in Britain before 1914. *Economic History Review*, 2nd series, **41**, 236–58.

Southall HR (1991): Poor Law statistics and the geography of economic distress. In J Foreman-Peck (ed.) *New Perspectives on the Late Victorian Economy*. Cambridge: Cambridge University Press, 180–217.

Southall HR and Gilbert DM (1996): A good time to wed: marriage and economic distress in England and Wales, 1839–1914. *Economic History Review*, 2nd series, **49**, 35–57.

Sparks R, Genn H and Dodd D (1977): *Surveying Victims*. Chichester: Wiley.

Stanley L (1992): *The auto/biographical I*. Manchester: Manchester University Press.

Statewatch (1997): *Key Texts on the European State*. London: Statewatch.

Stats Watch (monthly) available from J. Martyn, Department of Sociology and Social Policy, Roehampton Institute London, Southlands College, Wimbledon Parkside, London SW19 5NN.

Stedman-Jones G (1971): *Outcast London: A Study in the Relationship between Classes in Victorian Society.* Oxford: Clarendon Press.

Steedman C (1984): *Policing the Victorian Community.* London: Routledge & Kegan Paul.

Steering Group on Health Services Information (1982): *A Report on the Collection and Use of Information about Hospital Clinical Activity in the National Health Service.* London: HMSO.

Stern J (1983): Social mobility and the interpretation of social class mortality differentials. *Journal of Social Policy,* **12**, 27–49.

Stevens G (1992): Workplace injury: a view from HSE's trailer to the 1990 Labour Force Survey. *Employment Gazette,* December, 621–38.

Stewart A, Prandy K and Blackburn RM (1973): Measuring the class structure. *Nature,* **245**, 415–17.

Stewart DW and Kamins MA (1993): *Secondary Research: Information Sources and Methods.* Newbury Park: Sage.

Stigler SM (1986): *The History of Statistics: The Measurement of Uncertainty before 1900.* Cambridge, MA: Harvard University Press.

Stockholm International Peace Research Institute (various years): *SIPRI Yearbook: World Armaments and Disarmaments.* New York: Oxford University Press.

Stokes D (1995): Development of a model of local competitive forces acting on primary schools in the UK. Paper presented to the Twenty-First Annual Conference of the British Educational Research Association, Bath.

Stray SJ and Upton GJG (1989): Triangles and triads. *Journal of the Operational Research Society,* **40**, 83–92.

Susser E, Conover S and Struening EL (1989): Problems of epidemiologic method in assessing the type and extent of mental illness among homeless adults. *Hospital and Community Psychiatry,* **40**, 261–5.

Sweeting H and West P (1995): Family life and health in adolescence: a role for culture in the health inequalities debate? *Social Science and Medicine,* **40**, 163–75.

Task Group on Assessment and Testing (1987): *National Curriculum.* London: Department of Education and Science.

Temple B (1992): Household strategies and types: the construction of social phenomena. Unpublished PhD dissertation, Manchester University.

Tenner E (1996): *Why Things Bite Back: Predicting the Problems of Progress.* London: Fourth Estate.

Thatcher M (1995): *The Downing Street Years.* London: HarperCollins.

Third King's Fund Forum (1987): The need for asylum in society for the mentally ill or infirm. In *Consensus Statement.* London: King Edward's Hospital Fund.

Thomas G and Davis P (1997): Special needs: objective reality or personal construction? Judging reading difficulty after the code. *Journal of Educational Research,* 39, 263–70.

Thomas R (1997): The Labour Force Survey in the dock. *Radical Statistics,* **64/65**, 4–16.

Thomas R (1998): How enlarged Travel to Work Areas conceal inner city employment. *Radical Statistics,* **67**, 35–45. *See also* reply by Mike Coombes, pp. 46–9, and letter, pp. 62–4.

Thomas WI and Znaniecki F (1918) *The Polish Peasant in Europe and America*, 2 vols. Chicago: University of Chicago Press.

Thunhurst C (1985): The analysis of small area statistics and planning for health. *The Statistician*, **34**, 93–106.

Thunhurst C and Macfarlane AJ (1992): Monitoring the health of urban populations: what statistics do we need? *Journal of the Royal Statistical Society A*, **155**, 317–52.

Tickle L (1996): Mortality trends in the UK, 1982–1992. *Population Trends*, **86**, 21–8.

Times Educational Supplement (1997): A-level board gave 'unjustified' marks. 10 January.

Timms PW and Fry AH (1989): Homelessness and mental illness. *Health Trends*, **21**, 70–1.

Todorov T (1993): *On Human Diversity: Nationalism, Racism and Exoticism in French Thought*. Cambridge, MA: Harvard University Press.

Toro PA and McDonnell DM (1992): Beliefs, attitudes and knowledge about homelessness: a survey of the general public. *American Journal of Community Psychology*, **20**, 53–80.

Townsend P (1979): *Poverty in the United Kingdom*. Harmondsworth: Penguin.

Townsend P (1993): *The International Analysis of Poverty*. Brighton: Harvester Wheatsheaf.

Townsend P and Davidson N (eds) (1982): *Inequalities in Health: The Black Report*. Harmondsworth: Penguin.

Townsend P and Davidson N (eds) (1988): *Inequalities in Health: The Black Report*, 2nd edn. Harmondsworth: Penguin.

Townsend P, Phillimore P and Beattie A (1986): *Inequalities in the Health of the Northern Region*. Northern Regional Health Authority and the University of Bristol.

Townsend P, Phillimore P and Beattie A (1988): *Health and Deprivation: Inequality and the North*. London: Croom Helm.

Townsend P and Walker A (1995): *New Directions for Pensions: How to Revitalise National Insurance*. London: Fabian Society.

Tuan Y-F (1976): Humanistic geography. *Annals of the Association of American Geographers*, **66**, 266–76.

Tufte ER (1983): *The Visual Display of Quantitative Information*. Cheshire, CT: Graphics Press.

Tufte ER (1990): *Envisioning Information*. Cheshire, CT: Graphics Press.

Tufte ER (1997): *Visual Explanations: Images and Quantities, Evidence and Narrative*. Cheshire, CT: Graphics Press.

Turner J and Fichter R (1972) *Freedom to Build*. New York: Macmillan.

Turok I (ed.) (1997): *Travel to Work Areas and the Measurement of Unemployment*. Conference Proceedings, Centre for Housing Research and Urban Studies, University of Glasgow.

UK Government (1989): *Working for Patients*. London: HMSO.

UK Government (1991): *The Citizen's Charter*. London: HMSO.

UK Government (1996): Employment and unemployment statistics: response to the Employment Select Committee's report, and consultation on monthly publication of results from the Labour Force Survey. Reprinted in *Labour Market Trends*, **104**, 175–88.

United Nations (1993): *A System of National Accounts*. New York: United Nations Department of Publications.

United Nations Development Programme (1994): *Human Development Report 1994*. Oxford: Oxford University Press.

Upton GJG (1976): The diagrammatic representation of three-party contests. *Political Studies*, **24**, 448–54.

Upton GJG (1991): Displaying electoral results. *Political Geography Quarterly*, **10**, 200–20.

Upton GJG (1994): Picturing the 1992 general election. *Journal of the Royal Statistical Society A*, **157**, 231–52.

Vågerö D and Illsley R (1995): Explaining health inequalities: beyond Black and Barker. *European Sociological Review*, **11**, 219–39.

Valiente C (1997): Feminist activism and social policy towards male violence, the example of Spain. Paper presented at the ESF Network Gender Inequality in the European Regions Workshop on 'Analysing Gender Culture and Gender Divisions of Power in European Regions', Dublin, June 1997.

van de Water H, Boshuizen H and Perenboom R (1996): Health expectancy in the Netherlands 1983–1992. *European Journal of Public Health*, **6**, 21–8.

Veit-Wilson JH (1987): Consensual approaches to poverty lines and social security. *Journal of Social Policy*, **16**, 183–211.

Vincent J (1996): Who's afraid of an ageing population? *Critical Social Policy*, **16**, 3–44.

Wadsworth M (1986): Serious illness in childhood and its association with later life achievement. In R Wilkinson (ed.) *Class and Health: Research and Longitudinal Data*. London: Tavistock.

Wadsworth M, Butterfield W and Blaney R (1971): *Health and Sickness: The Choice of Treatment*. London: Tavistock.

Walker A (1990): The economic 'burden' of ageing and the prospect of intergenerational conflict. *Ageing and Society*, **10**, 377–96.

Walker A and Walker C (1997): *Britain Divided: The Growth of Social Exclusion in the 1980s and 1990s*. London: Child Poverty Action Group.

Walker E, Bruce I and McKennell A (1992): *Blind and Partially Sighted Children in Britain: The RNIB Survey*, vol. 2. London: HMSO.

Waller R and Criddle B (1996): *The Almanac of British Politics*. London: Routledge.

Ward C (1991): *Influences: Voices of Creative Dissent*. Bideford: Green Books.

Warde A and Hetherington K (1994): English households and routine food practices. *Sociological Review*, **42**, 758–78.

Warnock Committee (1978): *Special Educational Needs* (The Warnock Report). London: HMSO.

Watson S (1984): Definitions of homelessness: a feminist perspective. *Critical Social Policy*, **11**, 60–73.

Webb S (1994): *My Address Is Not My Home: Hidden Homelessness and Single Women in Scotland*. Edinburgh: Scottish Council for Single Homeless.

Webster C (1997): *The Health Services since the War*, vol. 2. London: The Stationery Office.

Webster D and Turok I (1997): The future of local unemployment statistics: the case for replacing TTWAs. *Quarterly Economic Commentary*, **22**, 36–40.

Webster F (1995): *Theories of the Information Society*. London: Routledge.

Weller BGA and Weller MPI (1986): Health care in a destitute population: Christmas 1985. *Bulletin of the Royal College of Psychiatrists*, **10**, 233–5.

Weller BGA, Weller MPI, Coker E and Mahomed S (1987): Crisis at Christmas 1986. *Lancet*, **i**, 553–4.

Welsh Office (1995): *Welsh Health Survey*. Cardiff: Welsh Office.

Welsh Office (1996): *Welsh Economic Trends*. Cardiff: Welsh Office.

Wenban-Smith A (1998): *Memorandum of Evidence*. Environment, Transport and Regional Affairs Committee Housing Inquiry, Session 1997–98.

West P (1991): Rethinking the health selection explanation for health inequalities. *Social Science and Medicine*, **32**, 373–84.

Westlake L and George SL (1994): Subjective health status of single homeless people in Sheffield. *Public Health*, **108**, 111–19.

White A (1901): *Efficiency and Empire*. London: Methuen.

White H (1987): *The Content of the Form*. Baltimore: Johns Hopkins University Press.

Whitehead C and Kleinmann R (eds) (1992): *A Review of Housing Needs Assessment*. London: Housing Corporation.

Whitehead M (1988): *Inequalities in Health: The Health Divide*, 2nd edn. Harmondsworth: Penguin.

Whiteside N (1991): *Bad Times: Unemployment in British Social and Economic History*. London: Faber & Faber.

Wilcox B and Gray J (1996): *Inspecting Schools: Holding Schools to Account and Helping Schools to Improve*. Buckingham: Open University Press.

Wilkinson M (1993): British tax policy 1979–90: equity and efficiency. *Policy and Politics*, **213**, 207–17.

Wilkinson RG (1990): Income distribution and mortality: a 'natural' experiment. *Sociology of Health and Illness*, **12**, 391–412.

Wilkinson RG (1992): Income distribution and life expectancy. *British Medical Journal*, **304**, 165–8.

Williams B (1997): Utilisation of National Health Service hospitals in England by private patients 1989–95. *Health Trends*, **29**, 21–5.

Williams BT and Nicholl JP (1994): Patient characteristics and clinical caseload of short stay independent hospitals in England and Wales, 1992–3. *British Medical Journal*, **308**, 1699–701.

Williams K (1981): *From Pauperism to Poverty*. London: Routledge.

Williamson J (1997): To be, or not to be. *Guardian*, 25 October, p. 8.

Willie C (1967): The relative contribution of family status and economic status to juvenile delinquency. *Social Problems*, **14**, 326–35.

Wilson A and Kirkby M (1975): *Mathematics for Geographers and Planners*. Oxford: Clarendon Press.

Wilson RA and Briscoe G (1991): *Employment Forecasting: A Brief Review*. Coventry: Institute for Employment Research, University of Warwick.

Wing JK (1990): Meeting the needs of people with psychiatric disorders. *Social Psychiatry and Psychiatric Epidemiology*, **25**, 2–8.

Wohl A (1983): *Endangered Lives: Public Health in Victorian Britain*. London: Methuen.

Women and Geography Study Group (1997): *Feminist Geographies: Explorations in Diversity and Difference*. Harlow: Longman.

Woodhead C (1995): *A Question of Standards: Finding the Balance*. London: Politeia.

Woodhouse G and Goldstein H (1989): Educational Performance Indicators and LEA league tables. *Oxford Review of Education*, **14**, 301–19.

Woolford C and Denham J (1993): Measures of unemployment, the claimant count and the LFS compared. *Employment Gazette*, **101**, 455–64.

Woolford C, Patel D and Evans A (1994): Characteristics of the ILO unemployed. *Employment Gazette*, **102**, 249–60.

Woolgar S (1988): *Science: The Very Idea*. Chichester: Ellis Horwood; London: Tavistock.

Working Party on the Measurement of Unemployment in the UK (1995): The measurement of unemployment in the UK (with discussion). *Journal of the Royal Statistical Society A*, **158**, 363–418.

World Bank (1994): *Averting the Old Age Crisis*. New York: Oxford University Press.

Woudhuysen J (1997): Before we rush to declare a new order. In G Mulgan (ed.) *Life after Politics: New Thinking for the Twenty First Century*. London: Fontana, 352–60.

Wragg EC and Brighouse T (1995): *A New Model of School Inspection*. London: Macmillan Education.

Wright DB (1998): Social science. In T Johnston and D Machin (eds) *Encyclopedia of Biostatistics*. Chichester: Wiley.

Wright EO (1985) *Classes*. London: Verso.

Wulf H (ed.) (1993) *Arms Industry Limited*. Oxford: Oxford University Press for SIPRI.

Yule GU (1900): On the association of attributes in statistics. Reprinted in *The Statistical Papers of George Uday Yule*, ed. A Stuart and MG Kendall. London: Griffin, 7–69.

Yule GU (1912): On the methods of measuring association between two attributes. Reprinted in *The Statistical Papers of George Uday Yule*, ed. A Stuart and MG Kendall. London: Griffin, 107–70.

Author Index

Abdul-Hamid, W, 194, 196, 197
Abramson, JH, 194
Ahmad, WIU, 128–9, 216, 218
Alcock, P, 244
Aldridge, J, 100, 101
Alen, J, 106
Allison, PD, 99
American Psychological Association, 66, 69
Anderson, C, 76
Anderson, M, 29
Andrews, AY, 139
Andrews, B, 66
Arber, S, 108, 117, 118, 119, 121, 122
Archard, P, 190, 191
Ardener, S, 91
Arms Control and Disarmament Agency, 380
Association of Metropolitan Authorities, 253
Atkinson, 30
Audit Commission, 254
Axelson, ML, 142

Bachrach, LL, 194–5
Baghurst, KI, 246
Bahr, H, 191, 196
Baker, RM, 61
Bakewell, J, 112
Balarajan, R, 218
Baldwin, J, 201
Baldwin, S, 167, 168, 170
Ball, SJ, 282, 288, 294, 296
Ballard, R, 130
Banister, P, 84
Barbesino, P, 149, 153
Barclay, GC, 198
Barnes, B, 61
Barot, R, 137, 138, 217
Bartley, M, 25, 236
Bassuk, EL, 190

Beardsworth, A, 143
Beatty, C, 318, 322, 325, 336, 339
Bebbington, PE, 196
Becker, J, 67
Beishon, S, 217
Bennett, SN, 298
Ben-Shlomo, Y, 165
Benzeval, M, 251
Beresford, B, 167, 171
Beresford, P, 167, 190
Bernstein, B, 292
Berrington, A, 20
Berthoud, R, 125, 129, 130, 131, 167, 170
Bettig, RV, 385
Bhaskar, R, 84
Bhopal, R, 218
Bhrolchain, NM, 127
Billig, M, 87
Blackburn, S, 23
Blair, PS, 229
Blalock, HM, 77
Blane, D, 239, 240, 242, 243
Blasi, GL, 181
Blaxter, M, 239
Block, FL, 343
Block, G, 141, 142
Bolling, GF, 347
Bone, M, 118, 162
Booth, C, 173
Boreham, J, 328
Bowling, A, 239
Box, S, 200
Boyle, J, 384, 385, 387
Braithwaite, G, 191
Bramley, G, 372, 373
Brandon, D, 190
Brandon, S, 66
Brannen, J, 25

Bravais, A, 58
Breakey, WR, 195
Brember, I, 292
Brighouse, T, 296
Brimblecombe, N, 296
Brimmer, MJ, 334
Brinberg, D, 142
Briscoe, G, 341, 347
Brittan, S, 339
Britton, M, 229
Brodzki, B, 96
Brown, M, 25
Bruegel, I, 106, 107
Bryman, A, 80
Bryson, A, 317
Budge, D, 292
Bulmer, M, 124, 127
Burghes, L, 25
Burrows, L, 187
Butler, DE, 395, 396

CACI, 200
Callan, T, 174
Calnan, M, 246
Caplow, T, 191, 196
Carstairs, V, 175, 248, 250
Cartwright, A, 239
Casey, B, 106
Cassidy, J, 343
Castillo, IY, 174
Central Statistical Office, 105, 112, 364, 365,
    367
    Social Trends, 106, 134, 135, 143, 145, 337
Chadwick, E, 173
Chamberlain, C, 182
Chambliss, W, 200
Chant, S, 107
Chapman, M, 79
Charles, N, 142, 145, 146
Charlton, J, 164, 230
Chartered Institute of Public Finance and
    Accountancy, 183
Chatfield, C, 76
Chesher, A, 143
Chetwynd, J, 170
Chief Inspector of Factories Annual Report for
    1945, 264
Chief Inspector of Factories Annual Report for
    1967, 264
Chief Medical Officer, 243, 251
Chitty, C, 287, 288
Choldin, H, 29
Clearing House on Child Abuse and Neglect
    Information, 67
Clemente, F, 198
Clinard, MB, 191
Clode, D, 190
Cockburn, C, 107
Cohen, J, 62, 63
Coleman, D, 126, 128, 130

Collins, J, 118, 190, 191
Commission for Racial Equality, 138
Commission on Social Justice, 116
Connelly, J, 190
Cook, D, 200
Cook, T, 191
Cooney, C, 197
Cooper, CL, 146
Cooper, L, 73, 74
Cooper, P, 191
Cornford, D, 386
Cowan, RS, 56
Cox, BD, 238
Cox, DR, 98
Cox, LH, 35, 36
Cramer, D, 80
Creeser, R, 234
Cresswell, M, 279
Criddle, B, 413
Crown, J, 190
Cunningham, P, 41, 42
Currie, E, 181
Curwin, J, 79

Dale, A, 20, 23, 33, 133, 164, 172, 176, 177
Darnton-Hill, I, 141
Davey Smith, G, 165, 215, 220, 239, 240, 243,
    246
Davidoff, P, 92
Davies, H, 123
Davies, J, 292
Davies, RB, 20, 99
Davidson, MJ, 146
Davidson, N, 23, 162, 215, 220, 244, 245,
    246
Davis, P, 290
Dawes, RM, 64, 65
Day, C, 298
Deane, P, 351
Dearing, R, 274
Deming, W, 74
Denham, J, 331
Denman, J, 326, 327
Denver, DT, 393
Denzin, NK, 96
Department for Education, 282, 303, 307
Department for Education and Employment,
    277, 283, 287, 304, 305, 309, 310
Department of Education and Science, 72, 289,
    307
Department of Employment, 21
Department of Employment and Productivity,
    351
Department of the Environment, 177, 179, 190,
    370, 372
Department of the Environment, Transport and
    the Regions, 185, 186, 189
Department of Health, 189, 225, 226, 227, 228,
    231, 232, 233, 234, 254, 255, 256, 258,
    259, 260, 261

Department of Health and Social Security, 116, 240, 242
Department of Social Security, 116, 250
Department of Trade and Industry, 384, 390
Derrida, J, 96
Dex, S, 101, 106
Dilnot, A, 120
Dineen, T, 65, 66, 69
Disability Now, 170
Disablement Income Group, 170
Dominian, L, 185
Dorling, DFL, 399, 403, 406
Downes, D, 198
Downs, C, 116, 123
Doyal, L, 73
Drever, F, 229, 234, 235, 242, 244, 245
Driver, F, 353
Duncan, S, 183
Dunne, G, 109
Dunne, P, 377, 380, 382
Dunnell, C, 239

Eagle, A, 269
Eagle, PF, 196
Ebel, R, 289, 290
Economist Intelligence Unit, 142
Edwards, D, 87
Edwards, G, 192, 193
Ehrenberg, ASC, 76
Elford, J, 165
Elliot, M, 36
Elliott, C, 291
Elpers, JR, 190
Employment Policy Institute, 317, 319
Enloe, C, 137
Equal Opportunities Commission, 106, 107, 112
Erikson, R, 238
Estaugh, V, 24–5
European Commission, 39
Eurostat, 154, 156
Evans, A, 183
Evans, C, 344
Evans, J, 71, 75, 76

Fagan, C, 322
Falkingham, J, 117, 120
Farr, W, 232
Farrall, LA, 56
Fazeli, R, 364
Fellegi, LP, 35
Fenton, S, 216
Fenwick, D, 326, 327
Ferri, E, 110, 111
Feuerstein, MT, 244, 246
Feyerabend, P, 90
Fichter, R, 92
Fiddes, N, 140, 143
Filakti, H, 238
Finch, S, 231

Fine, B, 143, 144
Fink, A, 79
Fischer, PJ, 195
Fisher, K, 190, 191
Fisher, RA, 55, 57
FitzGibbon, CT, 274, 295
Fitzhugh Directory, 258
Fitzpatrick, R, 239
Flanagan, K, 42
Fleming, PJ, 229
Forbes, WF, 238
Forrest, R, 164, 178
Foster, J, 198
Fothergill, S, 318
Fox, J, 236, 238, 242, 245, 246
Freedman, S, 79
Freeman, A, 364
Friedman, A, 386
Frisbe, D, 289, 290
Frow, J, 385
Fry, AH, 192, 195

Gaffney, D, 259
Gallie, D, 331
Galton, F, 55, 56
Gamble, A, 287
Gandy, OH, 385
Gani, J, 77
Gapper, J, 348
Gard, JF, 304
Garside, WR, 350
Gay, JD, 136
George, SL, 194
Georghiou, L, 43
Gershuny, J, 110
Gewirtz, S, 282
Giannelli, PC, 68
Gigerenzer, G, 63
Gilbert, B, 56
Gilbert, DM, 354
Gilbert, N, 79
Gill, R, 112
Ginn, J, 108, 117, 118, 119, 121, 122
Glendinning, C, 169
Glover, J, 108
Goddard, E, 231
Goering, P, 190
Goldblatt, PO, 200, 204, 205, 236, 242, 245, 246
Goldstein, H, 279, 280, 283, 284, 285, 286
Goldthorpe, JH, 238
Goodman, A, 30, 174
Gordon, D, 164, 165, 167, 177, 178, 199
Gough, D, 170
Government Statistical Service, 105, 110
Government Statisticians' Collective, 328
Graham, H, 146
Graham, S, 170
Gray, J, 296
Green, AE, 175, 317, 318, 319, 322

Green Balance, 371
Gregory, JR, 231
Greve, J, 181
Gross, E, 91
Grove, WM, 65
Groves, D, 121
Grunberg, J, 196
Gudgin, G, 406
Gunter, B, 112
Gupta, S, 217
Guy, W, 75

Hacking, I, 60
Hakim, C, 108
Hamnett, C, 256
Hampshire County Council, 303, 306
Hancock, R, 116
Hand, DJ, 77
Hannay, DR, 239
Hands, G, 393
Harding, S, 121, 220
Harré, R, 84, 85
Harrington, C, 257
Harris, P, 63
Hart, J, 337, 338
Hartman, C, 189
Hartmann, H, 108
Harvey, D, 180
Haskey, J, 22
Hasluck, C, 319, 322
Hattersley, L, 234
Haveman, RH, 344
Heady, P, 164, 398
Health and Safety Commission, 263, 265, 266
Health and Safety Executive, 264, 270
Heath, A, 220
Heath, I, 261
Heijken, JAM, 348
Hemingway, 239
Henry, N, 106
Her Majesty's Inspectorate, 304, 308
Hetherington, K, 146
Hickman, M, 138
Hills, J, 116, 119, 190
Hilts, V, 58
Himmelweit, S, 110
Hindess, B, 24
Hirdes, JP, 238
Hirsch, E, 387
Hoch, C, 182
Hodgson, GM, 342
Hoggart, K, 183, 184
Hollowell, J, 227
Holmes, T, 189
Hope, T, 198, 200
Horn, R, 170
Horwath, C, 141
House of Commons Employment Committee, 317

House of Commons Health Committee, 225, 256
Howitt, D, 80
Huff, D, 415
Hutchins, R, 142, 143
Hutson, S, 181, 182, 187
Hutton, S, 117, 118

Illsley, R, 165, 215, 245
Independent Healthcare Association, 257
Institute for Employment Research, 317
Institute of Statisticians, 73
Irvine, J, 4, 61, 75, 414

Jacobs, E, 198
Jarman, B, 176, 248, 250
Jarvis, H, 110
Jarvis, L, 231
Jeffreys, B, 297
Jenkinson, C, 239
Jessop, B, 112
Johnson, P, 116
Johnston, RJ, 396
Joiner, BL, 80
Jones, T, 198
Joseph Rowntree Foundation Income and Wealth Group, 31

Kalra, V, 130
Kamins, MA, 23
Kane, E, 79
Kass, G, 206
Kauppinen, L, 167
Kean, H, 107
Keenan, Michael, 42
Keighley, J, 120
Keil, T, 143
Kendell, RE, 196
Kerr, M, 142, 145, 146
Keys, W, 278
Kiernan, KK, 24–5
Kinsey, R, 198
Kirkby, M, 90
Klein, R, 245
Kleinman, M, 198
Kleinmann, R, 190, 191
Koegel, P, 191
Kotlikoff, L, 116
Kranzler, GD, 80
Kristeva, J, 91
Kuh, D, 165
Kunst, AE, 238, 240

Labov, W, 101
Laing, W, 256, 261
Lambert, C, 183
Lane, T, 116
Lanning, KV, 67
Lansley, S, 165, 174
Lapsley, DK, 63

Laux, R, 320, 321, 333
Lea, J, 198
Lee, MP, 80
Lee, P, 175, 177, 178, 179
Leech, K, 127, 128, 131
Lees, S, 113
Le Fanu, J, 246
Leicester, M, 257
Leighton, P, 107
Leon, DA, 227
Lévi-Strauss, C, 91
Levitas, R, 29, 75, 174
Lewis, P, 139
Lewis, T, 77
Liddiard, M, 181, 182, 187
Lidstone, P, 185
Lilley, P, 117, 118
Lincoln, S, 23
Lindsay, DS, 66
Linstead, S, 92
Lloyd-Sherlock, P, 116
Loach, I, 167
Lodge Patch, IC, 192, 193
Loftus, EF, 66
Loftus, GR, 63
Loveland, I, 191
Loynes, RM, 76
Luloff, AE, 181
Lundeberg, O, 215
Lupton, D, 140
Lynch, JW, 238, 246
Lyon, D, 385

Mabbett, D, 116
McCloskey, DN, 347
McCormick, A, 227
McCrossan, L, 20
McCulloch, A, 106
McDonnell, DM, 191, 195
McDowell, L, 107
Macfarlane, AJ, 115, 227, 230, 231
Macintyre, S, 215
Mack, J, 165, 174
Mackenbach, JP, 238, 240
McKay, S, 317
MacKenzie, D, 56, 61, 182, 240
McKenzie, J, 80
Mackintosh, NJ, 416
Macourt, MPA, 135, 136
McRae, S, 317
McWilliams, P, 170
Mahon, B, 79
Malpass, P, 190
Mamuya, WPM, 308
Mandelson, P, 175
Marchant, P, 79
Mar Molinero, C, 304, 306
Marmot, MG, 216, 220, 238, 243
Marsden, D, 342
Marsh, C, 33, 36, 79, 95, 164, 172, 176, 177

Marshall, EJ, 194
Marshall, M, 190, 192, 194, 195
Martin, B, 42
Martin, J, 162, 170
Matthews, R, 71
Matthews, S, 243
Maung, NA, 200
Maxwell, R, 220
Mayhew, PP, 199, 200
Meager, N, 344
Meehl, PE, 63, 65
Meier, RF, 191
Meltzer, H, 162, 169, 231
Mennell, S, 146
Metcalf, H, 107
Micklewright, 30
Middleton, D, 87
Milburn, N, 195
Miller, RJ, 196
Miller, WL, 406
Ministry of Agriculture, Fisheries and Food, 144, 145
Ministry of Education, 72
Ministry of Health, 225
Minkler, M, 116, 117
Mirrlees-Black, C, 198
Mischler, EG, 100
Mitchell, BR, 351
Modood, T, 128, 133
Montgomery, S, 243
Moore, DS, 79
Moore, J, 117
Moore, PG, 72–3
Morgan, K, 336
Morrell, D, 239
Morris, JN, 238
Morris, R, 175, 248
Morris, S, 80
Morris, T, 200
Mossman, D, 196
Motorola, 384
Moursund, JP, 80
Mugford, M, 230
Mullings, B, 190
Murgatroyd, L, 109
Myers, K, 285, 286

Naismith, DJ, 141
National Audit Office, 117
National Centre for Health Statistics, 238
National Consumer Council, 257
National Food Survey, 144
Naylor, K, 328
Nazroo, JY, 215, 216, 217, 218, 219, 220, 221, 222
Neale, J, 181, 182
Nelson, M, 141
Nesbitt, S, 116
Neuburger, H, 109
Newton, PE, 279

NHS Executive 227
NHS Management Executive 226
Nicaise, I, 317
Nicholl, JP, 258
Nichols, T, 263, 267, 268
Nightingale, DS, 344
Nord, M, 181
Norton, WW, 416
Norusis, MJ, 98, 99

Oakley, A, 246
O'Callaghan, B, 185
Office for National Statistics, 5, 30, 38, 113,
    133, 231, 260, 261, 325, 330
Office of Population Censuses and Surveys
    (OPCS), 22, 175, 178, 229, 234, 261
OFSTED, 275, 294, 296, 297, 298, 299, 300
OFSTIN, 296
Olney, J, 96
Open University, 76
O'Reilly, K, 239
Orwell, G, 192
Owen, C, 25, 26, 236
Owen, DW, 318, 322

Pahl, J, 107
Pantazis, C, 165, 177, 199
Parker, H, 218, 219
Parker, I, 85, 88
Paterson, G, 117
Paul, D, 61
Paulos, JA, 71, 73
Pearce, D, 164
Pearson, K, 55, 56, 57, 59, 60
Pease, P, 329, 330, 333
Peck, J, 112, 321, 342, 348
Pendergrast, M, 66, 67
Perlin, ML, 196
Perrons, D, 105, 106
Perry, K, 330
Petersen, W, 373
Phillimore, P, 200, 204, 250
Phillips, AW, 354
Phoenix, A, 26
Piachaud, D, 167
Plewis, I, 280, 283
Plummer, K, 96
Pocock, SJ, 238
Polanyi, L, 100
Polk, K, 202
Pollock, AM, 254, 257, 259
Poortvliet, W, 116
Popper, K, 84
Porter, K, 61
Power, C, 165, 242, 243, 245
Prescott-Clarke, P, 231, 237
Preston, S, 115
Priest, RG, 192, 195, 196
Primatesta, P, 231, 237
Pumfrey, P, 291

Purdie, E, 19

Quadagno, J, 116

Radford, J, 113
Radical Statistics Education Group, 75
Radical Statistics Health Group, 23, 75, 226,
    227, 231, 232, 254, 257, 260
Rahkonen, O, 165
Ramsay, M, 198, 200
Randall, S, 190
Ranson, S, 288
Read, JD, 66
Realeat Survey Office, 143
Redfern, P, 35
Reed, A, 192, 194
Reed, JL, 194
Rees, P, 133
Reicher, S, 87
Reichmann, WI, 415
Reid, F, 167
Rex, J, 137, 138
Rhondda, Cynon, Taff Economic Development
    Committee, 339
Ridge, J, 220
Ringen, S, 174, 200
Ritchie, C, 79, 415
Ritson, C, 142, 143
Roberts, HV, 71, 74
Robins, K, 90–91
Robinson, WS, 201
Robus, N, 167
Römkens, R, 113
Room, G, 174
Rose, D, 239
Rose, ME, 353
Rosenthal, R, 63, 87
Roseveare, D, 118
Ross, SM, 80
Rossiter, DJ, 393
Rowlands, O, 23
Royal College of Psychiatrists Working Group,
    66, 69
Royal Statistical Society, 30, 73, 317
Rubery, J, 105, 107, 322
Rubin, DB, 63
Rudat, K, 215, 217
Rumeau-Rouquette, C, 194
Ryan, BF, 80

Sachs, J, 116
Salt, J, 126, 128, 130, 149, 156
Salter, A, 42
Samet, JM, 141
Sammons, P, 284, 291
Scambler, G, 240
Schumpeter, JA, 361
Shorrocks, D, 288
Schenk, C, 96
Schols, CM, 58

Schools Curriculum and Assessment Authority, 282
Scott-Heide, W, 121
Screen Digest, 389
Secord, PF, 85
Secretary of State for the Environment, 370
Seidman, R, 200
Sen, A, 107
Senior, PA, 218
Serlin, RC, 63
SERPLAN, 370
Shaikh, AM, 364
Shanks, NJ, 193, 196
Shaw, M, 198
Sheaffer, R, 67
Sheldon, TA, 128–9, 218, 219
Shotter, J, 87
Sibley, D, 89, 93
Silver, H, 298
Silverstone, R, 387
Simpson, S, 74
Sinfield, A, 121, 123
Singleton, A, 149, 153, 156
Sivard, RL, 380
Slater, JM, 21
Smaje, C, 216
Smith, AM, 246
Smith, D, 220
Smith, K, 110
Smith, R, 380, 382
Smith, T, 295
Smyth, M, 167
Snow, DA, 191, 193, 196
Social Security Advisory Committee, 117
Social Trends, 106, 134, 135, 143, 145, 337
Soni Raleigh, V, 218
Soper, JB, 80
Sorlie, PD, 239
Southall, HR, 353, 354, 355
Sparks, R, 199
Spiegelhalter, DJ, 283
Stanko, E, 113
Stanley, L, 96
Statewatch, 148
Stedman-Jones, G, 56, 354
Steedman, C, 189
Steering Group on Health Services Information, 225
Stephenson, NJ, 295
Stern, J, 242
Stevens, G, 265
Stewart, A, 238
Stewart, DW, 23
Stigler, SM, 61
Stockholm International Peace Research Institute (SIPRI), 379
Stokes, DE, 288, 395, 396
Stray, SJ, 406
Susser, E, 195
Sweeting, H, 243

Task Group on Assessment and Testing, 281
Taylor, PJ, 406
Temple, B, 101
Tenner, E, 390
Thatcher, M, 327
Third King's Fund Forum, 190
Thomas, F, 164
Thomas, G, 290
Thomas, R, 327, 330, 333
Thomas, S, 284
Thomas, WI, 96
Thorstad, I, 75
Thunhurst, C, 175, 227
Tickell, A, 112
Tickle, L, 120
Tidmarsh, 192
Times Educational Supplement, 275, 297
Timms, PW, 192, 195
Todorov, T, 91
Tonak, EA, 364
Tonks, E, 333
Toro, PA, 191, 195
Townsend, P, 23, 116, 161, 162, 173, 174, 176, 177, 200, 204, 205, 215, 220, 244, 245, 246, 247, 248, 250
Tuan, Y-F, 136
Tufte, ER, 76, 400
Turner, J, 92
Turock, I, 327

UK Government, 226, 282, 325
United Nations, 362
Upton, GJG, 403, 406

Vågerö, D, 215
Valiente, C, 113
van de Water, H, 240
Veit-Wilson, JH, 165
Veltman, M, 280
Victor, C, 117
Vincent, J, 119, 120
Vincent, L, 274

Wadsworth, MEJ, 165, 239
Wale, C, 239
Walentowicz, P, 187
Waletzky, J, 101
Walker, A, 23, 116, 120
Walker, C, 23
Walker, E, 167
Waller, R, 413
Ward, C, 92
Ward, S, 123
Warde, A, 146
Watkins, C, 372, 373
Watson, S, 181
Watts, RJ, 195
Webb, S, 30, 174, 184, 187
Webster, C, 260
Webster, D, 327

Webster, F, 385, 387
Weir, P, 116
Weller, BGI, 190, 192, 193, 195, 197
Weller, MPI, 190, 192, 193, 195, 197
Welsh Office, 336, 337
Wenban-Smith, A, 373
West, P, 243
Westlake, L, 194
White, A, 56, 162, 170
White, H, 100
Whitehead, C, 190, 191
Whitehead, M, 162, 229, 234, 235, 242, 244,
    245
Whiteside, N, 356
Wilcox, B, 296
Wilkinson, M, 117
Wilkinson, RG, 247
Williams, BT, 258
Williams, K, 351, 353
Williams, S, 246
Williamson, J, 113
Willie, C, 202
Wilson, A, 90
Wilson, RA, 341, 347

Wing, JK, 196
Wohl, A, 244
Women and Geography Study Group, 105
Wood, L, 348
Woodhead, C, 297
Woodhouse, G, 283
Woods, P, 297
Woolford, C, 331, 333
Woolgar, S, 84
Worcester, R, 198
Working Party on the Measurement of
    Unemployment in the UK, 325
World Bank, 116
Woudhuysen, J, 387
Wragg, EC, 296
Wright, DB, 62
Wright, EO, 238
Wroblewska, A, 170
Wulf, H, 380

Young, J, 198
Yule, GU, 58, 59

Znaniecki, F, 96

Subject Index

Accountability
  Income Support *vs* Poor Law, 357
  and performance indicators in education, 282–3
ACORN system of neighbourhood classification, 200, 211
Action research, 87
'ActivStats', 80
Acute hospital care, privatisation, 257–9
Adam Smith Institute, 116
Addresspoint, 11
Administrative records, 31–2
Age issues
  eating habits, 143, 145
  employment, 337
  voting patterns, 404, 405
Agricultural sector
  injuries, 265
  unemployment and pauperage, historical records, 353, 354
Alternative Measures of Unemployment (AMU) series, 329
Alternative technology approach to statistics, 415
American Civil War, 353
American Psychological Association (APA), 64
Americans for Generational Equality (AGE), 116
Anthropologists, social, 91
Anti-science approach to statistics, 415
Arms trade, 378, 380–82
Armstrong, Hilary, 185
Artisans, 354
Aspirations, labour market, 320
Association for Research in the Voluntary and Community Sector (ARVAC), 80
Asylum-seekers, 155–6
Attainment Targets (ATs), National Curriculum, 288, 289, 290

Attribute disclosure, 36
Australia, 46
Autobiography in sociology, 96, 101–2

Babies, 17
Ball, Sidney, 56
Banks, 364–5, 367
Barefoot statisticians, 71, 77
  and community research, 73–4
  resources, *see* Lay statisticians, resources
  training, 76–7
Behavioural explanations, health inequalities, 242, 246
Benn, Tony, 406
Bevan, Aneurin, 261
Biography in sociology, 96, 101–2
Birth registration, 228–9, 232
Books for lay statisticians, 78, 79
Bottomley, Virginia, 258
Breadline Britain survey (BBS), 177
  crime, 199, 206–9
  disabled chidren, 165–6, 169
Breadline Index, 177–9, 180
British Aerospace, 382
British Crime Survey (BCS), 198, 199–200
  measurement and definition of crime, 199
  poor areas, 200, 202–6, 209
  violence against women statistics, 112
British Household Panel Survey (BHPS), 113
  attrition, 110
  life histories, 97
  paid and domestic work, linking, 110
British Official Publication Current Awareness Service (BOPCAS), 81
British Psychological Society (BPS), 64
British school of statistical theory, 55–61
British Society for Social Responsibility in Science (BSSRS), xxv, xxvi

British United Provident Association (BUPA), 258
Bureau of Statistics, Australia, 46
Burges, Patricia, 66–7
Bush, George, 189

Canada, 46, 93
Cancer registries, 227
Car ownership
    company cars, 366, 367
    and disabled children, 169
    as income proxy, 204–5
    and unemployment, 175–6
Cartograms, 401–4, 405
CASMIN schema, 238, 239
CBI/NatWest Innovation Trends Survey, 44
Census Act 1920, 32
Census Bureau, USA, 46
Census (Confidentiality) Act 1991, 32
Censuses, national, 9
    citizenship, 153
    Germany, 35
    Ireland, 2
    Netherlands, 35
    United Kingdom, 9–10, 18
        conduct, 10–11
        confidentiality, 31–5, 36
        content, 11–17
        coverage, 17–18
        deprivation, 172–3
        development, 2
        ethnicity question, 16, 126–7, 128–31
        gender differentiation, 110
        health questions, 16, 164, 230–31
        household projections, 370
        limiting long-term illness question, 16, 164, 230
        religion question, 17, 132–5
    United States of America, 29, 36
Censuses, one number, 17–18
Census Office, 11
Central heating, 16
Central Statistical Office (CSO, later Office for National Statistics), 267
Centre for Policy Studies, 116
Charities Evaluation Service, 80
Chemical industry, 43
Childbearing outside marriage, 25
Childcare, 109, 110
Children Act 1989, 171
Children
    disabled, 165–71
    sexual abuse, 66–7
Chi-squared Automatic Interaction Detector (CHAID), 206–9
Christianity, 132–3, 136, 137–8
Cigarette smoking patterns in pregnant women, 246
Citizen's Charter, 50, 282
Citizenship, 153–4

critical, 71, 75–7
    resources, *see* Lay statisticians, resources
Class, social, *see* Socio-economic group
Class size in schools, 298–300
Clinton, Bill, 343
Coal extraction industry, injuries, 266
    see also Mid Glamorgan, permanent sickness in
Cobwebs, 411–12
Codification of information, 385–6
Cohabitation before marriage, 25
Cold War, 380
Commission for Racial Equality, 128
Commodification of information, 385–6, 387–8
Communicable diseases, notification of, 227
Community
    and religion, 137
    research, 73-4
Community care, 253, 254
    and homelessness, 196
Community Innovation Surveys, 44
Community Operations Research Network (CORN), 82
Competitive tendering, 258
Computer-aided learning, 79–80
Computers
    and disabled children, 170
    see also Information society
Confederation of British Industry (CBI), 44
Confidence intervals, 250
Confidential enquiries into deaths, 229
Confidential enquiries into stillbirths and deaths in infance (CESDI), 229
Confidentiality of official statistics, 29–30, 37
    and availability of data, balance between, 35–7
    census, 32–5
    context of publication, 30–31
    Data Archive, 50
    range of official statistics, 31–2
    UK situation, 31
Conservative government (1979–97)
    dissemination of information, 48
    education, 287–8
        inspection of schools, 294, 297, 301
        league tables, 282
    employment income, 368
    housing expenditure, 190
    industrial injury statistics, 269
    National Health Service, 233
        'Health of the Nation' strategy, 230–31
        privatisation, 226, 257, 260
        pension privatisation, 116, 117, 118, 123
        unemployment statistics, 327, 350
Conservative Party, elections, 391, 393–5, 396
    graphical presentations of 1997 results, 400, 412
        cartograms, 401–3
        ring doughnuts, 410, 411
        triangles, 404, 406–8, 409

Constituencies, 391–2, 393
Construction industry, injuries, 266
Consumer durables, and disabled children, 169–70
Consumption expenditure, 161
  and disabled children, 169–70
Continuous Household Survey, 21
Copyright, 387–8
Coronary heart disease (CHD), 217–18, 219
Correlation, 56, 58
Count of Claimants (CoC), 324–9, 331, 332, 334
Courses for lay statisticians, 78–9
Covariates, 98, 101
Cox proportional hazard models, 98–9, 101
Crime and fear of crime, 198–9, 209–10
  ACORN system, 211
  Breadline Britain survey, 206–9
  British Crime Survey, 202–6
  and insurance, lack of, 209
  measurement and definition, 199–200
  and poverty, 200–202
  against women, 112–13, 199
Criterion-referenced interpretation of Standard Assessment Tasks, 289, 290, 291
Critical citizens, 71, 75–7
  resources, *see* Lay statisticians, resources
Cultural difference, 89, 94
  distanciation, 90–91
  ethics, 93–4
  muteness, 91–3

Darwinism, social, 57
Data Archive, 20, 28, 81
  confidentiality issues, 31, 32
  dissemination of information, 49–50
'Data Desk', 80
Davies, Frank, 263
Death certificates, occupational status exaggerated on, 220, 235
Death registration, 228–9, 232
Decennial Supplements on Occupational Mortality, 234, 235
Dentistry, privatisation, 260
Department for Education and Employment (DfEE), 325
Department of the Environment (DoE)
  Index of Local Conditions (ILC), 177, 178–80
  Z score, 176–7
Department of the Environment, Transport and the Regions (DETR)
  homelessness, 183, 184, 185
  Priority Estates Project, 198
Department of Health, 226, 231
Department of Social Security (DSS), 23
Department of Trade and Industry (DTI), 44
Depression, inter-war, 355–7
Deprivation, 172–3, 180
  conceptualising disadvantage, 173–4

and disabled children, 165–7
  indexes, 161, 176–8
  and crime, 204
  reliability, 178–80
  proxy indicators, 174–6
Deviance and homelessness, 191
Dietary research, *see* Eating habits; National Food Survey
Direct mail marketing industry, 385
Disability Discrimination Act 1995, 171
Disability-free life expectancy, 240
Disability Living Allowance, 168
Disability Surveys, 162–3, 165–6, 169, 170
Disabled people, poverty, 162–5
  children, 165–71
Disclosure of information, 35–6
Discouraged workers, 329, 331
'Discovering Statistical Concepts Using Spreadsheets', 80
Dissemination of official statistics, 46–52
Dissemination Standards Bulletin Board, 51
Dissociative identity disorder, 65–6
Distanciation, 86, 87, 89, 90–91
Domesday Book, 2
Domestic violence, 112
Domestic work, 108–11
  national income accounts, 362, 363
Dual-use civilian/military technologies, 378

Early retirees, 318
Eating habits, 140, 146–7
  measuring, 140–42
  National Food Survey, 142–6
Ecological fallacy
  educational attainment and teaching quality, 278, 279
  poverty and victimisation, 201–2
  proxy indicators of deprivation, 176
Economic activity rate, 313, 317
Economic and Social Research Council (ESRC), 50, 51
Economic inactivity rate, 313–14
Economic distress, historical statistics, 350–54
  inter-war Depression, 355–7
  unemployment and the labour market, 357
Economies
  employment forecasting, 340–49
  household projections, 373–4
  limitations, 361–2
Economy, measuring the UK, 361–2
  national income accounts, 362–8
Education
  census question, 17
  comparisons, 273–4, 279–80
    of different measures at a single point in time, 274–5
    of educational attainments at different ages, 276–8
    international, 278–9
    of performance across time, 275–6

Education (*Cont.*)
  and employment patterns, 317
  inspection system
    constructing the statistics, 295–6
    criteria setting, 298–301
    interpreting the statistics, 297–8
    philosophy and politics, 294–5
  performance indicators, 281–2
    and public accountability, 282–3
    schools, comparison of, 283–5
    social manipulation of statistics, 285–6
  as proxy indicator of socio-economic group,
    238
  reading standards, measurement of trends in,
    287–92
  special educational needs pupils, 302–3
    class size, 298
    Green Paper 1997, 310
    Hampshire review, 306–9
    social trends, 304–6
    statistics, 309–10
    Warnock Report, 303–4
  universities, research and development, 39,
    42–3
Education Act 1981, 303, 306
Education Reform Act 1988, 273, 281, 287–8
Education (Schools) Act 1992, 294
Elderly dependent ratio, 118–20
Electoral Register, 32
Electoral system and electorate, British, 391,
  397–9
  geography of, 391–7
  graphical presentations of 1997 results,
    400
    cartograms, 401–3, 404
    ring doughnuts, 410, 411
    triangles, 404, 406–8, 409
Embedded objectivity, 86–7
Empiricism, 84, 88
Employment
  definitions, 313
  and disabled children, 168
  disaggregation, 315, 316
  health inequalities, 235–6, 237
  household surveys, 25, 26, 27
  labour migration, 156–7
  life history data, 98–9, 100
  participation inlabour market, measurement,
    313, 323
    conventional approaches, 313–18
    inter-area comparisons, 321–3
    wider range of statistics, 318–21
  politics and reform of statistics, 324, 334
    Alternative Measures of Unemployment,
      329
    Count of Claimants, 324–7
    Labour Force Survey, 328–33
    political and statistical background,
      327–8
  rate, 315

women, 105–7
  return to work after childbirth, 98–9
Energy sector, injuries, 265
England
  elections, 392
  health statistics, 228, 231
    National Health Service, 225, 226, 227–8,
      232
    privatisation of long-term care, 254–5
Enumeration districts (EDs), 10–11
  ACORN system, 211
Enumerators, 10–11
Error theorists, 58
Ethical issues
  ethnography, 93–4
  qualitative research, 88
Ethnicity, 124, 131
  census question, 16, 126–7, 128–31
  defining, 124–7
  health inequalities, 215–16, 222
    class sensitivity, 219–21
    demarcating 'race' and ethnic groups,
      216–17
    hypothesised variables, measurement,
      219–21
    'untheorised' research, consequences,
      217–19
  Labour Force Survey, 22, 25–8
  and life expectancy, 121
  mixed, 130
  and religion, 125, 136–7
  and social policy, 127–9
  voting patterns, 404, 405
Ethnography, 91, 93–4
Eugenics, 55–61, 242
Eurofighter, 377
European Economic Area (EEA), 153–5
European Free Trade Association (EFTA), 149,
  155, 156
European Labour Force Survey (ELFS),
  322–3
European Union (EU)
  health and safety directives, 269
  migration, 148, 149, 155, 156
Eurostat
  dissemination of official statistics, 51–2
  migration statistics, 149
Event history analysis, *see* Life histories
Everyday life, use of statistics in, 71, 77
  barefoot statisticians, 73–4, 76–7
  critical citizenship, 75–7
  parastatisticians, 74–5, 76–7
  resources, *see* Lay statisticians, resources
  statistical skills, 72–3
Excellence for All Children Green Paper
  (1997), 310
Exclusion, social, 174–5
Expatriates, 399
Experimenter effects, 87
Eye tests, privatisation, 260

Fabianism, 57
False/recovered memory, 66–7, 68–9
Families
    definition, 20, 24
    homeless, 190
    household surveys, 24–5
Family and Working Lives Survey, 110–11
Family Expenditure Survey (FES), 20, 21
    and Family Resources Survey information,
        23
    gender differentiation, 107
    using the data, 23, 24, 30–31
Family Resources Survey (FRS), 20, 23
    gender differentiation, 107
    using the data, 24
Fatality, *see* Mortality
Fear of crime, *see* Crime and fear of crime
'Fictive devices' in life history data, 100–102
Field, Frank, 339
First-past-the-post electoral system, 391
Fisher, RA, 55–6, 57, 60
Flexible working patterns, 317
Flow-of-the-vote matrices, 396, 397
Food, *see* Eating habits; National Food Survey
Food diaries/records, 140, 141
Food frequency questionnaires, 140, 141
Foresight Programme, 39, 42
France
    arms production, 382
    citizenship and nationality laws, 153–4
    foreign population, 154
    research and development, 42
Freud, Sigmund, 63
Fringe psychologists, 62, 67–9
Full-time employment, 315, 317, 320

Galton, Francis, 55–6, 57–8
Gender issues, 105, 113
    eating habits, 143–4, 145–6
    education, 275
    health
        official statistics, 229
        mortality inequalities, 236
    homelessness, 194
    pensions, 118, 120–23
    see also Men; Women
General Household Survey (GHS), 20, 21–2
    childhood disabiity, 163, 164, 169, 170
    gender differentiation, 107
    health questions, 230, 234, 237, 242, 243
        private health care insurance, 261
    pension privatisation, 121
    using the data, 23–5
General practitioners (GPs), official statistics
        about, 227
Germany
    censuses, 35
    foreign population, 154
    research and development, 39, 42
Gini coefficients, 240–41

Global Plan for Action, 105
Governance, women's roles, 112
Government household surveys, *see* Household
        surveys
Government training schemes, 318, 326
Graphical techniques, 400, 412
    cartograms, 401–4, 405
    cobwebs, 411–12
    ring doughnuts, 408–11
    triangles, 404–8, 409
Gross domestic product (GDP), forecasting,
        344, 345
Gypsies, 89, 90, 92

Hampshire Local Education Authority, 302,
        306–9
Health and illness
    ethnic inequalities, 215–16, 222
        class sensitivity, 219–21
        demarcating 'race' and ethnic groups,
            216–17
        hypothesised variables, measurement,
            219–21
        'untheorised' research, consequences,
            217–19
    household surveys, 25, 27
    industrial injury statistics, 263–70
    inequalities, 234
        explanations, 240–43
        ill health, 236–8
        life-course approach to understanding, 243
        measurement, 240
        measures of health, 239
        socio-economic position, measuring,
            238–9
    measures, 239–40
    official statistics, 223, 224
        about the circumstances and health of the
            population, 228–31
        about the National Health Service, 223–8
        interpretation, 231–3
    and poverty, 244–5, 251
        explanations, 245–7
    research and development, 43
    see also Limiting long-term illness; Mental
        illness; Mortality; Permanently sick
        people
Health and Safety Executive (HSE), 263, 265,
        266, 267, 268, 269
Health Education Authority, 231
Health for All targets, 237
'Health of the Nation' strategy, 230–31
Health Surveys for England (HSE), 231, 234,
        237–8, 242, 243
Health Surveys for Scotland, 242
Her Majesty's Inspectors (HMIs), 294
Higher Education Funding Council, 42–3
Holmes, Thomas, 189
Home care, 256
Home computers, and disabled children, 170

Home help, 256
Homelessness, 188, 189–90, 196–7, 414
    defining, 181–2, 188, 194–5
    Housing Act 1996, impact of, 184–6
    'official', 183–4
    psychiatric surveys, 192–6
    quantification difficulties, 182
    social origins, 190–91
    'unofficial', 187
Home Office, 112
Hospital Activity Analysis (HAA), 225
Hospital Episode System (HES), 226
Hospital In-patient Enquiry (HIPE), 225, 226
Household projections, 369–70, 374
    alternative approaches, 374–5
    effects of reliance on, 372–4
    reliability, 370
    role in planning for housing, 371–2
Households
    definition, 20
    identification, 11
    relationships within, 16–17
    rooms, number of, 17
Household surveys, 19–23, 28
    using the data, 23–8
    see also Family Expenditure Survey; Family
        Resources Survey; General Household
        Survey; Labour Force Survey; National
        Food Survey
Housing, 189–90
    and ethnicity, 220
    graphical representation of types, 407–8,
        409–11, 412
    planning based on household projections,
        369–75
Housing Act 1985, 183, 185
Housing Act 1996, 183, 184–6
Housing (Homeless Persons) Act 1977, 184
Hypnotherapy, 68–9

Identity disclosure, 35
'Illegal' migrants, 157
Illness, *see* Health and illness; Limiting long-
        term illness; Mental illness
Immigration, *see* Migration
Inactivity, labour market
    Alternative Measures of Unemployment
        series, 329
    definition, 313
    disaggregation, 315, 316
    gender differences, 107–9
    measurement, 317
Incapacity benefit, 325, 339
Income
    car ownership as proxy measure, 204–5
    census question, 16
    and disability, 167–70
    and eating habits, 144, 145
    and ethnicity, 220
    gender differences, 106–7

inequality, 30, 31
    Mid Glamorgan, 336
    national statistics, 362–8
    older people, 117
        pensions, 118
    as proxy indicator for socio-economic group,
        238–9
    see also Poverty
Income Support, 118, 119, 357
Income tax records, 334
Index of Local Conditions (ILC), 177, 178–80
Indirect health selection, 243
Industrial injury statistics, 263–70
Inflation, 344, 345
'Information commons', 47
Information society, 384–5, 390
    codification and commodification, 385–6,
        387–8
    hype, 386
    information *vs* technology, 386–7
    Internet, 82, 388–90
Inheritance, 55, 56, 59–60
Injuries, industrial, 263–70
Inspection of schools
    constructing the statistics, 295–6
    criteria setting, 298–301
    interpreting the statistics, 297–8
    philosophy and politics, 294–5
Institute of Economic Affairs, 116
Institute of Statisticians (later Royal Statistical
        Society), 77
Insurance
    health care, 260, 261, 338
    home and contents, 209, 210
Intellectual property rights, 387–8
Intelligence, 55
Intelligence quotient (IQ) tests, 290
Interest payments, 364–5
Interest rates, 341–2, 344
Internal migration, 318, 370, 371, 372–3
International Labour Office (ILO)
    classification of employment/unemployment,
        313, 319
    gender differentiation, 106, 107
    Labour Force Survey, 22
    participation in labour market, measurement,
        322
    politics and reform of statistics, 329, 331
International Mathematics and Science Study
        Research, 278
International Monetary Fund (IMF)
    Dissemination Standards Bulletin Board, 51
    military spending, 380
Internet, 82, 388–90
    dissemination of information, 50–51
Interpretation in qualitative research, 83–4
Invalidity benefit, 325, 339
IQ tests, 290
Ireland, 2, 118
Irish people, 137–8

Islam, 138–9
Italy, 322

Jamaat-i-Islami, 138–9
Japan
  education, 278
  research and development, 39
  statisticians, 74
Jarman index, 176
Job Seeker's Allowance, 327, 336
Judiciary, and gender, 112

Key Stages (KSs), National Curriculum, 288,
  289, 291

Labour Force Survey (LFS), 20, 22
  EU legislative requirements, 48–9
  gender differentiation, 106, 110
  industrial injury statistics, 265
  participation in labour market, measurement,
    313, 320–21
  politics and reform of statistics, 324,
    328–34
  using the data, 24, 25–8
Labour government (1997–)
  devolution of power to the regions, 375
  education
    inspection of schools, 300, 301
    National Curriculum, 288
    performance indicators, 282, 283
  health and safety, 269
  National Health Service, 227-8, 233, 262
  pensions, 123
  unemployment as proxy indicator of
    deprivation, 174–5
  women MPs, 111
Labour migration, 156–7
Labour Party, elections, 391, 393–5, 396
  graphical presentations of 1997 results, 400,
    412
  cartograms, 401–3, 404
  ring doughnuts, 410, 411
  triangles, 404, 406–8, 409
Labour reserve, 321–2
Landlords, 362–3
Land Registry, 50
Land-use planning, *see* Planning, land-use
Language in research, 88
Lay statisticians, resources, 78, 82
  books, 78, 79
  computer-aided learning, 79–80
  electronic, 81, 82
  software for statistical practice, 80
  statistics courses, 78–9
  support groups, 80–82
  see also Barefoot statisticians;
    Parastatisticians
League tables, schools, 282
  comparisons, 283–5
  and special educational needs children, 306

Learning difficulties, people with, 255
Liberal Democrats, elections, 394, 395, 396
  graphical presentations of 1997 results, 400,
    412
  cartograms, 401–3
  ring doughnuts, 410, 411
  triangles, 404, 406–8, 409
Life expectancy, 120–21
  disability free, 240
Life histories
  and auto/biography in sociology, 96
  Cox proportional hazard models, 98–9
  'fictive devices', 100–102
  in survey research, 97
Limiting long-term illness (LLTI)
  census question, 16, 164, 230
  and income, 168, 169
  see also Disabled people, poverty
Literacy standards, measurement of trends in,
  287–92
Local authorities, 253–6
Local Base Statistics, 33
Local Government Board, 353
London Tin-Pan Workers, 351
Lone-parent families
  disabled children, 168
  employment, 110
  household surveys, 25, 26
Longitudinal Study (LS), 229, 234, 235–6
Long-term health care, privatisation, 253–9
Long-term illness, *see* Limiting long-term
  illness
Loyalty cards, supermarkets, 385
Luxembourg, 154–5

Mack, John, 65
Mailbase, 81, 82
Major injury rate, 264–5, 267, 269
Malingering, 338
Malthus, Thomas, 9
Management of Health and Safety at Work
  Regulations, 269
Manufacturing sector, injuries, 263, 264, 265–8,
  269, 270
Meat consumption, 143
Media, women and, 112
Members of Parliament (MPs), women, 111
Men
  eating habits, 143–4, 145–6
  employment, 106, 317
  'inactivity', labour market, 107–8
  pensions, 118, 120–23
  see also Gender
Mental Health Enquiry, 225
Mental illness
  homeless people, 189–90, 192–4, 196–7
    detection, 195–6
  official statistics, 225
  privatisation of long-term care, 255, 256
Metadata, 49

Metalliferous ores, extraction and preparation of, 265–6
MIDAS, 50, 107
Mid Glamorgan, permanent sickness in, 335–6
  coping with, 338–9
  health of the valley people, 337
  historical and political considerations, 338
  illness and despair, 337
Migration
  international, 148–9
    asylum data, 155–6
    construction and production of statistics, 149–53
    defining, 153
    European Economic Area, 153–5
    household projections, 370
    'illegal', 157
    labour migration, 156–7
  regional, 318, 370, 371, 372–4
Military sector, 376, 380–83
  finding and interpreting data, 379–80
  research and development, 38, 39, 41–2, 378, 379
  statistics, 376–9
Ministry of Agriculture, Fisheries and Food (MAFF), 21
Minitab, 80
MIQUEST, 227
Mixed ethnicity, 130
Models, statistical, 95
  Cox proportional hazard models, 98–9
  life histories
    and auto/biography in sociology, 96
    'fictive devices', 100–102
  in survey research, 97
Mortality
  death certificates, occupational status exaggerated on, 220, 235
  death registration, 228–9, 232
  industrial causes, 264, 265, 266–7, 269–70
  inequalities, 235–6
  and poverty, 247–51
Morton, John, 67
Multiple personality disorder, 65–6
Muslims, 138–9
Mutedness and cultural difference, 91–3

National Child Development Survey (NEDS)
  attrition, 110
  Cox proportional hazard models, 98
  gender differentiation, 110
  life histories, 97, 100
  social selection, 245–6
National Curriculum
  Education Reform Act 1988, 288
  reading standards, measurement of trends in, 287, 288–92
National Food Survey (NFS), 20, 21, 142–6
  health inequalities, 242
  using the data, 24

National Health Insurance (NHI) Act 1911, 338
National Health Service (NHS)
  privatisation, 252–3, 261–2
  and employment of staff, 259–60
  funding sources, 260–61
  insurance, private health care, 261
  long-term care, 253–9
  statistics about the, 223–8, 232–3
National Health Service and Community Care Act 1990, 254, 255
National Information Infrastructure (USA), 51
National Institute for Economic and Social Research (NIESR), 343, 344, 345, 349
National Insurance (NI)
  historical records, 350, 351, 355, 356, 357
  pensions, 117, 120
  unemployment statistics, 328, 330, 333, 334
National Spatial Data Infrastructure (USA), 51
National Survey of Ethnic Minorities, 110
NATO, 379
Netherlands, 35, 322
Neuro-linguistic programming, 69
Neutral positions, objectivity and subjectivity, 86, 87
New Cronos database, Eurostat, 52
New Earnings Survey (NES), 106–7
New Poor Law (1830s), 350, 353
New Right
  Education Reform Act 1988, 288
  pension privatisation, 116, 117, 123
NHS and Community Care Act 1990, 254, 255
Nomis, 48
Non-claimants, Labour Force Survey, 329–31, 332, 333
Non-employment rate, 315
Non-permanent employment, 317, 320
Norm-referenced interpretation of Standard Assessment Tasks, 289, 290, 291
North Atlantic Treaty Organization (NATO), 379
Northern Ireland
  elections, 392, 397, 398
  health statistics, 226, 228
  household surveys, 21
  religion, 134, 135–6
Northern Ireland Statistics and Research Agency, 21
North Korea, 47
Null hypothesis statistical testing (NHST), 62–4
Numeracy, 3, 72–3
Numerator–denominator bias, 235
Nurses, whole-time-equivalent qualified, 259–60
Nutritional research, *see* Eating habits; National Food Survey

Objectivity, 83, 85–8
Observation, eating habits, 142
Office for National Statistics (ONS, earlier Central Statistical Office; Office for

Population Censuses and Surveys), 81
census, 11, 18
  confidentiality, 35
  consultations, 132
  content, 16, 17
  coverage, 17
confidentiality issues, 31, 32, 35, 37
dissemination of official statistics, 46, 50
domestic work, 108
Family Expenditure Survey, 21
Family Resources Survey, 23
health statistics, 228, 231
  cancer, 227
  Longitudinal Study, 229, 234
unemployment, 29–30
Office for Standards in Education (OFSTED),
  see Inspection of schools
Office of Population Censuses and Surveys
  (OPCS, later Office for National Statistics)
  Disability Surveys, 162–3, 165–6, 169, 170
  Longitudinal Study, 229
  Omnibus survey, 108
Office of Science and Technology, 38–9, 42
Official Secrets legislation, 47
Omnibus survey, 108
One number censuses, 17–18
One-parent families, *see* Lone-parent families
Ordnance Survey, 11, 50
Organization for Economic Co-operation and
  Development (OECD), 149
Orr, Marjorie, 67
Out-commuting, 318
Outdoor Relief Regulation Orders, 353
Out-migration, 318
Output, measurement of, 362–8
Over-three-day injury rate, 265, 270

Parastatisticians, 71, 74–5, 76–7
  resources, *see* Lay statisticians, resources
Parliamentary Constituencies Act 1986, 391
Parmenides, 90
Part-time employment, 315, 317, 320
Patient's Charter, 232
Pay as you earn (PAYE) records, 334
Pay beds in NHS hospitals, 258, 261
Pearson, ES, 60
Pearson, Karl, 55–7, 58, 59–60
Pensions
  national income statistics, 362, 363–4
  privatisation, 115, 123
    actuarial tables, 120–21
    apocalyptic demography and, 115–17
    gender impact, 121–3
    legitimating statistics, 117–20
Permanent employment, 320
Permanently sick people, 318, 325
  Mid Glamorgan, 335–6
    coping with, 338–9
    health of the valley people, 337
    historical and political considerations, 338

illness and despair, 337
Pharmaceuticals industry, 43
Planning, land-use, 92–3
  household projections, 369–70, 374
    alternative approaches, 375
    effects of reliance on, 372–4
    reliability, 370
    role, 374–5
Poisson regression, 250
Policewomen, 112
Policy Studies Institute (PSI), 128, 130
Poor Britain survey, 165
Poor Law, 338, 350, 351, 353, 354
  inter-war Depression, 355, 356–7
Popper, Karl, 84, 85
Portfolio careers, 317
Positivism, 57, 85, 88
  migration statistics, 149
Poverty
  absolute *vs* relative, 244
  and crime, 198–9, 200–202
    Breadline Britain survey results, 206–9
    British Crime Survey results, 202–6
  and death in Yorkshire, 247–51
  defining, 161, 200
  disabled people, 162–5
    children, 165–71
  and economic distress, historical statistics,
    250–54
    inter-war Depression, 355–7
    unemployment and the labour market, 357
  gender differences, 106–7, 108
  and ill health, 244–5, 251
    explanations, 245–7
  line/threshold, 161, 174
  see also Deprivation; Homelessness
Practising psychologists, 62, 64–7
Pregnant women, smoking patterns, 246
Prescribing analysis and cost (PACT) data, 260
Prescription charges, 260
Primary Reading Test, 292
Primary schools, 297–8
Priority Estates Project, 198
Private Finance Initiative (PFI), 258, 259
Privatisation, *see* National Health Service,
  privatisation; Pensions, privatisation;
  Social services, privatisation
Problem-solving strategies, 76–7
Production, measurement of, 362–8
Productive activities, 363–4
Profits, banks, 364–5, 367
Proportional hazard models, 98–9, 101
Psychologists, 62, 69–70
  fringe, 67–9
  practising, 64–7
  scientific, 62–4
Psychology industry, 65, 69
Psychotherapy, 68, 69
Public accountability
  Income Support *vs* Poor Law, 357

Public accountability (*Cont.*)
  and performance indicators in education,
    282–3
Public Assistance Committee, 338
Public Health Acts (1870s), 227
Public Records Office, 32

Qualifications and Curriculum Authority
    (QCA), *see* Schools Curriculum and
    Assessment Authority
Qualitative research, 83
  interpretation in, 83–4
  life histories, 102
  'objective' facts, 87–8
  reflexivity, 85
  subjectivity, 85–7
Quangos, 112
Quantitative research, 83
  life histories, 96, 102

Race, 126, 216–17
  see also Ethnicity
Racialisation, 129
  of ethnic inequalities in health, 215–16, 222
    class sensitivity, 219–21
    demarcating 'race' and ethnic groups,
      216–17
    hypothesised variables, measurement,
      219–21
    'untheorised' research, consequences,
      217–19
Radical science approach to statistics, 415
Radical Statistics, xxv–xxvii, 4, 80, 81, 82
Rape, 112–13
Rayner, Sir Derek, 327, 328
Reading standards, measurement of trends in,
    287–92
Reagan, Ronald, 189
Recall method, eating habits, 140, 141
Recovered/false memory debate, 66–7, 68–9
Reflexivity, 84, 85, 86–7
Regional migration, 318, 370, 371, 372–4
Regional Planning Guidance, 371, 374
Registration, electoral, 397–8
Regression, 56, 58
Religion, 139
  census question, 17, 132–5
  Christian and Irish categories, 137–8
  and community, 137
  and ethnicity, 125, 136–7
  Northern Ireland, 135–6
  South Asians, 138–9
  and the state, 136–7
Rent, 363
Reporting of Injuries, Diseases and Dangerous
    Occurrences Regulations (RIDDOR),
    265, 270
Research and development (R&D), 38, 39–41,
    44–5
  data sources, 38–99

funding, 39–41
  business, 43–4
  public, 41–3
  military sector, 38, 39, 41–2, 378, 379
Research Councils, 39, 42
Resource Centre for Access to Data on Europe
    (r•cade), 51–2
Ring doughnuts, 408–11
Rooms, number in household, 17
Royal Statistical Society (earlier Institute of
    Statisticians), 77
Runnymede Trust, 128

Safety, industrial, 263–70
Sample of Anonymised Records (SARs), 164
Satanic ritual abuse (SRA), 66–7
Schools, *see* Education
Schools Curriculum and Assessment Authority
    (SCAA, later Qualifications and
    Curriculum Authority), 276, 277
Scientific psychologists, 62–4
Scotland
  education, 296
  elections, 392
  health statistics, 228
    National Health Service, 226, 232
Seasonally Adjusted Count Consistent with
    Current Coverage (SAUCC) series, 327
Service sectors, 43–4
Sex Discrimination Act 1975, 120
SF-36 measure, 239
Shocks, economic, 341, 342
Short-time working, 355–6
Sickness benefit, 325, 339
Singapore, 46, 47, 278
Single-parent families, *see* Lone-parent families
Single Regeneration Budget, 172
Small Area Statistics, 33, 34
Smith, Cyril, 406
Smoking patterns in pregnant women, 246
Social and Community Planning Research
    (SCPR), 23, 231, 234
Social anthropologists, 91
Social Change and Economic Life Initiative, 97
Social control of homeless people, 191
Social Darwinism, 57
Social exclusion, 174–5
Social responsibility approach to statistics, 415
Social Science Information Gateway (SOSIG),
    81
Social Security Act 1986, 117, 253
Social Security Act 1989, 327
Social services, privatisation, 252–3, 261–2
  and employment of staff, 259–60
  funding sources, 260–61
  insurance, private health care, 261
  long-term care, 253–9
Socio-economic group (SEG)
  childhood disability, 162–3, 164–5
  and education

class size, 299
  inspection of schools, 295–6
  reading standards, 290, 291
eugenics, 55–6
health inequalities, 237–9
  and ethnicity, 215–16, 219–21
  explanations, 241–3
  mortality, 235–6
  and poverty, 245–51
  private health care insurance, 261
homelessness, 191
household surveys, 25, 27
and life expectancy, 121
pensions, 122
Software
  for learning, 79–80
  for statistical practice, 80
South Asians
  health inequalities, 217–18, 219
  religion, 138–9
Spain, 322
Special educational needs (SEN) pupils, 302–3
  class size, 298
  Green Paper 1997, 310
  Hampshire review, 306–9
  social trends, 304–6
  statistics, 309–10
  Warnock Report, 303–4
Special needs assistants (SNAs), 302
Speculation, 85
Spin driers, and disabled children, 169
SPSS, 80
Standard Assessment Tasks (SATs), 287, 288, 289–92
Standard deviation, 250
Standardised mortality ratios (SMRs), 121, 240
State earnings-related pension scheme (SERPS), 116–17
Statements of Attainment, National Curriculum, 288–9
Statements of special educational need, 303, 304, 306–8, 309, 310
'Statistical Education through Problem Solving (STEPS)', 79–80
Statistical News, 28
Statistics, 1–4, 414–16, 419–20
  assembly, 417
  dissemination, 418–19
  interpretation, 417–18
  purpose, 416–17
Statistics Canada, 46
'Statistics for the Terrified', 80
Statistics Singapore, 46
StatsWatch, 82
'STEPS (Statistical Education through Problem Solving)', 79–80
Stockholm International Peace Research Institute (SIPRI), 379, 380
Structure Plans, 371
Subjectivity, 83, 85–7

Sudden Unexpected Death study, 229
Suicide, 218
Supermarket loyalty cards, 385
Support groups, 80–82
Survey Methodology Bulletin, 28
Sweden, 278
Swing, inter-election, 395–6

Tactical voting, 396
Tatchell, Peter, 406
Taxation
  employment statistics, 334
  national income accounts, 364
Teaching, quality of, 275, 278
  see also Education
Temporary employment, 317, 320
Tetrachoric method, Pearson's, 59, 60
Textbooks for lay statisticians, 78, 79
Thatcher, Margaret, 327
Time-varying covariates, 98, 101
Trade union membership, 351–2, 353–4, 355, 356
Training
  employment, 339
    government schemes, 318, 326
  for lay statisticians and critical citizens, 76–7
Training and Enterprise Councils, 348
Tramping system, 351
Transnational companies, 41
Travellers, 89
Triangles, 404–8, 409
Tumble driers, and disabled children, 169

Unemployment
  definitions, 29–30, 313
  disaggregation, 315, 316
  dissemination of statistics, 48
  and ethnicity, 220
  eugenics, 56
  forecasting, 340–49
  gender differences, 107–8
  historical statistics, 350–54
    inter-war Depression, 355–6
    and labour market, 357
  measurement, 317–21
  and permanent sickness in Mid Glamorgan, 335–6
    coping with, 338–9
    health of the valley people, 337
    historical and political considerations, 338
    illness and despair, 337
  politics and reform of statistics, 324, 334
    Alternative Measures of Unemployment series, 329
    Count of Claimants, 324–7
    Labour Force Survey, 328–33
    political and statistical background, 327–8
  as proxy indicator of deprivation, 174–6
  rate, 314–15
  volatility, 340–41

see also Employment
Unemployment Assistance Board, 357
Unemployment benefit, 325, 339
    historical statistics, 352
Unemployment Unit, 331
United Nations
    international migrant, definition, 153
    System of National Accounts, 362
United Nations Economic Commission for
        Europe (UNECE), 149
United States of America
    censuses, 29, 36
    Civil War, 353
    dissemination of official statistics, 47, 48, 52
        Census Bureau, 46
        National Information Infrastructure, 51
    employment forecasting, 343
    health care corporations, 257, 258
    homelessness, 196
    military sector, 380
    pensions, 115–16
    psychologists, 65
    research and development, 39, 42
    statisticians, 74
    unemployment measures, 319
Universities, research and development, 39, 42–3
Unmatched aspirations, labour market, 320
Unproductive activities, 363–4

Vacancy rates, housing, 371
Vagrancy Act 1824, 189
Value-adding
    dissemination of official statistics, 51
    education statistics, 282, 283–4, 285
        inspection of schools, 295
    national income accounts, 364, 365, 366
Vegetarians, 143
Victim blaming, health inequalities, 246
Violence against women, 112–13
    see also Crime and fear of crime
Visual sense, domination of, 90–91

Wales
    elections, 392
    health statistics, 226, 228
Warnock Report, 303–4, 306, 307, 308
Washing machines, and disabled children, 169
Welfare-to-work, 344–5
Women, 113
    crime against, 112–13, 199
    eating habits, 142, 143–4, 145–6
    in the economy, 105–6, 110–11
        domestic work, 108–11
        employment patterns, 315–17
        incomes and poverty, 106–7
        return to work after childbirth, 98–9
        unemployment and 'inactivity', 107–8
    health
        and childhood disability, 162
        disability-free life expectancy, 240
    and the media, 112
    pensions, 118, 120–3
    in power and decision-making, 111–12
    smoking in pregnancy, 246
    suicide rates, Indian immigrants, 218
    see also Gender
Women in Employment Survey, 97
Workforce in Employment (WiE) series, 326,
        327, 328, 330
World Bank, 380
World Health Organization (WHO), 46
World Wide Web, 82, 388–90
    dissemination of information, 50–51

Years of potential life lost (YPLL), 240
Yugoslavia/former Yugoslavia, 155
Yule's Q, 58–9, 60

Zero sums, objectivity and subjectivity, 86
Z score, 176–7